LMKH
36⁵⁰

To my children, from whom I have learned a great deal:
Marsha
Stephen
Lawrence

Marshall B. Clinard

To my parents, who made my learning possible:
Frank
Eileen

Robert F. Meier

Preface

Sociology of Deviant Behavior presents a theoretical overview of the nature and meaning of deviance and examines in detail a number of forms of behavior that are commonly regarded as deviant behavior. Throughout the book, sociological concepts and processes underlie the presentation. We have attempted to identify and explain the leading theories and sociological orientations of deviant behavior: anomie, control, labeling, conflict, and learning. When appropriate, we have also attempted to be sensitive to other perspectives where they apply. The theoretical frame of reference throughout the book is socialization, or learning theory, together with a normative perspective. The reader will see that we find the most meaning of deviant behavior in the context of the acquisition of any (normal) behavior. As such, the central theme in the book is that understanding deviant behavior is no different from understanding any other behavior; deviant behavior is human behavior, understandable within a context of socialization and role playing. Such a frame of reference furnishes theoretical continuity throughout the book, although other viewpoints are recognized as well. Where possible, we have attempted to illustrate sociological ideas from the deviant's own perspective through the use of case histories or personal accounts.

It is not easy to define deviant behavior. Where a consensus has appeared to exist, it usually has been the result of political, social, and economic powers of certain interest groups that have succeeded in imposing on others the views they hold of what constitutes deviance. Here, we examine the merits of four definitions of deviance: the statistical, absolutist, reactivist, and normative. We have adopted the normative definition as best fitting the complex, largely urbanized societies that are characterized by a high degree of differentiation and, as a result, a high degree of deviance.

As with previous editions, this eighth edition is a complete revision that incorporates the most recent theoretical developments in the field and the latest research findings. Reviewers of previous editions have suggested changes. In the eighth edition, we have placed more emphasis on some forms of deviance and issues of social control that are of great contemporary concern, for example, drugs and violence, both personal and family. We have augmented the material throughout the book with first-person accounts to illustrate some of the sociological concepts and theories discussed. We have attempted to devote attention to "newer" forms of deviance, including eating disorders such as anorexia and bulimia. We have attempted to integrate the material on the urban context of much deviance into the remainder of the book. Every chapter has been updated and some, such as the material on heterosexual deviance, have been enlarged.

We are aware of the political nature of some material. There are those, for example, who would disagree on whether a book on deviance should have a chapter on homosexuality, which they regard not as deviant but merely as an alternative lifestyle. Here, we do not take a position on that issue, but have decided to retain the chapter on the basis that regardless of our personal predilections, most people do appear to regard homosexuality as deviant. Similarly, a chapter on physical disabilities might, at first, appear odd in a book on deviant behavior since physical disabilities imply involuntarily conditions. There is no attempt here to "blame" persons with physical impairments for their condition; these individuals are not voluntary deviants. There are, of course, other involuntary deviants, such as persons with mental disorders. It is significant sociologically to note that these persons experience social reactions very similar to those experienced by voluntary deviants (e.g., criminals), and the chapter serves to point to the fact that deviance is a concept that can apply to conditions as well as behavior.

Chapters 1 and 2 deal with the nature and definition of deviance. They introduce the sociological concepts necessary to understand the processes as well as the theories of deviance that follow. Chapter 3 represents a discussion of general sources or contexts of deviance, with a particular focus on urbanization as an important context in which to view deviance. Chapter 4 examines and contrasts two major sociological theories of deviance, anomie and learning. Chapter 5 examines three other theories: control, labeling, and conflict. We shift then to an in-depth examination of various forms of deviant behavior. Chapters 6 and 7 identify the processes involved in crimes of interpersonal violence and crimes against the economic and political order. Chapters 8 and 9 deal with drug usage, with one chapter devoted to the drug most widely used: alcohol. Chapter 10 focuses on forms of heterosexual deviance, and Chapter 11 is on homosexuality. Chapters 12 and 13 analyze mental disorder and suicide from a sociological point of view. We conclude with Chapter 14 in which we discuss certain physical disabilities, including eating disorders, as examples of conditions that are often regarded sociologically as deviant, with profound social and personal implications for the self-concept of the individual.

This book first appeared in 1957. It pioneered a major shift from the then characteristic approach to deviance, termed "social disorganization" or "social problems," to a more basically sociological orientation built around the concept of normative deviance and deviant behavior. Subsequently, the conceptual framework of deviant behavior has received wide acceptance and greater use in sociology. In the eighth edition, we continue the tradition begun in the first edition of attempting to understand deviance in its social context. This edition also emphasizes that deviance is an inescapable feature of modern, complex societies because such societies are characterized by a system of ranked social differentiation (stratification) that is the basis for social deviance. We also wish to affirm in this edition the obvious relationship between deviance and social order, and the necessity for a sociological understanding of all aspects of society in order to comprehend the nature and complexity of social deviance.

Over the years, numerous sociologists and friends have contributed the basic data for this book through their theoretical writings and research on deviance. The references in this book acknowledge some, but not all, of them. At various times, other sociologists have criticized various editions, including the present one, and they have thus contributed valuable ideas and suggestions. We are grateful to all of them.

We wish specifically to thank the following reviewers who provided valuable suggestions for this edition of the book:

Roger C. Barnes
Incarnate Word College

Antonia Keane
Loyola College

Will C. Kennedy
San Diego State University

Bettie M. Smolansky
Moravian College

An Instructor's Manual is available. The manual may be obtained through a local HBJ representative or by writing to the Sociology Editor, College Department, Harcourt Brace Jovanovich College Publishers, 301 Commerce Street, Suite 3700, Fort Worth, Texas 76102.

M.B.C.
R.F.M.

Brief Contents

PART THREE
Forms of Deviance   123

Contents

PART THREE
Forms of Deviance   1 2 3

6   Crimes of Interpersonal Violence   1 2 5

13   Suicide   390

The Nature of Deviance

PART ONE

The Nature and Meaning of Deviance

Deviance is found everywhere people are, but there is no agreement on what behavior constitutes deviance. Our everyday conversations show that what some people regard as deviant, others regard as virtuous; what some might praise, others condemn. This disagreement may inhibit a social and political consensus about deviance but not a full discussion about the important dimensions of deviance: what it is, how we can define it, what explains it, and what a social group can do to reduce deviance. To say that deviance exists does not specify which acts are considered deviant by which groups in what situations and at a given time.

To understand deviance, one must first understand this paradox: there is no consensus on which behavior, people, or conditions are deviant. Yet, most persons would say that they know deviance when they see it. Mental disorder, suicide, crime, homosexuality, and alcoholism would be on many persons' lists. Yet, even with this list of "generally accepted" forms of deviance, there is disagreement. To some, for example, homosexuality is not at all deviant. What one person considers problem drinking may pose no such problem to another. The "wrongfulness" of certain crimes, such as prostitution and the use of marijuana or cocaine, is currently disputed among different segments of the population. As such, deviance may be similar to what St. Augustine said about time: we know pretty much what it is—until someone asks us to define it.

Even scholars who study deviance are not in agreement on which people, acts, or conditions are deviant. Cohen (1966: 1) says his deviance book is about "knavery, skulduggery, cheating, unfairness, crime, sneakiness, betrayal, graft, corruption, wickedness, and sin." Gouldner (1968) complained that the empirical literature on deviance has been limited largely to "the world of the hip, night people, drifters, grifters, and skidders: the 'cool world'." Howard Becker (1973), a writer we will encounter again, limited his influential study on deviance to jazz musicians and marijuana users. A British collection of papers on deviance dealt with drug users, thieves, hooligans, suicides, homosexuals and their blackmailers, and industrial saboteurs (Cohen, 1971). Lemert (1951) illustrated his theoretical position on deviance with reference to, among others, the blind and stutterers. Dinitz, Dynes, and Clarke (1975) find the following types of persons deviant: midgets, dwarfs, giants, sinners, heretics, bums, tramps, hippies, and Bohemians. Becker (1977) finds the "genius" deviant. Liazos (1972) attempts to capture the essence of deviance by claiming that the study of deviance has traditionally been concerned with "nuts, sluts, and preverts." Henslin (1972) discusses four types of deviants to illustrate research problems in the field: cabbies, suicides, drug users, and abortionees. Stafford and Scott (1986: 77), in a contemporary list, offer the following list of disapproved conditions: "old age, paralysis, cancer, drug addiction, mental illness, shortness, being black, alcoholism, smoking, crime, homosexuality, unemployment, being Jewish, obesity, blindness, epilepsy, receiving welfare, illiteracy, divorce, ugliness, stuttering, being female, poverty, being an amputee, mental retardation, and deafness." Davis (1961) talks about blacks as deviants and both Davis (1961) and Schur (1984) discuss women as deviant. Many persons would include witches on their lists of deviants (see Geis and Bunn, 1990).

These lists include behavior (e.g., crime), conditions (e.g., ugliness), and types of persons (e.g., bums). The acts are both voluntaristic (e.g., smoking) and involuntary (e.g., stuttering). It is difficult to imagine what is common about these lists, particularly when one such deviant—genius—appears to represent a quality that is positive-valued. This book examines a number of forms of deviance, including some of those mentioned on the preceding lists. Not everyone will agree that each form is "deviant" and there will be disagreement on the extent to which some acts are deviant, but, we feel, many persons will so regard these forms as deviant. One thing that is clear is that if persons are asked for a list of things they consider deviant, these lists would vary from one another, sometimes drastically. Yet, deviance is not an individualistic judgment because

The Concept of Deviance

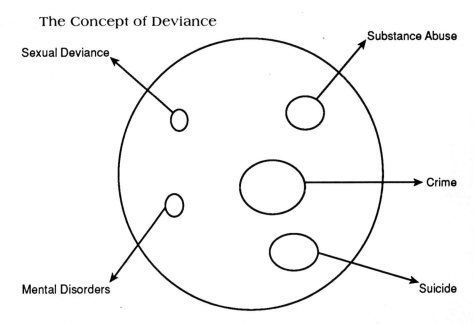

Deviance is a general behavioral category that includes many different, specific kinds of subcategories, as illustrated in the few <u>examples</u> above.

Substance Abuse	Involves violations of norms surrounding the use of alcohol and other drugs
Sexual Deviance	Involves violations of norms concerning sexual behavior, including appropriate partners, times, acts, places, etc.
Crime	Involves violations of legal norms from legislatures and other governmental agencies
Suicide	Involves violations of norms concerning the value of human life and circumstances under which humans can end a life (in this case, their own)
Mental Disorders	Involves violations of norms concerning expected or "normal" behavior and thoughts

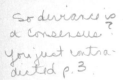

*So deviance is
a consensus?
You just contra-
dicted p. 3.*

many persons agree on the deviant nature of a wide number of acts and persons. There are some acts and actors that do, in fact, share something called "deviance." There is, in other words, a degree of consensus about the deviant nature of some acts and conditions. To identify what characteristics these judgments share, a more formal definition is necessary.

WHAT IS DEVIANCE?

Some sociologists conceive of deviance to be those conditions, persons, or acts that are either disvalued by society (Sagarin, 1975: 9) or simply offensive (Higgins and Butler, 1982: 3). Such a conception, however, begs the question of how or why persons find an act or individual offensive and, hence, on what basis that person is disvalued. Nor does such a conception recognize the distinct possibility that deviance might be highly valued, that there can be "positive" as well as "negative" deviance, as illustrated in the examples of the genius (see Dodge, 1985; see also Sagarin, 1985). Another example of positive deviance might be contained in a United Nations report on exceptional children who grow and develop well in impoverished environments (Zeitlin, Ghassemi, and Mansour, 1990). But defining deviance by examples is not sufficient, and only a more explicit definition of deviance can identify fully examples that might be considered deviant, either in a positive or negative sense.

Deviance may be defined in four different ways; this variety of definitions is one of the reasons that lists of examples of different types of deviance and deviants differ. The four ways of defining deviance are the statistical, absolutist or violation of values, reactivist, and normative.

Statistical *(falls under normative)*

One of the most common definitions of deviance heard in everyday conversations is to regard it as variations or departures from the "average." In this conception, deviance is behavior that is not average; it is behavior that is rare or infrequent. This approach assumes that whatever it is that most people do is "correct."

This definition faces immediate difficulties because it can lead to some confusing conclusions if, for example, the minority is always defined as deviant. With a statistical definition of deviance, those who have never stolen anything or never violated the law, those who have never used marijuana, those who do not drink alcoholic beverages, and those who have never had premarital sex might be considered deviant (Rushing, 1975: 3–4). The meaning of deviance is not to be found in the statistical regularities of behavior, but rather in the fact that deviance connotes some difference or departure from a standard of behavior or what "should" or "should not" be rather than "what is."

Absolutist

Some persons view social rules as "absolute, clear and obvious to all members of society in all situations" (Scott and Douglas, 1972: 4; see also Hawkins and Tiedeman, 1975: 20–41). This absolutist conception of deviance assumes that the basic rules of a society are obvious and its members are in general agreement on what constitutes deviance, because the standards for acceptable behavior are laid out in advance. Deviance is taken for granted, as though everyone agreed that certain violations of rules are abnormal while others are not. Everyone is presumed to know how to act according to universally held values; violations of these values constitute deviance. The sources of these universal standards have usually been identified as moral values of the middle class and the personal biases of some writers who, coming from rural, traditional, and religious backgrounds, have viewed many forms of behavior related to urban life and industrial society as being destructive of what they thought was moral (Ranulf, 1964). Still another version of this definition asserts that conceptions of what is deviant stem ultimately from preferences and interests of the elite segments of society (Schwendinger and Schwendinger, 1977).

The absolutist definition of deviance is still supported today, particularly by psychiatrists and some psychologists who think of deviance in terms of a "medical model" as a form of "sickness." Thus, crime, mental disorder, suicide, alcoholism, drug addiction, and so on, become absolutes much like diseases such as cancer, and they are universal expressions of individual maladjustment regardless of differences in cultural and subcultural norms. But the absolutist conception of deviance ignores too many facets of social life:

> The absolutist asserts that, regardless of time and social context, certain culture-free standards, such as how fully persons develop their innate potential or how closely they approach the fulfillment of the highest human values, enable one to detect deviance. Thus suicide or alcoholism destroys or inhibits the possibility of the actor's developing his full human potential and is therefore always deviant. . . . The absolutist believes that he knows what behavior *is,* what people *should be,* and what constitutes full and appropriate development. (Lofland, 1969: 23–24)

Reactivist

While the statistical and absolutist definitions are used in everyday conversations, the two major competing definitions of deviance are the reactivist and the normative. The reactivist conception defines deviance as behavior or conditions labeled as deviance by others. As one reactivist puts it: "The deviant is one to whom that label has successfully been applied; deviant behavior is behavior that people so label" (Becker, 1973: 9). Thus, in the reactivist definition, acts can be identified as deviant *only* with reference to the reaction to the acts through the labeling of a person as deviant by society or by its agents of social control. Once behavior has been labeled deviant, it is an easy step to consider the actor as deviant as well.

The reactivist conception of deviance has been very influential, and the reasons for its popularity are easy to identify. The reactivist conception attempts to concentrate on what is truly social about deviance—the interaction between the deviant and society (really, the agents or representatives of society in the form of social control agents)—and the consequences of that social relationship. Thus, reactivists reject the notion that what is considered deviant depends on some innate quality of the act; rather, they claim that what is and what is not deviant depends exclusively on the actions of the social audience to the act.

Critics of the reactivist definition of deviance have pointed out that while the interaction between deviant and social control agents is an important process, it does not *define* deviance. The illogical nature of this view can be illustrated by a case in which a man is engaged in an act of burglary and is not discovered; because he is not discovered, he is not reacted to and, thus, not regarded as deviant. Furthermore, even those acts that do elicit a social reaction do so on some basis. That is, there must be something about the act that prompts others to react against it in the first place, and that quality (such as the "innate" wrongfulness of the act, or that the act violates some agreed-upon standard of behavior) is what really defines deviance.

Normative

A normative definition of deviance claims that deviance is a violation of a *norm*. A norm is a standard about "what human beings should or should not think, say, or do under given circumstances" (Blake and Davis, 1964: 456; see also Birenbaum and Sagarin, 1976, and Gibbs, 1981). Violations of norms often draw a reaction or *sanction* from the social audience to the violation. These sanctions constitute the pressures most people feel to conform to norms. There are two common conceptions of norm: norm as an evaluation of conduct, and norm as expected or predictable conduct (Meier, 1981). The former conception recognizes that some conduct (behavior or beliefs) "ought" or "ought not" to occur, either in specific situations (for example, no smoking in public elevators) or at any time or place (for example, no armed robbery, ever). The latter conception points to regularities of behavior that may be based on habit or traditional customs—a child is expected to act a certain way in church, another way on the playground. Norms are not necessarily rules: norms are social properties because norms are shared; rules can be formulated individually and imposed on others (such as the laws of a monarch or despot). Rules and norms do share many characteristics, however, not the least of which is the ability of rules and norms to direct behavior; they are necessary components of social order (Bryant, 1990: 5–13).

One virtue of the normative conception is that it offers an answer to a question generated by the reactivist conception: on what basis do people react to behavior? If deviance is indeed identified through the reactions of others, how do people know whether to react to or label a given instance of behavior? The only obvious answer to this question is norms. For this reason, the reactivist and normativist conceptions may be similar in that norms provide the basis for reacting to deviance, but it is through social reactions that norms are expressed and deviance is identified.

Here we adopt a normative definition of deviance. *Deviance constitutes only those deviations from norms in a disapproved direction such that the deviation elicits, or is likely to elicit if detected, a negative sanction.* The key element in this conception is the idea of "norm." Norms do not simply exist in society. They are created, maintained, and promoted against one another. In addition to being properties of small groups, norms, and hence deviance, are also related to certain structural features of society. The larger meaning of deviance is found in the context of social differentiation and stratification. It is also found in the properties and nature of norms, social groups who subscribe to those norms, and the degree to which those norms are influential over behavior. It is to an examination of these concepts— differentiation, norm, sanction, and social control—that we now turn.

DEVIANCE AND SOCIETY

At the simplest level, deviance refers to something that is "different" from something else. "Deviants" are not like "us"—"they" behave differently. Or, so many people think. But deviance goes beyond the simple and everyday

Shorthand Definitions of Deviance

Statistical	What everyone does is "normal" or nondeviant. Whatever is in the statistical minority is deviant.
Absolutist	What is deviant is a value judgment based on absolute standards. Certain things are deviant because they have always been deviant (because of tradition or custom).
Reactivist	Whatever is reacted against (or labeled) by a social audience to behavior is deviant. If there is no reaction, there is no deviance.
Normativist	What is deviant depends on a group's notion of what "ought" or "should" not be. This is a situational conception, meaning that the conception can change as situations do.

observation that people and behavior are different from one another. We may observe differences in styles of dress, for example, that are not deviant. Persons who wear the same clothing may wear different colors that are not in any way deviant from one another.

Beyond the idea of differences, deviance refers to something that is evaluated negatively or disvalued (see also Terry and Steffensmeier, 1988). Returning to the fashion example, the colors that someone wears may be different but they may be considered deviant only if they are so different as to be considered "in bad taste," such as when the colors clash with one another. Or when the clothing is not suitable to the occasion—wearing a bathing suit to a funeral—it would be considered deviant. Some people would never wear red and orange or black and blue in the same outfit. They, and perhaps you, would consider it deviant in the context of fashion.

But colors that clash is a problem only for some persons and not others. Youth who prefer to dress in punk rock styles seem to value highly clashing colors, unusual styles of clothing, and atypical hair styles. This suggests that *deviance is a relative notion* that depends upon the audience to the behavior or condition that defines something as deviant. These three ideas, differentness, judgment, and relativity, are each important to a sociological understanding of deviance. These ideas let us understand the meaning of individual deviant conduct, as well as the connection of that individual conduct to the larger social community.

Norms

Because deviance is here defined in terms of norms, it is necessary to identify the importance of norms to everyday life. Human social relations and behavior are regulated through social norms. *Norms are expectations of conduct in particular situations.* Norms can be classified according to the degree of their acceptance, the mode of any norm's enforcement, the way a norm is transmitted, and the amount of conformity required by the norm. Some social norms may require considerable force to ensure compliance; others may require little or none. Some norms are fairly stable in nature, others are more transitory (Gibbs, 1965). Rarely are individuals in a group consciously aware of the often arbitrary nature of the social norms in their group since they have been introduced to them in the ongoing process of living. Norms are learned and transmitted in groups from generation to generation. In this way, individuals have incorporated into their own life organization the language, ideas, and the beliefs of the groups to which they belong. Human beings, thus, see the world not with their eyes alone, for if they saw only with their eyes, each would see the same thing; rather, they see the world through their cultural and other group experiences.

Even moral judgments are generally not those of an individual alone, but of the group or groups to which the individual belongs. The significance of the group nature of norms and of understanding the world through norms has probably never been stated more cogently, even poetically, than by Faris many years ago when he stated:

> For we live in a world of 'cultural relativity' and the whole furniture of earth and choir of heaven are to be described and discussed as they are conceived by men. Caviar is not a delicacy to the general [population]. Cows are not food to the Hindu. Mohammed is not the prophet of God to me. To an atheist, God is not God at all. (Faris, 1937: 150–151)

Norms are crucial in the maintenance of order. They may be regarded as cultural ideals, or they may be expressed in terms of what we expect in certain situations. For example, sexual behavior may be examined as cultural ideals or in terms of specific expectations for persons associated in certain situations, such as a married couple on their honeymoon. Ideal cultural norms can be inferred from what people say, or observing what they sanction or react against. *Proscriptive norms* tell people what they "ought not" to do; *prescriptive norms* tell them what they "ought" to do. Not only are norms social or group standards for conduct, but they also provide categories through which we interpret our experiences. Norms provide us with a means by which to interpret both actions ("He should not have laughed at the funeral") and events ("Funerals are certainly sad").

The social norms and behavior of social classes in the United States vary greatly with respect to many attitudes and values. The norms of long-shoremen differ from those of doctors and professors; construction workers display markedly different attitudes than do college students. Child-rearing patterns have been shown to differ from one social class to another. Lower-class parents, for example, tend to use physical punishment more often as a disciplinary measure than do middle-class parents, although not as much as some expect (Erlanger, 1974). Most crimes of violence, such as murder, aggravated assault, and forcible rape, are committed by lower-class persons, and the existence of a lower-class "subculture of violence" (to be discussed in Chapter 6) may offer a partial explanation. Child-rearing patterns are also influenced by how religious the parents are, more so than the religious affiliation of the parents (Alwin, 1986). This suggests that belief in some norms effects belief in others and that this, in turn, is made part of the socialization process.

Norms are an integral part of the organization of all societies, from small tribal groups to modern industrial societies. In complex modern societies, group norms may differ radically or only slightly from one another; in other cases, the norms simply differ in emphasis. As a result, persons who belong to a number of groups, with each group having either different norms or emphasizing them differently, may experience personal conflict. We are often expected to act in different ways according to which role we are performing at the time. A *social role* is a collection of norms conveying expectations about appropriate conduct for persons in a particular position (Biddle, 1986). Thus, the norms governing the behavior of husbands are often different from those governing the behavior of bachelors; the role of a shopper is different from that of a sales clerk; and so on. The norms and roles a person acquires from the family group do not necessarily always agree with the norms and social roles of the play group, the age or peer group, the work group, or political group. Certain groups may become more important to an individual's life organization than others, and he or she may, as a result, tend to conform more closely to the norms of the groups with which the individual feels more closely identified. Although the family group is important, it is only one of several groups related to a person's behavior, whether deviant or nondeviant. Many other sources of norms in modern societies are important: social class, occupation, neighborhood, school, church, and one's immediate friends.

Among more homogeneous peoples, such as primitive or folk societies, most norms and values are perceived in a like fashion by group members, although certainly not entirely so (Edgerton, 1976). Members of such societies thus come to share many common objectives and meanings, in contrast to more modern, complex societies where social groups arise out of many attributes of race, occupation, ethnic background, religion, political party affiliation, residence, and many more. Particularly important in the development of this differentiation are social class and age or peer groups.

In spite of the importance of norms, there is no single agreed-upon research strategy for studying them. One strategy is *inferential* and regards the existence of norms as detectable only in social control efforts. That is, norms are observable only after their violation by observing patterns of sanctions (Kitsuse, 1972). A second strategy regards norms as *qualitative* properties of social life that are not amenable to either direct measurement or analysis. Thus, some sociologists indicate that norms are so deeply embedded in social situations, they cannot be analyzed apart from those situations (Douglas, 1970). Still a third strategy regards norms as *cognitive* properties that can be inferred from responses by articulate persons to questions that ask for normative opinions (Rossi, et al., 1974). This strategy requires a sensitivity to the ways in which norms differ from one another and an awareness of how many different situations apply to norms (Meier, 1981).

Regardless of the strategy, most persons are aware of the importance of norms in their everyday lives. Norms define acts, actors, and conditions as acceptable or deviant and unacceptable. While norms can and do change through the actions of individuals and groups who are able to promote their norms over others successfully, the idea of norm provides the foundation of the meaning and explanation of deviance.

Differentiation and Deviance

There are a number of ways in which people are different from one another, including age, sex, race, educational achievement level, and occupational status. Differentiation is the sociological term that refers to such differences. At the most general level, deviance refers to differentness. For this reason, the concept of deviance would have no meaning in an undifferentiated society. Since, however, a society where everyone is the same does not and can not exist (Durkheim, 1982; 1895), deviance occurs everywhere. Durkheim said that deviance can be found even in a society of saints where the differences among them would be small but they would be morally magnified. Some saints, in other words, would still be literally "holier" than others.

A recent example illustrates the wisdom of Durkheim's observation. In 1987, religious broadcasters approved a code of ethics to avoid scandals like the one that forced Jim Bakker, a well-known evangelist, to resign as head of a popular television ministry (*New York Times*, September 13, 1987, p. 18). Bakker was forced to vacate his ministry after a sex scandal and subsequent questions of financial impropriety in spending millions in solicited money for high living finally ended in a 40-year federal prison term. These highly visible evangelical TV preachers, considered by many to be among the most moral persons in society, found it necessary to adopt a special code after a series of scandals raised the issue of ethics as applied to these moral leaders. The code of ethics also spoke to the issue of fund raising, a topic brought about by questions generated when Rev. Oral Roberts declared God would take his life if he did not raise $8 million by the end of March, 1987. Mr. Roberts later said his life had been spared because he attained that goal. Subsequently, another evangelist, Jimmy Swaggart, was said to have engaged in sexual improprieties with a prostitute. Because of such activities, viewership of these "televangelists" declined significantly, and the revenues generated by television evangelists were cut in half by the early 1990s.

The conditions that promote social differentiation in society also promote deviance (Meier, 1989). Those conditions that increase differentiation are likely to increase the degree and range of social stratification by increasing the number of ways in which people can be compared among one another. Those comparisons are often invidious or ranked, with some characteristics being more highly valued than others. To the extent that the bases on which persons are stratified increases, so too does the range of conditions that are disvalued or not as highly ranked (see also Cohen, 1974).

The more people are different from one another, the more they are likely to be stratified with respect to one another. Modern, industrial societies are extremely complex in the ways in which people are differentiated from one another. In addition to such characteristics as age, sex, and race, persons in modern societies display greater diversity than persons in more homogeneous societies in terms of behavior, dress and clothing styles, attitudes, and interaction patterns. Even within modern societies, there are differences between urban and rural areas regarding such matters. Sometimes, what some people mean by "deviance" is diversity, or behavior that is the result of social differentiation.

Beyond this, however, it seems clear that the more persons are stratified with respect to one another, the more likely some of these rankings will reflect disvalued characteristics. That is, not only will some persons be ranked lower than others but they may also be disvalued as well. To the extent that education is valued, undereducation is not; to the extent that having an occupation with much prestige is valued (Supreme Court Justice), having one with little or no prestige (ditch digger) is not. When we make judgments about "better" or "worse" we are beginning to make judgments about deviance.

It is in these ways that deviance is linked with the stratification system in a society. The range of statuses that are ranked to form a stratification system have roughly the same range from top to bottom as the range of ranked negative statuses that comprise a structure of deviance. A more highly differentiated society should have a greater number of ranked statuses, while a less differentiated society should have a smaller number of ranked statuses. Similarly, a more stratified society should have a greater number of negatively ranked statuses than a less stratified society. And, just as the number of bases on which judgments of positive and negative status are made increases, so, too, should the correspondence between the two systems. So, a simpler society should have both a simple structure of social stratification and a simple or narrow structure of deviant statuses.

There are many bases of differentiation. Age, sex, status, occupational achievement, race, and occupational prestige are only some bases on which persons are differentiated. A definition of deviance would make clear which kinds of differentiation would be regarded as deviant and which as just "different" without any moral connotations attached to the conduct. Some sociologists, however, have recommended an alternate strategy: leaving deviance undefined, and proceeding with research on "matters dealing with deviance." Lemert (1982: 238), for example, has suggested that ". . . the study of deviance can best proceed by identifying bodies of data through primitive, ontological recognition rather than by formal definition." But only some statuses and acts are considered morally inferior or are evaluated negatively, and a definition of deviance would make clear which kinds of differentiation would be regarded as deviant and which as just "different" without any moral connotations attached to the conduct.

Deviance is not static or constant. Deviance changes constantly in terms of the forms it takes and the degree of disapproval it elicits. And, frequently, to understand which conduct or conditions are disapproved, an understanding of social power is necessary. Power can be defined as the ability to make choices by virtue of political, economic, or social resources. Persons who have money, education, and social influence are generally more powerful than those who do not. Persons of power, by virtue of their influence, are often able to define what is deviant and, unexpectedly, they often find more deviance among persons of lesser power than themselves. White-collar and corporate crimes are often seen by the general public as less serious than ordinary street crime. This perception exists in spite of the fact that crimes by the powerful are as dangerous as and involve more financial loss than ordinary crime. The reasons for this disparity reside in the fact that these crimes are often not handled under the criminal law like other crimes, and most people are not used to conceiving of powerful persons as "evil" or "depraved" as they do lower-class persons. Crimes by the powerful have the appearance of not being real crimes. Put another way, crimes committed with a pen, such an embezzlement, appear to many people to be less serious than those committed with a gun, such as robbery.

The importance of social power can also be expressed in terms of social differentiation. Deviance is relative not because no trait or act is everywhere and for all time deviant, but because the processes of social differentiation and social change produce alterations in social judgments. The key question is how some conditions come to be ranked the way they are (i.e., why are some acts and actors deviant but not others?). This is a question sociologists frequently answer in terms of power—powerful groups expand the range of stratified social phenomena by engaging in a process of definition and influence (Chambliss, 1976). A generic term for this process is "norm promotion," that is, being able to promote successfully particular norms to the exclusion of other, competing norms. Regardless of the answer to the question about how specific acts come to be defined as deviant, it can hardly be denied that social judgments of disvaluement represent a core component of the concept of deviance. This is presumably why some sociologists use the phrase "moral differentiation" to refer to deviance (Lemert, 1982). Deviance judgments are moral judgments.

Subcultures

Since norms are properties of groups, it is not surprising that different groups have different norms. One is expected to behave differently according to the group to which one belongs. What is deviant in one group may be perfectly acceptable behavior in another group. Sociologists often refer to such differences as subcultural differences.

Sometimes social groups share a set of values and meanings not shared by the society of which they are a part. When this occurs, we may speak of a *subculture.* A subculture is a "culture within a culture." More specifically, a subculture is a collection of norms, values, and beliefs whose content is distinguishable from that of the dominant culture. This implies that the persons who subscribe to the subculture participate in and share the "larger" culture of which the subculture is a part. At the same time, it implies that the subculture has some norms and meanings that are peculiar to it. A subculture is not necessarily in opposition to the larger culture, although if it is, the term *counterculture* is more appropriate (Yinger, 1982).

An example of a counterculture is outlaw motorcycle gangs whose members refer to themselves as "one percenters." When the American

Motorcycle Association condemned the activities of outlaw bikers, it claimed that these cyclists were only "one percent" of the organized motorcycling population. The term was adopted as a symbol of distinction by Hell's Angels and other such groups (Thompson, 1966: 13, 18). One-percenters live a hedonistic life and they often see themselves as social pariahs, often by engaging in totally outrageous behavior for the benefit of onlookers. The values of the subculture include mobility, mechanical ability, skill at fighting and riding very large Harley-Davidson motorcycles, and the ability to manipulate or "con" others (Watson, 1982). Crime is often a part of the life of these cyclists and their street lifespan is claimed to be only about five years (Quinn, 1987). After that, the effects of the law, brawls, or crashes all take their toll and the one percenters move out of the gangs, usually into working-class occupations.

A variety of subcultures and countercultures characterize modern industrial societies. Some of these subcultures are deviant. Cohen has suggested that subcultures arise in highly differentiated, complex societies when a number of persons have similar problems with the prevailing culture. In his view, subcultures represent collective solutions to shared problems posed by the dominant culture (Cohen, 1955: 14). According to him, the delinquent subculture comes about as a result of the frustration of the lower-class boy trying to meet the middle-class expectations of him in the school. The delinquent subculture provides an alternative status system, one in which the boy is more equipped to compete. Lewis (1961) has adopted a similar view in his description of a subculture of poverty.

This is the same process that criminologists have described for the origin of subcultures within institutions for deviants, such as prisons (Johnson, 1987). In prison, subcultures represent social alternatives to the prison world. Composed of opposing norms and values, these subcultures may be affiliated with a gang in the prison that provides support and protection for inmates who are members of that gang. Prison subcultures are not only different from the larger culture (i.e., the culture of the prison) but they are also different from one another. Racial and ethnic conflict among inmates in prison is now a common feature of many maximum-security prisons because the content of these subcultures conflict with one another.

Large city slum areas are more than overcrowded, congested, run-down physical areas. Sociologically, a slum represents a subculture with its own set of norms and values which are reflected in poor sanitation and health practices, a lack of interest in formal education, and characteristic attitudes of apathy and social isolation from conventional institutions. Inner-city areas are also characterized by subcultural norms conducive to violence, theft, delinquency, vandalism, selling and taking illegal drugs, and the presence of street addicts. In this sense, a "slum way of life" born from a combination of economic-specific cultural characteristics and wider social and economic opportunities frequently characterizes high-rise slum clearance projects in our major cities (Wilson, 1987).

The subcultural dimension of many forms of deviance is quite evident and, in the chapters that follow, subcultural influences and contexts to drug use, homosexuality, skid row drinking, delinquency and crime, and even suicide will be noted. There are even descriptions of subcultures for discharged, chronic psychiatric patients. Deinstitutionalization of large numbers of chronically mentally disordered persons has resulted in their becoming street people in large cities. The problems they face in meeting the demands of modern urban society are partially solved by subcultures that have arisen to perform this function by providing social support for members, enhancing self-esteem by providing rationales for their conditions, and making available practical suggestions for "making it" on the outside. The ex-mental patients may engage in several deviant activities, including selling their legally-obtained medications, shoplifting and even prostitution, but the norms of the subculture provide limits on these activities and, at the same time, the subculture provides a justification for deviance. As one patient phrased it:

> We're not doing anything that's really wrong. We don't murder or rob or things like that. We only take a few groceries once in a while from the A&P store. And we only do that when it's absolutely necessary. Other people who have lots of money do it all the time, and they take things much bigger than we do. We do it for medical reasons—our health, but they just do it for greed. (Herman, 1987: 252)

To summarize the importance of group norms in modern, complex societies, the following generalizations may be helpful. (1) In modern societies, there may be almost as many pronounced differences among the groups within the society with respect to the norms of accepted behavior as there are differences between cultures themselves. (2) Any logical explanation of why members of certain deviant subgroups act as they do must trace the development of the behavior in the same way that any member of any cultural group learns to act, for example, how Eskimos learn through their culture to be an Eskimo (e.g., how to act, think, and interpret the world like Eskimos). (3) Finally, it must be remembered that even when the norms of any given family are discussed, it is probable that one is actually discussing the social class, the occupational group, or some specific subcultural group to which the family happens to belong.

The Relativity of Deviance

Defining deviance in terms of norms does not identify any particular types of conduct as deviant. Deviance is not static or constant. Deviance changes constantly in terms of the forms it takes and the degree of disapproval it elicits. In this sense, deviance is not a unique type of behavior but, rather, common behavior that happens to offend some group. Because norms are relative (to groups, to places, to times), deviance is relative. This is why

The Relativity of Deviance

Activity	Probably Not Deviant For	Probably Deviant For
Drinking beer	Fraternity members celebrating a football victory	Baptist deacons and elders celebrating a successful church fund raising campaign
Asking someone of the opposite sex out on a date	Unmarried pesons	Persons married to someone else
Being able to set one's own bedtime	parents	children
Sexual intercourse	married couple	Catholic priest
Selling drugs	pharmacist	illicit drug dealer
Consuming alcohol	Priest during religious ceremony	someone with a desire to get drunk
Acting "weird"	Someone who just won the state lottery	An older person who has no reason to act differently

there is an almost endless variety of acts and characteristics that are considered deviant depending on the conditions and circumstances. Debates over such acts as prostitution, gambling, nudism, cheating, medical quackery, and marijuana use are based on conflicting norms pertaining to these acts. And, just as some acts are considered deviant by some, so too are various kinds of people. Social types perceived by some as deviants include reckless drivers, pacifists, racists, "hippies" and radicals, "squares" and conservatives, the very rich and the very poor, old people, drinkers and nondrinkers, and motorcycle gang members. Conservatives, for example, are considered deviant by liberals, and vice-versa.

The fact that deviance is a violation of a norm does not identify whose norm it is. When we ask ourselves what deviance is and who is a deviant, we are faced with specifying which groups within society define certain behavior as deviant and which groups do not. We are asking, in other words, whose norms are being violated. In this sense, deviance must be viewed from the perspective of the social audience of the act. Take, for example, the designation of sexual "promiscuity."

Suppose that a particular unmarried woman maintains an active and varied sex life. While some people may condemn her as "promiscuous" others may view her and her behavior as "liberated." Not that these highly divergent designations do not stem from differences in the sexual behavior itself. On the contrary, the behavior has been the same; it is only the evaluation of it that has varied. (Schur, 1984: 5)

Those who regard the woman's behavior as promiscuous might not permit her to reenter conventional social sexual roles, even after a long period of "nonpromiscuous" behavior.

Sociologists often maintain that there is nothing inherently deviant in any act; deviance is a judgment that is made with reference to some norm. It is the norms that "create" deviance by creating social differentiation and attaching a moral ("ought" or "ought not") quality to the act. This is not to say that there is not widespread agreement on the wrongfulness of certain acts, such as the deliberate killing of a person, the physical assault of a very old person, or sexual intercourse with a young child; but it does suggest that moralities differ because norms differ.

While norms are relative, it is also the case that some norms are given more attention than others, and it often matters how much power a group has in enforcing its norms over group members as well as over others. What is considered deviant may depend on the relative power of groups to enforce their norms on others. Social power, then, is important in understanding why deviance is relative. For example, strong negative attitudes toward taking one's own life through suicide, the practice of prostitution, homosexuality, drunkenness, and other means of expressing personal choice have stemmed, for the most part, from the actions of certain church groups (see Greenberg, 1988). Opposition to the use of marijuana, nudity, and the distribution of pornographic materials rests with other "moral entrepreneurs" who attempt to impose their norms on others (Becker, 1973: 147–163). Some criminologists have maintained that what constitutes a crime and the severity of the penalty for that crime is specified by segments of society that have the power to shape criminal policy (Quinney, 1981). Thus, a burglar who netted $200 may be treated more harshly by a criminal court than a corporate executive who caused consumers to be defrauded of millions of dollars. The burglar may be sentenced to a long prison term, while the executive may be put on probation, given a small fine, or told to perform some public service for the community.

The Example of Nudism.

Nudity, like all behavior, is governed by norms. These norms are relative to groups, times, and social situations and it follows that nudity is sometimes considered deviant, sometimes not. To many people, nudity outside of one's bathroom or bedroom is deviant, but nudists do not have such a conception.

Whispering Pines is a nudist resort in North Carolina. There are a number of conditions that one might think would be conducive to sexuality:

nudists camps are located in relatively private settings, members are there voluntarily and, presumably, these are people with more "liberal" attitudes about being naked; and, perhaps above all, people are unclothed and have plenty of opportunity to engage in sex. Many people have a strong association between nudity and sex.

But like most nudist resorts, there are a number of norms at Whispering Pines that restrict sexual behavior. For example, attendance at the resort is legitimated by a "club" atmosphere. The fees at Whispering Pines, a nudist resort in North Carolina, are $250 per year plus $32.50 for annual American Sunbathing Association dues. Members then pay a per-day site rental charge. Whispering Pines allows no more than 10 percent single males and 10 percent single women at one time, thus restricting the proportion of sexually unattached persons. Children, whose socialization to the sexuality is incomplete, must be supervised at all times in the camp. If one acts rude, insists on having binoculars or cameras, or acts lewdly, that person's name and picture could be put on an American Sunbathing Association list which would bar visits to nudist resorts around the country. First-timers, called "Cottontails," must learn these norms.

In addition to these restrictions, there is an explicit rule that sex in public is forbidden. Perhaps, because of all these norms, the owners of Whispering Pines indicate that open sexuality just doesn't happen. Attendance at Whispering Pines is defined as nonsexual. As one owner put it:

> There are no bathing beauties. People are just people. Women are always worrying about fat knees, legs, that sort of thing. You come out here a while, you wouldn't worry about stuff like that. They accept themselves as what they are. (Hill, 1990)

Persons who attend nudist camps go to some length to dissociate the relationship between nudity and sexuality. Nudity is considered natural and not "dirty." But such a conception would be considerably more difficult to maintain if nudist resorts condoned or encouraged more profuse sexuality.

But there are other nudists who find the participants at Whispering Pines deviant. Naturists are persons who wish to be nude in public places, not just nudist camps. They view nudity as natural in virtually all social situations. Naturists have disdain for nudist camp attendees because the nudist resort attendees display their nudity only in these situations and because they divorce nudity from sexuality, another natural process. Jeannette, a naturist nudist, says:

> We went to a camp where we were the first naturists to visit. The first thing we did was to introduce ourselves by telling everyone our first and last names. They jumped all over us because we gave out our last names. They told us that we should never tell anyone our last name at a nudist camp because we would be giving someone the opportunity to do terrible things with the information that could hurt our reputations. It is pretty obvious to us that they didn't believe that nudity is alright. (Cox, 1989: 123)

To the general public, persons who attend nudist resorts are engaging in deviance because there is a commonly held norm that restricts nudity to "private" places or relationships. To the naturists, persons who attend nudist resorts are engaging in deviance because they are not "true" nudists. True nudists are not afraid of associating sex with nudity. One of the naturists said in his experience the nudist camp people didn't even talk about sex: "I find them to be nauseatingly sterile" (Cox, 1989: 123).

Who is deviant? The nudists in the resort who dissociated nudity from sex, the naturists who feel comfortable with that association, or non-nudists? It is all relative.

Creating Deviance

Deviance is often a created entity. The process of defining of an act as deviant is a political act that involves power and is related to some symbolic or moral context (Ben-Yahuda, 1990). Groups whose interests are threatened may attempt to promote those interests by attempting to persuade others of the legitimacy of those interests.

Social issues can be created and maintained by successful social promotion. Such processes can be identified for a number of forms of deviance, including homosexuality, drunk driving, and public concern about the use of certain drugs. Trebach (1987), for example, relates public attitudes about cocaine to the national attention focused on the deaths of two well-known athletes in 1986, specific television specials about "crack" and other drugs, and the statements of political leaders about starting a "war on drugs" that would include drug testing and greater legal penalties.

Missing Children. Deviance can be created when groups are able to persuade others of the legitimacy of their norms. The problem of missing children illustrates how the concerns of one group can be enlarged to that of other groups as well. Children in the United States have historically been the object of a number of concerns, including delinquency, poor school performance, and neglect. In recent times, the list of concerns has broadened to include issues of child prostitution, child pornography, negative consequences of listening to rock music, Halloween sadism, incest, child molestation, involvement with religious cults, and drug use. The most recent concern is over missing children, most of whom are incorrectly presumed to have been kidnapped (Best, 1987). The missing children's movement has been generated by individual parents of missing children, as well as groups that have been formed to help locate missing children, such as Child Find, a national organization. The promotion of the missing children has included bringing the issue to the attention of legislatures and the public through television programs and movie adaptations of actual missing children cases. Public service announcements on television and publishing the pictures of missing pictures on grocery bags and milk cartons nationwide testifies to the success of this movement. The tactics of those promoting this issue include using a broad definition of the problem, using horrific examples, and using large estimates of the scope of the problem. While the movement has implied that the missing children are the result of the actions taken by strangers against the children, it usually involves a kidnap by an estranged parent. The numbers of such children taken annually under these circumstances is relatively small, and the kidnapping of children by strangers is even more rare.

Satanic Cults.   Some believe there is a large amount of Satanic worship and cult activity in the United States at present. Media descriptions of unknown groups engaging in animal slaughter, late-night rituals, and human sacrifice have been presumed by some to indicate a growing incidence of this type of behavior. An increase in books, articles, and talk show topics on Satanism could lead one to conclude that this type of behavior is increasing.

Are Satanic cults on the increase? Although there are, clearly, some groups which have a concern in the occult, witchcraft, and Satanism, it is not clear that this type of activity has increased. Victor (1990: 288) has reported that one claim of ritualistic killing of animals in New Hampshire was "later determined to be only road kills cleaned up by state road workers and deposited in the woods." Hicks (1990) observes that a comprehensive investigation of cult-mutilation claims concluded that they are due to the natural actions of scavengers and predators.

In this sense, without denying the existence of some Satanic groups, it appears that Satanism can be socially created (see also Forsyth and Oliver, 1990). The activities of a few can be overemphasized to suggest that there are many such groups and that these persons pose a significant threat to others. Such a view might be beneficial to religious persons and groups who use the idea of the devil to scare persons into the safety of the Christian church.

Determining Norms and the Content of Deviance

A difficult sociological problem in the study of deviance is determining how strongly certain norms are opposed by various groups within a society (Sagarin, 1975: 222). Some norms are more easily determined than others. The criminal law contains a body of legal norms that all can see; norms regulating sexual behavior, on the other hand, are often ambiguous, changing from group to group and situation to situation over a period of time.

Some normative change may be predictable. Some observers have argued that there is a cycle, or stages, in which some forms of behavior that are disapproved move to greater tolerance (see Winnick, 1990). At one time, a behavior may be widely socially disapproved, but some group may call for a change in that disapproval. The group may be aided in their efforts to promote a new norm by claims that many persons are being victimized by a current policy or the publication of a major research study (such as a public opinion poll). Over time, others may come to share the norm, thereby changing the social designation of what is and what is not deviant. This life-cycle approach to deviance may explain changes in such things as cigarette smoking, the consumption of alcohol, the use of heroin, and homosexuality.

The problem is made even more difficult by the fact that many norms change. Cigarette smoking, for example, has undergone a number of changes in acceptability in the United States since the 1800s (Troyer and Markle, 1983). In the 1870s, cigarette smoking was strongly condemned by many groups and by many individuals, in part because it was most common among urban immigrants who had lower social status and who were often characterized by heavy drinking. At that time, too, smoking by women was considered to be particularly deviant, being associated with prostitution. In spite of its being considered deviant, cigarette smoking increased in the United States, and attitude began to change following World War I. By the end of World War II, it was considered acceptable behavior, socially desirable, and, in some circles, even necessary.

Attitudes began to change again in the 1960s as evidence began to accumulate that tobacco smoking was linked with a number of serious physical illnesses. In the 1970s, cigarette advertising was legally prohibited on television and radio. Smoking is increasingly banned in public places because many nonsmokers find the odor objectionable and because medical evidence shows that inhaling someone else's cigarette smoke is dangerous to the inhaler even if that person is a nonsmoker. Indeed, citizen groups have been campaigning for even stronger measures, and the 1980s have witnessed several cities enacting ordinances prohibiting cigarette smoking in certain places, for example, elevators, public meeting rooms, and certain areas in restaurants. And cigarette smoking has again become deviant as norms regulating this behavior have changed. Anti-smoking is a norm that has been successfully promoted in recent years.

By 1989, 42 states had enacted legislation restricting smoking in public places. Most of this legislation has been based on the idea that smoking, regardless of its effects on the smoker, also adversely affects others (Goodin, 1989). Clearly, the norms on smoking have changed.

Normative change sometimes takes place for very complex reasons (as in the case of cigarette smoking), sometimes for more identifiable reasons (as in the case of Prohibition). Linking cigarette smoking with health hazards and an increasing emphasis on self-control and physical fitness have helped produce a change in many peoples' evaluations of smoking. Even more important has been linking smoking with the notion of individual rights of non-smokers to avoid the smoke of smokers. With respect to Prohibition, some groups were instrumental in creating the legislation that outlawed the production, sale, and consumption of alcoholic beverages in 1920. Alcohol offended their sense of morality and they were successful in convincing others of that evaluation (Gusfield, 1963). Still, this is not to say that everyone considers cigarette smoking or the consumption of alcohol (or even marijuana) deviant. In fact, there is great dispute over how deviant each of these activities is, depending on who is making that judgment.

SOCIAL CONTROL

Many scholars regard the problem of social order as the fundamental question for all of the social sciences (Rule, 1988: 224). Why do people conform to rules and norms, even when it is not in their interest to do so? How is it that some persons come to violate laws and others to violate deeply held social understandings about appropriate conduct? Most sociologists regard such questions as involving social control.

All social groups have means to deal with behavior that violates social norms. These methods, taken together, are called *social control* (Meier, 1982). The notion of control is broad but might be conceived to be "overt behavior by a human in the belief that (1) the behavior increases or decreases the probability of some subsequent condition, and (2) the increase or decrease is desirable" (Gibbs, 1989: 23). To speak of social control is to speak of deliberate attempts to change behavior. The purpose of social control measures is to insure, or at least attempt to insure, conformity to norms. In some situations, people conform to norms because they know of no alternative. In other situations, they do so because they have been offered an inducement of some kind to do so. These inducements may be through informal social control mechanisms, such as ridicule, or through formal agencies, such as the church or government.

Processes of Social Control

Two basic processes of social control are distinguished by sociologists: (1) the *internalization of group norms* wherein conformity to norms comes about through the socialization of both knowing what is expected and being desirous and able to conform to that expectation (Scott, 1971); and (2) social reaction through *external pressures in the form of sanctions from others* in the event of anticipated or actual nonconformity to norms. These are not mutually exclusive processes and, in fact, they can and do occur together.

Internalization.   In the internalization of group norms, social control is the result of a person learning and believing in the norms of his or her group. This process is a result of the overall socialization process whereby persons acquire a motivation to conform to group expectations regardless of other external pressures. No conscious effort is required on the part of society to secure compliance with norms, for they are the spontaneous and unconscious ways of acting that characterize the bulk of customs of any culture.

Mechanisms of social control, like customs, traditions, beliefs, attitudes, and values are generally taught through prolonged interaction between persons. The fact that most wives do not murder their husbands is due not entirely, or even mostly, to the severe legal penalties for criminal homicide; most automobile drivers stay on the right side of the road not entirely because, otherwise, other drivers will regard their driving as deviant; and drinkers of alcoholic beverages do not all become drunks simply because the neighbors will gossip. Rather, most persons conform to most norms most of the time because, first, they have been taught the content of those norms and, second, because they have accepted the norms as their own and operate as though they are taken for granted.

A great deal of conformity to norms takes place because persons have been socialized to believe they *should* conform, regardless and independent of anticipated reactions from others. In this sense, socialization can be referred to as "self-control" because this is often the result of the socialization process. Social control consists, in a sense, of processes which teach the person not to embark on processes of deviance; they consist of how not to, rather than how to, engage in deviant behavior.

Sanctioning.   Sanctions are reactions to behavior and can be classified according to their content. Social controls through external pressures include both negative and positive sanctions. A *negative sanction* is a punishment meant to discourage deviant conduct. A *positive sanction* is a reward meant to encourage conduct to conform to a norm. Sanctions can also be classified according to the source of the sanction, or who is doing the reacting. *Informal sanctions*, such as gossip and ostracism, are unofficial actions of groups or individuals, while *formal sanctions*, such as the use of the criminal sanction, are official group expressions meant to convey collective sentiments.

Examples of Different Kinds of Sanctions

		Source of Sanction	
		Formal	Informal
Content of Sanction	Positive	Raise in salary Medal in the army Certificate	Praise Encouragement Smile
	Negative	Imprisonment Being fired from a job Excommunication	Criticism Spanking a child Withholding affection

Formal and informal sanctions are not independent from one another. In one study, for example, it was found that among a sample of 800 teenage boys, more concern was indicated for what their families would think of them than about having been arrested by the police and having been subjected to formal penalties (Willcock and Stokes, 1968). Yet, the fear of formal penalties, such as arrest and incarceration, was important too. This suggests that a combination of both informal and formal sanctions is of great importance in influencing behavior.

Informal Social Controls

Informal social sanctions consist of the reactions to behavior of people who know one another on a personal basis. Informal sanctions are used to enforce informal norms, often in small groups.

Informal sanctions, such as gossip and ridicule, may be more effective in smaller social groups where everyone knows everyone else and the same people are brought into continual face-to-face contact with one another. One example of an informal sanction, in this case gossip, has been reported from such a society.

> Early this morning, when everyone was still around the village, Fokanti began loudly complaining to an affine (who was several huts away and was probably chosen for that reason) that someone was "killing her with broken promises." Who? Asibi. He promised to help Fokanti with her rice planting today and now he's reneged. At this point, Asibi appeared and tried to explain how something else had come up which required his attention. This cut no ice with the woman, who proceeded—her voice still at a high volume—to attribute Asibi's unreliability to his "just wanting to go to dances all the time, like last night!" None of this public broadcasting was helping Asibi's reputation any, so he promised to change his plans and make good his original promise. (Green, 1977: 10)

Clearly, the effect of informal sanctions, as in this example, can be quite powerful. Asibi kept his original promise because he cared about what others thought of him. His reputation in the group was sufficiently important that he did not wish it damaged. He wished, in other words, to avoid the shame of community embarrassment for not keeping his word.

Informal sanctions' reactions may be observed in specific behavior such as ridicule, reprimands, criticism, praise, gestural cues, glances, and other mechanisms of "body language" meant to convey approval or disapproval, denial or bestowal of affection, or verbal rationalizations and expressions of opinion. "Frequently, the penalty consists of verbal expressions of displeasure; even a glance of annoyance on the face of a friend is often enough to inhibit deviant acts or to arouse feelings of guilt or shame" (Shibutani, 1986: 218). Gossip, or the fear of gossip, is a very effective sanction among people who have close personal relationships.

Braithwaite (1989) believes that an important dimension to crime control generally is the ability to use informal social controls such as shame to prevent and reintegrate illegal behavior. He notes that most persons refrain from crime not because they fear the legal sanctions, but because their consciences do not permit legal violation. Most people do not contemplate doing "bad" things because they believe them to be bad and should be avoided. For those persons who do violate the law, in addition to the formal sanctions of fines, jail, and prison, an appeal to shame should be made where their consciences are aroused and they control themselves. The use of shame, a common sanction in such countries as Japan, bridges the informal power of the individual conscience and the formal power of the state in the criminal sanction.

> When a young constable raped a woman in Tokyo several years ago, his station chief resigned. In this way, junior and senior ranks express a shared commitment to blameless performance. This view of responsibility is part of the Japanese culture more largely. When a fighter aircraft struck a commercial airliner, causing it to crash, the minister of defence resigned. Parents occasionally commit suicide when their children are arrested for heinous crimes. . . . Japanese policemen are accountable, then, because they fear to bring shame on their police "family," and thus run the risk of losing the regard of colleagues they think of as brothers and fathers. (Bayley, 1983: 156)

Formal Social Controls

Formal controls involve organized systems of reactions from specialized agencies and organizations. The two main types are those instituted by the political state and those imposed by agencies other than the state. The latter groups include churches, business and labor groups, educational institutions, clubs, and other organizations.

The development of formal systems of control may be related to conditions that weaken informal systems of control (see Horwitz, 1990: 142–149). When the control of family, church, clan, or community is largely unavailable, as is the case in the process of urbanization, alternative forms of control are necessary. This alternative form of control involves the use of third parties—such as the state in the form of police, courts, and correctional systems—to enforce various norms and regulations.

Because formal sanctions are incorporated into the institutional systems of society, they are administered by persons who occupy particular positions or roles within those institutions. These persons are commonly known as "agents of social control" since the administration of sanctions is part of their official duties. In the most general sense, anyone who attempts to manipulate the behavior of others through the use of formal sanctions can be considered an agent of social control. The police, prosecutors, and judges in the criminal justice system obviously qualify, but so too do employers, psychiatrists, teachers, and ministers and priests who promise heaven and threaten hell to believers. In each instance, the persons who occupy these roles are charged with making reactions (sanctions) to the behavior of others. The behavior of social control agents can be viewed as comprising a system of social control where the control efforts from different sources—police, judiciary, corrections, juvenile justice, psychiatry, welfare, the family, and other agencies of the state and civil society—form a network of control (see Lowman, et al., 1987).

Nonpolitical groups impose penalties, some of which may be more severe than punishments imposed by the political state for crimes. A business concern may fire a person, even after long years of employment, for an act of deviance, such as employee theft or embezzlement. A professional group or union may suspend or even expel the individual from the group, which may mean a loss of livelihood (see Shapiro, 1984: 135–166).

Professional athletes who do not obey the rules of the club or league may be fined several hundred or several thousands dollars for an infraction,

Examples of Institutions of Social Control

Institution	Agent	Deviance	Sanction
Religion	Minister, priest	Sin	Penance, withholding rites, excommunication
Business	Employer	Absence, laziness, violation of work rules	Dismissal, suspension, fine
Labor union	Labor representative	Failure to obey union rules	Expulsion from union, fine
Professional group	Officer	Ethical violations	License revocation, expulsion from group
Political state	Police, prosecutor, judge	Violations of administrative, civil, or criminal law	Fine, probation, imprisonment, damage suit
Clubs and social organizations	Officers	Violations of club rules	Fines, suspension of privileges, expulsion
Family	Parents	Youthful disobedience	Spanking, "grounding," withholding privileges

such as insulting a spectator, not obeying club rules, or losing the club's playbook. Violations of such norms may result in a fine or even being suspended, meaning the loss of even more pay. Religious organizations may demand penance or withhold certain religious services, such as the wedding privilege or a religious service at death. They may even use what is, to members of a particular faith, the most drastic punishment of all—excommunication from the church. Clubs and similar groups generally utilize a scale of fines, temporary suspension of membership privileges, or even expulsion as a means of controlling their members.

A series of specific actions is established not only to punish the transgressor, but to reward those whose compliance with norms equals or goes beyond the expectations of the group. Curiously, nonpolitical agencies, such as businesses and professional, religious, or social groups, probably use rewards more than punishments to manipulate the behavior of their members (Santee and Jackson, 1977). Through promotions, bonuses, or some token of merit, business organizations frequently reward those who have made an outstanding contribution to the firm. Professional groups reward faithful members with election to honored offices or a special citation. Religious groups reward their members with promises of a future state of euphoria, by positions of leadership within the church organization, and by pins and scrolls for exemplary service and commitment. Clubs, lodges, and fraternities and sororities likewise offer a large number of prestige symbols for those who walk the path from initiate to full-fledged member without reflecting dishonor on the group. Recognition of a type similar to military rewards is given to a small number of civilians each year in the United States through the Carnegie awards for outstanding heroism.

Unlike many other kinds of organizations, the political state seldom exercises control through the use of positive sanctions, or rewards. Citizens who go through life systematically obeying most requirements put on them by the law seldom receive rewards or commendations for such behavior. It is simply not practical for the state of, say, Iowa to award a certificate to all those who did not commit a burglary in the past year, even if such persons could be reasonably and practically identified. Some states and cities occasionally give publicity to long-term safe or courteous drivers, but this is one of the few exceptions. The fact that the state is unable to make use of positive sanctions has important implications for our expectations regarding the effectiveness of the social control efforts of the state. We might expect that official social control from the state would be less effective because state sanctions are only negative in content, and therefore limited.

The political state can impose a variety of penalties upon state or legal norm violators and some observers have noted that the power of the political state has increased over time (Lowman, et al., 1987). Law violators below the legal age of adulthood come under the jurisdiction of juvenile courts; those who have attained adulthood are subject to punishment under the criminal law. Offenders can be fined, imprisoned, placed on probation, or in some states, even condemned to death. The state also has a means of controlling law violations by business groups and corporations in addition to the criminal law. It can use administrative sanctions to compel adherence to law. These civil actions include monetary payments, court injunctions, and license revocations.

Deviant Contacts with Social Control Agencies

The deviant interacts with associates, victims and others in the commission of a deviant act. The deviant, however, also may interact with agents of social control. Agents of social control represent the "community" or "society," and that interaction has important consequences for the deviant. The application of social control measures, under certain circumstances, may intensify or reinforce deviant acts in ways that had not been intended. Participation in a drug treatment program, for example, may intensify the drug user's self-conception or identity as an addict. The person may develop the feeling that continued association with other addicts and participation in the addict subculture is necessary or even "natural" in light of this self-conception.

Whether a person is directed toward or away from deviance as a result of contact with a social control agency depends on many factors. Certainly, to the extent that persons have contact with social control agencies, the feeling of differentness and apartness that most deviants experience will be enhanced. Some deviants will continue their association with other deviants and their deviant conduct primarily as a result of their contact with agencies and agents of social control. In this sense, rather than solving them, social control agents and agencies can contribute to deviance problems.

A Final Note

Deviance related to social behavior and deviance related to social problems are not necessarily the same, even though the two kinds of deviance overlap. Not all social problems are instances of deviance. For example, unemployment is considered by many to be a major social problem, as are population control and the provision of adequate medical care to poor people; yet, these can hardly be considered instances of deviant behavior. The same could be said about other conditions, such as aging and racial discrimination (See Manis, 1976; and Spector and Kitsuse, 1979). Consequently, the forms of deviance selected for study here are those over which there is, for the most part, much contemporary interest, debate, and concern. In the past, different types of behavior might have been discussed as deviant. Three hundred years ago, or even less, blasphemy, witchcraft, and heresy might have been included because they were then regarded as serious forms of deviance often punishable by death. In the future, some forms of behavior regarded today as deviant may well disappear as new norms arise and new issues replace old ones.

Obviously, space limitations preclude a book such as this from analyzing all forms of social behavior that might possibly be termed deviant; thus, one must be selective. Those forms of behavior designated here as deviance in terms of the criteria stated above are certain types of crimes, including those of personal and family violence and those against property, crimes against the political state, and those committed in connection with an occupation such as white-collar and corporate crime; illegal drug use; deviant alcohol use and problem drinking; prostitution; co-marital sexual relations; homosexuality; mental disorders; suicide; and severe physical disabilities, such as those experienced by the crippled, the obese, the mentally retarded, and the blind.

Considerable consensus exists about the deviant nature of some of these acts, as in the case of murder and alcoholism. In others, such as homosexual behavior, the use of marijuana, and certain types of heterosexual acts, less agreement is evident. Even when some measure of agreement exists that certain behavior is deviant, there may be strong disagreement regarding approaches to its social control.

SUMMARY

The notion of deviance most generally refers to differentness and applies to behavior, conditions, and people. Deviance can be defined in statistical, absolutist, reactivist, or normative terms, although there may be less difference between the reactivist and normativist conceptions than some believe. Deviance is here defined as those deviations from norms in a disapproved direction such that the deviation elicits, or is likely to elicit if detected, a negative sanction. The amount and kind of deviance in a society is related to the degree of social differentiation in that society.

Deviance is relative to the norms of a group or society. Just as norms change, so too does deviance. At times, it is difficult to identify norms prior to their violation and, because not everyone subscribes to a given norm, there may be disagreement on what is and what is not deviant. Deviant acts are a necessary, but not sufficient, condition for becoming a deviant. One cannot become a deviant by committing deviant acts alone because, if that were true, we would all be deviants. Deviance is linked to the stratification system of a society. The greater the differentiation in society, the greater the potential for deviance.

Deviants are persons who come to adopt the role of deviant. Just as persons learn conventional norms and social roles, so too do they learn deviant roles and patterns of behavior. The relationship between adopting deviant roles and the commission of deviant acts is complicated. A full understanding of deviant acts requires knowledge of the process of committing deviant acts, and the role and actions of victims.

Social control efforts are usually administered through sanctions, or specific reactions to behavior. The internalization of norms is probably the most effective form of social control because it makes sanctions unnecessary. Social control can be either formal or informal. There are different types of sanctions and their effectiveness varies. The social control process is part of the deviance definition process. Some people control others by defining their conduct as deviant. In this way, the definition of deviance serves the same function as specific sanctioning efforts—keeping people "in line" or in their "place."

There is some overlap between the notions of deviance and social problems, but they are not the same thing. The range of deviance is enormous and not every instance of deviance can be discussed here. Only those instances of deviance about which there seems to be strong consensus or over which there has been strong normative dispute are discussed in this book.

SELECTED REFERENCES

Becker, Howard S. 1973. *Outsiders: Studies in the Sociology of Deviance,* enlarged edition. New York: Free Press.

> A classic statement on the sociology of deviance that proposes an interactionist or labeling perspective (see also Chapter 5). Becker's observations in this book on the nature of deviance may be the most widely cited of any contemporary book.

Bryant, Clifton D., ed. 1990. *Deviant Behavior: Readings in the Sociology of Deviant Behavior.* New York: Hemisphere.

> A collection of recent papers on different aspects of deviance. The breadth of the book is impressive. Useful for both classroom and research work on different forms and process of deviance.

Gibbs, Jack P. 1989. *Control: Sociology's Central Notion.* Urbana: University of Illinois Press.

> An extended discussion of the idea of "control" generally with particular attention to social control. Gibbs maintains that control is a unifying concept in social science in general and sociology particularly.

Horwitz, Allan V. 1990. *The Logic of Social Control.* New York: Plenum.

> A systematic examination of the nature and meaning of social control. Well written and very informative. The author examines social control as something itself to be explained.

Sagarin, Edward. 1975. *Deviants and Deviance: An Introduction to the Study of Disvalued People and Behavior.* New York: Holt, Rinehart and Winston.

An older but not dated discussion of the concept of deviance and processes of becoming deviant. Written in textbook form, this book contains many sociological insights on deviance and social control.

Schur, Edwin M. 1984. *Labeling Women Deviant: Gender, Stigma, and Social Control.* New York: Random House.

This book makes a cogent and forceful argument applying the concepts of deviance to women. Excellent discussion of appearance norms that identify "ideal" physical appearance and the consequences for violating those expectations.

Becoming Deviant

2

Up to this point, the discussion has centered around the definition of deviance, norms, subcultures, societal reaction, and the social control of deviance. In this chapter, the emphasis will be on how an individual becomes deviant. People do not become deviant by simply committing a deviant act. If the sociological criterion for being a "deviant" were simply the commission of a deviant act, we would all be deviants and the term would have little meaning. Sociologically, deviance involves playing a social role that exhibits this behavior. The way in which people come to occupy deviant roles is here referred to as the process of becoming deviant. To understand processes of acquiring deviant roles, it is necessary to examine the social nature of human beings, including the self and identity of deviants, and the process of socialization into a deviant role. It is also important to be able to empathize with deviants, to see the world the way they do. We gain important insights into the processes of becoming deviant when we understand the meaning of deviant acts for such persons. In this chapter, various ideas on the origin of deviant behavior are critically examined, including biological, psychiatric, psychoanalytic, and a generalized perspective called the medical model.

It is important at the outset, however, to make one fundamental point: The belief that deviants are inherently different from nondeviants is built upon a series of false assumptions, for *all deviant behavior is human behavior*, and the basic processes that produce social behavior are the same for the deviant and the nondeviant. The subprocesses affecting deviants operate within the general framework of a theory of human behavior. The units of analysis, as well as the fundamental social processes, in all human behavior are the same whether the end products are inmates in correctional institutions or wardens, mental patients or psychiatrists, corrupt businessmen or ministers.

Deviance is neither unique nor clear-cut; there is no explicit way to distinguish deviance from nondeviance without reference to norms, and, as we have seen, norms change. The relativity of deviance means that behavior that is not deviant at one time or in one situation may be considered deviant at another time or in another situation. For much the same reason, deviance is ambiguous. Human beings must live with changing norms, which means that the expectations or norms that govern behavior are constantly shifting and their applicability must be continuously assessed. As children mature, for example, the expectations of the parents may change. Even daily rules change. Bedtimes, being able to travel to certain parts of the neighborhood, telephone calls, and other privileges are renegotiated as the child grows older, and what was a "given" at one time is not clear anymore. Human beings live their lives in situations with conflicting expectations and demands. Teenagers, for example, who follow the expectations of their parents may violate the expectations of their peers. Similarly, employees who follow the expectations of their employers may violate the expectations of their fellow workers.

Since deviant behavior is human behavior, the same general explanations of one should apply to the other. It is necessary to discuss the social nature of humans to see how deviant and nondeviant conduct stems from the same basic social processes. Deviance may become the role behavior of an individual. Just as there are conventional social roles, so too are there deviant roles to which one can be socialized.

SOCIALIZATION AND SOCIAL ROLES

In a sense, deviants are hypocrites. They will violate some norms but conform to others. There does not appear to be any general behavior pattern of conformity and nonconformity with social norms. Persons may deviate from certain norms and comply with others. A criminal may break the law by extorting money from people, but he would never steal from friends or cheat on his wife because he thinks that marriage is sacred. Those who deviate from sexual norms may not steal, for example, while many white-collar criminals have a rigid sexual code and are largely ethical in their dealings with neighbors. Top management of criminal corporations may be highly dedicated citizens of their local community. In most cases, strongly disapproved deviations may be but a small proportion of a person's total life activities. Even where deviations constitute a more organized subculture, as among heroin addicts, accepted conduct may coincide at many points with norms and values of the larger community (Levison, et al., 1983). No one is deviant all of the time, and even the most committed deviant will engage in deviant acts only at some times.

Social behavior is acquired. It is not present at birth; it develops through socialization. Behavior becomes modified in response to the demands and expectations (norms) of others. Practically all behavior is a product of social interaction. Words like "honesty," "friendliness," and "shyness" have meaning only in relation to other people. Even expressions of emotionality, such as anger or depression, although they may have physiological components, are mostly the expressions of social reactions. Individual emotions are social products too (Scheff, 1983).

As a result of group experiences, a human being becomes dependent upon others for human associations, conversation, and social interaction. The importance of this dependence on groups can be demonstrated in situations where group contacts are removed. Admiral Byrd, the first person to fly over the North and South Poles, voluntarily isolated himself for several months in unhabitable polar regions more than a hundred miles from the nearest human being of his expedition. Byrd was interested in discovering his reactions to such isolation. His diary described his experiences of being alone and vividly showed how dependent the individual is on social groups when such contacts are removed.

> Solitude is an excellent laboratory in which to observe the extent to which manners and habits are conditions by others. My table manners are atrocious—in this respect I've slipped back hundreds of years; in fact, I have no manners whatsoever. If I feel like it, I eat with my fingers, or out of a can, or standing up—in other words, whichever is easiest. What's left over, I just heave into the slop pail, close to my feet. Come to think of it, no reason why I shouldn't. It's rather a convenient way to eat. I seem to remember reading in Epicurus that a man living alone lives the life of a wolf.
>
> My sense of human remains, but the only sources of it are my book and myself and, after all, my time to read is limited. [W]hen I laugh, I laugh inside; for I seem to have forgotten how to do it aloud. This leads me to think that audible laughter is principally a mechanism for sharing pleasure. . . . My hair hasn't been cut in months. I've let it grow because it comes down around my neck and keeps it warm. I still shave once a week— and that only because I found that a beard is an infernal nuisance outside on account of its tendency to ice up from the breath and freeze on the face. Looking in the mirror this morning, I decided that a man without women

around him is a man without vanity; my cheeks are blistered, and my nose is red and bulbous from a hundred frostbites. How I look is no longer of the least importance; all that matters is how I feel. (Byrd, 1966: 139–140)

Deviants and nondeviants perform a variety of social roles which represent the behavior that is expected of a person in a given position or status with reference to a particular group (cf. Heiss, 1981). The activities of a human being in the course of daily life can be regarded as the performance of a series of roles that have been learned and which others expect the person to fulfill. A person learns to play such roles as son, daughter, man, woman, father, mother, husband, wife, old person, doctor, lawyer, or police officer. Similarly, a person may learn to perform the deviant roles of gang member, professional thief, drug addict, alcoholic, or that of a mentally disabled person.

Although a great deal of socialization in role-playing and role-taking occurs in childhood, it also continues in later life. Individuals learn new roles and abandon old ones as they pass through the life cycle and encounter new situations. Adolescence represents a period of adjustment to new roles (Hogan and Astone, 1986). Marriage represents the acquisition of new roles, as does entrance into the working world in the form of one's new professional or occupational roles. Old age also often requires a major role adjustment as persons must leave behind old roles and assume other ones (such as that of a "retired person" or "senior citizen").

Social behavior develops not only as we respond to the expectations of others, as we confront their norms, but also through social interaction as we anticipate the responses of other people to us and incorporate them into our conduct. When two or more persons interact, for example, all are more or less aware of the fact that each is evaluating the behavior of others; in this process, we also evaluate our own behavior in relation to others. Orienting one's behavior to a set of expectations found in a role is called "role-playing." A *role set* is a complement or collection of role relations which persons have by occupying a particular social status, such as the role of a teacher to his or her students and to all the others connected with the school. Put another way, a role "is a set of expectations attached to a particular combination of actor-other identities (for example, father-son, father-daughter), and all the roles associated with one of the actor's identities is that identity's role set" (Heiss, 1981: 95). Social control becomes possible and effective through the fact that persons acquire the ability to behave in a manner consistent with the expectations of others. Even self control—an individual decision to engage in some behavior—is social control in that persons come to develop self concepts in reaction to group expectations (Gecas, 1982).

Socialization as Role-Taking

Socialization largely represents the learning of norms and roles. Put another way, socialization refers to the process by which persons acquire the skills, knowledge, attitudes, values, and motives necessary for the performance of social role. It involves a process of learning in which an individual is prepared to meet the requirements of society in a variety of social situations. The required behavior (habits, beliefs, attitudes, motives, and actual conduct) are an individual's *prescribed roles*, the requirements themselves are the *role prescriptions*. The role prescriptions, or norm requirements, are learned in interaction with others. What roles a child learns in the family, such as a male or female sex role, are largely dictated by the social structure or society itself. Groups, then, are multidimensional systems of roles; a group is what its role relations are. The individual members of a group may change, but the group may continue, as in the case of a delinquent gang, where the roles of leader and other required roles in the gang may continue despite changes in gang membership. In fact, much deviant behavior can be directly expressive of roles:

> A tough, bellicose posture, the use of obscene language, participation in illicit sexual activity, the immoderate consumption of alcohol, the deliberate flouting of legality and authority, a generalized disrespect for the sacred symbols of the "square" world, a taste for marijuana, even suicide—all of these may have the primary function of affirming, in the language of gesture and deed, that one is a certain kind of person. (Cohen, 1965: 13)

Professional thieves, for example, perform a variety of roles. Punctuality in keeping appointments with partners and the code of not "squealing" on another thief are of great importance in their profession (Sutherland, 1937). Social status or position among thieves is based on their technical skill, connections, financial standing, influence, dress, manners, and general knowledge. The professional criminal may play different roles toward victim, friend, spouse, children, father, mother, grocer, or minister (Inciardi, 1984).

Actual role behavior may be somewhat different from the role prescriptions because it is affected by a variety of influences, such as the behavior of others in the situation, membership in groups where the role prescriptions are different and confusing, and so forth. *Role strain* may arise in situations requiring complex role demands and where a person is required to fulfill multiple roles (Heiss, 1981; Parsons, 1951: 280–283). Many of the problems arising in systems of roles are due to the fact that (1) the role prescriptions are unclear and the person has difficulty in knowing what is expected, (2) the roles are too numerous for the individual to fulfill adequately, with a resulting "role overload," and (3) the roles may conflict or be mutually contradictory, so that the individual must perform a role for which that person is unprepared. The diversity of social roles in modern, urban society is an important factor in the extent of social deviation in society.

Keeping "Role" Terms Straight

Role Term	Meaning
Role	A set of expectations for persons occupying a particular social position (social positions are called "statuses"). Also, the behavior that is expected of a person in a given status with reference to a particular group.
Prescribed Roles	Required behavior (habits, beliefs, attitudes, motives, and actual conduct) of a status.
Proscribed Roles	Roles not permitted an individual because of other roles the person occupies (e.g., "bachelor" is a proscribed role to a "husband").
Role Playing	Orienting one's behavior to a set of expectations found in a role.
Role Taking	The decision to adopt a particular social role.
Role Set	A set of expectations attached to a particular combination of actor-other identities (for example, father-son, father-daughter), and all the roles associated with one of the actor's identities is that identity's role set.

Deviant Role-Taking

One can speak of deviant roles in the same way one can speak of any other social roles. Some criminals perform criminal roles, as do many deviants who have physical disabilities, such as the obese, the crippled, the blind, the retarded, who often come to occupy the role expected of them in their physical condition. In fact, such persons come often to perform social roles that can not be explained entirely by the physical disability itself. The behavior of much mental disorder can be viewed in terms of social roles, as will be shown later, as is the behavior of the homosexual and the organized criminal offender. Even suicide is often the enactment of a social role to its final and ultimate conclusion. Much deviance that appears to be "irrational" or "senseless" is interpretable when seen as an effort to proclaim or test a certain kind of identity or self.

The use of marijuana and heroin, especially the early experimental stages; driving at dangerous speeds and "playing chicken" on the highway; illegal consumption of alcoholic beverages; participation in illegal forms of social protest and civil disobedience; taking part in "rumbles"—all of these are likely to be role-expressive behavior. In order to recognize this motivation, however, one must know the roles that are at stake, and what kind of behavior carry what kinds of "role-messages" in the actor's social world (Cohen, 1966: 99)

There are a number of compelling reasons for viewing deviant behavior in terms of roles (Turner, 1972). One such reason is that it brings diverse actions together into a particular category or style of life, such as the "homosexual," "drug addict," or "criminal." In so doing, examining each type from the perspective of the deviant role facilitates identifying those common dimensions among each type. Many people have, at one time or another, engaged in homosexual acts. But homosexuality is more than engaging in homosexual acts. It is a role that is performed to some degree by persons who identify with homosexuality. It may involve a style of dress, gestures, certain language, knowledge of homosexual meeting places, and how to react to heterosexuals. Similarly, many people have been drunk at some time in their lives, but only a few people come to perform the role of alcoholic or problem drinker. There are great differences among people who drink, but many fewer differences among alcoholics. Once a deviant role is assumed, deviants become more like one another.

One characteristic of deviant roles is that they are powerful roles, both for the person performing the role and for others. Once a person is considered an "alcoholic," a "homosexual," a "criminal," or a "mentally disordered" person, other social roles are organized around the deviant role. The deviant role thus becomes a *master role* for the individual. The deviant individual develops a deviant self-conception through selective identification with the deviant role out of the many roles he or she plays. When other people come to stress a particular deviant role a person performs, it is difficult for this person not to regard his or her deviance as central to that person's identity, as in the case of the physically handicapped.

The roles to which people are assigned cannot be easily changed by their own desires. Whether a person plays the role that society has assigned or not, the person's behavior is still often interpreted by society as consistent with this role and its corresponding status. For example, the behavior of the former inmate of a prison in his home community may be interpreted in a manner consistent with real or imagined "criminal tendencies," even if he or she is making a determined effort to go straight. The deviant may be blocked from reentering conventional social roles while, at the same time, having to deal with social rejection and exploitation. As one ex-mental patient put it:

Since I was let out, I've had nothing but heartache. Having mental illness is like having the Black Plague. People who know me have abandoned me—

my family and friends. And the people who find out that I was in the mental [hospital] . . . treat me the same way. . . . And, at the boarding home where I was placed, I hardly get enough to eat. For lunch and supper today, all we got was a half a sardine sandwich and a cup of coffee, and they take three hundred and fifty dollars a month for that kind of meals and lousy, overcrowded, bug-infested rooms to live in. (quoted in Herman, 1987: 241)

The power of community interpretations in perpetuating a person's occupancy of a criminal status and role may have several consequences. Sometimes such persons will "give in" to the societal definition and actively play the part expected of them. The treatment of a person as if he or she were generally, rather than specifically, deviant may produce a "self-fulfilling prophecy," setting in motion several mechanisms that "conspire to shape the person in the image people have of him" (Becker, 1973: 34).

Deviant behavior can affect the selection of other roles the deviant acquires in life. For example, family life and deviant behavior as the result of the performance of deviant roles are often incompatible and in conflict with one another. Marijuana use, for example, seems to be associated with postponing becoming a mother among women, and increasing the propensity to dissolve marriage among men and women (Yamaguchi and Kandel,1985).

Deviant Acts and Deviant Roles: The Example of Heroin Addiction

Assuming deviant roles takes place over a period of time. Opiate addiction illustrates this point with behavior that some people think of as simply physical addiction, as something over which the addict has no control once addicted. Bennett (1986) studied 135 English addicts between 1982 and 1984 to identify the stages of their drug careers. The majority of addicts began their drug-taking with marijuana or amphetamines. They later turned to heroin after considerable drug experience, usually when a friend offered them the opportunity.

The process of becoming an addict was usually a slow one. The majority of Bennett's addicts took more than one year to become addicted. A number of heroin users continued for many months taking the drug only occasionally. Once addicted, some addicts discontinued their use—sometimes for as long a period as a year or two. As one of the addicts phrased it: "I usually use every day for a couple of months and then I start cutting down. I have occasions when I dry myself out for three or four months. I don't want my habit to get too big" (Bennett, 1986: 96). In other words, these addicts performed the role of "addict" more at some times than others and they were able to perform their other, conventional roles as well.

Addiction careers vary both in terms of the amount of heroin consumed during that career and one's socialization to the drug subculture. Although one can continue to maintain an addict self-conception without daily heroin use, one requires contact with the drug subculture to ensure future supply and support. Contact with that subculture greatly increases the chances that the drug user will develop a deviant self-concept and begin to adopt the addict role. In this sense, the adoption of a deviant role is a matter of degree. Most persons are neither totally conforming nor deviant in terms of roles, but somewhere in between these extremes.

Bennett's research was of current heroin users. When asked whether they would abstain from heroin in the future, about one-half reported they would like to continue to use heroin. These addicts felt comfortable with their addiction and they thought their lives were better with than without heroin. The other half indicated they would like to terminate their use within the next decade, usually citing other people's expectations (e.g., spouse, other family member, friend, employer) as the main reason. Thus, the expectations of others are important both in occupying and in leaving deviant roles.

SEEING THE DEVIANT'S PERSPECTIVE

It is easy to condemn the norms and values of others because they are not our own. Ultimately, it is necessary to comprehend the world of the deviant as that individual experiences it, at the same time remaining sufficiently detached to understand and analyze the interrelationships of the deviant world and the larger social order. All too frequently, only the world of the observer is considered. Categorizations of "senseless," "immoral," "debauched," "brutal" and so on, are often bestowed on the deviant by outside observers with no awareness that deviant actions might have a different meaning to the actor. To develop an "appreciation" for deviance is not to approve of it but to approach it from the standpoint of the deviant, to try to see the world and the meaning of deviance from the perspective of the deviant (Matza, 1969). Much research on deviance has approached the subject with an eye toward correcting, not understanding it. On the other hand, an excessive emphasis on the appreciation of the deviant's world can lead to an overly romanticized view that obscures a meaningful, honest appraisal of deviant life-styles. Clearly, both correctional and appreciative perspectives are necessary to provide a balanced view of deviance.

Understanding Deviant Worlds

Part of the difficulty has been that most observers have had little firsthand contact with the deviant world. To share the perspective and definition of the situation of the deviant does not mean the observer "always concurs with the subject's definition of the situation; rather, that his aim is to comprehend and to illuminate the subject's view and to interpret the world as it *appears to him*" (Matza, 1969: 15). The initial problem, therefore, involves gaining access to deviant worlds.

Much material on deviant perspectives in now being obtained by sociologists and others through in-depth interviewing, by "insider" reports of deviants, and by participant observation (Douglas, 1970 and 1972). Through such sources, we have learned much about such things as nudist camps, drug users, call girls, homosexuals, youth gangs, pool hall hustlers, Hell's Angels, and topless bar maids. Some of the information was collected by insider participant observation which is the result of becoming a member of a group. This research is usually done secretly; the persons being studied do not know they are being studied. The members of the group treat the researcher as one of their own. There are obvious ethical and practical limitations to the use of insider participant observation, as well as the use of what Douglas (1970: 6–8) has called "fictitious membership." In this technique, the group knows who the researcher is, but they also know that the researcher will not report them to the police or other officials. This specific method raises ethical questions relating to the obligations of the researcher to contact authorities about "serious" acts of deviance, such as serious crimes.

Other firsthand material can be obtained from life histories, diaries, and letters of deviant persons. In the chapters that follow, much use is made of such material from the perspective of the deviant, such as the analysis of suicide notes to understand the meaning of suicide to the participant.

The sociologist wishing to comprehend the world of the deviant, or even to gain access to those who are deviant, need not, of course, be a deviant. In fact, there are disadvantages to being "coopted" into the deviant life-style and becoming "one of them." Deviants do not have an exclusive claim to knowledge of "things deviant." Drug addicts are not necessarily experts on the addiction process, though they have experienced it personally. Homosexuals are not experts on the social dynamics of homosexuality, although they must "manage" (a term to be explained shortly) their identities and must have gone through the process of becoming a homosexual. No one would claim that only mentally ill persons can understand that condition, although they have intimate knowledge of what it is like to be one. Thus, being an "insider" does not mean that the knowledge one has acquired in this manner is reliable for others who have undergone roughly the same experience, or that it is the type of information that is valuable to have about deviance. No one but a heroin addict can know personally what it feels like to really experience the severe pains resulting from withdrawal from that drug. The important questions about heroin addiction and the withdrawal process are not what it feels like, but rather the role of the withdrawal process in continued addiction and the importance of the drug subculture in defining that experience.

Deviants, naturally, see the world differently from those who are not "insiders." One must balance a sensitivity to the perspective of the deviant with a concern with objectivity. Drug addicts know where to obtain their supply of drugs and from whom; they do not know, as addicts, the process whereby they came to be addicts. They know which of their acquaintances are "for" and "against" them; they do not necessarily know the extent and types of influences each exerts on them or their deviance. In other words, insiders do not know all that is known or should be known about deviance; in fact, what they know often does not provide reliable basis for generalized knowledge about one type or deviance or another (Merton, 1972). Deviants can not be expected to provide all worthwhile information about deviance merely because they are deviants. The virtue of being an insider is that one has certain kinds of information. The limitation of being an insider is that the person has knowledge only with narrowly prescribed boundaries. An alcoholic may be a good source of information about the person's experiences; but if the issue is the nature and extent of alcoholism in the United States, the social processes that generate and inhibit alcoholism, or the most effective means of treatment for the widest variety of alcoholics, one of the most unreliable sources of information would be the minute questioning of local skid row alcoholics.

In spite of these limitations, information about deviants from firsthand observation or from deviants' own accounts is important for a fuller understanding of the phenomena. While personal experience as a deviant is not necessary to formulate sociological questions about deviance, processes of deviance are interactive processes, and to exclude information from the deviant is to ignore one side of that interaction (Skipper, et al., 1981).

The benefits of a greater appreciation or understanding of deviance in this manner can be illustrated by example of heroin addicts. Often the addict is seen only as an emotionally ill retreatist who has dropped out of many social relationships and resorted to the use of heroin. Ethnographic research by urban anthropologists and sociologists that approaches the lives of addicts in a more empathetic manner portrays addict life differently (Agar, 1973; Hanson, et al., 1985). Addicts are engaged in what, to them, is a meaningful way of life, in spite of its deviant nature. Urban heroin addicts adopt to a greater or lesser extent a master social role, the "street addict role," that dominates their relationships and life activities (Stephens, 1987: 77–79). This role comes to form a personal identity for the addict and it is learned in association with other addicts who participate in the drug subculture. The subculture also provides access to drugs and support from other addicts. Even persons who use heroin only occasionally for recreational purposes adopt a set of attitudes and norms from others that inhibits their becoming addicted (Zinberg, 1984). But without firsthand information about these processes, they would not be understood, and effective efforts to deal with heroin addiction would be diminished.

The Management of Deviance

One of the most valuable aspects of seeing the deviant's perspective is that we become sensitive to some of the problems deviants face. The imposition

of negative sanctions from others poses obvious difficulties th
would like to avoid. In addition to specific negative sanction
must also deal with the *stigma* of being deviant.

It is understandable that social groups stigmatize so
Stigma functions for the group; it "reaffirms the rule, reaffirms the conformists as conformists, and separates off the wrongdoer who has broken
the rule" (Harding and Ireland, 1989: 105). But if stigma serves the punishing
group, it represents a problem for the deviant. The deviant must learn to
live with the fact that others think the deviant is "odd" or "strange" compared to "normals." Through the use of a number of techniques, the deviant
is able either to manage (or cope with) the stigma, prevent the stigma altogether, or lessen what stigma may be present. In other words, such techniques
"save face" for the deviant and ward off social rejection.

Management techniques are used that suit the particular form of rejection the deviant encounters, but a number of techniques may be common
to many forms of deviance (Elliott, et al., 1982).

1. Secrecy. If others are not aware that an act of deviance has been
committed or that a person occupies a deviant role, there will be no negative
sanction. Homosexuals may hide this fact from their families and employers;
obese persons may avoid social gatherings and maintain an isolated existence; heroin addicts may wear clothing that hides the needle marks on their
arms and legs; and, criminals attempt to elude the police through planning
and careful execution of their crime. "Secrecy is [often] urged upon deviants
by their in-the-know friends and family among normals: 'That is what you
want to do, okay, but why advertise it'?" (Sagarin, 1975: 268).

2. Manipulating the Physical Setting. Often negative sanctions
can be avoided if a deviant situation or act has the appearance of legitimacy,
regardless of the situation's true nature. Persons who are embezzling funds
from their employers attempt to continue to give the appearance of an
honest, trustworthy employee. Problem drinkers may turn down a drink
when with friends to lessen suspicion. A homosexual may continue his marriage to a woman because it "looks good," even though his sexual preference
is for other males. Prostitution is sometimes undertaken under the guise of
massage parlors and escort services (Prus and Irini, 1980: 65–68). Here, the
outward appearance of the setting in which deviant acts are committed is
made as legitimate or normal as possible, which is made easier because there
really are, for example, some legitimate massage parlors and escort services.

3. Rationalizations. A deviant may "explain away" the deviance by
justifying it in terms of the situation, the victim (if there is one), or some
other factor usually beyond the deviant's control. Cheating on one's income
taxes may be justified by the offender in terms of the already excessive taxes
he or she pays. Shoplifting may become acceptable because "the store can

afford the loss and insurance will cover it anyway." Obese persons may falsely attribute their problem to a physiological or glandular condition that they do not have. If an act is justified after it is committed, the term *rationalization* is appropriate; if the act is justified before it is committed, the term *neutralization* is more appropriate. The use of neutralizations weakens the strength of the norm by placing the deviance in a more acceptable framework or by convincing deviants that the norm does not apply to them for some reason. They also provide an effective means by which persons can maintain "face" when confronted with a troublesome or embarrassing situation. Singles dances, for example, which are often seen by both participants and others as lacking full respectability, generate pressures toward the use of rationalizations either to account for one's presence at the dance ("A friend made me come because she/he was recently divorced") or rejection at the dance ("I haven't danced yet because I had a tiring day at work and want only to stand around for a while") (Berk, 1977).

Similarly, it is often thought that the condition of divorce is less deviant than in previous times. Divorce, however, still contains a good deal of stigma that is felt and must be managed by divorced persons. Gerstel (1987), in interviews with 104 divorced men and women, found that it was not the actual experience of divorce that created stigma, but the circumstances surrounding the divorce. Some conditions, such as where one of the parties committed adultery, are associated with greater blameworthiness than others, such as where the divorce took place because of simple incompatibility. The divorcees also reported that they viewed marriage as a "normal" condition and divorce as not normal, thus suggesting that these people stigmatized themselves.

4. Change to Nondeviance. Another technique involves the movement from deviant to nondeviant status. For some criminals, this technique is usually referred to as "going straight" or being rehabilitated. For example, the obese person loses weight, the prostitute marries and settles down with a family, and the problem drinker shuns alcohol. It is often difficult to determine whether or not someone has terminated deviance since this judgment is often a social one. The heroin addict who no longer uses heroin may take methadone, itself an addicting drug, though a more socially acceptable one. Or, the addict will turn to the heavy use of alcohol. The change to nondeviance is troublesome for some stigmatized people. Some deviants are unable to use this technique, such as the physically disabled. There are also some deviants who have no motivation, even if they could, to change, such as a homosexual who wishes to remain homosexual.

The change to nondeviance can also be seen on a group level with the affirmation of deviance by militants. Homosexuals, for example, in some communities have publicly proclaimed their status and have pressured legislatures to change laws concerning this behavior, while at the same time asking the public to be more tolerant of homosexuality. Kitsuse (1980) has

Techniques of Managing Homosexuality

Sometimes management techniques are specific to the kind of deviance at issue. Homosexuals and lesbians, for example, have used a number of management devices that are geared toward reducing the stigma they experience as a result of their sexual preference. Troiden describes four such techniques.

1. *Capitulate.* To capitulate to homosexuality is to refrain from engaging in it and from openly expressing it.

2. *Minstrelization.* This term, meaning to act like a minstrel like in an old-time show, refers to the the person acting in accord with popular stereotypes of homosexuality, dressing and walking in certain ways and affirming a homosexual lifestye.

3. *Passing.* This is probably the most common adaptation for homosexuals. It means for many to lead a double life by trying to control information between the straight and gay worlds. Persons who pass do not deny their sexual preference but neither do they publicize it.

4. *Group Alignment.* This refers to belonging to and participating in homosexual subcultures. This adapation is used by homosexuals who have admitted to themselves and others that they are openly homosexual.

Adapted from Richard Troiden. 1989. "The Formation of Homosexual Identities." *Journal of Homosexuality,* 17: 43–73.

suggested the term "tertiary deviant" (in contrast with primary and secondary deviant) to denote the deviant who presses for the transformation of deviant conduct into something more acceptable. Similarly, militant prostitutes have taken like public stands advocating the decriminalization of this offense. In each offense, the idea is to change to nondeviance by redefining the behavior itself rather than changing the individuals.

5. Deviant Subcultures.

Participating in a subculture is helpful for deviants in managing their deviance: It lessens the chances of receiving a negative sanction from others by protecting them from contact with "normals" (Troiden, 1989). The subculture may also facilitate deviant acts by providing a necessary commodity, for example, drugs, and by reinforcing deviant attitudes. Homosexuals who frequent gay bars and maintain interaction with other homosexuals, at least during that time, decrease the chance that outsiders will stigmatize them. Gay bars can also facilitate the maintenance of a homosexual identity by managing interactions with non-gays in a situation that gays control; the effect is that gay life can be better reinforced and perpetuated (Reitzes and Diver, 1982). Subcultures offer the deviant sympathy, support, and association with other deviants. They help the deviant cope with social rejection while, at the same time, providing the opportunity to commit deviant acts (Herman, 1987).

INDIVIDUALISTIC THEORIES OF DEVIANCE

A sociological theory of deviance explores the social conditions that underlie deviance—how it is defined, how group and subcultural factors are related to deviance, and what reactions there are to deviations from norms. Individualistic theories of deviance, on the other hand, seek to explain deviance by conditions or circumstances uniquely affecting the individual, as for example biological inheritance or early family experiences. Such theories based on the individual largely disregard both the learning of deviant norms and group and cultural factors affecting deviance. The explanations to be critically examined here include biological, psychiatric or medical model, the psychoanalytic, and the psychological.

Biological Explanations

Human beings possess a biological nature and a social nature, and it is obvious that without a biological nature there could be no human nature. There is an interplay, rather than opposition, between the two. Humans are animals who must breathe, eat, rest, and eliminate wastes. Like any other animals, they require calories, salt, and other chemicals, and a particular temperature range and oxygen balance. Humans are animals that are dependent on their environment and limited by certain biological capacities.

Some scientists and practitioners claim that certain forms of deviant and antisocial behavior can be traced to certain physical anomalies, body chemistry, or heredity (Fishbein, 1990). These beliefs, in turn, have important consequences for the nature of some suggested preventive and "treatment" programs. Those, for example, who believe in the sterilization of certain types of deviants are advocating a biological view of human nature. More specifically, some biologists believe that crime, alcoholism, drug addiction, certain types of mental disorders, and certain sexual deviations can be carried as a specific unit in biological inheritance. The evidence with respect to such a view is limited. While there is the possibility of the inheritance of alcoholism (Secretary of Health and Human Services, 1990), recent research has failed to identify a specific "alcoholism gene" which predisposes individuals to heavy drinking. The National Institute of Alcohol Abuse and Alcoholism reported in a 1990 study that it had found a so-called "alcoholism gene" to be no more common among 40 alcoholics than among 127 non-alcoholics. There is no evidence that criminality or drug usage is inherited, and most mental disorders appear to be social, not biological, in origin.

The notion of inherited tendency or biological predisposition figures prominently in the theory of crime developed by Wilson and Herrnstein (1985). These authors attribute criminality to the acquisition of criminal attitudes and to a biological tendency in some offenders to violate the law. Wilson and Herrnstein say that crime is the result of a choice that people make that is determined by the consequences of that behavior. The consequences of committing a crime include both rewards (or "reinforcers") and punishments; the consequences of not committing a crime also entail costs and benefits. While the rewards of crime are immediate, the rewards of noncrime are all in the future, and some people seem better able to anticipate future events than others. But our ability to determine these choices is limited by biological constraints. Our biological makeup, for example, may influence the kinds of social interaction we have and, therefore, the kinds of learning experiences we have. While many sociologists would agree with this conception, it is a far cry from claiming that offenders have a "biological predisposition" to crime.

Biology is of little relevance to the social or symbolic behavior of humans or to deviant behavior in general, with some exceptions. There are no physical functions or structures, no combination of genes, and no glandular secretions that contain within themselves the power to direct, guide, or determine the type, form, and course of the social behavior of people (Fishbein, 1990). Physical structures or properties set physical limits on the activities of people, but whether such structures also set social limits depends on the way in which cultures or subcultures symbolize or interpret these physical properties.

Deviant behavior can not be inherited since social norms that are directly related to deviance can not be inherited. While individuals can inherit a particular way of looking or, sometimes, acting, whether that appearance or behavior will be considered deviant is a social, not biological, matter. With respect to crime, it is "obviously impossible for criminality to be inherited as such, for crime is defined by acts of legislatures and these vary independently of the biological inheritance of the violators of law" (Sutherland and Cressey, 1978: 123). As a result, few people today support the view of the direct inheritance of deviance. Instead, some have substituted the "inherited tendency" for such behavior. In many ways, this is even more unscientific and vague, since the nature and physiological location of this tendency is usually left unspecified.

What many persons may confuse as inheritance in behavior is the social transmission of somewhat similar ways of behavior from one generation to another in a culture, or from one family to another. Actually, none of this is heredity, for there is no way in which so-called family behavioral traits or culture can be inherited through the genes. The complexity of gene structure which would be required to transmit a culture or family attitudes and values as part of the biological heritage would be inconceivable. Behavioral traits can be passed on in a family through the sharing of common experiences and attitudes. It is in this fashion that persons who know one another, or who are related to one another, also may act similarly.

From time to time, various explanations of certain forms of deviance in terms of biological characteristics arise, but sooner or later all of them disappear as individual valid theories. Explanations have been offered in terms of body type, glandular disorders, brain pathologies, and, in the 1970s, the relation of chromosomes to criminality (see Meier, 1989: Chapter 4). While there may indeed be particular offenders who possess an abnormal chromosome pattern, this explanation, like other biological explanations, fails to take into consideration the relativity of "deviance" and the essentially socially acquired nature of human behavior. Even more recent theories that have focused on combining biological and nonbiological explanations have failed to explain how physical and, say, social dimensions come together to form one unitary explanation.

Psychiatric or Medical Model of Deviance

Psychiatrists regard deviants as psychologically "sick" individuals. This view is sometimes also called the "medical model" of deviance. In this view, deviant behavior is a product of something within the individual, such as personal disorganization or "maladjusted" personality. Culture is seen not as a determinant of deviant and conforming behavior, but rather as the context within which these tendencies are expressed.

It is common for psychiatrists to emphasize that all persons at birth have certain inherent basic needs, in particular the need for emotional security. Furthermore, deprivations of these universal needs during early childhood leads to the formation of abnormal personality patterns. Childhood experiences, such as emotional conflicts, largely, but not exclusively, determine personality structure, and thus the pattern of behavior in later life. The degree of conflict, disorder, retardation, or injury to the personality will vary directly with the degree of deprivation. By affecting personality structures, childrens' family experiences largely determine their behavior in later life, whether deviant or nondeviant. The need for the mother to provide maternal affection is particularly stressed in this view.

According to this theory, a high degree of so-called general personality traits, such as emotional security, immaturity, feelings of inadequacy, inability to display affection, and aggression, characterize the deviant but not the nondeviant. These traits are the products of early childhood experiences in the family. It is argued that because a child's first experiences with others are within the family group, traits arising there form the basis for the entire structure of personality. Deviant behavior is often a way of dealing successfully with such personality traits; for example, so-called immature or emotionally insecure persons may commit crimes, or emotionally insecure persons may drink excessively and become alcoholics.

The psychiatric position implies that *certain childhood experiences have effects that transcend all other social and cultural experiences.* These proponents suggest that certain childhood incidents or family relationships lead to the formation of certain types of personalities that contain within themselves seeds of deviant or conforming behavior, irrespective of culture.

Thus, childhood is the arena in which personality traits toward or away from deviance are developed, and a person's behavior after the childhood years is fundamentally the acting out of tendencies formed at that time.

The psychiatric view is often referred to as the medical model of deviance. It is a perspective that views the causes of deviance as residing within individual deviants in the form of defective biology or psychology. It equates deviance in many instances as examples of "sick" behavior and usually calls for medical intervention, either in the form of a physician using medical treatments or persons identified with the medical profession using psychiatric or psychoanalytic methods. Many persons, for example, view mental disorders within the larger context of "illness" that requires medical intervention. Many persons also view homosexuality as an "illness," and some persons regard alcoholism the same way. Frequently, we hear that some spectacular crime was the result of a "sick" mind, and that someone who committed suicide did so because he or she was "obviously" disturbed. These are all examples of the application of the medical model.

The medical model views deviance as merely a symptom of some underlying individual "psychological sickness" which can be detected and treated only by professionals. Those who take this view regard most forms of deviance as a form of "mental illness" or psychological disorder. What is significant about a criminal act, for example, is not the behavior itself, as serious as it may be, but the fact that the criminal act is a symptom of the "real" problem which lies deep within the personality structure of the individual.

The medical model is effective in the treatment of physical illnesses where agreement on what constitutes health and what constitutes illness is more widely extended. A fever is indeed symptomatic of an infection somewhere in the body which the physician attempts to locate and to treat. In the area of social deviance, no such agreement about normative behavior exists. A number of more specific problems have become apparent with the medical model as it is applied to deviance. Criticisms of the psychiatric or medical model explanation largely involve a confusion about "sickness" and norms, the lack of objective criteria for assessing mental health, an overstatement of early childhood experiences, and the lack of scientific verification of their claims.

1. Psychiatric explanations of deviant behavior blur the line between "sickness" and simply deviations from norms. Thus, deviant behavior is itself made the criterion for the diagnosis of mental abnormality. In this sense, deviations from norms, or illegal behavior such as delinquency or crime, are used as the basis for inferring the presence of "sickness" or mental aberration. This tendency is similar to older attempts to link behavioral deviations with "possession by devils." Moreover, the varied nature of deviance and its relation to groups and the power structure are not considered. "When we relate crime to mental illness, we do not know whether the crime produced the mental illness, the mental illness the crime, or both were a product of a third factor" (Jeffery, 1967: 212–213).

2. Psychiatric diagnoses are often unreliable and psychiatrists often can not agree among themselves concerning what objective criteria are to be used in assessing degrees of mental well-being or mental aberration. To a great extent, it is the very absence of objective criteria of either mental disorder or mental health that is responsible for the tendency of psychiatrists to equate "sickness" with, for example, delinquency and crime (Hakeem, 1984). Even within broad diagnostic categories, there is little agreement on the nature of psychiatric disorders. This is what leads some critics, such as the psychiatrist Thomas Szasz, to claim that psychiatry is more religion than science and that psychiatrists have too much power. Szasz (1987) and his followers have claimed since the early 1960s that the entire concept of mental illness be abandoned, since they feel it is a value-laden, relativistic view of people's adjustment problems to living.

3. Early family influences are overemphasized in this view, sometimes to the virtual exclusion of the effect on personality of other groups, such as the youth peer group, and of occupation, neighborhood, marriage, and other social institutions. Even in early life, the socialization of the child is greatly influenced by the play group, by street play in urban areas, by preschool and kindergarten activities, and by neighbors and others such as relatives. The rigidity of character structure during the first five years of life has been exaggerated, for life must be regarded as a continuous experience of social interaction.

4. As applied to the study of deviance, the medical model is particularly prone to a logical fault called "tautology," or circular reasoning. Consider the following example. In 1966, Richard Speck entered an apartment in Chicago and brutally killed eight student nurses. In this nationally publicized and sensational case, the immediate reaction of most persons was that Speck must be "crazy" because of the particularly shocking nature of the crime. The evidence offered in support of this claim was the behavior itself, since one *had* to be crazy to commit such an atrocious act. The model, therefore, has two uses for the term "illness." One is to take as evidence the deviant act of the illness, while the other is to use the concept of illness to explain the deviant act. In this case, Speck's actions were taken as evidence of his illness, and that some illness was said to be the cause of the behavior. This circular reasoning is a common problem with the model. Often the deviance is taken as evidence of some underlying problem, and this presumed problem is then stated as reason the behavior occurred. Obviously, what is required to break the tautology is to have evidence of the "problem" that is independent of the deviant act which is said to be simultaneously cause and effect of the problem.

The Psychoanalytic Explanation

Psychoanalysis is closely related to general psychiatric thought, but it has a particular system of its own in explaining deviant behavior. Psychoanalysis was founded by Sigmund Freud, a Viennese physician who died in 1939 (see Gay, 1988). According to psychoanalytic writers, the chief explanation of behavior disorders must be sought in an analysis of the *unconscious mind*, which is said to consist of a world of inner feelings that are unlikely to be the obvious reasons for behavior or to be subject to recall at will. Antisocial conduct is a result of the dynamics of the unconscious mind rather than of the conscious activities of mental life. Much of the adult's behavior, whether deviant or nondeviant, owes its form and intensity to certain instinctive drives, particularly sexual ones, and to early reactions to parents and siblings.

Psychoanalysis assumes that the conscious self is built over a great reservoir of biological drives. In the psychoanalyst's scheme, personality is thought of as composed of three parts: the primitive animal *id*, *ego*, and the *superego*. The id is the buried reservoir of unconscious instinctual animal tendencies or drives. The ego is the conscious part of the mind. Freud postulated here a dualistic conception of mind: the "id" or internal unconscious world of native or biological impulses and repressed ideas, and the "ego," the self, operating on the level of consciousness. There may be constant conflict between the ego and the id. The superego, on the other hand, is partly conscious; it is the conscious part which corresponds to the conscience (see Lilly, Cullen, and Ball, 1989: 38). It is man's social self, derived from the cultural definitions of conduct.

Some psychoanalysts have made almost synonymous with criminal behavior the unresolved conflicts between the primitive id and its instinctive drives and the requirements of society. According to this view, crime arises out of inadequate social restrictions which society has placed on what psychoanalysts assume to be the original instinctive, unadjusted nature of man, which is savage, sensual, and destructive. Psychoanalytic writers stress the strong desire in the id for self-destruction. The superego, in turn, contains various social and moral restrictions on personal violence and self-destruction. The forces pulling toward self-destruction and self-preservation are in constant interaction, and when the former overcomes the latter, suicide ensues.

Psychoanalysts also think of a normal personality as having *developed through a series of stages*. The development of personality involves shifting interests and changes in the nature of sexual pleasure from the "oral" and the "anal" preoccupation of infant life to love of self, love of a parent of the opposite sex, and, finally, love of a person of the opposite sex other than one's parent. Some of these stages overlap and may go on simultaneously.

Some persons do not progress through all of them; consequently they have conflicts and develop personality difficulties. The activities of deviants unconsciously represent unresolved infantile desires. Some psychoanalysts believe, for example, that the type of crime and the types of objects involved in the crime often indicate infantile regression. Others have concluded that the etiology of schizophrenia, a form of mental disorder, is a retreat to a form of infantilism. The alcoholic has often been characterized by psychoanalysts as a passive, insecure, dependent, "oral" stage personality whose latent hostility has been obscured. Drug usage has been likened to infantile masturbation (Rado, 1963).

Numerous criticisms can be made of the psychoanalytic explanation of deviant behavior.

1. Contrary to psychoanalysis, human behavior is a product of social experience and is not determined by an innate reservoir of animal impulses termed the id. Depending upon social and cultural experiences, a person can be either cruel or gentle, aggressive or pacific, sadistic or loving. One can be either a savage Nazi or a compassionate and tender human being like Albert Schweitzer or Mohandas Gandhi.

2. The psychoanalytic emphasis on sexual eroticism is a great overstatement of an important aspect of human behavior. Conflicts can arise in many other areas of human experience, for example, through economic competition, the achievement of status, and religion.

3. The entire psychoanalytic scheme is too bodily conscious rather than sufficiently socially conscious; the child's development is greatly influenced by social relationships that have little or no connection with bodily functions.

4. Psychoanalysis is not a scientific explanation of human behavior. Most psychoanalytic claims have not been scientifically verified. It has not proven possible to measure or otherwise verify the existence of the id, ego, and superego, and many other claims of psychoanalysis must be taken on faith.

Psychological Explanations of Deviance

Many efforts have been made, primarily by psychologists, to isolate by various tests those personality traits that would distinguish deviants from nondeviants. Such an approach assumes that the basic components of any personality are individual personality traits or generalized ways of acting. Many personality traits have been identified, such as aggressiveness or submissiveness, intense display of emotions or the lack of any such display when appropriate, suspiciousness or the lack of it, self-centered reactions as opposed to those directed toward the welfare of others, withdrawal of contacts with other persons, and feelings that one is regarded with affection or dislike. At one time, the term "temperament" was used to encompass all such personality traits. Many researchers used to believe that personality traits were hereditary and that some persons were "naturally" aggressive or "naturally" shy. Substantial research has shown that such behavior patterns are primarily developed out of social experiences. Other research that has attempted to link crime with such psychological characteristics as feeble-mindedness, insanity, and stupidity (measured by I.Q. tests) has been disappointing since it has failed to find a strong relationship (Lilly, Cullen, and Ball, 1989: 39).

Dozens of personality tests, rating scales, and other psychological devices have been used to try to distinguish deviants from nondeviants. Some of the more widely used tests have been used for many years. Traits are often ascertained and measured by a variety of pencil-and-paper type tests, such as the MMPI (Minnesota Multiphasic Personality Inventory) and the CPI (California Personality Inventory). Projective tests are also used, such as the TAT (Thematic Apperception Test) which consists of a series of pictures about which the subject comments, and the Rorschach which uses cards containing standardized inkblots, to which subjects respond by telling what they mean to them.

Psychologists often seek to explain nearly all forms of delinquent and criminal behavior in terms of abnormalities in the psychological structure of the individual. They believe that inadequacies in personality traits interfere with the individual's adjustment to the demands of society. Eysenck's (1977) interactionist theory is perhaps the most broad of the personality test theories. Eysenck claims that criminal behavior is the result of both certain environmental conditions and inherited personality traits. Persons are genetically predisposed toward crime at birth and this, coupled with adverse environmental conditions such as poverty, poor education, and unemployment, creates criminality.

There are several major difficulties in distinguishing the personality traits of offenders with those of nonoffenders, not the least of which is that the tests have not been able to distinguish the personality traits of criminal offenders from those of nonoffenders. Psychologists have been unable to identify a set of personality traits that consistently differentiates deviants from nondeviants.

A widely held belief is that differences in personality traits or the need to escape explains addiction to opiates. There is, however, no evidence of anything approaching an "addict personality" or a cluster of personality traits that are consistently associated with addiction. To some persons, alcoholism is the result of personality maladjustment. In this view, early childhood experiences produce feelings of insecurity which, together with difficulties in interpersonal relations of adult life, produce tensions and anxieties. The use of alcohol reduces anxiety and persons may come to depend on it for this purpose. However, efforts to document the existence of an "alcoholic personality" have not succeeded. Moreover, there is no reason to believe that persons of one type of personality are more likely to become alcoholics than persons of another type. The view that alcoholism is the result of particular personality traits often fails to take into account that prolonged use of alcohol may itself change aspects of the drinker's personality. Efforts to identify the personality traits that would distinguish homosexuals from heterosexuals have also been unsuccessful.

The psychological literature of many types of deviant behavior contains reference to a deviant personality type termed a *criminal psychopath* or a *psychopathic personality*, a habitual antisocial deviant. Although there has been considerable dispute over the meaning of the term "psychopath," some of the characteristics of a so-called psychopath are said to be demonstration of poor judgment and an inability to learn from experience, as is seen in "pathological lying," repeated crime, delinquencies, and other antisocial acts. The lack of precision describing psychopathic traits has been shown by the wide difference in the diagnoses of "psychopathic" criminals in various institutions and by research on the traits. Furthermore, the view that persons are psychopaths merely because they are repeaters or are persistent in their behavior is circular reasoning. Writing on the characteristic of persistent antisocial behavior as a criterion of a sexual psychopath, Sutherland (1940: 549) stated: "This identification of a habitual sexual offender as a sexual psychopath has no more justification than the identification of any other habitual offender as a psychopath, such as one who repeatedly steals, violates the antitrust law, or lies about his golf scores."

The general criticisms of the psychological approach to deviant behavior include the following.

1. Human behavior consists primarily of social roles that are variable and socially determined, and not static entities like so-called personality traits. Psychological theory also does not explain how specific behavior, such as techniques of stealing, are acquired.

2. It is almost impossible to isolate effects of societal reaction on the behavior of deviants. One is never sure whether given personality traits were present before the deviant behavior developed or whether experiences encountered as a result of the deviation produced the traits. An alcoholic or a drug addict may have developed certain personality traits as a result of a long period of alcoholism or drug addiction, and consequent rejection and stigma, rather than having had the trait prior to the deviance.

3. Finally, no evidence has been produced that so-called personality traits are associated with deviations from disapproved norms. Comparisons with control groups have revealed that no series of traits can distinguish deviants from nondeviants in general. The studies do not show that all deviants have particular traits and that none of the nondeviants have a given trait. Some deviants, for example, are "emotionally insecure," but so also are some nondeviants. On the other hand, some deviants are "emotionally secure." It is difficult to interpret such mixed results without accounting for the fact that, though the proportions may vary, the same characteristics may be present in both deviant and nondeviant groups.

SUMMARY

Deviant behavior is human behavior and is understandable only within the larger framework of other human actions and thought. One becomes deviant just as one becomes anything else—by learning the values and norms of one's group and in the performance of social roles. Some values are conventional, some deviant; some norms are conventional, some deviant; some roles are conventional, some deviant. The difference is the content of the values, norms, and roles. Viewing deviance in the sociological context of roles enables us to interpret the meaning of deviance both for the deviant and for others. Deviant roles may be powerful ones because they tend to overshadow other roles the person may play. Insofar as deviance at least partially expresses role behavior, it conforms to certain expectations about behavior in particular situations. A drug addict may conform to the demands of an addict role, just as a criminal may be conforming to those of a criminal role. But, most deviants perform deviant roles infrequently and then usually only for short times.

Deviance is usually approached from the perspective of someone who is not a deviant. But a full understanding requires an attempt to understand the meaning of the deviance for the deviant. Observational studies can provide some insights into deviance not found in other methodologies. To appreciate deviance means to understand, but not necessarily agree with, the world as the deviant sees it. The ways in which deviants handle the rejection or stigma of nondeviants are called management techniques. No one technique guarantees that deviants will be able to manage living in a world that rejects them, but not all techniques are available to each deviant. These techniques include secrecy, manipulating aspects of the physical environment, rationalizations, participation in deviant subcultures, and changing to nondeviance.

Individualistic approaches to deviance attribute the process of becoming deviant to something within people—their biology or psychology. Individualistic theories are consistent with a medical model view that likens deviance to illness in need of treatment and correction. The psychiatric and psychoanalytic viewpoints are similar in finding the roots of deviance in early childhood experiences, but the psychoanalytic perspective puts more stress on inadequate personality development, sexual conflicts, and the unconscious mind. Neither view can be established scientifically. In spite of years of psychological testing, no method has proven capable of consistently distinguishing deviants from nondeviants in terms of their personality traits.

SELECTED REFERENCES

Biddle, B. J. 1986. "Recent Developments in Role Theory." *Annual Review of Sociology*, 12: 67–92.

An overview of social-psychological role theory concentrating on extensions of the basic idea. Biddle is a leading figure in the sociological literature on the concept of role.

Fishbein, Diane H. 1990. "Biological Perspectives in Criminology." *Criminology*, 28: 27–72.

An extensive review of biological approaches in the study of criminality. If interested in this subject, this is the place to start.

Kitsuse, John I. 1980. "Coming Out All Over: Deviants and the Politics of Social Problems." *Social Problems,* 28: 1–13.

A statement that attempts to connect social problems theory and social deviance theory by an influential theorist.

Skipper, James K., Jr., William L. McWhorter, Charles H. McCaghy, and Mark Lefton, eds. 1981. *Deviance: Voices From the Margin.* Belmont, Calif.: Wadsworth.

A collection of papers written by and about deviants that describes the social worlds of deviants from their own perspective. An introductory essay sketches possible theoretical frameworks within which to view deviance.

Szasz, Thomas. 1987. *Insanity: The Idea and Its Consequences.* New York: Wiley.

The most recent theoretical statement by a psychiatrist who denies that mental illness has a reality apart from a particular social context. Szasz denies the "illness" metaphor as applied to behavior problems. This is Szasz's most extensive discussion of his position in many years.

Explaining Deviance

PART TWO

Urbanization, Urbanism, and Deviant Attitudes

Almost everywhere, the spread of urbanization—in the United States, Europe, Latin America, Africa, and Asia— has been accompanied by a marked increase in various forms of deviant behavior. Regardless of the country, the rates for juvenile delinquency, prostitution, drug use and addiction, alcoholism, mental disorders, and suicide are generally far greater in the cities than in rural areas. Something about urban life accounts for this situation. This chapter examines the relationship between urbanization and deviance, as well as identifying major sources of deviant attitudes. These sources include associates, community or neighborhood factors, mass media, occupation, and the family. While such sources are not in themselves theories of deviance, they can be important elements of a theory of deviance.

THE PROCESS OF URBANIZATION

While city living has characterized some areas for centuries, urbanization has increased at such an accelerated rate over the past century that today it encompasses hundreds of millions of people throughout the entire world. Urban life has produced what some people have called the mass society. It has greatly increased social differentiation, the clash between different norms and values, and the breakdown of interpersonal relations (see Spates and Macionis, 1987). The effects of urbanization are vast and these effects ultimately can be seen in conceptions and rates of deviance.

Many forces are responsible for the growth of cities, only some of which can be mentioned here. These include initially the breakdown of feudalism in Europe and the loss of prescribed duties and obligations, and an integrated village way of life; the growth of trade and commerce and later the Industrial Revolution, which produced a dispersion of the population, particularly to cities; the development of the factory method of production and, with it, an extensive division of labor; new forms of transportation; and improvements in agriculture that freed some people from the land. It was also necessary that agricultural production be high enough to sustain an increasing number of people in nonagricultural production. All of these forces, and more, sped the urbanization process. In the last few centuries, technological changes have increased this process even more.

The United States has been no exception to this worldwide trend. Although some people live in rural areas, the United States is an urban society. About 75 percent of the U.S. population reside in metropolitan areas (Population Reference Bureau, 1987). A metropolitan area is a big city and its independent suburbs. Nearly half (48 percent) of the U.S. population live in metropolitan areas of 1 million or more, although suburban growth in most areas has been rapid. Six of the ten fastest growing metropolitan areas in the United States are in Florida, three are in Texas (Anchorage, Alaska is the 10th). Generally, the Sun Belt states have experienced the greatest growth, while parts of the Midwest and eastern U.S. have experienced the greatest population declines. It is neither possible nor desirable to review here the scope of urbanization in the United States, let alone throughout the world. Suffice it to say that after the Civil War, the rate of urbanization in the U.S. increased rapidly. The Table on page 63 shows the growth of the urban population in the U.S. from 1790 through the 1980 census. The increase in the proportion of the population who lived in cities is unmistakable, from about 5 percent in 1790 to 75 percent in 1980.

Recent Patterns in Urbanization

Changes in urbanization have been related to economic changes in recent years and these changes have, in turn, influenced some forms of deviance. Two important trends have occurred during the past few decades: a countermigration to rural areas and the growth of suburbs near large cities. The countermigration to rural areas was evident during the 1970s. By 1975, rural areas were growing in population while urban areas were declining. The

Growth of Urban Population in the United States,
1790–1980

Year	Percentage Urban	Percentage Rural
1790	5.1	94.9
1850	15.3	84.7
1900	39.7	60.3
1910	45.7	54.3
1920	51.2	48.8
1930	56.2	43.8
1940	56.5	43.5
1950	64.0	36.0
1960	69.9	30.1
1970	73.5	26.5
1980	75.0	25.0

Source: Bureau of Census, *United States Census of Population, 1970, Summary of Inhabitants.* (Washington, D.C.: Bureau of the Census, Government Printing Office, 1971), pp. 3–6 and Population Reference Bureau, "Where is the Metropolitan U.S.?" *Interchange,* no volume: (1987): 1–4.

change reflected the movement of retired persons relocating to warmer climates, blue- and white-collar workers leaving the cities for more pleasant living, and younger persons trying their hands at subsistence living. The resulting decline of urban populations, particularly in the eastern U.S., took place at the same time suburban growth was taking place.

During the 1970s, suburban areas of large cities were growing rapidly. The increase in suburban population paralleled the exodus from central cities. The growth of suburbs meant not only that people had moved from the city to suburban areas, but also that manufacturing and service occupations left the central city. Factories, office building, banks, and retail establishments went to where the workers were located. At present, it is no longer necessary for many suburban residents to commute to work in the central city. Because of improvements in communication and transportation, corporations—especially those dependent on white-collar workers—have been able to move to suburban areas. This has produced a different pattern of urbanization than found previously. Cities and suburbs are becoming separate entities, both socially and economically. The build-up of malls and other shopping services has increased the economic split between suburbs and the central city. Consumers no longer have to commute to the city for their purchases, and retail establishments in cities have suffered.

Social Consequences of Shifts in Urbanization

The social consequences of this shift from cities to suburbs have similarly been great. The service sector or blue-collar jobs previously offered to inner city residents have largely disappeared or are shrinking, moving to the suburbs. Therefore, many cities, especially the slum or inner city areas, have very high unemployment rates particularly among teenage and youthful persons. There is also a great deal of instability in employment, so that if a person is able to secure employment, it may be short-term or temporary.

These changes have had a serious impact on black youth. In the early 1950s, black youth had an unemployment rate almost identical to that of whites (Murray, 1984: 72). Beginning the late 1950s, the unemployment rate among young blacks began to increase. Much of this initial increase was due to the loss of agricultural jobs in the South (Cogan, 1982), but subsequent increases in unemployment were the result of the loss of manufacturing jobs in the North. The unemployment rate stabilized during the 1960s but at a very high rate; nearly one-quarter of the potential black teenage labor force was unemployed. The 1970s were worse, and by 1980, nearly 40 percent of 16 and 17 year old black youth were out of work. Actually, it appeared that many of this age group had employment but the work was transitory, did not pay much, and failed to provide skills, work habits, and a work record necessary to move into the adult world of work.

By the 1980s, jobs were leaving the city and going to the suburbs, where many of the people had moved. Many of the better educated and occupationally trained blacks moved too, leaving mainly unskilled and untrained workers behind with few prospects for steady employment. This *underclass* has had largely to subsist on government programs and with what few jobs they could find and keep (Wilson, 1987). Between 1970 and 1986, for example, the city of Chicago lost 211,000 jobs that did not require a high school diploma, while gaining 112,000 jobs that required a college degree. While 25% of all job holders in Chicago have a degree, only 7% of black men do (cited in *The Chronicle of Higher Education*, Vol. 36, no. 43, July 11, 1990, p. A6). Clearly, the kinds of skills needed for employment in modern cities are not the ones possessed by urban workers.

With many of the unskilled jobs no longer available, many black youths have had to find employment where they can, and whenever they can. The migration of middle-class blacks—those most likely to be able to exploit the shifting economic opportunities—has had marked differences in the inner-city communities. Among the more pronounced consequences is a leadership vacuum in these communities that affects most institutions, including politics, education, welfare, and community services.

The effect of these population and economic shifts on females has been serious as well. Rates of black teenage pregnancy are very high compared to those of white teenagers, and the the percentage of teenage and young women who have babies out of wedlock is high compared to whites. The reasons for this include higher rates of premarital sexual intercourse among blacks as compared to whites, a lower percentage of the use of effective contraception, and a lesser tendency to abort a pregnancy (see the summary in Reiss and Lee, 1988: 147).

Another important reason is that there are fewer "eligible" black male marriage prospects. Some have suggested that there are three times as many marriageable black women as men (Pins, 1990). The black males who remain in the inner city are generally either unemployed or do not have steady employment. This results in a smaller marriageable pool of males from which to choose husbands (Wilson, 1987).

These social consequences are reflected in the nature of the minority populations of large cities. Just 38 percent of the U.S. black children lived with both parents in 1989, less than half the proportion of white children who lived in traditional homes (Pins, 1990). The National Center for Health Statistics reported that in 1987, the most recent figures available, 62 percent of all black babies were born to unwed mothers, up from 55 percent in 1980. In comparison, 17 percent of all white babies were born to unwed mothers in 1987. The composition of black families reflects these demographic trends. Because of the economic hardships and the absence of an adult male, many black children are cared for by grandparents: 1.2 million, or 13 percent of black children under the age of 18 live with their grandparents; in more than half of those cases, the mother also lives with the grandparents.

Other minority groups have not been immune to the debilitating effects of these urban processes. Moore (1987: 7) describes a similar outcome for Chicano gang members in Los Angeles: "Youngsters look ahead to a dreary round of welfare payments, living off relatives, street hustling (big-time and small-time), dehumanizing jail terms and short-lived dead-end jobs."

To make matters worse, as the suburbs expanded and job opportunities followed the population, middle-class blacks who could obtain those jobs also migrated to the suburbs. In this sense, the shift of economic opportunities produced a change in the composition of inner-city populations. For example, large numbers of homeless people live in big urban areas. Some of these persons are ex-mental patients for whom continued mental hospitalization is not warranted (Herman, 1987). Others are former inmates from correctional institutions and ex-residents of drug rehabilitation institutional programs. There are also individuals who have been unable to find and keep steady employment for a number of reasons, including poor job skills or work habits. Most homeless persons appear to have a long history of evolving personal problems, as reflected in personal histories of illegal drug use, alcoholism, crime, and psychiatric symptoms (Benda, 1987). For many of these individuals, homelessness and participation in the urban welfare system is simply the next step in a downward spiral of social debilitation.

It is difficult to estimate with accuracy the number of homeless people because it all depends on one's definition of "homeless." Sleeping in a motor vehicle might define some homeless people, but it also might define the activities of long-distance truckers who make $40,000 per year as well. In spite of difficulties in definitions, one estimate puts the figure at about one-half million homeless persons in the United States (Wright, 1989: 24). These persons are the poorest of the urban poor, persons who are unable to compete any longer in a marketplace of dwindling low skill-low pay jobs found in urban areas.

Rates of deviance are high in most inner city areas, and have been traditionally high in slum communities. Deviance is reflected in a variety of criminal activities, drug usage and trafficking, prostitution and commercial vice, and high rates of alcoholism and problem drinking. Suicide rates and the rates of various mental disorders are also high in these areas. For some, deviance becomes a way of life, much like the city becomes a way of life. To see how, it is necessary to examine the nature of city life and the life styles found there.

THE URBAN CONTEXT OF DEVIANT BEHAVIOR

A close relationship exists between types of deviance and city life. Modern cities, as compared to rural areas, appear to have higher rates of crime, illegal drug usage, heavy drinking and alcoholism, homosexual behavior, mental disorder, and suicide. This relationship is evident in the United States as well as other countries. With respect to crime specifically, almost without exception, it is the developing countries that report that crime is most rapidly increasing. This increase is almost all due to the accelerated urbanization that has accompanied industrialization (Shelley, 1981; Clinard, 1976). In fact, most crime is concentrated in larger cities which account for a relatively small proportion of the total population in these developing countries. The relationship between crime and urbanization is evident in European, South American, African and Asian nations as well (Johnson, 1983). A similar relationship has been observed in such diverse cities as Bogotá, Tokyo, Bangkok, Singapore, and Warsaw (Buendia, 1990). The relationship between city life and specific crimes, such as rape (Baron and Straus, 1989), has also been well documented.

While some criminal acts committed in rural areas are informally handled and not officially reported, and while cities undoubtedly offer more opportunities for crime, the differences between rural and urban crime rates are so great that differential reporting and opportunity account for only part of the differences. Moreover, there is mixed evidence on whether the city attracts deviants from rural areas. Studies of the relationship between geographic mobility and crime show that mobility has no direct effect on crime (Tittle and Paternoster, 1988). It appears that there are certain conditions and processes in the city that generate crime from its residents.

There are some exceptions to these general relationships. One example in Europe is Zurich, a city of about half a million persons and a total metropolitan population of about 850,000. Due to a number of reasons, the crime rate in Zurich is comparable to that for much smaller cities (Clinard, 1978). One of the world's largest cities, Tokyo, with nearly 12 million people, is another example of a very large city with a low official crime rate (Citizen's Crime Commission, 1975). Even in the United States and Canada, the relationship between crime and urbanization is such that there are exceptions. Population density, for example, has been implicated in the deviance process, but it appears that population density is not strongly related to crime rates or several other forms of deviance (Choldin, 1978). The physical crowding of persons together, either into neighborhoods or into houses within neighborhoods, is not in itself crucial. What matters are the social accompaniments to those kinds of crowded conditions. It is the social, not the physical, environment that matters. It is not the case that crime is always highest in the largest cities. Some studies report that robbery and auto theft are the only officially recorded offenses that appear to increase consistently as city size increases (Brantingham and Brantingham, 1984: 153). While the general relationship between crime and other forms of deviance and urbanization still holds, the relationship is a complicated one.

As with crime, other forms of deviance are found disproportionately in urban as opposed to suburban or rural communities. The greater the size of the community, for example, the greater the proportion of alcohol users it contains. This relationship relates to the larger proportion of middle- to upper-status, white, and Protestant alcohol users in the more urban areas, as well as to a value system that is conducive to alcohol consumption (Peek and Lowe, 1977). Suicide rates are also higher in cities than in suburban or rural areas (Stack, 1982), and the rates of many different types of mental disorders are higher in cities (Dohrenwend and Dohrenwend, 1969). The use of opiates and cocaine is more common in cities, and the existence of drug subcultures is almost exclusively an urban phenomenon (Stephens, 1987). The existence of deviant street networks is also related to cities (Miller, 1986). These groups that are organized for illicit purposes, including crime, prostitution, and drug selling, survive in an urban environment that provides a market for their services. Gay and lesbian communities are found in large cities throughout the world where they find a more tolerant environment than in smaller towns and rural areas (Reiss, 1987).

The effects of urbanization are often found concentrated in the inner-city areas of large metropolises. It is here that we often find the highest rates of such deviant acts as crime, illicit drug usage, problem drinking, and suicide. Not all forms of deviance, of course, are found in inner-city areas. White-collar and corporate criminality, for example, are upper-class forms of deviance. The upper and middle classes also have many of the other forms of deviance found in the lower classes, although generally not to the same extent. Similarly, rural areas are also characterized by many forms of deviance, although they are not as common as in urban areas.

Urbanism as a Way of Life

One must examine the effects of urbanization in order to explain differences in the incidence of deviant behavior in rural and urban areas, as well as the differences in rates by city size. Urbanism brings about pronounced changes in the way of life, the patterns of population distribution, contacts with ever-wider circles of people, with subsequent increases in impersonal relationships, differences in leisure-time pursuits, and the great diversity of subcultures and countercultures. Urbanism results in increased social differentiation, the clash of norms and social roles, and a breakdown in informal, traditional controls. Among other things, urbanism means a more complex life, impersonality of relationships, the existence of many different subcultures, and less direct behavior controls.

Urbanism is different from urbanization. Urbanization refers to the process of growing concentration of people in cities, while urbanism refers to the cluster of social qualities and characteristics that distinguish the city from rural areas. As such, "urbanism" is not synonymous with "city;" whereas "city" refers to an area distinguished principally by population size, density, and heterogeneity, "urbanism" refers to a complex of social relations that is embodied in a particular way of life or way of perceiving the world. Rural areas in urban-industrial societies may become "urbanized" as the way of life there reflects urban characteristics. It is for this reason that the relationship between deviance and the city can not be explained only with reference to size alone. The terms "urban" and "rural" do not refer exclusively to the relative size of a community, but to the social accompaniments of the development process. It is not the absolute number of people in a community that leads to predictions about the extent of deviance, but the social dynamics among those persons.

The diversity of urban life, and the conditions that lead to that diversity and the tolerance of alternative life-styles, is intimately related to deviance. Consider the following scene:

> Imagine sitting in Union Square in downtown San Francisco noting social differences in people who walk by. A well-dressed young woman carrying an attaché case hurries past, presumably on her way to her next appointment. An old man, unshaven and in shabby dress, reclines on the grass enjoying the sun; occasionally he takes a brown bag from his coat pocket, unscrews the top of a bottle inside, and has a drink. Four long-haired young men in bright floral shirts and dungarees are engaged in a serious discussion about the state of the nation. A short distance away, a black woman plays a guitar as she offers Christian messages to anyone who will listen. Few do. A group of Chinese children, 10 or 11 years old, playfully skip their way through the square. A middle-class couple emerges from the City of

Paris department store carrying a large assortment of packages. They
buy an ice cream from a street vendor, stroll into the square, and sit wearily
but happily on the grass (Spates and Macionis, 1987: 286).

Similar scenes are available in many large cities and they all illustrate
the great social differences in our cities. Diversity in cities arises from having
a concentration of people with similar social characteristics: age, sex,
income, education, ethnic or racial background, and political and religious
beliefs. People from like backgrounds and with similar characteristics often
reflect similar life-styles and interests. And the city itself often shapes and
determines these life-styles by bringing people into contact with one another.
The processes of describing and explaining the social diversity in cities is
called urbanism.

Wirth (1938) provided the first systematic description of urbanism.
For Wirth, there is an interaction between aspects of the physical environment in cities and aspects of the social environment. The size of the city
increases the range of human interaction but may also make those contacts
more superficial. The high density of people in cities leads to the creation
of subgroups and subareas with, inevitably, some people not belonging anywhere. The city reflects its heterogeneous population. Social roles are often
segmented and membership in groups is often fluid. As a way of life, urbanism is often characterized by extensive conflicts of norms and values, rapid
social change, increased mobility of the population, emphasis on material
goods and individualism, and an increased release from informal social
controls.

Certain cultural values in a society may increase the effects of urbanization. If a culture emphasizes material possessions as a central value, as
is largely true of the United States, the impersonality of urban life will tend
to increase that emphasis. City living does not, of course, directly result in
deviant behavior; it is just that many of the conditions associated with city
life are, to a preponderant degree, conducive to deviance. It is also a statistical fact that with more groups there is a greater chance of deviating from
some group's norms. Some characteristics of urban life that are conducive
to deviance include the following.

1. Norm Conflicts Diversity of interests and backgrounds is a major
characteristic of urbanism. Urban residents vary in age, race, ethnic background, and occupation, as well as interests, attitudes, and values. Large
cities are really cities within cities, each community with its own subculture,
religious affiliation, history, or racial characteristic. Often groups have different customs as well as separate languages (Hunter, 1974). The
heterogeneity of the population, the complex division of labor, the class
structure, and, apparently, the simply physical dimension of population size,
generally produce divergent group norms and values as well as conflicting
social roles (Mayhew and Levinger, 1976).

Urbanites have more opportunities to meet people and make friends. Urbanites are also more likely to be involved in more and different activities. For this reason, urban environments reduce residents' involvement with people from the traditional complex of kin and close neighbors. The importance of urban environments is found in the kind of social opportunities they provide, opportunities for deviance as well as conformity. In urban areas, people are more likely to find others like themselves with whom they can build networks and get incorporated into subcultures (Fischer, 1984). It is not just the fact that there are more people in cities that matters, but the fact that there is greater social diversity because such conditions are conducive to maintaining the social differentiation in cities.

It happens that the ends sought by different groups become so differentiated and so conflicting that individuals often do not know the conventional ways of behaving and suitable social roles in many areas of their lives. The simple fact that there are more people providing social expectations results in difficulties in fulfilling role obligations. Thus, urban residents tend not to agree about norms, and show a tendency to ignore traditional norms. There is not one central value system, but many such systems, and they may conflict at various points, leading to role confusion or conflict. Even values that are strongly maintained or brought to the city are subject to change (Fisher, 1975). One consequence of this heterogeneity is that urban residents are more tolerant of behavioral diversity than are nonurban residents (Wilson, 1985; Tuch, 1987).

As normative consensus decreases, alternative norms are available to learn, and these norms may permit deviance. This important consequence of urbanism refers to the creation and strengthening of urban subcultures, and the norms from these subcultures are often conducive to deviance. In this sense, the city creates social groups differentiated from one another by distinctive sets of beliefs and ways of living. Many of them are conventional, but others are considered deviant because of their lack of commitment to the dominant social order, their particular normative structure, and their unconventional types of behavior. Not all deviance in cities is the result of subcultures (Tittle, 1989), but the relationship between the overall rates of deviance and the conditions that create and maintain subcultures is strong.

With a proliferation of different groups and subcultures comes about the opportunity of learning norms that will be considered deviant to other groups. Alternative normative systems also provide the opportunity to associate with and receive support from others in the group. The large city populations provide what is termed the "critical mass" of deviants, as well as the opportunity to commit deviant acts.

Large population size provides a "critical mass" of criminals and customers for crime in the same way it provides a critical mass of customers for other services. The aggregation of population promotes "markets" of clients—people interested in purchasing drugs or the services of prostitutes, for example; and it provides a sufficient concentration of potential victims—for example, affluent persons and their property. (Fischer, 1984: 93)

2. Rapid Cultural Change   Rapid social and cultural change, features of urbanization, have the effect of reducing the importance of long-held traditions and customs. Social changes are more likely to occur in urban than rural areas because of the diversity of urban life. There are more sources—ideas, different groups, and values—for change in the city, and this gives the city a certain unpredictable quality. The effects can produce political changes in the form of new leadership, social changes in the form of new fashions, music, or life-styles, or technological changes. Sometimes the practical exigencies of urban life produce these changes. At other times, the changes seem to be outgrowths of the failure of informal controls to uphold and to maintain the older values and ideologies. The reduced size of the modern family is both a characteristic and a result of urban life. Urban life has also led to the development of the concept of equality of the sexes, both in marriage and outside it. The structuring of urban society into often fairly distinct peer groups has resulted in the magnification of age differences and the widening of the communication and status gaps between teenagers and adults, a situation that greatly increases the possibility of delinquency, crime, drug use, and different sexual behavior (Friday and Hage, 1976; Glassner and Loughlin, 1987).

3. Mobility   It has been said that less than a century or more ago a person might live a lifetime without ever going far from home and without seeing more than a handful of strangers; today, the picture is quite different. Modern transportation, particularly in urban areas, enables persons to move about rapidly and to have frequent contacts with many different people. Urban societies generally tend to regard mobility favorably, but frequent moves may have unsatisfactory effects. They tend to weaken attachments to the local community, to make persons less interested in maintaining certain community standards, and to increase contact with secondary groups of diverse patterns, thus weakening bonds which help to provide the basis for social control among members of local groups. As persons become more mobile, they come into contact with many different norms, and they begin to understand that other codes of behavior differ from their own. Moves necessitate changes in friendships, social roles, and adjusting old norms to new ones.

Some forms of deviance may be the outgrowth of mobility, such as suicide (Stack, 1982) and crime (Reiss and Tonry, 1986) in areas where people have few ties to the community and other residents. The rates of most crimes are higher in those areas of the city that have the highest transient population (Brantingham and Brantingham, 1984). Tittle and Paternoster (1988) found that mobility appears to reduce moral commitment which, in turn, increases the chances of criminality, but that mobility had little relationship with marijuana use. Some forms of deviance, therefore, may be more strongly related to mobility than others.

4. Materialism External appearances and material possessions have become of primary importance in modern urban society, where people are more often known for their gadgets than for themselves. People increasingly come to judge others by how well they display their wealth, a display Veblen termed "conspicuous consumption." Under urban conditions, the type of clothes people wear or the automobiles they drive, the costliness of their homes and furnishings, the exclusiveness of the social clubs to which they belong, and the amount of their salaries are often the sole means others have of making judgments about a person's success in life. If there are too many people to know personally, one can at least know other peoples' possessions.

By the 1980s, the emphasis on possessions had become associated with a type of urban resident, the "Yuppy," or Young Urban Professional. Partly because of the significant expense associated with life in many urban communities, another urban type has arisen, called a "Dink," or Double Income-No Kids. Both of these types are characterized by a concern over possessions and a life-style that emphasizes high-paying employment to help accumulate possessions. While not associated with a specific life-style, the 1980s was termed by many the "greed decade," given the high cultural emphasis on materialism.

At the other end of the income ladder are individuals and groups that are trapped in a cycle of poverty in urban areas. While urban areas often contain significant economic opportunity, that opportunity can not be enjoyed by all. Large cities in the United States contain very wealthy people but they also contain very poor people. As pointed out above, Wilson (1987) has observed that cities now contain a permanent "underclass" of individuals who have been unable to break from the chains of poverty in spite of sustained governmental efforts to facilitate this break. As pointed out earlier, decades of black emigration from the rural South and shifts in the national economy have left many black males without employment as manufacturing has shifted, ironically, to the South as well as overseas. Many jobs are also shifting to what previously had been considered suburban areas. The centers of major cities have been changing as suburban growth and urbanization in these areas have altered the economic structure of many areas. Many cities are now centers for financial and other professional services, thereby decreasing employment opportunities for unskilled workers even more. As a result, such persons may be disenfranchised from the larger culture as much by their feelings of alienation as the lack of economic resources to enjoy the material benefits of urban life.

5. Individualism Individualism may come to play a more important role in life in cities than in other communities. Urban persons may come to regard their own interests and self-expression as paramount in their social relations. The "I" feelings come to replace much of the feeling of cooperation characteristic of rural life. Competition may also intensify in urban settings where there are not only more opportunities to achieve goals, but more persons in competition for those same goals. This may lead to a reduction of close social contacts with persons with whom one is in competition, such as fellow employees.

This feature of urbanism may have intensified in recent years with the popularity of various psychologies that stress individual self-satisfaction over social relationships and obligations. To be primarily interested in one's own happiness, self-concept, and other interests is to lose sight of existing and potential social relationships and responsibilities. As Richard Sennett (1977: 259) has observed:

> The reigning belief today is that closeness between persons is a moral good. The reigning aspiration today is to develop individual personality through experiences of closeness and warmth with others. The reigning myth today is that the evils of society can all be understood as evils of impersonality, alienation, and coldness. The sum of these three is an ideology of intimacy: social relationships of all kinds are real, believable, and authentic the closer they approach the inner psychological concerns of each person.

Of course, there may be nothing wrong with the pursuit of individual desires, except perhaps if it is at the expense of the larger culture—that is, at other people's expense. It is sometimes difficult to find the intersection of self-interest and community interest, but individualism asserts that individual interests take precedence over social obligations and responsibilities.

6. Increasing Formal Social Controls As urbanism increases and conformity to social norms becomes less affected by informal group controls, greater opportunities and inducements develop for behavior that deviates from that of others or the norms of others. As impersonality increases and intimate communication declines, normative violations produce less and less informal censure of the kind seen in rural areas. The conflicting normative experiences in an urban setting tend to weaken parental authority and other traditional controls over youth and also over all individuals. Alternative normative systems provide an alternative to traditional authority that is appealing to some people. As a result, responsibility for controlling behavior in cities is shifted more and more to the police, courts, and other agencies of government that tend to enforce the norms of certain groups (Meier, 1982).

Urbanization throughout the world is now becoming associated with specialization where cities come to lay claim on certain products or services. Thus, in the United States, New York has become a major center for financial and professional services, whereas other cities, such as Detroit and Houston, are centers for the production of certain commodities, automobiles and oil

What is it about Cities that Make Them Conducive to Deviance? Six Sociological Reasons

1. Norm Conflicts	4. Materialism
2. Rapid Cultural Change	5. Individualism
3. Mobility	6. Increase in Formal Social Controls

respectively (Feagin, 1985). Just as cities have become specialized, so too have individuals. Systems of social control associated with interpersonal relations have come to be replaced by systems of control that utilize third parties (see the general discussion in Gibbs, 1989). The importance for interpersonal relationships of such roles as marriage counselors, police officers, judges, psychiatrists and other therapists, and mediators has increased as social life has moved to cities. People used to make do for themselves in solving interpersonal problems, but they now rely more on others for expert assistance.

A diversity of groups in the city also means that persons can belong to more groups. As a result, as Simmel pointed out, each individual is part of a larger network of group affiliations, but each group may have little influence over the individual. Membership in occupational, civic, religious, political, and recreational groups, as well as other voluntary associations, may be combined with association with other groups composed of particular sets of persons one meets in daily life. For many persons, while they belong to and associate with a wide variety of groups, any single group may be able to exert relatively little control over their behavior. As these informal controls dissipate, other kinds of more formal controls arise to replace them.

SOURCES OF DEVIANT ATTITUDES

Urbanism is not, of course, the only source of deviant attitudes. Other sources, such as the family, associates, and mass media, are also sources of such attitudes. The city is often the context in which these sources of deviance operate, but it is necessary to understand the nature of these other sources of deviant attitudes.

Specific theories of deviance, to be discussed in the next chapter, address the origins of deviance, but certain general observations are possible regarding the sources of deviant attitudes in society. A central theme of this book to this point has been that deviance is behavior or conditions that violate norms, and that much deviance is learned. It is not surprising that in a pluralistic society like the United States norms regarding a variety of behavior and conditions differ from group to group, and that adherence to one set of norms may mean that someone else's norms will be violated. People acquire their norms in a process of socialization or learning. This learning takes place in interaction with persons of one's own group. Intimate and personal associations probably have the greatest influence in the acquisition of deviant norms and attitudes. Most of these associations are likely to be of a group nature, such as through certain companions, gangs, families, occupational colleagues, or persons in local neighborhoods. Some norms that are learned in such interactions tend to push the individual away from deviance ("It is wrong to steal") while others pull the person toward it ("Everyone cheats on his income taxes; the government expects it"). Persons learn that marijuana use may violate a norm of appropriate behavior of a church group at a picnic, but not a college fraternity at a private picnic. Persons come to learn the content of the norms of the groups they belong to; some of those norms prescribe and some proscribe conduct that might be evaluated as deviant by another group. When people learn the norms of group A, and base their behavior upon them, they may violate the norms of group B.

The learning of both deviance and nondeviance are "natural" in the sense that they are the outgrowths of processes of social definitions and social processes of the types of groups to which one belongs. There are a number of sources of deviant norms and attitudes that affect these learning processes. In the discussion that follows, the relationships between deviant attitudes, associates, neighborhood, family, mass media, and occupation will be identified.

Associates

Attitudes favorable to deviance are primarily acquired through companions and by participation in small, intimate groups, such as gangs or families. Thus, deviant attitudes are learned in much the same manner as law-abiding norms are transmitted. One of the most consistent findings in criminology is the importance of companions in the commission of illegal acts by juveniles and adults. Most forms of criminality, and virtually all forms of juvenile delinquency, are committed in groups or have a group context (see the review on delinquency in Short, 1990). This is the case not only in the United States but in all other countries where the situation has been studied. The number of companions may vary from two or three "best friends," with whom acts of adolescent shoplifting or vandalism are committed, to a larger number of companions in the illegal acts of juvenile delinquent and young adult gangs. For those crimes that are committed alone, the offender's acquisition of attitudes and values toward criminality depended upon others.

Companions are found in the places where social life exists: schools, work places, and recreational facilities. It is in such social settings that deviant norms are learned as well as deviant roles. Deviant companions actively teach others attitudes, norms, and skills conducive to deviant acts (e.g., Akers, 1985). On the other hand, some claim that the influence of deviant companions is more indirect and that they merely fail to control by not providing negative sanctions to others who commit deviant acts for other reasons (Hirschi and Gottfredson, 1980). Since most delinquency is committed in groups, it is important to understand the role of peer influence in adolescent groups (Dornbusch, 1989). Violations of law by corporations are concentrated more in some industries than others, suggesting that these industries constitute more effective learning environments for law violations than other industries where violations are low (Clinard, 1990).

The role of companions can not be overstressed for many forms of deviance because they are the important source of deviant attitudes. Virtually all illicit drugs require not only a learning context for users, but also access to the drug. One must know from whom one can obtain the drug, and access to that world is necessarily limited. Only participation in a drug-using community will yield that information and, for that reason alone, it is usual that drug dealers have often been users themselves (Adler, 1985). In fact, one can not become a regular marijuana user (Becker, 1973) or a heroin user (Stephens, 1991) without the help, encouragement, and support of others.

Various sexual acts, whether considered deviant or not, obviously require other persons, and some acts are such that they require the support of others who may compose a community, such as those found among homosexuals. Homosexual communities vary in size and activity throughout the U.S., and their activities include not only interpersonal support and sociability but also, in some cities, political activism. Similarly, gambling, some of which may be illegal, can not be done alone and requires the active participation of others with whom one is gambling. The other party may be an individual, a gaming casino, or a group of undetermined numbers who "control" a particular form of gambling, such as policy or numbers gambling. While it is a popular view to think that compulsive gamblers have a form of addiction, it may be just as plausible that "problem gamblers" (those who repeatedly lose a lot of money gambling) suffer more from defective gambling strategies than a psychological "illness" (Rosecrance, 1988).

The production of pornography requires large numbers of persons who may reinforce in each other the feeling that they are merely providing a demanded service (Attorney General's Commission on Pornography, 1986: 284–291). The production of X-rated (NC-17) movies and video tapes requires a crew of several people and a cast; most adult books are not written by one author, but a staff of writers who guarantee that a pornographic book can be written quickly on a specific topic; pornographic magazines also require a staff for their production. And, all pornographic materials require an extensive and sophisticated distribution system that results in retail sales.

Gangs in Des Moines

Nathaniel Eugene Terrell and James Fleming, two graduate students in Sociology at Iowa State University, began studying urban gangs in 1989. No urban area appears to be completely free of gangs, even in a rural state like Iowa. Terrell and Fleming drove the city streets of Des Moines (population 200,000) as part of an urban project on homelessness. Indeed, most of the gang members were "homeless," or runaways. They were able to obtain estimates of the number of members of four gangs.

In addition to these, Terrell and Fleming found evidence of other groups, including Crippes, Bloods, BDG (Black Disciples Gang), Lords of Disciples, Vice Lords, and chapters of other national gangs.

Estimated Sizes of Selected Gangs, Des Moines, 1990

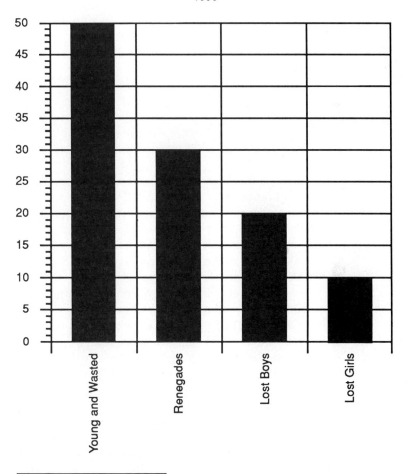

Estimated Members

Neighborhoods

The neighborhood or local community is primarily one of personal relationships, where people live and where their local institutions are situated. It is a world of meaningful experiences of the individual, a place where intense social relationships with others occur. It is also, at the same time, a place that often reflects different normative standards that may lead to deviant behavior. Neighborhoods reflect different social communities, where people feel a sense of belonging. Neighborhoods also provide the physical context in which much deviance takes place, and there are differences among neighborhoods that contribute to this deviance.

Neighborhoods differ as to social class, in the variety of the composition of racial, ethnic, and religious groups, and in the stability of the population. Even more important, there may be pronounced differences in the presence of norms supporting, for example, criminal activities in the neighborhood. Persons in high crime rate areas may have opportunities for many close associations with people who engage in or even encourage them to engage in crime. In this sense, crime may be self-perpetuating in certain areas of the city, especially those characterized by low-income, rundown housing, poverty, and the presence of certain ethnic groups (Brantingham and Brantingham, 1984: 297–331). Persons who live in middle- or upper-class neighborhoods may express approval of violations of laws that govern their own businesses while condemning ordinary crime, such as burglary and theft.

Such observations have led to the common sociological observations that communities differ in their social structure and social control, and that these differences are associated with different rates of crime and other forms of deviance (Reiss, 1986). Communities characterized by racial income inequality, poverty, and low occupational status—inner city or slum areas—have higher rates of many property crimes such as robbery and burglary than do other communities (Sampson, 1986).

Neighborhoods contain members of the same social class and are thus susceptible to class influences. The rates of conventional (meaning not white-collar or corporate) crime appear to be higher in the lower classes, regardless of whether we examine official police statistics, reports from victims of crimes, or self reports of illegal acts (Tittle, Villemez, and Smith, 1978 and Elliott and Huizinga, 1983; but compare Tittle and Meier, 1990). This suggests that different neighborhoods will have higher rates of deviance depending on the social class of the residents of that community. We might expect that persons residing in high crime and delinquency rate communities, for example, would have more opportunity to acquire criminal attitudes. But it is also the case that even in high crime rate areas, there are noncriminal norms and values (Kobrin, 1951).

The effects of neighborhood influences can be seen with other forms of deviance, such as teenage pregnancy. Black teenagers living in metropolitan areas of the United States have sexual intercourse at earlier ages than other teenagers, and thus they have higher rates of premarital pregnancy. Certain neighborhoods present a higher risk for teenage pregnancy than others. Hogan and Kitagawa (1985: 852) concluded that teenagers who are from high-risk environments—lower-class, residents in a ghetto neighborhood, nonintact family, five or more siblings, a sister who became a teenage mother, or lax parental control of dating—have rates of pregnancy that are 8.3 times higher than those from low-risk neighborhoods. Clearly, high rates of teenage pregnancy are the result of the unfavorable social circumstances in which these black teenagers are growing up. But such conditions are also related to various neighborhood norms pertaining to sexual activity and pregnancy. The prevalence of early sexual activity in the neighborhood legitimates the acceptability of this practice, in spite of the costs associated with teenage pregnancy. Having a sister who has experienced teenage pregnancy and the inability of parents to control the dating behavior of the teenage girl further contribute to and reinforce a normative climate that is conducive to norms that permit early sex. If such behavior is not rewarded in the neighborhood, neither is it strongly condemned.

The Family

The family, as an institution, has been undergoing great social changes which have resulted in the decline of its importance in general social life (Reiss and Lee, 1988). Increasingly, the socialization of young children has been taken on by other groups, such as the school and even the street gang. With weakened kinship ties, and with more mothers employed outside the home, the urban child may spend even less time with immediate family members.

Some people believe that unsatisfactory family influences constitute a chief source of deviant behavior. A number of possible family influences might be related to deviant conduct, particularly crime, including the one-parent (or broken home) situation and the family itself as a source of deviant norms. For example, one extensive review concluded that socialization variables, such as a lack of parental supervision and parent-child involvement, are among the most powerful predictors of juvenile misbehavior (Loeber and Stouthamer-Loeber, 1986). Furthermore, the presence of one child with delinquency or other behavior problems increases the probability that other children in the family will exhibit these problems. Attributing delinquency or crime simply to family influences, however, leaves out the very important source of deviant attitudes in the neighborhood and peer group.

Persistent efforts have been made to link youth crime to homes where there is only one parent (usually the mother), on the assumption that such a break in family ties may lead a young person to commit delinquent or other deviant acts. Some studies do show a high percentage of delinquents (between 30 and 60 percent) come from one-parent households, but such figures are misleading because the relationship between delinquency and broken homes is confounded by a selection process in the juvenile justice system; it appears that juveniles from broken homes are more likely to be arrested, convicted, and sentenced to a juvenile institution (Schur, 1973: 121). It is possible, however, that broken homes may influence the acquisition of deviant attitudes because of the reduced degree of supervision over youth, thus permitting adolescents to come into contact with deviant norms in the community (Rankin, 1983; Grinnel and Chambers, 1979). Research does suggest, however, that family and marital disruption is an important influence over crime rates in some cities. Cities with a high percentage of black and white households comprised of a married couple family have low rates of black and white juvenile offenses, while the divorce rate appears to have a significant effect on adult criminality (Sampson, 1986).

Since most one-parent households are the result of divorce, the consequences of marital disruption on children are important. It appears that the process of adjusting to divorce—which is most intense in the first year after the divorce—is more difficult and takes longer for boys than for girls (see Hodges, et al., 1983; Lowry and Settle, 1985). This may reflect the fact that most children of divorced parents live with their mothers and the absence of the father may present greater adjustment problems for boys.

The Mass Media

Persons still regard as controversial whether we learn deviant attitudes from the mass media, particularly television and motion pictures. At an earlier time, the issue of the effects on youngsters of reading crime comic books was raised with respect to delinquency. But since the 1950s, the influence of television has far surpassed any presumed influence from comic books. Without question, television and motion pictures are important media in the United States and they present a version of our culture that emphasizes wealth, materialism, and conflicting conduct that furnishes juveniles with models conducive to many forms of deviance. One is also occasionally able to read accounts from deviants that suggest that they were presented with an idea they had encountered from the media, and they acted upon this idea.

For example, one set of crimes has been attributed to a video. Terry Lee Corron had a video that described in vivid detail the strangulation of six women by a killer who wore camouflaged clothing, lifted weights, and kept a pornographic gallery (Spokane *Spokesman Review*, November 8, 1987, p. A2). Shortly after 3 a.m. on May 20, 1987, Corron, a security guard, crept up behind a co-worker at the switchboard of the hospital that employed him. He wrapped his hands around her neck and threw her to the ground, breaking her nose and shoulder. He then choked her for a while before fleeing. That attempted murder paralleled the first murder in the video. Both victims wore white nurses' uniforms, both were attacked from behind, picked up by the neck, and thrown down. Many felt that Corron had learned to kill from the video. It was not so.

In spite of such spectacular revelations, a more realistic appraisal of both television and motion pictures in relation to crime specifically indicates that on the whole their direct influence on viewers serves only to aggravate or reinforce whatever deviant attitudes and subcultural roles already exist (Lowery and DeFleur, 1988). There is no reason to believe that if all the media were to disappear from a society such as the United States, there would be any less delinquency and crime than there is now. Certainly, extensive delinquency and crime existed before the present mass media were considered to be of any social consequence. Children both learn and are influenced by the various media of mass communications, but all of what they receive passes through another set of influences, such as the family, school, relatives, and church before it becomes a real guide to action. No doubt viewers can learn different ways or techniques of acting out preexisting attitudes, but such persons may have acted upon those attitudes in any case, although in a different way.

The role of the media with respect to violence has received special attention, and there are three different views have been proposed on the issue of whether or not television promotes aggressive acts by viewers: (1) that television teaches aggressiveness; (2) that it reduces aggressiveness by serving as an outlet; and (3) that it has no demonstrable influence on viewers with respect to aggression. Some researchers have found a more direct role for the television media in the learning and modeling of aggressive behavior (Liebert, et al., 1973; Bandura, 1973), while others have failed to document such a role (Feshbach and Singer, 1971). One sociologist, for example, believes that "violent entertainment increases the likelihood of aggressive behavior among some categories of children" (DeFleur, 1983: 588).

Media effects can be either short-term or long-term. A media task force in the 1960s looking at the scientific evidence regarding media influences on violence concluded that the major short-term effects included that audience members who are exposed to violence will learn how to perform violent acts, and that audience members are more likely to exhibit that learning if they expect to be rewarded (Baker and Ball, 1969). The longer-term effect, the task force concluded, was that "exposure to mass media portrayals of violence over a long period of time socializes audiences into the norms, attitudes, and values for violence contained in those portrayals" (Baker and Ball, 1969: 376).

Media influences are often not as direct as such conclusions suggest. Some writers, for example, have claimed that media influences are mediated by certain psychological characteristics that exist in the viewer prior to media exposure (Feldman, 1977: 86). An extensive review of studies examining the relationship between television viewing and violent conduct concluded that there was no evidence of any significant effects on aggressive behavior (Kaplan and Singer, 1976).

There is no question that the mass media do present an image of the United States as a crime-ridden, violent society. The newspapers' newscasts of major cities contain graphic examples daily of such incidents, and they give the impression that deviants conform to certain stereotypes, such as the image that all criminals are lower-class citizens, rather than from white-collar occupational groups. The impact of such misleading stereotypes may be severe in the sense that persons may come to view upper-status deviance as inconsequential. Although many persons in the media may think that they are simply telling people what they need to know, public opinion polls indicate that the public thinks the media spend too much time on crime (Harris, 1987: 184–187). In any case, much more work, both theoretical and empirical, needs to be done before we can draw firm conclusions about media influences (DeFleur and Ball-Rokeach, 1982).

Occupation

Occupations may also serve as the opportunity for deviant acts. Techniques of law violations in business, for example, may be picked up from conversations with business associates. Advertising specialists, for example, learn how to prepare misleading advertising. In such instances, the line between ads that are false and "just" misleading may be the size of the fine print. Illegal practices are often diffused among corporations and among the industries of which they are a part. A systematic examination of violations and sanctions among the largest corporations in the United States found that three industries—the pharmaceuticals, auto, and oil industries—had particularly high violation rates compared to other industries (Clinard and Yeager, 1980).

As another example, physicians and nurses have access to many kinds of drugs that other people do not. It is not surprising, therefore, that addiction rates are relatively high among physicians. A study of 98 physicians who either were or had been opiate addicts found that the physicians began their drug using careers later than other opiate addict, and that they almost never associated with other physician addicts (Winick, 1961). Their addictions were usually discovered by the indirect evidence of checking prescription records. The level of general drug use among physicians, except for tobacco and alcohol, is very high compared to other occupational groups (Vaillant, et al., 1970). Similarly, performers in the entertainment business, such as jazz musicians and performers in rock bands, are more likely to use marijuana and/or cocaine largely because such drug use appears to be a part of the music subculture. Drugs may be used for prestige among other musicians, and because "one-nighters" are tiring and drugs provide recreational outlets as well as stimulation.

SUMMARY

Much deviance has an urban context. The strong relationship between urbanization and deviance has been noted for virtually every form of deviance, including crime, delinquency, alcoholism, drug addiction, mental disorders, and suicide. Not surprisingly, the development and growth of deviance is related to the general process of urbanization found in all parts of the world. In the United States, processes of urbanization resulted initially in shifts of the population to urban areas, but more recently it has resulted in a shift of population, as well as economic opportunities, to suburban sectors of metropolitan areas. In some cities, an "underclass" of persons remains because they have been unable to follow those opportunities. They are poor and, many, persons of color who experience very high rates of deviance in their daily lives.

The urban context of deviance is related more to certain social characteristics of cities (called *urbanism*) than to the simple effects of living with many people. Urbanism is the way of life that develops in cities and is reflected in a number of social features of city life, such as norm conflicts, rapid cultural change, mobility, materialism, individualism, and an increase in formal social controls. In cities, people must learn new ways of thinking and doing things, and they must do so in a setting of great diversity of people, ideas, and events. It is the greater differentiation in the cities that is related to deviance.

The city is also conducive to deviance by being the setting in which deviant attitudes are found. These attitudes are related to the structural feature of cities that permits a number of different subcultures and patterns of behavior. City dwellers' attitudes are related to patterns of the number and kind of associates they have, the neighborhood they live in, characteristics of their family life, their exposure to the mass media, and in the context of their employment. In cities, there is greater opportunity that one will be exposed to deviant attitudes from such sources.

An understanding of deviant behavior must take into account the role of associates, neighborhoods, the family, mass media, and occupation in describing how persons and groups come to commit deviant acts. These broad influences are important because they are closely related to many forms of deviance. But, as important as these sources of deviant attitudes are, they do not in themselves provide an explanation of deviance. They do not, in other words, constitute sociological theories of deviance, even though they may be incorporated into theories. Theories are specific explanations of the deviance process. It is this topic to which we now turn in Chapter 4.

SELECTED REFERENCES

Brantingham, Paul and Patricia. 1984. *Patterns in Crime.* New York: Macmillan.

A collection of research findings concerning the relationship between crime and city characteristics. Well-written and informative, this book summarizes much of relevant literature.

Fischer, Claude S. 1984. *The Urban Experience,* 2nd ed. New York: Harcourt Brace Jovanovich.

Fischer is one of the best known urban sociologists. His work has concentrated on the nature and meaning of urban communities and the importance of subcultures that form there.

Harris, Louis. 1987. *Inside America.* New York: Vintage.

The results of a collection of Harris Polls in recent years. An indispensable reference for those interested in what representative samples of Americans think about a number of topics. Short and interesting.

Reiss, Jr., Albert J. and Michael Tonry, eds. 1986. *Communities and Crime.* Chicago: University of Chicago Press.

A recent collection of papers from sociologists interested in a communities orientation that characterized work done in the "Chicago School" in the 1920s. This is a theoretical perspective that has been revived in recent years.

Wilson, William Julius. 1987. *The Truly Disadvantaged: The Inner City, The Underclass and Public Policy.* Chicago: University of Chicago Press.

An important statement by a leading sociologist on the nature of the inner city for its black residents. Debilitated by shifting economic opportunities and the departure of more skilled blacks who have migrated to the suburbs, those who remain in the slum suffer from a number of intense problems, including poverty, teenage pregnancy, and high rates of deviance.

General Theories of Deviance

Most sociologists would agree that processes of urbanization have been linked with deviance and that there are a number of sources of deviant attitudes. In fact, the importance of the link between deviance and urban context can hardly be overstated. But, there is no single agreed-upon sociological theory to explain deviance. The absence of a single theory might have been anticipated from the fact that, as was seen in Chapter 1, sociologists do not agree on a definition of deviance. Several controversial theories have been developed, in fact, and it is important to understand the various approaches that have been used in studying deviance. It is also important to understand the context in which these theories were developed. Two earlier perspectives that are examined here—social pathology and social disorganization—no longer have a sizable sociological following, but they are important because they set the intellectual stage for subsequent thinking and theorizing about deviance, particularly in the form of those theories that are more contemporary— anomie, labeling, conflict, control, and socialization theories.

Sociological thought on deviance coincides with the development of that discipline in the United States in the late 1800s. The social pathology perspective was the first orientation among sociologists, and this gave rise to the social disorganization viewpoint.

SOCIAL PATHOLOGY

The concept of "social pathology" was developed in the late 1800s and continued in use until the end of the 1930s or so. The social pathology perspective emphasized the debilitating effects of city life but it sketched its portrait in large, wide strokes, linking deviance to pathological conditions in cities and the large society. It used an *organic analogy* that likened society to the functioning of a biological organism. Social pathology in society was the direct counterpart to physical illness in the individual, and the social pathology perspective was the sociological counterpart to the medical model thinking found in individualistic theories of deviance.

Social pathology attempted to apply a biological model to deviance in social settings. Social pathologists believed that in a healthy society some universal criteria were to be found, but that, at the same time, societies could develop pathologies or abnormalities such as crime. The conditions considered to be deviant were those that interfered with the "normal" or "desirable" workings of society. Conditions like crime, suicide, drunkenness, poverty, mental illness, prostitution, and so forth, were deviant because they were known to be "bad" or pathological (Higgins and Butler, 1982: 20). Even today, it is not uncommon to hear such phrases as a "sick society" or that "society is suffering from a sickness." Within this framework, deviance was somewhat on the order of a universal disease or an "unhealthy" deviation from some assumed universally accepted norm of conduct. With the conception of society as an organism, it is important that it be healthy, thus good, while sickness is undesirable and, therefore, bad. Persons or situations diverging from expectations that have been formulated in these terms are "sick." For social pathologists, a social problem or an instance of deviant behavior was, in the end, a violation of *moral* expectations (Rubington and Weinberg, 1981: 19). In this sense, the social pathologists interchanged and confused the concepts of sin, sickness, and deviance.

Social pathologists worked from two perspectives—first, they believed that there was a certain "sickness" in society, and, second, that certain individuals were also "sick." Increasingly, sociologists developed the view that personal maladjustment or "sicknesses" such as "bad" heredity, physical illnesses, mental deficiency, mental disorder, drunkenness, lack of education, and personal immorality due to poverty, lay at the heart of what they then termed social pathology, later becoming known as deviance. All this was the forerunner somewhat later of the "medical model" approach of the psychiatrists. Just after the turn of the century, one writer wrote graphically about this organic analogy to social science:

> Pathology in social science has a certain parallel to pathology in medical science. As the study of physical disease is essential to the maintenance of physical health, social health can never be securely grounded without a wider and more definite knowledge of social disease. General pathology in medicine teaches that many diseases have much in common and there are morbid processes which may be discussed, as well as particular diseases. In social pathology, the interrelation of the abnormal classes is one of the most impressive facts. Paupers often beget criminals; the offspring of criminals become insane; and to such an extent is the kinship of the defective, dependent, and delinquent classes exhibited, that some have gone so far as to hold that under all the various forms of social pathology, there is a common ground in the morbid nervous condition of the individual. (Smith, 1911: 8–9)

The social pathology approach was largely consistent with the personal ideologies and social backgrounds of its advocates. Many early American sociologists interested in social pathology were recruited from small, mid-western rural communities where they had been raised with a sense of the importance of traditional religion, as well as a general distrust of social change and city life (Mills, 1943; Schwendinger and Schwendinger, 1974). As a result, they had developed an attitude, which one might describe as "sacred provincialism," that displayed a moralistic approach to deviance and to social problems. The social pathology approach came at a time of relatively rapid urbanization and increasing social differentiation as a result of extensive immigration. From their own points of view, the social pathologists had no need to understand the relative nature of the behavior they condemned because it was obviously wrong and must be changed. One observer wryly noted: "Like General Custer's, their [the social pathologists] tactics were simple; they 'rode to the sound of the guns' " (Lemert, 1951: 3).

The social pathologists employed a universal moral standard—rather than a normative one—against which all behavior and persons could be judged, and it was against this standard that the determination was made of which behavior was to be considered deviant and which acceptable (Meier, 1976). The social pathologists were social reformers who wished to remove the "evils" of society and to "salvage" it for the middle class (Davis, 1980: Chapter 3). Their reform efforts dealt principally with such issues as poverty, child labor, divorce, ordinary crime, and associated "vices" such as drinking alcoholic beverages and prostitution. They took these vices as city behavior and in need of change. In spite of their good intentions, the views of the social pathologists are not taken seriously today for a number of reasons.

Deviance and Illness

Deviance is not an individual or social "illness," but rather a departure from norms. It is not the case that "deviance and crime are reflections of fundamental disorders in society or in the individual or both" (Aday, 1990: 64). Individuals, in the social pathology model, can not be divided into the normal and the pathological (Matza, 1969: 41–46). What is pathological in a social sense is relative to norms, while what is pathological in the physiological sense is universal and unchanging.

The Universality of Norms

Norms are actually changeable entities, and as norms change, so too do forms of deviance. As Matza (1969: 43) has pointed out, "the idea of diversity contested pathology." Social pathologists failed to recognize the important point that deviance varies over time and from group to group as norms change, and it can not be compared to a disease, like cancer, which is universally designated as an illness. While there is little disagreement on what constitutes a healthy state of an organism, much less agreement is seen on what constitutes a "healthy" or "sick" behavior; what is deviant is a function of norms. People who agree on what constitutes a healthy organism do not similarly agree on what constitutes a healthy society. In fact, "it is impossible to find [a criterion] that people generally accept as they accept criteria for health for the organism" (Becker, 1973: 5).

SOCIAL DISORGANIZATION

The growth of cultural relativity (the recognition that cultures are not better or worse than one another, only different) in sociology and the questioning of the existence of "universal" values brought an end to the social pathology perspective. Particularly instrumental were the pronounced social changes following World War I and the Great Depression, along with extensive immigration, urbanization, and industrialization in the United States. These were accompanied by the crowding of large numbers of newly arrived persons from a diversity of cultures into urban areas and the rapid development of conditions conducive to deviance. Some sort of different explanation with new concepts was necessary.

The need for a new framework was satisfied with the evolution of the concept of *social disorganization*, which was developed originally by Thomas and Znaniecki (1918), and by Cooley (1918). The term "social disorganization" is used to refer to both an explanation of deviance and a state of society that produces deviance. The concept of social disorganization is associated with the "Chicago School" of sociology because most of the persons originally using the concept came from there (Bulmer, 1984). Social disorganization was the result of the intellectual development that had taken place in sociology since about 1910 in both theory and method. Social disorganization "rooted" the problem of explaining deviance in the social norms and activities of communities.

Deviance, in the context of social disorganization, was seen as a product in society of uneven development, with much social change and conflict affecting the behavior of individuals. Social disorganization theory emphasized that society was organized when people are presumed to have developed common agreement about fundamental values and norms, as reflected in a high degree of behavioral regularity. In other words, social organization (social order) exists when there is a high degree of internal cohesion binding together individuals and institutions in a society. This cohesion consists largely of consensus about goals worth striving for (values) and how or how not to behave. When consensus concerning values and norms is upset and traditional rules no longer appear to apply, social disorganization results.

In contrast to social disorganization, successful social organization, it was thought, involved an integration of customs, effective teamwork, and high morale; all this led to harmonious social relationships. A well-integrated social group showed group solidarity, and modes of behavior were more homogeneous and traditional. There was little unconventional behavior and deviance, and informal controls usually sufficed to regulate behavior (Meier, 1982).

The United States, particularly the urban areas, did not fit well into this idyllic set of characteristics, which made it clear to social disorganization theorists that a state of social disorganization existed in much of city life. While the social pathologists were well aware of the relationship between city life and deviance, the social disorganization theorists made that relationship unmistakable. They did so by proclaiming the city their laboratory in which to study deviance. It was in the cities where they concentrated their research, on what they termed "disorganized local areas," generally slum or inner city areas of high crime, prostitution, mental disorder, suicide, and other forms of deviant behavior. It was assumed that changes were taking place in the ecological or spatial growth of the city and that these changes resulted in the general deterioration of group solidarity in certain areas (Davis, 1980: 67–69).

Social disorganization theorists singled out the city to study what they saw as the "disorganizing" aspects of life there, and the effects of rates of deviant conduct in urban areas. In their own theoretical framework, the social patterns of the urban environment were conducive to social disorganization, which then led to the deviant behavior. Social disorganization theorists were aware of the great heterogeneity of people living in the same geographic area and the vast differences in values and norms of these people; in addition, they noted that people moved into and out of the areas without ever developing a sense of neighborhood social organization. Lacking a real social organization, certain neighborhoods were, they considered, "socially disorganized," and this condition itself resulted in high rates of deviance, particularly when examined in terms of the more organized suburban and rural neighborhoods. Within these "disorganized" parts of the city, certain areas, particularly inner city slum neighborhoods, appeared to have even more deviant behavior. It was generally felt that the higher rates of deviance, particularly rates of high delinquency, were to be found in the central inner-city core where basic social solidarity was particularly lacking. By this, social disorganization theorists meant the original group norms that had been brought to the United States from the old country had collapsed in the new land and that no new effective standards had become solidified; in sum, a true situation of "social disorganization" existed (Kornhauser, 1978: Chapter 3).

An important implication of this view for the control of deviance was that deviants should be integrated into conventional groups and that the integration process should focus on the local community. The most typical example of a large-scale program designed to implement these ideas was the Chicago Area Projects, initiated in the 1930s and continuing to this day. This series of neighborhood programs was meant to organize the change in local communities with high rates of crime and delinquency.

A number of problems became evident with the social disorganization concept to explain deviance, and these problems contributed to the decline in the use of this concept.

Subjectivity

A fundamental problem is that it is subjective and judgmental while it masquerades as an objective conceptual framework (Gibbons and Jones, 1975: 19). The concept of social disorganization seemed almost as subjective as social pathology. In the social disorganization perspective, the concept of pathology was simply applied to the group instead of to the individual. No longer were persons pathological; communities were now disorganized. The designation of phenomena as deviant and the equation of deviance with disorganization were the focus of the sociological analyst or observer rather than findings derived from actual studies of what some people would term a state of social disorganization. Social disorganization was usually thought of as something "bad," and what was bad was often the value judgment of the observer and the members of his or her social class or other social groups.

Deviants and the Lower Class

Concentrating as it did on inner city areas, the concept of social disorganization tried to explain deviance almost entirely as a lower-class phenomenon, to the exclusion of middle- and upper-class deviance. In other words, it was biased in favor of middle-class values and standards. The lower class was assumed to have higher deviance rates because its members lived in the most disorganized areas of the city. This argument appears to suffer from circular reasoning: the lower class has the most deviants because it is the most disorganized, and it is the most disorganized because it contains the most deviants (Traub and Little, 1985: 43). Of course, if the lower-class does contain the most deviants, it may be for reasons other than social disorganization.

Deviance or Change?

Social disorganization implies the disruption of a previously existing condition of organization, a situation that generally can not be established. Social change was often confused with social disorganization, and little attention was paid to explain why some social changes are disorganizing and others are, in fact, organized.

What Is Disorganized?

What may seem like disorganization may actually at times be quite highly organized systems of competing norms and values. Many subcultures of deviant behavior, such as youth gangs, organized criminal syndicates, homosexuality, prostitution, and white-collar crime, including political corruption as well as corporate crime, are highly organized. Even the norms and values of the slums are highly organized, as Whyte (1943) showed in his classic study of a slum area, *Street Corner Society*. In fact, the social disorganization theorists often were describing not disorganization but diversity (Scull, 1988). Such confusion regarding what is and what is not "disorganized" reflects the basic problems of this approach. The problem was in the ambiguity of the concept of social disorganization itself (Liska, 1987: 88).

GENERAL SOCIOLOGICAL THEORIES OF DEVIANCE

General sociological theories of deviance can be divided into two main types: structural and processual. Structural theories emphasize that deviance is related to certain social structural conditions in society, while processual theories describe the processes by which individuals come to commit deviant acts. Structural theories are also more interested in explaining the *epidemiology* of deviance—that is, the distribution of deviance in time and space— while processual theories are more interested in *etiology*—that is, the specific causes of deviant acts. For these reasons, structural theories often attempt to explain such things as why certain forms of deviance are more prevalent in the lower classes (an aspect of epidemiology) while processual theories attempt to explain how specific persons came to commit deviant acts (etiology). Because they operate on different levels of analysis, structural theories are sometimes called sociological theories, and processual theories are called social psychological theories. Actually, there is a good deal of overlap between structural and processual theories since many of these theories have implications both for epidemiology and etiology. Still, the distinction is a useful one in terms of the ability of a given theory to account for these two dimensions of deviance.

The two remaining theories discussed in this chapter illustrate each of these types of theories. Anomie is a structural theory that attempts to explain group differences in the rates of deviance, while learning or socialization theory is a processual or social psychological theory that concentrates on the processes by which individuals come to commit deviant acts.

ANOMIE

Anomie is a sociological perspective that is related to the social disorganization view previously mentioned. Anomie is a general perspective on deviance because it is meant to explain a number of forms of deviance, including crime, alcoholism, drug addiction, suicide, and mental disorder. Anomie is both a theory of social organization as well as a theory of the origins of deviant motivation, but the social organization component of the theory has received less attention (Messner, 1988).

Anomie theory contains the core idea that elements in the structure of society promote deviance by making deviant behavior a viable adaptation to living in the society (see Aday, 1990: 63–64). The theory of anomie claims that deviance is the result of certain social structural "strains" that place pressure on individuals to become deviant. This view was originally proposed as a general theory by sociologist Robert Merton in the 1930s (Merton, 1968: 185–248; see also Clinard, 1964: 1–56). Modern industrial societies, like the United States, emphasize the acquisition of material success in the form of wealth and education as accepted status goals, while simultaneously limiting institutional access to certain segments of society who are able to use them legitimately. The segments of society that are generally denied access to these status goals are the poor, those who belong to the lower class, and persons of certain racial and ethnic groups who are discriminated against, such as African-Americans and Chicanos. As a result, a situation of *anomie* takes place when there is an acute dysjuncture between the cultural goals and the legitimate means available to certain groups in society to achieve those goals.

Success goals in cultural terms are generally presumed to be achieved by *legitimate means* through regular employment, in higher paid occupations, and through access to further education. These channels, however, are not available to certain persons, such as those in the lower class. Thus, while everyone learns to aspire to the "American Dream" where success can come to everyone, the reality is that this dream is reserved for only a few since the social structure can actually provide opportunities for only a small number of persons. Anomie is the name of the social condition where success goals are emphasized much more than the acceptable means by which to achieve those goals. Consequently, some persons are forced to achieve them through *illegitimate means,* through such forms of deviance as crime, prostitution, drug use, alcoholism, and mental disorder. In attempting to explain these forms of deviant behavior, anomie theory has relied on the fact that official rates of deviance are highest among the poor and the lower class where the greatest pressure for deviation occur and opportunities to acquire both material goods and a higher level of education are limited (Clinard, 1964).

Adapting to Strain

According to the anomie perspective, several adaptations are available to persons in an anomic society. The most common adaptation is to continue to conform to society's norms and not become deviant. This, according to Merton, is the most common adaptation. Some individuals may adapt by becoming ritualists, or persons who conform to society's norms without any expectation of achieving the goals or values of society.

There are also several *illegitimate adaptations* that can be used by poor, lower-class persons where legitimate means to achieve the culturally prescribed goals of success have been blocked. The adaptations that are relevant to the study of deviance include rebellion, innovation, and retreatism. The particular adaptation is dependent on the individual's acceptance or rejec-

Anomie: Adapting to the Strains Between Values and Norms

Modes of Adaptation to an Anomic Society		
	Values (Goals)	Norms (Means)
Conformity	+	+
Innovation	+	−
Ritualism	−	+
Retreatism	−	−
Rebellion	±	±
Legend: + = Acceptance − = Rejection ± = Substitution		

Adapted from Robert K. Merton, *Social Theory and Social Structure* (New York: Free Press, 1968).

tion of cultural goals and the adherence to, or the violation of, accepted norms. Persons may turn away from conventional cultural goals and rebel against them. Through this *rebellion* they may seek to establish a new or greatly modified social structure. They try to set up new goals and procedures to change the existing social structure instead of trying to achieve the goals traditionally established by society. This type of deviant adaptation is represented by political radicals and revolutionaries.

Innovation is an adaptation involving the use of illegitimate means such as theft, burglary, robbery, organized crime, or prostitution to achieve culturally prescribed goals of success. This response is "normal" where access to success through conventional means is limited (Merton, 1968: 199). As evidence, Merton has maintained that unlawful behavior such as crime and delinquency are most common in the lower strata in society. The poor are largely restricted to manual labor, which is often stigmatizing. As a result of the low status and the low income, they can not readily compete in terms of established standards of worth, and therefore they are more likely to engage in crime.

Retreatism, according to Merton, represents the substantial abandonment of the cultural goals that society esteems and of the practices that have become institutionalized to achieve these goals (Merton, 1968: 203–204). The individual has fully internalized the cultural goals of success, but has not found them available through the institutional means of obtaining them. Being held back from achieving the goal, through internalized pressures which prevent innovative practices, the individual becomes frustrated and handicapped, often defeated and even withdrawn. Retreating from cultural goals, the person becomes addicted to drugs, becomes an alcoholic, or may completely "escape" through a mental disorder or even suicide. Retreatism tends to be a *private* rather than a group or subcultural form of adaptation, even though the person may have contact with others in a similar fashion. In addition to becoming withdrawn in an individual sense, the retreatist also withdraws from social life, or, in the case of suicide, from physical life.

Reformulations

Cloward and Ohlin (1960) point out that illegitimate means are not equally available to all persons, for much the same reasons that legitimate means vary by social strata. The lower-class poor are provided greater opportunities for the acquisition of deviant roles, largely through access in inner city areas to deviant subcultures and the opportunity for carrying out such deviant social roles once they have been acquired. Delinquency arises from the disparity between what lower-class youth are led to want and what is actually available to them. Desirous of such conventional goals as economic and educational success, many persons are faced with limitations on the legitimate avenues to success. Since they are unable to revise their goals downward, they become frustrated and turn to delinquency if the norms and opportunities are not available to them. A similar position has been taken by Agnew (1985) who argues that delinquency may actually result from an inability to avoid negative or painful situations in life. The lack of opportunity in this case may lead adolescents to feel they are trapped with few prospects for the future.

Another reformulation of anomie theory has been proposed by Simon and Gagnon (1976). They point out that Merton had originally formulated his theory in the 1930s, during a period of chronic economic depression, but that the economic affluence of the 1970s had had a substantially different impact on deviance than had been the case several decades earlier. Simon and Gagnon (1976: 370) expand the number of adaptive responses to nine based on (a) commitment to approach cultural goals, and (b) the degree to which progress is achieved toward the realization of these goals. These responses are optimal conformist, detached conformist, compulsive achiever, conforming deviant, detached person, escapist, conventional reformer, missionary, and total rebel. With their added adaptations, Simon and Gagnon argue that the revised model appears to account more adequately for deviance at higher socioeconomic levels than does Merton's original model, a contention of course that still awaits empirical confirmation.

Evaluating Anomie

Explanations of deviance in terms of anomie tend to oversimplify what is a far more complex problem and, probably for this reason, it seems that anomie theory is almost ritually constructed and demolished every school term in courses on deviance and crime. Only a few of the more important inadequacies can be pointed out here (Clinard, 1964; Cohen, 1965; Liska, 1987: 54–55).

The Assumption of Universality. Anomie theory assumes a universality of what constitutes "illegitimate means" that is not the case because delinquent and criminal acts vary in time and place. Deviance is a relative concept; it is not the same in all groups. The use of drugs, for example, such as marijuana and even cocaine and opium, are not deviations in many parts of the world today. It was not a century ago in Western societies, including the United States, that the use of opiates was generally made illegal. Even in the United States, what is considered deviant depends on the norm being violated, which group subscribed to that norm, and with what intensity.

Class Bias. Anomie theory rests on the assumption that deviant behavior is disproportionately more common in the lower class. This assumption is made because the lower class is where the gap between pressures to succeed and the reality of low achievement is the greatest. Considerable deviance can be cited since lower class persons and members of minority groups are more likely to be detected and labeled as delinquents, criminals, alcoholics, drug addicts, and mental patients than persons who belong to the middle and upper classes who may engage in the same behavior. Furthermore, studies of occupational, white-collar, and corporate crime have shown that crime also occurs in the highest social strata (Clinard and Yeager, 1980; Coleman, 1985), although the pressures from an anomic society should not be very great there. Even conventional offenders are not exclusively found in the lower classes. Sykes (1971) points out, for example, that middle-class persons and college students apprehended for shoplifting do not fit the "poverty syndrome" where crime is a means of breaking free from severe material deprivation.

Simplicity of Explanation. While it is possible for an individual in some cases to be subject to something that resembles the strain of anomie, there are clearly many other factors that influence deviant acts. Although some deviants undoubtedly experience frustration regarding their inability to achieve success goals legitimately, most deviant acts actually arise out of a process of interaction with others who may serve as a reference group for the individual, and whose advice is important to the individual. Many deviant acts, in fact, are part of role expectations rather than representing a dysjuncture between goals and means (Cohen, 1965). Anomie does not recognize the importance of deviant subcultures, deviant groups, the role of characteristics of urban life, and processes of interpersonal influence and control. Many forms of deviance, such as drug addiction, professional theft, prostitution, and white-collar crime, are collective acts in which association with group-maintained norms explains the behavior.

The Trouble with Retreatism. The theory of the adaptation of means to goals through "retreatism" lacks precision and oversimplifies what is actually a much more complex process of how alcoholism, drug addiction, mental disorder, and suicide develop. As will be shown later, the process by which one becomes addicted or mentally ill is much more complex than a simple process of retreating from success goals. It involves normative actions and role-playing. Drug addicts are not retreatists in any conventional sense of that term. Rather, they are active participants in their social worlds (e.g., Hanson, et al., 1985). It is also difficult to perceive the many physician addicts as "retreatists" or suffering from a general inability to achieve culturally prescribed goals (Vaillant, et al., 1970). An additional major criticism of retreatism as an adaptation is that it fails to distinguish the origins of the deviance from the actual effect produced. Long periods of excessive drinking or drug use may impair a person's social relations and ability to achieve certain goals in society; in this way, anomie may confuse what is cause and what is effect.

Alternative Perspectives. Because of a broad, social structural approach, acts of deviance are presented as having only one social meaning.

Thus, while drug use is alleged to represent an escape from economic failure in anomie theory, it may actually serve different purposes. Drug use may be a form of innovative behavior, such as risk taking or "getting kicks," a ritual act, such as the American Indian's use of peyote, an expression of rebellion, an act of peer conformity, or an act of social consciousness as reflected in instances of medical experimentation (Davis, 1980: 139).

LEARNING OR SOCIALIZATION THEORY

A second general theory of deviant behavior can be called "socialization," or learning theory, and this perspective is adopted as the central frame of reference throughout the remainder of this book. The socialization approach has certain weaknesses, to be pointed out from time to time, but it seems best suited to account for the facts about deviance that require explanation. Deviant behavior is here approached as a learned phenomenon according to the same basic processes through which conformity is learned. Although the basic processes are the same, the direction and content of the learning may differ. The process of acquiring norms, social roles, and self-conceptions has been discussed in previous chapters and, thus, the details are omitted here. The central contexts of learning deviant behavior, such as urbanism, have also been discussed and will not be repeated here.

Sutherland's Theory: An Example of a Socialization Theory

Deviance is the consequence of the learned acquisition of deviant norms and values, particularly those learned within the framework of subcultures and among peers. The best-known socialization or general learning theory in these terms is Edwin H. Sutherland's (1947) theory of *differential association*, one of the most widely known theories in sociology. Sutherland's theory was developed to account for crime, but actually it is a perspective that accounts as well for both the etiology, or "cause" of an individual's deviance, and epidemiology, or the "distribution" of deviant behavior as reflected in various rates. This combination requires an analysis of conflicting deviant and nondeviant social organizations or subcultures (differential organizations) and a social psychological approach to deviation at the individual level in terms of conflicting deviant and nondeviant associations of deviant norms (differential associations).

At the group level, Sutherland argued that deviant behavior was a consequence of normative conflict. Conflict among norms is related to deviance through differential social organization, which consists of neighborhood structures, peer group relationships, and family organization. At the individual level, normative conflict results in deviant behavior through differential association, the learning of definitions of behavior from primary groups.

In propositional form, Sutherland's theory is stated in terms of its application to crime and criminal behavior, but the concept is modified here to apply to other forms of deviant behavior, such as prostitution, drug addiction, alcoholism, and homosexual behavior, along with other forms of deviance. The propositions of the theory of differential association (from Sutherland and Cressey, 1978: 80–82), with the adapted amplification from McCaghy (1985: 66–67), are as follows:

1. Deviant behavior is learned. This means that deviance is not inherited; nor is it the result of low intelligence, brain damage, and so on.

2. Deviant behavior is learned in interaction with other persons in a process of communication.

3. The principal part of the learning of deviant behavior occurs within intimate personal groups. At most, communications such as the mass media of television, magazines, and newspapers play only a secondary role in the learning of deviance.

4. When deviant behavior is learned, the learning includes (a) techniques of deviance, which are sometimes very complicated, sometimes quite simple; and (b) the specific direction of motives, drives, rationalizations, and attitudes.

5. The specific direction of motives and drives is learned from definitions of norms as favorable or unfavorable. This proposition acknowledges the existence of conflicting norms. An individual may learn reasons for both adhering to and violating a given rule. For example, a person might argue that stealing is wrong—that is, unless the goods are insured and, of course, nobody really gets hurt.

6. A person becomes deviant because of an excess of definitions favorable to violation of norms over definitions unfavorable to violation of norms. This is the key proposition in the theory. An individual's behavior is affected by contradictory learning experiences, but the predominance of deviant definitions leads to deviant behavior. It is important to note that the associations are not necessarily only deviant persons, but also definitions, norms, or patterns of behavior. Furthermore, in keeping with the notion of a learning theory, the proposition can be phrased: A person becomes nondeviant because of an excess of definitions unfavorable to violation of norms.

7. Differential associations may vary in frequency, duration, priority, and intensity. Frequency and duration are self-explanatory, referring to how long one is exposed to particular definitions and when the exposure began. Priority refers to the time in one's life when exposed to the association. Intensity concerns the prestige of the source of the behavior pattern.

8. The process of learning deviant behavior by association with deviant and nondeviant patterns involves all of the mechanisms that are involved in any other learning. Again, there is no unique learning process associated with acquiring deviant ways of behaving.

9. While deviant behavior is an expression of general needs and values, it is not explained by those general needs and values since nondeviant behavior is an expression of the same needs and values. A "need for recognition" can be used to explain why certain persons commit mass murder, run for President, or work to maintain a .320 batting average, but it really explains nothing since it apparently accounts for both deviant and nondeviant actions.

Other Learning Theories

Ronald Akers (1985) has attempted to explain deviance on the basis of learning principles in a differential association-reinforcement theory of deviance. Similar to Sutherland's theory, Akers (1985: 51) claims that deviance results from the learning of definitions that portray some conduct as desirable, even though deviant. "Definitions are normative meanings which are given to behavior—that is, they define an action as right or not right." This is what is learned; this motivates or makes the deviant willing to violate norms. Akers has applied this perspective to a wide variety of different forms of deviance, and we will have occasion to make reference to his insights in subsequent chapters. Wilson and Herrnstein (1985) also present a theory of criminality that appears to be relevant to other forms of deviance. The theory asserts that criminality is essentially learned behavior within certain biological constraints, some of which may be considered to be "predispositions" to crime.

The notion of socialization is central to virtually all of the perspectives we have discussed so far, even though these other perspectives are not learning theories. In anomie theory, it is necessary that persons learn success goals and agree upon general social values; in control theory, it is necessary that persons are socialized into a conventional value system with which one can develop bonds; in conflict theory, to be discussed below, persons are socialized to the interests of their groups; and, in labeling theory, persons are socialized by the societal reaction to a deviant role and status. While the extent to which each perspective stresses socialization varies, clearly the concept of socialization is indispensable to a full understanding of deviance.

Some Evidence

Learning theories, such as Sutherland's differential association, have generated long-term acceptance among sociologists. The appeal of differential association has largely been based on its flexibility in meeting simultaneously both the sociological and social psychological aspects of deviance. At the social psychological level, the theory can account in large part for the processes by which one individual, not a deviant at time A, becomes one at time B as a result of the learning process described in Sutherland's nine propositions.

The theory is equally explanatory at the sociological or group level, thus accounting for the differential rates of deviance among groups, some of whom contain or exhibit higher rates of deviance than others. Arrest and conviction statistics reveal, for example, a ratio of males, urban residents, persons of lower socioeconomic status, and some minorities disproportionately higher than their distribution in the general population (Federal Bureau of Investigation, 1987; Brantingham and Brantingham, 1984). Sutherland's theory explains this disequilibrium on the basis that these groups are more exposed to deviant norms and, thus, have a higher probability of learning, internalizing, and acting on these norms. Even changes in the official rates of deviance for one group, women, would tend to support the differential association theory. Official crime rates for women have traditionally remained low, except for a short period of increased rates during World War II. Even with this increase, however, rates of female criminality generally declined after the war, only to increase again in recent years (Simon, 1975). In terms of differential association theory, this can be explained by the increased opportunities for women during World War II to participate more fully in the general society, and thus be more exposed to deviant norms. The more recent rate increases have been due to increased learning opportunities for women, together with changes in traditional sex roles, which previously had often placed a high premium on female submissiveness and a "stay-at-home" attitude.

Socialization theory may also describe in general terms any learning process that ultimately leads to deviance. Forcible rape, for example, may result from the separate and unequal socialization process for both males and females through which traditional masculine qualities (for example, aggressiveness, power, strength, dominance, and competitiveness) may be translated into aggressive sexual behavior (Randall and Rose, 1984). Similarly, the role of peers and drinking companions may greatly influence attitudes toward, and behavior with, alcohol (Downs, 1987). One study reported that particular value orientations learned by lower-class youth are related to delinquency (Brownfield, 1987).

Other studies applying the principles of differential association have found support for the theory with respect to a variety of forms of deviance, from crime (Matsueda and Heimer, 1987) to adolescent drug and alcohol use (Akers, et al., 1979), as well as in diverse settings, such as counseling groups in correctional programs (Andrews, 1980). These studies suggest that socialization principles are important to a full understanding of the deviance processes.

Contrary Evidence

Socialization theory, a term we use to denote the learning of deviant norms and values, as well as the more specific applications, as in the theory of differential association, has been subject to some criticism. The most common criticism is that the theory tends to present an oversocialized conception of human beings in which the differential response in the form of individual motivation and rational actions are not sufficiently considered (for others, as well as responses to them, see Sutherland and Cressey, 1978: 83–95). Dennis Wrong (1961) has argued, for example, that it is common in socialization theory to claim that people internalize social norms and seek a favorable self-image by conforming to the expectations of others.

Other criticisms of socialization theory include logical problems, such as the tendency for socialization theories to commit the logical error of "tautology." Socialization theories attribute deviance to deviant norms, and then often take as evidence of those deviant norms the fact that the person committed a deviant act. This circular reasoning requires independent evidence of the existence of deviant norms aside from the deviant conduct (Kornhauser, 1978).

The learning of deviant norms and behavior patterns parallels the learning of nondeviant norms and behavior patterns. What differs, of course, is the content of the learning. The oversocialized conception of human beings sensitizes us to the dangers of claiming that all deviant acts were the result of learning. This is not the case, and we have included a discussion of one form of deviance that does not require learning: physical disabilities (see Chapter 14). In spite of criticisms, however, socialization theory seems to offer the most adequate perspective to explain deviance.

Learning theories require a clear explication of their causal arguments. Even Sutherland's long-standing theory could benefit from a clearer discussion of the principal components of the theory—definitions favorable to crime, differential social organization, and normative conflict (Matsueda, 1988). Even without such a specification, however, the theory appears to be well supported.

SUMMARY

General theories of deviance attempt to explain virtually all instances of deviance, regardless of its form (e.g., crime, mental disorder, suicide) or frequency. Earlier theoretical perspectives paved the way for the two major general theories of deviance discussed in this chapter. The social pathology perspective likened society to a biological organism and deviance to some illness, or pathology, in that organism. It represented a sociological counterpart to the medical model thinking of some psychologists and psychiatrists discussed in the last chapter. The social disorganization perspective, another early view, sought the meaning of deviance in the malfunctioning of local community institutions. Each of these views was important for leading to the development of subsequent theoretical insights on deviance.

Anomie theory has been a major structural theory of deviance for over fifty years. Anomie locates the cause of deviance in the imbalance of values and norms in society (the social structure) where culturally induced goals are emphasized more than socially approved means to achieve those goals. Individuals and groups in such a society must adapt to this situation, and some of those adaptations may be in a deviant manner. Most persons conform to society's norms most of the time. Some persons and groups, however, may move toward deviance. Groups that experience more strain from this imbalance (e.g., lower-class persons) are more likely than others to make a deviant adaptation.

Socialization or learning theory asserts that deviant behavior arises from normative conflict where individuals and groups learn norms that permit or condone deviance under some circumstances. This learning may have a subtle content, such as when persons learn that deviance is sometimes unpunished, but learning can also include the acquisition of deviant norms and values that define deviance as either necessary or desirable under certain circumstances, such as when one is around certain people. Sutherland's theory of differential association is one of the most well-known learning theories of deviance and, although his theory is oriented around the general explanation of criminality, there is no reason it could not be applied to other forms of deviance as well. Virtually every sociological theory of deviance assumes that persons are socialized to become members of groups or the general society. Some theories emphasize this learning process more than others, as we shall see in the next chapter where more specific theories of deviance are discussed.

SELECTED REFERENCES

Akers, Ronald L. 1985. *Deviant Behavior: A Social Learning Perspective*, 3rd ed. Belmont, CA: Wadsworth.

> More than a textbook on deviance, Akers' book is a systematic exposition applying learning theory to different forms of deviance. This book is valuable because it identifies both a content (what is learned) and a process (how it is learned) of socialization to deviance.

Clinard, Marshall B., ed. 1964. *Anomie and Deviant Behavior*. New York: Free Press.

> A collection of papers that explores the theoretical and empirical status of anomie theory with respect to different forms of deviance, including alcoholism, crime, gang delinquency, mental disorder, and drug addiction.

Kornhauser, Ruth Rosner. 1978. *Social Sources of Delinquency: An Appraisal of Analytic Models*. Chicago: University of Chicago Press.

> A conceptual and logical (rather than empirical) evaluation of major theoretical perspectives in the sociology of deviance. The discussion focuses

on the kind of assumptions that are made in various theories, including anomie and learning theories.

Liska, Allen E. 1987. *Perspectives on Deviance*, 2nd ed. Englewood Cliffs, NJ: Prentice-Hall.

> A summary of major theoretical perspectives in the sociology of deviance. Particular attention is devoted to the causal structure underlying each perspective.

Matsueda, Ross L. 1988. "The Current State of Differential Association Theory." *Crime and Delinquency*, 34: 277–306.

> A thorough assessment of differential association theory with reference to both the theoretical and empirical literature.

Sutherland, Edwin H. and Donald R. Cressey. 1978. *Criminology*, 10th ed. Philadelphia: Lippincott.

> The last edition of this classic text that contains Sutherland's theory of differential association. The theory is systematically applied throughout the book to all topics in criminology, making it still one of the most theoretically coherent textbooks on crime available.

Labeling, Control, and Conflict Theories of Deviance

5

The anomie and learning, or socialization, perspectives are broad, and such general theories attempt to explain all instances of deviance. There are, however, other sociological perspectives that are more specific in scope. A theory can be limited to explain only certain types of deviance or a particular form of deviance. An example of a theoretical perspective that attempts to explain only certain types of deviance is labeling theory, which attempts to explain "secondary" deviation, or deviation that represents role behavior for the deviant. This theory does not attempt to explain "primary" deviation.

A theory can also be geared specifically to the explanation of a specific form of deviance, such as crime, mental disorder, or suicide. Control theory is an example of a theory that has been applied in a systematic fashion to only one form of deviance (crime). In subsequent chapters, we will have occasion to identify other specific theories of various form of deviance, such as alcoholism, drug addiction, suicide, and mental disorder.

There is still a third way in which a theory can be limited. A theory may be specific to explaining or describing the processes by which some acts come to be deviant. Conflict theory is a perspective that attempts to describe how some acts, but not others, come to be considered deviant. It does so by exploring the origins of rules (norms, or laws) and the processes by which those rules come to be shared in a society.

To say that a theory is "limited" to certain instances of deviance does not mean that the theory might not be applied to other forms of deviance. For example, while control theory has been applied most directly to crime, as mentioned above, the main ideas of control theory, or any other theory, might be applicable to a wider range of deviance than previously thought. Some observers think that deviance may be a general rather than specific tendency, and that an explanation for one form of deviance might explain other forms as well (Osgood, et al., 1988). Whether this is true remains to be seen. In any case, the identification of certain theoretical perspectives as being limited in scope is useful for teaching purposes.

LABELING THEORY

Labeling theory is a processual theory. Processual theories concentrate on the social psychology of deviance, that is, the conditions that bring about deviance at the individual and small group level. Labeling theory, or the interactionist approach, is interested in the consequences of interaction between the deviant and conventional society, particularly with representatives of society such as official agents of social control. The major conceptualizations of this perspective are based on the writings of Lemert (1951 and 1972) over 30 years ago, although similar ideas had been expressed previously by others, particularly Mead (see Blumer, 1969: 62, 65–66), Tannenbaum (1938), and Schutz (1967). More recently, Becker (1973), Garfinkel (1967), Goffman (1963), Scheff (1984), Erikson (1962), Kitsuse (1962), Schur (1979), and others (e.g., Plummer, 1979) have contributed to the literature on interactionist approaches.

In the labeling perspective, no attempt is made to explain why certain individuals engage initially in deviance; instead, labeling theorists stress the importance of both the social definitions and the negative social sanctions as they relate to the pressuring of an individual to engage in more deviant acts (Traub and Little, 1985: 278). Attention is shifted from individuals and their actions to the dynamics involved in socially defining particular activities or persons as deviant. The developmental process in which deviance results is emphasized, a process with "varying stages of initiation, acceptance, commitment, and imprisonment in a deviant role because of the actions of others" (Traub and Little, 1985: 277). This analysis of the process is centered on the reactions of others (termed "definers" or "labelers") to individuals or acts that these "evaluating others" perceive negatively.

There are actually two important components to the labeling perspective, a particular conception or definition of deviance (the reactivist conception), and a concern about the consequences of social control efforts (the theory of secondary deviation).

Deviance as Reaction

Labeling theorists claim that since the concept of deviance is relative and ambiguous, the only way one can understand what is meant by it is to examine the reactions of others to the behavior. Perhaps Becker's definition of deviance is the most well-known in terms of this approach. Becker (1973: 9) claims that deviance is a "consequence of the application by others of rules and sanctions to an 'offender.' The deviant is one to whom the label has successfully been applied; deviant behavior is behavior that people so label." Thus, the crucial dimension is the societal reaction to an act, not any quality of the act itself. Here, deviance is not defined by any reference to norms, but by reference to the reactions (sanctions) of the social audience to the act. In this view, deviance does not bring forth social control efforts, but the reverse: social control efforts "create" deviance by defining it and making it known to others (see Rubington and Weinberg, 1987).

Labels Create Types of Deviants

Since the emphasis is placed on the label, interest shifts from the origin of the deviant behavior (1) to characteristics of the societal reaction that is attached to persons and (2) to consequences of this "labeling" attachment for the individual's further deviation. The official labeling of a person as a delinquent, criminal, homosexual, drug addict, prostitute, or "lunatic" may have serious consequences for further deviation. Schur (1971: 27) believes that the emphasis on labeling signifies a shift from efforts to distinguish what "caused" individuals to be offenders to a more intensive study of the processes that have produced the deviant outcomes. Lemert (1972: ix) has particularly stressed this viewpoint and its consequences for deviance, since labeling represents a big step away from older sociology which tended to rely heavily on the idea that deviance leads to social control. "I have come to believe that the reverse idea (that is, social control leads to deviance) is equally tenable and the potentially richer premise for studying deviance in modern society."

According to labeling theorists, the deviant label may produce a basic change in the nature and meaning of deviance for the individual who is labeled. They distinguish *primary deviance*, or behavior that has arisen for a number of reasons, including risk-taking, chance, and situational factors, and *secondary deviance*, which Lemert describes as the behavior of a person when using his or her deviant behavior or a role based on it as a means of defense or adjustment to the problems caused by the label (Lemert, 1951:

76). The deviant label produces a deviant social role and is the basis upon which further social status is conferred on the deviant. As two observers put it:

> The idea of a master status is the end result of the entire model. It refers to a dominant status either socially conferred by rule enforcers and the audience or individually by the deviant actor. The emphasis here is on the development of a deviant self-concept and the consequent probability of a deviant career. (Dotter and Roebuck, 1988: 28)

This is a subtle process and it takes place over an extended period of time. For example, the individual who engages in a homosexual act for money may adopt a homosexual identity if others react toward him as though he were a homosexual; an individual who performs an "eccentric" act may enact the role of a mentally disturbed person if he or she is formally treated by a psychiatrist or admitted to a mental hospital; and a drinker who has been labeled as a "drunk" by his or her family may drink excessively to cope with the rejection found at home. In each case, the master status of deviant (or "homosexual," "mental patient," or "alcoholic") is developed and others react toward the individual in a manner consistent with the master status. All other statuses become secondary or less important.

Once a label has been attached by an arrest, confinement in a mental hospital, or other action by an official agency, it is claimed by labeling theorists, a spiraling action is initiated that sets off a sequence of events leading to further deviance because of the stigmatizing effect of the label. In a sense, the labeling of a person as a deviant may result in a self-fulfilling prophecy. Persons labeled as deviants continue to commit acts of deviance and can develop deviant careers by becoming secondary deviants. The person tends to be cut off from participation in conventional groups and thus moves into an organized deviant group.

Kitsuse (1980) has suggested further that some deviants, because of this labeling process, rebel against their labels by attempting to reaffirm their self-worth and lost social status. These *tertiary deviants* may join social movements to combat negative images and, in effect, to deny that they are deviants. Kitsuse distinguishes these deviants from secondary deviants

The Labeling Process

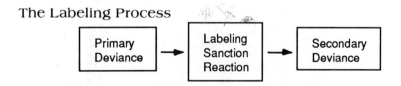

Derived from Edwin M. Lemert, *Social Pathology*. New York: McGraw-Hill, 1951.

because tertiary deviants are active in protesting their labels, whereas secondary deviants are passive recipients of their labels. While secondary deviants adapt to the labeling process, tertiary deviants "reject the rejectors" and attempt to neutralize the label. Recent activities of groups of homosexuals, prostitutes, and physically handicapped persons reflect this movement to tertiary deviance.

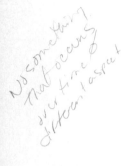

The process from primary to secondary deviation may be a lengthy one that requires many labelers and labels. People whose behavior is considered "odd" by some may be defended for many years by family members as simply "eccentric" rather than mentally disordered. The reactions of school officials, employers, and psychiatrists may take some time to move such people to the status of secondary deviation. Similarly, the adoption of a homosexual role is a complex process that involves the acquisition of a homosexual identity earlier in life and perhaps the reactions of family, friends, and others over a period of time.

The Power to Label

The labeling perspective has also correctly focused on the significant role played by social control and social power considerations in a society's determination of what is deviant. Certain groups may influence the administration of criminal law, for example, through agents such as the police and courts. Similar processes have been documented in other areas of deviance. For example, the purposeful actions of other agents of social control have been influential in determining which persons are to be considered mentally ill and the appropriate manner in which they are to be dealt with, either in an institution or the community.

The labeling perspective, because of its emphasis on the importance of rules, social control efforts, and the effect of stigma on deviants, is interested in the nature of deviant labels—who creates the rules that define deviance and how certain individuals and groups are singled out for labeling. Labeling theory, in short, is interested in power and the politics of deviance. Schur (1980) has suggested in this regard that we conceive deviance in terms of "stigma contests" where different groups have competing rules and definitions of what is deviant, and the determination of what is deviant is always a matter of the relative power of these groups. Thus, those persons who are most likely to be labeled as deviant are those from the most powerless groups, such as drug addicts, alcoholics, mental patients, and, as Schur (1984) has argued, even women.

Labeling theorists claim that the disproportionate representation of powerless groups in official statistics on deviance reflects class bias in the actions of agents of social control. The over-representation of these groups also reflects the fact that more powerful persons and groups were able to make deviant the behavior of other less powerful groups. Furthermore, since the distribution of deviance is more widespread than suggested in such statistics, it appears that characteristics other than the deviant act, such as the deviant's age, sex, race, social class, or characteristics of the social control agency, elicited the label (Box, 1981).

Evaluating Labeling Theory

In spite of the popularity of the labeling perspective and the intuitive appeal of many of its ideas, it has been subjected to various criticisms. A number of the criticisms relate to the problem that labeling theory is not precisely stated. There is ambiguity about key points in the theory and these confusions have not been clarified.

Where Is the Behavior? The label does not create the behavior in the first place. The labeling perspective, in this sense, denies the reality of the deviant act and the basis on which the societal reaction is made. The basis of the reaction, ultimately, is some normative standard or expectation that is being violated. "People can, and do, commit deviant acts because of the particular contingencies and circumstances in their lives, quite apart from or in combination with labels others apply to them" (Akers, 1968: 463). The majority of persons who are deviant are not officially labeled, whether their acts have involved stealing, homosexual behavior, marijuana use, drunk driving, or crimes committed by persons in business or politics.

Who or What Labels? Three groups can engage in labeling—official agents of social control, society at large, and the immediate group one participates in, the significant others for whom one receives cues as to role performance. One must define specifically which of these groups is doing the labeling, since the effect of labeling from each group may be quite different. In general, the labeling approach has emphasized almost exclusively formal agency labeling, with minor importance accorded informal sanctioning from family, friends, employers, and others. While it is often assumed that formal sanctions create more significant stigma problems for deviants, there is reason to believe that informal sanctions are powerful stigmatizers too; yet, these are often ignored.

How Much of a Label? Writers who have adopted a reactivist conception of deviance are vague about how much societal reaction constitutes effective labeling. In other words, how harsh must the reaction be to result in a person's being labeled or defined as deviant? One might ask if labeling is to be only by formal social control agencies and, if this is the case, how severe is the penalty to be? Is it to be arrest, conviction, imprisonment, mental hospitalization, and so on, and what effect, on the other hand, will result from informal social sanctions, such as those exercised by family and neighbors? Those who define deviance by the reactivist conception have not, according to one critic, been specific in the kinds of reactions that identify behavior as deviant (Gibbs, 1981: 497–504).

Who Is Deviant?   A major consequence of a reactivist conception of deviance is that it largely restricts the concept of deviance to the lower classes, since the acts for which one is labeled are far more numerous in this group. Persons who engage in acts that are largely not labeled, because of social and economic position, escape the appellation of deviant.

> Because of these biases, there is an implicit but very clear acceptance by [labeling theorists] of the current definitions of "deviance." It comes about because they concentrate their attention on those who have been successfully labeled as "deviant" and not those who break laws, fix laws, violate ethical standards, harm individuals and groups, etc., but who either are able to hide their actions or, when known, can deflect criticism, labeling, and punishment. (Liazos, 1972: 109)

What Are the Effects of Labeling?   The evidence contradicts the claims that the application of formal sanctions, even when severe and frequent, always strengthens deviant conduct patterns. The argument of labeling theorists is that persons assume deviant roles primarily because they have been labeled by others as deviants, and because they are excluded from resuming nondeviant roles in the community. In spite of such claims, labeling is not a necessary and sufficient condition for all secondary or career deviance, even though this is more likely to be the case in certain instances. In this connection, one can distinguish between achieved and ascribed rule-breaking (Mankoff, 1971). In ascribed deviance, the rule-breaking is characterized by particular, visible, physical disabilities such as mental retardation; achieved deviance involves activities such as those of a professional criminal, on the part of the rule-breaker. As Mankoff (1971: 207) says: "Ascribed deviance is based upon rule-breaking phenomena that fulfill all the requirements of the labeling paradigm: highly 'visible' rule-breaking that is totally dependent upon the societal reaction of community members while being totally independent of the actions and intentions of rule-breakers."

The severely crippled, the blind, the obese, the spastic, the mentally retarded, and those with severe facial disfigurements may encounter labeling because of imputed undesirable differences from what other persons regard as normal or appropriate. As a result of their being stigmatized, the social identity of the person is affected and, for these persons, the physical condition constitutes a necessary condition for labeling and career deviance (see generally Stafford and Scott, 1986).

On the other hand, labeling does not appear to be necessary for many forms of achieved deviance. The achieved types of deviants can have deviant careers without ever having been "forced" into them by agents of social control (Mankoff, 1971: 211). Many choose deviance as a way of life and are not forced to remain deviant because of the effects of any stigma; rather, they simply do not wish to conform. Ample empirical evidence supports the conclusion that a deviant career, or secondary deviation, can develop without arrest or other sanctions, as has been supported in studies, for example, of embezzlement (Cressey, 1971), and homosexuality. Most persons who come to develop a homosexual identity do so without having had contact with a police officer or psychiatrist (Langevin, 1985). Delinquent gang behavior may occasionally become highly sophisticated, yet with minimal or no contacts with the law; and offenders who have legitimate occupations, such as white-collar criminals, may pursue careers in deviance without ever having been sanctioned and often without fear of sanctioning (Coleman, 1989). Women alcoholics are persons whose problem drinking is not known to many persons and many of these women perpetuated their drinking careers without being labeled (Wilsnack, et al., 1987). Physicians have a high rate of narcotic addiction compared to persons in other occupations, yet they are seldom detected and labeled as drug addicts (Vaillant, et al., 1970). Thus, to say that "most deviantness is ascribed, not 'achieved' " (Schur, 1979: 261) surely goes too far.

It has also been claimed that once a person is labeled by commitment to a mental hospital, breaking out of the deviant status is difficult. Studies indicate that the extent of rehospitalization because of poor adjustment due to the effects of labeling is not as great as the labeling position has assumed. In one study, two-thirds of the patients had not been hospitalized for seven years, and even in those rehospitalized cases, few of the readmissions had been urged by relatives (Gove, 1970). Hospitalization may actually be positive in the sense that it may improve family relationships; the label "illness" may remove alienation of others, help erase the effects of long-term personal quarrels, and give the patient an opportunity to make new adjustments. In fact, one researcher maintains that the behavior in much mental disorder determines the expectations of others to a much greater extent than does the reverse situation (Gove, 1982).

CONTROL THEORY

The perspective known as "control theory" is one of the most popular general perspectives in the field of deviance. Like anomie theory discussed in the last chapter, control theory is generally consistent with some of the main ideas of the social disorganization approach. In fact, some of the sociologists associated with the social disorganization perspective, such as Frederick Thrasher, Clifford Shaw, and Henry McKay, have been called early control theorists (Kornhauser, 1978). Control theory, however, has been limited in its application mainly to explain crime and delinquency rather than other forms of deviance.

The central idea behind control theory is that deviance is the result of an absence of social control or restraint. Control theorists differ on the sources of this lack of control but they agree that a reduction of control—for whatever reason—will generate more deviance because people will be freed to do what comes "naturally." Reckless (1973) calls his version a "containment theory." He argues that controls over behavior can come from interpersonal, political, and legal sources. Two basic types of containment are *inner containment*, which is found within the person, and *outer containment*, which are controls that arise from forces in the individual's environment. These sources combine to keep most people from deviating from social norms most of the time. But the idea of control theory is not new.

Control theory has its origins in the emphasis on social integration in the pioneer work of Emile Durkheim. Durkheim was interested in how social order could be maintained in complex societies with a sophisticated division of labor and substantial social differentiation, both of which appear to be more conducive to social disorder. Durkheim sought the answer in the notion of "integration" and the bond of commitment that develops between persons and their larger social group (Durkheim, 1933; see also Fenton, 1984). Durkheim ambitiously studied what was thought to be a highly individualistic behavior, suicide, to show how this form of deviance was a social phenomenon that was related to different degrees of integration in social groups. His analysis indicated that, as predicted, suicide rates varied inversely with the degree of a person's social integration; for example, rates among Catholics were found to be lower than among Protestants because the Catholic Church provides its members with a greater sense of group belonging and participation.

Contemporary control theories predict that deviance will be greatest among those groups and individuals who are the most poorly integrated with conventional society. Conformity is assumed to be nonproblematic in most other theories of deviance, such as the theory of anomie. That is, conformity is the "natural" order of things that requires no explanation. Control theory, on the other hand, reverses this approach by indicating that it is conformity, not deviance, that requires an explanation. "The important question is not 'Why do men *not* obey the rules of society?' " (Hirschi, 1969: 10). Rather, the important question is why persons conform. This is the question control theorists ask.

Control theory attempts to combine theories of conformity with theories of deviance, and it is in this context that control theorists assert that deviance is not caused as much by forces that motivate persons to deviate as it is simply by the fact that deviance is not prevented (Nye, 1958: 3–9). Anomie theory posits that a dysjuncture between goals and means produces motivation within individuals to take a deviant adaptation, such as innovation. Control theory, on the other hand, assumes that everyone is motivated to commit criminal acts and that no special motivation is necessary to explain deviant behavior. Conversely, it is not necessary to explain motivation. It is simply there. The reason that some persons act on that motivation is that they are temporarily released from restraints that hold the behavior of others in check.

Assumptions and Structure of Control Theory

Obviously, in order to justify a view of deviance as the absence of controls, control theorists must make certain assumptions about the nature of human beings. A recent application of control theory to juvenile delinquency makes two of these assumptions explicit:

> (1) That human nature is on the "bad" side of a neutral position; that is, humans are naturally egocentric and seek to satisfy their wants and needs by the easiest means available, even if those means are illegal.
> (2) That decreases in prolegal controls (internal and external) allow delinquent behavior. (Arnold and Brungardt, 1983: 398)

In other words, control theorists make assumptions that are related to those made by psychoanalysts regarding the nature of human beings and the importance of controlling "innate" tendencies. The nature of this control leads to a commitment on the part of the person to conform (Reckless, 1973: 55–57).

Hirschi has provided what is probably the most clear statement of control theory, identifying four components of the person's bond with society that tend to prevent deviance (Hirschi, 1969: 16–26). *Attachment*, the first element of the bond, refers to the extent to which persons are bound to their groups through feelings of affection and respect and through socialization to group norms. *Commitment* is that element that describes the degree to which persons develop a "stake" in conforming behavior so that acts of deviance jeopardize other, more valued, conditions and activities (Toby, 1957). Concern over one's reputation, or being expelled from school or losing one's job are examples of commitment. *Involvement* refers to physical activities of a nondeviant nature; at the simplest level, little time is left for delinquency if one spends much time playing basketball, for example. Continued involvement in conventional activities also leads to strengthened commitment. *Belief* refers to persons' allegiance to the dominant value system in their group. These values may assume the nature of moral imperatives for the individual, and it would be unthinkable to violate them. Hirschi (1984: 51) has indicated that the general logic of control theory might be applicable to other forms of deviance as well, since delinquency results from "the tendency or propensity of the individual to seek short term, immediate pleasure," a tendency that may relate to other forms of deviance as well.

Other than Durkheim, most control theorists conceive of their perspective to be a processual rather than structural theory. Hirschi, Reckless, and Nye, for example, all talk about the process by which certain individuals are freed from controls and come to commit acts of deviance. Sykes and Matza (1957), too, conceive of delinquency as a process of weakening controls over deviant impulses, a process that the deviant may aid by "neutralizing" the restraining effect of norms and laws. Shoplifters, for example, may persuade themselves that "the store will never miss the item," or "they really owe me this item for overcharging me on other items all these years" (Meier, 1983). In this way, deviance is sometimes permitted for the person because normal restraints are neutralized and the individual is no longer held in check by these restraints.

The termination of crime can also be thought of in terms of the reestablishment of bonds and greater integration between deviants and conventional society. Many drug addicts reject treatment, even if they wish to quit their addiction, because they feel they could take care of it themselves or they thought treatment would not help them (Biernacki, 1986: 74). But that rejection did not lead automatically to recovery without getting out of the drug using world and reestablishment of ties with conventional society. Similarly, many criminals report they leave lives of crime because of a resolve to support themselves in a less risky manner (Shover, 1985: 94–97). Different life contingencies may account for this resolve, including the development of ties to a noncriminal of the opposite sex and to noncriminal activities.

Evaluating Control Theory

In spite of its widespread appeal, control theory suffers from a number of inadequacies that limit its role as a complete explanation of deviance.

Where Is Deviant Motivation?
Control theorists assume that all persons are equally motivated to commit deviant acts and the only reason some do not is the nature of the control over their conduct at any given moment. But there is reason to believe that some persons are more strongly motivated than others to engage in deviant acts. Because control theorists assume that everyone would be deviant given the chance, they must, like psychoanalysts, also assume that these tendencies are present at birth and invariant. Thus, if humans have any "natural" state, it is one that is self-seeking, harmful to others, and generally "evil." This perception does not conform to the view of many sociologists and anthropologists who find the most striking characteristic of humans to be their diversity (according to cultural background and normative circumstances), not any common tendency for wrongdoing.

Disagreement Among Control Theorists.
Control theory predicts that the rates of deviance are higher in those groups where controls are least effective. In Hirschi's language, the rate of deviance should be greatest among those groups that have the least attachment, commitment, involvement, and belief with the values of conventional society, such as the lower class and particular racial and ethnic groups. These groups have par-

ticipated least in the "American Dream" and would be expected to have developed the weakest bond with a society that has not been especially kind to them. Yet, Hirschi's (1969) own data find no strong relationship between the probability of delinquency and one's social class, and probably for this reason, Hirschi and other social psychological control theorists do not make the same kind of predictions as those control theorists who stress the structural relationship between groups and society, such as Durkheim (1933). And, even when a specific prediction of greater deviance among, say, lower class persons is made, this does not explain the extensive deviance among other social classes.

The Assumption of a Central Value System. Most versions of control theory assume the existence of a central value system that makes explicit to all in society what is and what is not deviant. In fact, without such an assumption, control theory would assert that deviance is the result of learning different moralities. But control theory is not a learning theory. Control theory must assume that there is one central value system so that variations in deviance can be attributed to variations only in controls, not beliefs. Sociologists do not agree, however, that a society like the United States can be said to have one central value system, and it appears that this may be the case for all modern societies. Rather, there are many value systems, some of which condone certain acts that would be considered deviant in other value systems.

Some Empirical Evidence. The issue of empirical evidence applied to control theory is particularly important because competing theories may present stronger theoretical arguments. In fact, as Hirschi has put it:

> The primary virtue of control theory is not that it relies on conditions that
> make delinquency possible while other theories rely on conditions that
> make delinquency necessary. On the contrary, with respect to their logical
> framework, these theories are superior to control theory, and, if they
> were as adequate empirically as control theory, we should not hesitate to
> advocate their adoption in preference to control theory. (Hirschi, 1969: 29)

But, the empirical evidence regarding the adequacy of control theory is mixed. Hirschi's (1969) own study, with some exceptions, supports the basic ideas of control theory as do some more recent studies (see Wiatrowski, et al., 1981 and Gibbs, 1982). There are other studies, however, that question the validity of control theory as applied to such areas as criminality and problem drinking. Matsueda (1982) and Matsueda and Heimer (1987) found that crime was more a function of the acquisition of criminal norms than the weakening of controls. The empirical evidence in areas of deviance other than crime is also mixed. Seeman and Anderson (1983) found that social integration was not a buffer against heavy drinking. In fact, they report that high social, conventional involvement was related to heavier drinking, not the reverse as predicted by control theory.

CONFLICT PERSPECTIVES

The conflict approach is a theory of deviance rather than a theory of deviant behavior. That is, it is more a theory or perspective about the *origin of rules or norms* than a theory about the origins of rule-breaking behavior. Most writings within the conflict perspective pertaining to deviance have been related to criminality, but this approach also appears to be relevant to a number of other forms of deviance as well (Spitzer, 1975). The conflict view stresses the pluralistic nature of society and the differential distribution of power among groups. Because of the power they possess, some groups can create rules, particularly laws, that serve their own interests, often to the exclusion of the interests of others. In this respect, the conflict perspective conceives of society as groups with competing interests in conflict with one another, and those groups with sufficient power will create laws and rules that guarantee their interests will be served (Quinney, 1979: 115–160).

Of particular interest to conflict writers is the origin of norms that define certain acts as deviant. Strong negative attitudes toward suicide, prostitution, homosexuality, drunkenness, and other behavior have stemmed, for the most part, from certain church groups who regard such behavior as immoral (Greenberg, 1988; Davies, 1982). Opposition to the use of marijuana, nudity, and the distribution of pornographic materials rests with other "moral entrepreneurs" who attempt to impose their views on others (Attorney General's Commission on Pornography, 1986). According to the conflict view, deviance represents behavior that conflicts with the interests of the segments of society with the power to shape public opinion and social policy.

Deviance and Marx

Many of these contemporary ideas on the importance of social conflict in society generally can be traced to such older sociological theorists as Marx, Simmel, and more recently to Coser and Dahrendorf, to whom society is not characterized by a consensus over values, but a struggle between social classes and class conflict between powerful and less powerful groups. In fact, most conflict writers on deviance and on crime identify themselves as "Marxists," although Marxists disagree on the extent to which crime should form the basis of a Marxist view of society (see O'Malley, 1987). Marx viewed society as composed primarily of two groups with incompatible economic interests: the bourgeoisie and the proletariat. The "bourgeoisie" are the ruling class—they are the wealthy, they control the means of economic production, have inordinate influence over the society's political and economic institutions, and have at their disposal great power to serve their interests. The "proletariat," on the other hand, are the ruled—they are the workers whose labor the bourgeoisie exploits.

The state is not a neutral party to the inevitable conflicts that arise between the two groups; it serves mainly to cushion the threats of the ruled against the rulers, and to foster the interests of the rulers. Marx foretold that as capitalism developed there would be a proliferation of criminal laws, since laws are considered important mechanisms by which the rulers maintain order (Cain and Hunt, 1979; Beirne and Quinney, 1982). First, the laws can define certain conduct as illegal, particularly conduct that might pose a threat to the rulers' interests. Second, a law legitimizes the intervention of society's social control apparatus through the police, the courts, and the correctional systems, all forces to be used against the ruled, whose behavior is most likely to be in violation of the law. In this sense, the criminal law comes to "side" with the upper classes against the lower classes. Marx's conception of conflict is ultimately tied to a particular economic system: capitalism. In this system, it is inevitable that a major division will develop based on the means of production and the economic interests of persons, depending upon whether they own these means or work for those who own them (Inverarity, et al., 1983: 54–99).

Other Conflict Theorists

The conflict approach to crime and deviance is generally identified with the works of Vold (1958), Quinney (1980), Turk (1969), Taylor, Walton, and Young (1973), Platt (1974), Takagi (1974), and Chambliss (1976), along with others. These writers view criminal behavior as a reflection of power differentials in the sense that crime comes to be defined as a function of social class position. Since the elite and the powerless have different interests, whatever benefits the elite will work against the interests of the powerless. It is not surprising, therefore, that the officially recorded crime rates are substantially higher in the lower classes than in the more privileged classes from which the elite are recruited. Since the elite also control the law-making as well as the law-enforcement process, the nature and content of the criminal laws will coincide with their interests (Krisberg, 1975). Laws relating to theft, for example, are said to have been enacted by persons in positions of power who have more to lose from theft. It is no social accident, moreover, that these particular laws are invariably broken by persons in the lower, less powerful classes whose temptation toward theft is greater.

Conflict theorists often regard crime as a rational act (Taylor, Walton and Young, 1973: 221). Persons who steal and rob have been forced into these acts by social conditions brought about by the inequitable distribution of wealth, while corporate and white-collar crime is directed at protecting and augmenting the capital of owners (Simon and Eitzen, 1987). Organized crime is a rational way of supplying illegal needs in a capitalist society (Block and Chambliss, 1981). Because lower-class membership is often accompanied by a lessening of commitment to the dominant social order, it is not surprising that one analysis combined a conflict orientation with a version of control theory to explain persistent crime and delinquency among working-class youth. Colvin and Pauly (1983) argue that economic repression of workers creates alienation from society and this alienation, which is manifested in weakened bonds to the dominant social order, produces criminality.

Law has also become a tool by which the ruling class exercises its control over the ruled. In addition to protecting the property of the elite, it serves to repress other political threats to the position of the elite through the coercive response of the criminal justice system. This is important for the elite because conflict over life-chances and power are inevitable, and the law will be used by some groups to the detriment of others. In many respects, political criminality is synonymous with being a member whose life-chances are so poor as to invite a challenge to the authority of the powerful group (Turk, 1984).

Conflict theorists perceive crime as an unchangeable feature of capitalist society. The United States is one of the most advanced capitalist societies, and its overall crime rates are among the highest in the world today. Since the country is organized to promote capitalism, it is organized to serve the interests of the dominant economic class, the capitalist ruling class. Recent developments in criminal justice, including rising imprisonment rates, increasing penalties for crimes, and more interest in retribution, reflect the influence of the bourgeoisie in manipulating laws and penalties (Horton, 1981). Access to criminal opportunities vary by class; the poor can hardly engage in embezzlement or corporate crime, so they must instead choose burglary and mugging.

Evaluating the Conflict Model

The conflict model has made an important contribution to the study of deviance. It has focused attention on the role of political, economic, and social structure in the definition of deviance, particularly through laws of the political state. Conflict theorists point out basic problems and contradictions of contemporary capitalism. They note that much crime is a reflection of societal values and not merely a violation of those values (Friday, 1977). The basic issue is how values are translated into crimes and other rules, and it is on this point that conflict theory is focused. Several problems are associated with the conflict view, however.

Explanation of Rules or Behavior?

Conflict theory does little to inform us about the process by which a person comes to commit crimes or to develop deviance. It does raise pertinent questions about the origin of laws and norms, but it is essentially an explanation for the formation and enforcement of certain rules and laws (Akers, 1968: 29). To conflict theorists, the basic structure of a society, both economic and social, shapes the behavior of individuals and not socialization processes or peer-group and subcultural patterns (Thio, 1973). When the conflict approach does deal with the individual, it is assumed that deviance is a rational and purposive activity. Because the socialization process is ignored, it is assumed that deviance is motivated only by political considerations.

Who Benefits?   Not all laws are necessarily devised by and operated for the advantage of one particular group. The conflict approach may be more applicable to those acts that generate disagreement about their deviant nature, such as political crime, prostitution, the use of certain drugs, and homosexuality, rather than acts that reveal no such disagreement. In fact, it would appear that most acts presently defined in the United States as conventional or ordinary crime have general consensus in regard to both the illegal nature of the behavior and the seriousness of the act (see Hamilton and Rytina, 1980). Laws against homicide, robbery, burglary, and assault benefit all members of society, regardless of economic position. Any statement that the elite alone benefit from such laws neglects the fact that most victims of these offenses are other poor, lower-class urban residents and not members of any elite, however broadly defined. Although certainly the elite have more property to lose from theft or robbery, persons who actually lose the most are those who are the least able to afford it. If, on the other hand, one regards the operations of the criminal justice system, one sees considerable validity in the conflict perspective in the sense that persons who commit conventional crimes (generally lower class citizens) are much more likely to be arrested, convicted, and sentenced to longer prison terms than persons who commit white-collar and corporate crimes (Reiman, 1984).

The Powerful Make the Rules Everywhere.   The assumption that powerful groups dictate the content of the criminal law, as well as other rule-making processes, and their enforcement for the protection of their own interests is too broad an assumption. All types of groups are involved in law-making, each having specific interests and concerns. Powerful groups do have substantial input into the legal structure, but this would appear to be the case in any social system, whether capitalist, socialist, or communist. By penalizing those who violate it, the criminal law always defends the existing order and those holding power in it. It means little to say that the rules are made by those who have something to gain from those rules. This leaves unanswered important questions related to the characteristics of the "powerful," the process whereby some norms are made into law and others are not, the selective enforcement of those laws, and differences in law-making and enforcement processes in different economic and political systems.

Law Does Not "Cause" Behavior.   Although the conflict perspective points to the criminal law, supported by certain interest groups, as the ultimate "cause" of criminal behavior, it does not follow logically that the law is responsible for the behavior. In referring to the labeling perspective, which generates similar confusion with its emphasis on rule-making and deviance by interest groups, Sagarin (1975: 143–144) observes that "without schools, there would be no truancy; without marriage, there would be no divorce; without art, there would be no art forgeries; without death, there would be neither body-snatching nor necrophilia. Those are not causes; they are necessary conditions." There would be no crime if there were no laws to prohibit some behavior, but the existence of a law is not sufficient to account for the behavior.

Theory as Ideology. The ultimate acceptance of the conflict view, particularly the Marxist perspective, depends only upon the acceptance of its ideological base (see Buraway, 1990). Other sociological perspectives are not completely free from ideology, but the conflict theorists' emphasis on combining theory with practice in a socialist framework makes more obvious and explicit the political connotations of its explanatory scheme. The movement toward socialism is the end product of a fully developed conflict theory. If the elimination of deviation and crime through the dissolution of capitalism and the transition to socialism are perceived to be too costly, however, the appeal of the conflict view diminishes considerably. It is not sufficient merely to analyze the conditions under which deviance develops; one must also be willing to change those conditions in a political sense. We are, thus, talking about someone who is not only committed to science as a means to discover the "real world" but someone who is also a political being committed to a political ideology, an ideology that, it is believed, can eradicate deviance. Thus, appeals to scientific evidence alone leaves untouched the ideological component to conflict theory (Gouldner, 1980: 58–60).

SUMMARY

Whereas general theories of deviance attempt to explain all instances of deviance, limited theories have a more narrow scope. Some theories might be limited to certain types of deviance, to a particular substantive form of deviance (such as alcoholism or homicide), or be limited to explaining the origin of deviant acts, not the origin of deviant behavior. In this chapter, labeling, control, and conflict perspectives as examples of limited theories were discussed.

The labeling perspective emphasizes an interactionist approach by concentrating on the consequences of the interaction between a deviant and agents of social control. Labeling theory predicts that social control efforts cause deviance in some instances by pushing people toward deviant roles. Closed off from conventional roles by negative or stigmatizing labels, people may become secondary deviants partly in defense from labeling efforts. Movement back to conventional, nondeviant social roles is also troublesome, and the individual may come to feel like an outsider. In this sense, sanctioning or labeling efforts that are designed and supposed to control deviance instead, according to labeling theory, amplify it.

Control theory is a perspective that has been usually limited to the understanding of delinquency and crime. It locates the cause of crime in the weakening of an individual's bonds or social ties with society, or a general lack of integration. Those groups who are less integrated into conventional society (such as lower-class individuals) are more likely than others to violate the law because they feel less committed to the conventional order. When one feels close to conventional groups, that individual is less likely to deviate from that group's rules, but when there is some social distance as a result of broken bonds, the individual is considered more free to deviate.

Conflict theory is an approach to deviance that is found most fully developed as applied to criminality, although it applies to other forms of deviance as well. It is more a theory of the origin of norms, rules, and laws than the origin of rule-breaking behavior. Rules come from individuals and groups who have the power to influence and shape public policy through the law. Groups of "elites" exert influence on the content of law and the processes by which persons who violate the law are processed through the criminal justice system. The origin of other social norms may follow this pattern. Some groups may be sufficiently powerful to have their norms dominate, such as those norms that prescribe heterosexual relations, abstinence from the drinking of alcoholic beverages, or refraining from suicide for moral or religious reasons.

SELECTED REFERENCES

Chambliss, William J. 1976. "Functional and Conflict Theories of Crime: The Heritage of Emile Durkheim and Karl Marx." Pp. 1–28 in *Whose Law? Whose Order? A Conflict Approach to Criminology*. Edited by William J. Chambliss and Milton Mankoff. New York: Wiley.

> A concise comparison between functional and conflict approaches to crime and law. This is almost essential reading on these general perspectives as applied to deviant behavior.

Hirschi, Travis. 1969. *Causes of Delinquency*. Berkeley: University of California Press.

> The most influential recent version of "control theory." The theory is identified and then tested with respect to delinquency. This is required reading for anyone who desires a more complete understanding of this perspective.

Lemert, Edwin M. 1951. *Social Pathology*. New York: McGraw-Hill.

> After forty years, this book still retains a freshness not found in many books on deviance, despite its misleading title. This book contains Lemert's

theory of secondary deviation and bears no relationship to the social pathology perspective discussed in the last chapter.

Quinney, Richard. 1980. *Class, State and Crime*, 2nd ed. New York: Longman.

An example of the conflict perspective applied to the problem of crime and crime control. Many of the observations Quinney makes with respect to crime could be extended to represent the conflict position on other forms of deviance as well.

Social Justice: A Journal of Crime, Conflict, and World Order. (Formerly, *Crime and Social Justice.*)

This is a journal devoted to publishing articles on crime and politics, topics that are closely related within the conflict perspective.

Taylor, Ian, Paul Walton, and Jock Young. 1973. *The New Criminology: For A Social Theory of Deviance.* London: Routledge and Kegan Paul.

This is a lengthy discussion and evaluation of major theoretical perspectives written from a conflict perspective. The "social theory of deviance" promised in the subtitle never quite materializes, but the journey through and around other theories is very rewarding nevertheless.

Forms of Deviance

PART THREE

Crimes of Interpersonal Violence

6

The scientific study of crime is only little more than one hundred years old (Meier, 1989). Before that time, crime was largely conceived as behavior that resulted from the moral inadequacies of the offenders or the depraved behavior of under- or dangerous-classes. The scientific study of crime has produced a great deal of knowledge about the nature of criminal activity and characteristics of persons who commit crime. Crime is one of the most widespread forms of deviance. This chapter deals with major forms of illegal interpersonal violence: murder, assault, and rape. The subsequent chapter covers crimes that are concerned mainly with the theft of property or threats to the political order.

There is great consensus about the deviant nature of some crimes, such as murder, forcible rape, burglary, and assault. But there are some crimes about which there is more dissensus, such as prostitution, pornography, and the use of certain drugs, such as marijuana. As this suggests, crime is a heterogeneous form of deviance. In order to understand fully the diversity of criminality, it is necessary to discuss the nature of crime as a form of deviant behavior.

CRIME AS DEVIANCE

Criminal behavior is deviant behavior. Crime violates a particular kind of norm, a law. Actually, the nature of a crime may be examined two ways: as a violation of the criminal law or as a violation of any law punished by the state. Sociologically, a crime is any act that is considered to be socially injurious and that is punished by the state, regardless of the type of punishment. Certainly, behavior that violates a specific criminal statute, such as the legislative prohibitions against robbery or fraud, is a crime and would be studied as such. But this conception of crime also permits us to consider as crimes the violations of other bodies of law as well, such as regulatory law created by agencies of the federal government, such as the Federal Trade Commission (FTC), the Securities and Exchange Commission (SEC), and the Occupational Safety and Health Administration (OSHA). These latter violations, usually called white-collar or corporate criminality, figure prominently in the everyday behavior of certain individuals and groups, and the impact of these crimes is great.

The Origin of Law

Sociologists are interested in the origin of norms, in this case, legal norms or law. The content of law is often tied to the conditions of the society in which it is found. In the United States, the issues of state's rights, slavery, economic development and the regulation of monopolies, and the role of the Constitution in protecting individual rights were important themes in the development of the content of law (Abadinsky 1988: 25–51). But, however detailed such accounts might be, they do not provide a theory of the origin of law.

There are two major views of the origins of law: one asserts that law comes about because it reflects the strong, majority sentiment of the population; the other asserts that law comes about because certain groups have enough power to legislate their interests into law. The *consensus* and *conflict* models, respectively, have been general orientations of law-making. It is the case that law is a product of government or governmental agencies. But before there was government or state law acts were considered wrong and punished by a central authority, such as a monarch. Such acts as murder, robbery, and assault have long been considered illegal and violations of what is called common law, an Anglo-Saxon legal tradition where law was reflected in judicial precedent rather than statutes. With the advent of state law, these common law crimes were simply incorporated into the legal system in a formal way by codifying the prohibitions (Thomas and Bishop, 1987). These laws could be said to have come about because of the strong consensus regarding the wrongfulness of the acts they prohibit. Many other criminal laws where there is more disagreement on the wrongfulness of the act develop from conflict among groups in society (Chambliss and Seidman, 1982). Conflicts are inevitable in any society—conflicts between states, between groups, and between cultural units. Conflict is a normal feature of social life, and its presence in the law-making process is well documented. To illustrate, historical analyses of the origins of the laws against embezzlement and vagrancy indicate that the statutes evolved through a conflict process motivated by various economic, political, or social interests. In the case of vagrancy, the new law was a device to protect the development of industrial interests in English society at the time by forcing people into the cities to work (Chambliss, 1964). With respect to embezzlement, the protection of foreign trade and commerce required more stringent measures against persons who might be retained by others as agents (Hall, 1952: Chapters 1 and 2). Under previous understandings, these agents who had come into possession of property legally were not guilty of a crime if they then turned this property into their own use. The first embezzlement statute overturned this older idea and made the agents responsible for that property. Without such a law, such persons as bank tellers, bookkeepers, and others trusted with other people's money could take that money for their own use without legal risk.

It is within this general framework that some criminologists have regarded virtually all crime as behavior that conflicts with the interests of those segments of society that have the power to shape social policy (Bierne and Quinney, 1982). Although this definition applies to much crime of the type regulating political behavior and personal morality, the definition is too broad to include all crimes. Crimes such as burglary, larceny, and robbery would be regarded as crimes by all social strata; they would remain crimes no matter who has power in the social structure and would more properly be regarded as having originated from a consensus in society. Moreover, Hagan (1980) has concluded after an extensive review of studies of historical analysis of laws that many interest groups were involved in the passage of most legislation, and to assert that only vested business or political interests benefited would not be accurate. It is clear, however, that a full understanding of the origin of laws requires sensitivity to the role of various interest groups both in the formulation of new legislation and in changes in the penalties for existing legislation (see Berk, et al., 1977).

Types of Crimes

Criminologists often distinguish between common law crimes (conventional or street crimes), white-collar crimes, and adolescent violations. Common law crimes are those offenses that virtually everyone would regard as criminal, such as murder, rape, robbery, burglary, and assault. Lawyers often refer to these violations as *mala in se*, meaning they are bad in themselves. These acts were judged illegal prior to written state laws when there was only common law, a term that refers to legal traditions in the form of judge-made decisions. At some time or another, a variety of behavior was covered under common law, including engaging in recreational activities on the Sabbath, practicing witchcraft, cigarette smoking, selling alcoholic beverages, wearing a one-piece bathing suit, and many others.

Certain other types of behavior have no such basis in common law and lawyers refer to them as *mala prohibita*, or as bad simply because they are prohibited by law. Most of these latter offenses have grown out of technological and social changes in society. Many are associated with the automobile, building codes, the manufacture and sale of impure foods and drugs, and sales of fraudulent securities.

Conventional crimes must be separated from white-collar crimes, or perhaps more accurately, occupational crimes. Conventional crimes are found in criminal law, but occupational crimes are seldom punished under the criminal law. They include violations of the law by persons in businesses, leading corporations, employees, politicians, government workers, labor union leaders, doctors, and lawyers in connection with their occupations (Braithwaite, 1985). Because criminologists wish to study these violations, many believe that a crime should be defined not only in terms of the criminal law but in broader terms as any act punishable by the state, regardless of whether the penalty is criminal, administrative, or civil in nature. The state has many ways of compelling individuals, business concerns, and labor unions to obey regulations under administrative law. It may withdraw a doctor's, lawyer's or druggist's right to practice, and it may suspend a tavern or restaurant owner from doing business for a few days or even permanently. Clearly, these can be very severe sanctions, even though they are not precisely criminal in nature.

A person below the age of 18 who commits a crime is generally regarded as a "delinquent" rather than a criminal, but one must not assume that *juvenile delinquency* is comparable to adult criminality in all respects but age. It is the case that many offenses committed by juveniles are also crimes when committed by adults. But it is difficult for juveniles to commit acts of white-collar or corporate criminality, and delinquency in some jurisdictions results from offenses that only juveniles can commit. These violations are called *status offenses,* a term that includes being unmanageable at home, running away from home, and truancy. For this reason, one should not assume that adolescents who are apprehended for delinquency and sent to state training schools are always "junior criminals." The great majority of all adolescents who commit acts of delinquency do not "graduate" to adult criminality. Many delinquency offenses are committed by only a few offenders, suggesting that most of the youth who engage in this type of behavior do so only on an experimental basis (Blumstein, et al., 1986).

Legally, only behavior (as opposed to thoughts or beliefs) deserve to be punished as criminal. Moreover, only persons who are culpable can be punished for crimes. These principles, called *actus reus* and *mens rea*, form the basis of the criminal law (Thomas and Bishop, 1987). These elements can not exist without one another for there to be crime. Behavior that violates the law (*actus reus*) is not criminal unless the actor had criminal intent (*mens rea*). Behavior alone is not enough; the mental element must also be taken into account. The criminal law contains a multitude of different kinds of prohibited behavior ranging from very minor, petty acts to major acts with enormous social, political, and economic implications. Because the law is so varied, it is inconceivable that persons can reach adulthood without at some time violating the law. If we regard the term "criminal" to refer to persons who have ever violated the law, we are all criminals.

The term "criminal" obviously does not refer to a homogeneous group. Instead of referring to criminals, it is more meaningful to refer to *types* of criminals because there are often more significant differences between different types of criminals than between criminals and noncriminals.

Types of Criminals

The notion of classifying criminals is not new. In everyday discourse, we commonly refer to "robbers," "burglars," and "rapists" as well as other criminals. In so doing, we have classified offenders according to the legal offense category of their behavior. This classification scheme, however, is not very useful since offenders frequently commit different kinds of crimes, making the classification of any one offender troublesome.

A more useful way of distinguishing between criminal offenders is a typology based on *behavior systems*. This involves making distinctions between offenders based on the extent to which crime is a career for the offender. The term career is perhaps most often used in the context of an occupation, but it would be a mistake to think that the term as used here refers to crime as an occupation. The notion of criminal career refers to the extent to which the individual comes to be committed to crime as an activity. Career in a sociological sense can refer to an action or activity that is more or less a pattern for an individual.

A criminal career can be distinguished from a noncriminal career by the extent to which the person has developed criminal norms that lead to criminal acts and how that person views the criminal behavior. A criminal career involves a life organization of roles built around criminal activities, such as (1) identification with crime; (2) commitment to crime and criminals as a social role and activity; (3) extensive association with criminal activities and with other criminals; and finally (4) a progression in crime that involves the acquisition of more complex criminal techniques and more sophisticated criminal attitudes. As offenders identify themselves more with crime, as they become more committed to crime, and as they progress in crime they develop a criminal self-concept. They also associate more with other criminals. A career criminal is one who identifies with crime and has a self-concept as a criminal, who demonstrates a commitment to criminality by committing frequent offenses or crimes over a long period of time, and who has progressed in the acquisition of criminal skills and attitudes. A noncareer criminal displays no such identification or commitment, has no criminal self-concept, and has not progressed in techniques or attitudes.

Offenders can be divided into behavior system types along a continuum from noncareer at one end to career at the other end. The distinction between career and noncareer offenders is not a precise one, but it does capture a major difference between types of offenders. Most offenders who engage in interpersonal violence are of the noncareer type, or primary criminal deviants, whereas property criminals are more likely to be of the career, or secondary, type. This chapter discusses the former, while the next chapter deals more with career offenders. At one end of the continuum are violent offenders and occasional property offenders; at the other end are organized and professional offenders. In between are political offenders, occupational and corporate violators, and conventional criminal offenders (Clinard and Quinney, 1973: 18–20). In other chapters, an analysis will be presented of

Behavior System Approach to Deviance

1	2	3	4	5	6	7	8

Noncareer   Career
Non-Role Behavior   Role Behavior
Primary Deviance   Secondary Deviance

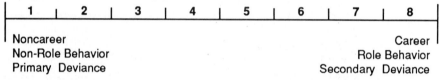

A behavior system perspective views deviance as a continuous rather than categorical variable. It differentiates deviants on the extent to which they perform the role of deviant. Some persons perform this role infrequently; they can be considered noncareer deviants. Other persons perform this role frequently; they can be considered career or secondary deviants. Still others fit somewhere between these two extremes. This view is consistent with a number of theoretical perspectives, including virtually all that have been discussed in class.

Defining Characteristics

1. Commitment   The extent to which the individual is committed to deviance as a behavior pattern.

2. Identification   The extent to which the individual identifies with deviance or other deviants.

3. Progression   The extent to which the individual progresses in the acquisition of deviant skills or attitudes.

public order types of behavior that are often criminal—illegal drug use, drunkenness, prostitution, and sexual deviance.

ASSAULT AND MURDER

All violent personal crimes involve the accomplishment of an objective through violence, whether it is the closure of an argument, winning a personal dispute, or sexual intercourse. Offenders who commit these crimes generally do not have criminal careers in such offenses. In fact, most murderers and assaulters do not conceive of themselves as being "criminals" because there is seldom an identification with crime and criminal behavior, as such, is not a meaningful part of their lives. Most murderers do not progress in the acquisition of criminal techniques or attitudes. The recidivism rate (the return to crime rate) for murderers, as shall be seen, is very low compared to some other offenders.

The type of criminal homicide discussed here consists of both murder and non-negligent homicide, but not justifiable homicide, accidental death, or negligent manslaughter. Technically, "murder" is determined by a criminal court through a legal process. In aggravated assault, there is an attempt to use physical force to settle a dispute or argument. Nearly all criminal homicide represents some form of aggravated assault, the chief difference being that the victim died. Most criminal homicide, like assault, is an outgrowth of personal disputes and altercations, some immediate and some long-standing in nature. A few of these offenses are associated with the commission of other crimes, such as robbery or drug trafficking, and some homicides are actually assaults in which the victim dies.

Frequency of Assault and Homicide

It is not possible to know about every crime that is committed. Not all crimes are reported to the police by victims and witnesses. Other major techniques of estimating crime—asking persons if they have committed a crime regardless of whether the police knew about it, and asking persons if they have been the victim of a crime. Usually victimization surveys identify more crime than is known to the police. But such techniques suffer from problems of selective memory, sampling, and other methodological problems that may limit the information from these sources.

Official information, such as police records of homicide, however, appear to be reasonably accurate because sooner or later the police do learn of most murders. Police information about assault is more limited since many assaults, like property crimes, are not reported to the police.

Incidence of Homicide. The Uniform Crime Reports (UCR), the major source of statistics on crime known to the police in the United States,

recorded 21,500 murder and nonnegligent manslaughters in the United States during calendar year 1989 (Federal Bureau of Investigation, 1990: 7). This represents one percent of all violent crimes committed during that year. The UCR regards murder and nonnegligent manslaughter as the willful killing of one human being by another. Aside from some annual variation, this figure has not changed much during the last decade. The comparable number in 1980, for example, was 23,040 (Federal Bureau of Investigation, 1990: 48). The official murder rate has been quite constant during this period.

Incidence of Aggravated Assault. There are many more assaults than homicides. Looking just at aggravated assaults—the unlawful attack by one person upon another for the purpose of inflicting severe or aggravated bodily injury—the police recorded 951,707 in 1989 (Federal Bureau of Investigation, 1990: 22). This represents more than a 42 percent increase since 1980. While there is reason to believe that the police eventually learn of most murders, this is not the case with assault, even the aggravated kind. Many assaults are not reported because victims regard the assault to be a "private matter" between the victim and assailant.

Group Variations in Homicide

Homicide is not just individualistic behavior. Rather, homicide is structured or patterned in societies in certain ways. The rates of homicide are higher in some groups than others, at some times more than others, and in some situations more than others. The acceptance of murder as a method of solving interpersonal conflicts varies a great deal in time, from country to country, from region to region, and by local areas, race, social class, and age. These variations provide us with clues as to potential explanations of this type of crime.

Variations by Country. The use of personal violence to settle disputes, even though it results in assault and murder, appears to have been common in nearly all of Europe a few centuries ago, even among the upper classes. Whereas at present most European countries, particularly most Scandinavian countries, the United Kingdom, and Ireland have low rates, the Latin American and African countries generally have high rates (Johnson, 1983). The higher homicide rates in Latin American countries are related to the attitude of masculinity or "machismo" involving particularly the customary recourse to the use of violence when one is personally insulted or when one's honor is challenged. The rates of violence in Turkey and Finland are higher than those in other European countries, and the rate in Sri Lanka is particularly high for Asia (Ferracuti and Newman, 1974: 194–195). The United States ranks about in the middle. A common variation among countries has been documented by Archer and Gartner (1984) who report that homicide rates increase in most countries after wartime.

Regional Variations. In the United States, homicide rates in the southern states are considerably higher than those in other regions, although the rates in the West are increasing rapidly (Federal Bureau of Investigation, 1990). The differences may be due largely to the fact that cultural definitions demand personal violence in certain situations and that weapons are more available in some areas than others.

Homicide rates also show variations within many other countries (Clinard and Abbott, 1973). For example, the rates are higher in Sardinia than in any other part of Italy; here the use of violence, particularly of vendetta (homicides), is regulated by a set of norms or a "code" that supersedes Italian criminal law. This code is learned and socially reinforced by others (Feracutti, et al., 1970).

Local Areas. Rates of interpersonal violence are highest in the inner-city areas of large cities (Brantingham and Brantingham, 1984). These high rates are related to the slum way of life in which the use of force may be approved to settle disputes. Studies of homicide in Houston found these crimes to be concentrated in a relatively small area of largely lower-class residents (Bullock, 1955; Lundsgaarde, 1977: 47–50. 105–106), while another study of these crimes in Cleveland found that two-thirds of the homicides occurred in 12 percent of the city, primarily in black inner-city areas (Bensing and Schroeder, 1960). Similar patterns have been documented in other large cities in the United States (Wolfgang and Zahn, 1983), as well as other countries such as England (McClintock, 1963: 44–45).

Race. African Americans are consistently more likely to be involved in homicide and assault, both as offenders and victims, than one might expect from their proportion in the population. Racial disparity is greatest in arrest rates for crimes of violence, due largely to inner-city living. In a well-known study of homicide in Philadelphia, Wolfgang found that the homicide rate for blacks was four times that of whites (Wolfgang, 1958). A similar finding has been reported for other cities. Blacks comprised three-fourths of the offenders in Cleveland where they were only 11 percent of the population (Bensing and Schroder, 1960: 22). In Houston, African Americans made up only one-fourth of the population, but two thirds of the offenders (Pokorny, 1965). And, in Chicago, the homicide rate for nonwhites was about 10 times that of whites (Voss and Hepburn, 1968: 501).

Nearly all crimes of violence in the United States involve persons of the same race. This is true of homicide, aggravated assault, and rape. Crimes of violence are *intraracial* rather than interracial. The majority of these crimes involved blacks murdering or assaulting other blacks; most of the rest involved whites victimizing other whites. The homicide information collected by the Federal Bureau of Investigation (1990: 10) for 1989 showed that 95 percent of the black murder victims were slain by black offenders, and 88 percent of the white murder victims were slain by white offenders.

Social Class.   Crimes of violence are found almost entirely in the lower class (Luckenbill, 1984). Ninety percent of the homicides in Wolfgang's (1958: 37) study in Philadelphia involved persons in lower-class occupations, and the majority of offenders of homicide and assaults in London were found to be lower class workers (McClintock, 1963: 131–132). Similar results have been reported elsewhere. For example, four-fifths of a sample of assaulters in St. Louis, as well as a like percentage of victims, were from the working class (Pittman and Handy, 1964; see also Hepburn and Voss, 1970).

It appears that different kinds of murders are committed in different social classes. A study of middle and upper class homicide found that the patterns for these crimes were different from the vast majority of all homicides which occur among lower class persons. For the middle and upper class murders, almost three-fourths of them were premeditated, the offenders were largely out for personal gain, and alcohol was not present in the great majority of these crimes (Green and Wakefield, 1979). In contrast, Wolfgang (1958) found that alcohol was present but not necessarily a cause in almost two-thirds of the Philadelphia lower-class murders. In fact, alcohol is a frequent accompaniment to both homicide and assault.

Age and Sex.   Personal violence in urban areas is generally higher among young age groups and among males. Specifically, homicide rates are much higher in the United States among males aged 18 to 24 than for any other group (Wolfgang and Zahn, 1983; Luckenbill, 1984). For aggravated assault, the 15–24 age group generally has the highest rate. In fact, the heaviest incidences of assaultive behavior, including murder, has been found to occur in many studies of criminal violence during later adolescence and early adult years (Weiner, 1989: 118). Most murder victims and assault victims are similarly young. People tend to murder and assault those who are similar to them in age, sex, race, and social circumstance. Concern that rates of female criminal violence would escalate during the 1980s because of larger social changes concerning the status and role of women seemed to be unfounded. Women play a relatively small role in the overall rates of violent crime, not only in the United States but in other countries as well (Simon and Baxter, 1989).

Interaction Between Offender and Victim

Most murders and aggravated assaults represent a response, growing out of social interaction between one or more parties in which a situation becomes defined as requiring the use of violence (Wilbanks, 1984). The violence may result from a single argument or dispute, or it may be the result of a long series of disputes between intimates, such as husband and wife, lovers, close friends, or fellow employees.

Homicides and assaults are often "precipitated" by the victims themselves. In these homicides, the victim was the first to draw a weapon, strike the first blow, or in some other way precipitate his or her own victimization. In one study, more than one in four homicides were precipitated by the victim in that the victim was the first to strike a blow or to use a deadly weapon (Wolfgang, 1958: 252). In another study, it was found to be one case in three (Voss and Hepburn, 1968: 506).

Assaultive crime, whether or not it leads to homicide, is intergroup behavior. Studies indicate that *close* relationships usually exist between offenders and victims of homicide and assault. These crimes usually involve relatives, close friends, or acquaintances. Males kill males, females kill females, and persons over 25 kill others of the same age group. In other words, persons kill and assault those persons with whom they are likely to have interaction. It is not surprising that domestic violence should also be a situation where homicide can result (Gelles, 1985). While many women report fear of criminal victimization of assault and homicide while away from home, "traditional domestic, especially marital, contexts . . . still prove the most lethal for women" (Zahn, 1975; also see Wilbanks, 1983).

Most cases of violence develop from what outsiders might regard as trivial disputes, but what is considered "trivial" by one person is related to his or her judgment derived from age, social class, and other background factors. Homicide may involve the nonpayment of a small debt, a petty jealousy, or a small neighborhood quarrel—events that may be extremely important to the persons involved.

As these factors suggest, those conditions that are conducive to social interaction in general are also related to interpersonal violence. The Federal Bureau of Investigation (1990) reports that homicides and assaults are higher on weekends than during the week, and during the summer than the winter months. Such times are more amenable to social interaction and much socializing takes place during these times. The presence of alcohol in many homicide and assault cases also reflects the fact that these crimes are tied to particular social situations. Violence is a kind of interaction just as loving is a form of interaction. It should therefore be unsurprising that those conditions that are conducive to people getting together for social purposes are also conducive to violence.

Understanding and Explaining Violence

In addition to the relationship between violence and social situations, one must also understand another characteristic of offenders and victims: differential power. Violence is frequently an attempt of one party to establish or to reestablish a more powerful position over the other party (Hepburn, 1973). Criminologists have pointed to the importance of *asymmetrical power relationships* in crimes of violence. Acts of violence can occur between husbands and wives, business partners, parents and children, and siblings. In each instance, the form of the violence may differ (homicide, aggravated assault, child abuse, spouse abuse), but the relationship between the participants can best be described as one of one party having more power relative to the other party. These crimes may be most likely to occur when the powerful person feels his or her power threatened or shifting to the less powerful person. Under these circumstances, violence may be used to reestablish control of the relationship. This view is similar to that offered by Daly and Wilson (1988) that homicide is conceived to be behavior that grows out of particular competitive relationships among people—usually males. "[T]wo individuals will perceive themselves to be in conflict when the promotion of one's expected fitness entails the diminution of the other's" (Daly and Wilson, 1988: 293). These conflicts usually involve young males because they often compete for women and status.

Many persons do not resort to violence to correct power imbalances in interpersonal relationships. Most acts of interpersonal violence appear to grow out of an interactive situation in which the acts come to be defined as expected or required of the offender. The fact that crimes of violence occur in some groups and in some places more than others suggests that it is related to social characteristics. Some criminologists have suggested that subcultural patterns determine the frequency of crimes of violence, and the extent to which violence is used varies by neighborhood within cities and by social class, occupation, sex, race, and age. These variations have been explained as representing *subcultures of violence*, or the normative systems of a group or groups. According to Wolfgang and Ferracuti (1982), specific populations, for example social classes or ethnic groups, have different attitudes toward the use of violence. These favorable attitudes are organized into a set of norms that are culturally transmitted. Such norms have to do with expected conduct in specific situations, the value of human life in the scale of values, and how situations are to be perceived and interpreted. Proponents of a subculture of violence explanation based their conclusions on differences in *rates* of violence among groups. This does not mean, of course, that all persons in that group share the values or act violently.

Persons who learn these violent norms interpret some situations as requiring the use of violence, such as when one's honor or reputation is belittled. Sometimes, the failure to use violence to, say, defend a male's girlfriend's sexual reputation, would incur an immediate and perhaps irrevocable loss of status in a group. Under such conditions, not only is violence appropriate within the subculture, but rewarded. Failure to act violently in such situations can bring an immediate loss of social status and, in the extreme, banishment from the group.

The distribution of violent crimes suggests that the subculture is found in the inner city regions of urban areas. In these places, the persons who come to subscribe to the violent norms or values are lower-class, minority, young males. Some sociologists, however, do not agree that subcultures of violence explain patterns of homicide and assault. Significant regional differences in homicide, for example, are not easily attributable to the subculture of violence (Dixon and Lizotte, 1987). Other studies have challenged the theory more directly. One study concluded that there was little empirical evidence for the subcultural theory based on the degree of self-esteem, violence in the form of fighting, and the conferral of esteem by others for using violence, thereby casting doubt on whether violence is actually rewarded in lower-class groups (Erlanger, 1974). Another study failed to find value differences among persons who reported different involvement in violence at different times of their lives (Ball-Rokeach, 1973). And, still a third study reported at best only partial support for the theory in data from adolescents (Hartnagel, 1980). It may be that violent values are important, but only at certain times and under certain conditions, and until those times and conditions can be identified the theory needs still to be evaluated (Erlanger, 1979).

Violence can also be learned in other, less direct, ways. Archer and Gartner (1984) claim that soldiers can learn to be more accepting of violence because of their war experiences. Phillips (1983) finds that homicide rates increase after publicized championship heavyweight fights. Phillips believes that such a relationship supports the notion that violence can represent an "imitative" effect of other forms of violence. These findings suggest that some instances of criminal violence can be modeled after other forms of violence.

Homicide is related to certain features of cities, particularly the degree of income inequality. Income inequality can produce hostility and frustration because of one's economic and social position. It has been reported that rates of interpersonal violence are related to the degree of racial income inequality in cities (Blau and Blau, 1982). Tests of this idea have found support in a number of countries when the relationship between homicide and income discrimination was examined (Messner, 1989). This is a structural theory; that is, it attempts to explain the violence rates of aggregates rather than individuals. How are these ideas acted upon at the individual level? One possibility is this: Often, this does not lead to revolutionary behavior, such as committing violent acts against rich persons, but, rather, to striking out at convenient targets, those who are physically and emotionally close to the violent person. The income inequality theory does not at first appear to be close to the subculture of violence theory and perhaps subsequent research will explore ways in which the two might be expressing similar social processes.

The Development of Dangerous Violent Offenders

While much violence is tied to specific social situations and power differentials, there are some persons who resort to violence in a variety of situations and circumstances. These persons are likely to have extensive criminal records and a lengthy history of antisocial conduct in other areas, such as school maladjustment and family problems. One study of 50 such persons concluded that they experience an increasing sense of violence as a solution to a problem (for example, an argument of feeling being pushed around) and in so doing come to affirm themselves as violent persons (Athens, 1989).

Dangerous violent offenders go through a relatively unique series of four stages. In the first stage, *brutalization*, these offenders, in their formative years, are coerced physically to submit to authority. They are witness to the brutalization of others and they are taught by others in their primary group to use violence to accomplish objectives. As an outgrowth of these experiences, these offenders move to a stage of *belligerency*, having concluded that they must resort to violence in future relations with people. In the third stage, they set out on a series of *violent performances* in which they intentionally and gravely injure another person. Finally, because of their violent performances, others in their primary group come to see them as really violent people instead of persons who are merely capable of violence. Others come to confer a sense of power on them that reinforce the use of violence. The manner in which they undergo virulency is illustrated by a person in his late teens recently convicted of aggravated assault (Athens, 1989: 76–77) who said:

> After the stabbing, my friends told me, "Hey man, we heard about what you did to Joe. It's all over school. Everybody's talking about it. You must really be one crazy ass motherfucker." My girlfriend said, "Wow, you stabbed that dude." Finally, things came together and hit right for me. My girlfriend and all my other friends were impressed with what I had done. I didn't really care what my parents thought. Everybody acted like nobody better piss me off any more unless they wanted to risk getting fucked up bad. People were plain scared to fuck with me. My reputation was now made.
>
> I was on cloud nine. I felt like I climbed the mountain and reached the top. I had proven to my friends and myself that I could really fuck somebody up. If something came up again, I knew I could hurt somebody bad. If I did it once, I could do it again. . . . I knew I could fuck somebody's world around, send them sideways, upside down and then six feet under. There was no doubt at all in my mind now that I was a bad son of a bitch, a crazy motherfucker. I could do anything, kill or murder somebody.
>
> Now that I had reached the top of the mountain, I was not coming down for anybody or anything. The real bad dudes who wouldn't associate with me before because they thought I was a nobody, now thought I was a somebody and accepted me as another crazy bad ass.

Some may actually go on to perform violent acts and learn from the reactions of others after such instances of violence. Some persons may assume a violent role and an important part of this process is the reactions of others who regard the person as powerful and dangerous, perhaps even as unpredictable. Others, in other words, will confer a sense of power on the individual and that will reinforce the use of violence.

FAMILY VIOLENCE

Domestic, or family, violence has been recognized increasingly as a serious social problem in the United States. Patterns of assaultive behavior committed within families, by nature, tend to be hidden from official view. But increased sensitivity in recent years has lead to a greater awareness of the existence of this problem and how pervasive it truly is in the American family structure. Any family member can be the victim of family violence, although children and wives are more apt to be victimized than are fathers or husbands.

Child Abuse

Child abuse, in general, can be defined as "nonaccidental physical injury" (Helfer and Kemp, 1974), although this is a broad conception. Beyond this, it is difficult to define physical child abuse precisely since physical measures are widely used to discipline children, although there is no dispute about abuse cases where the injuries are clearly excessive such that any reasonably strict person would agree that abuse has taken place. Children with broken bones, bruises, cuts, and burns would, by most people's definition, qualify for being abused. It is more difficult to get people to agree on a definition of "psychological" or emotional child abuse, except in very extreme instances. It is also difficult to get agreement on the abusing nature of lesser injuries that might have resulted from more or less reasonable parental measures. Because there is no universal standard for judging the best or most desired child rearing, there is no universal standard for judging what is child abuse and neglect.

Yet, there are some definitions that guide parents and policy makers. The official federal definition of child abuse, stated in the Federal Child Abuse Prevention and Treatment Act of 1974 (PL 93–237) states that child abuse is to be considered:

> . . . the physical or mental injury, sexual abuse, negligent treatment, or maltreatment of a child under the age of eighteen by a person who is responsible for the child's welfare under circumstances which would indicate that the child's health or welfare is harmed or threatened thereby (quoted in Gelles, 1985: 351).

The clearest example of child abuse takes the form of child battering, or the physical assaulting of children. The range of injuries suffered by children from battering includes relatively minor scratches and scrapes to life-taking injuries. Child battering includes virtually every kind of assault that can be inflicted upon an adult. Child abuse includes not only child battering, but also the exploitation of children through pornography and sexual assault, malnutrition, educational neglect, medical neglect, and medical abuse.

What is Deviant Violence?

We begin to answer the question of which violence is "deviant" in the same manner as determining whether anything is deviant—with reference to norms. What are the norms in the following situations?

The Situation	Meaning #1 Non-deviant Interpretation	Meaning #2 Deviant Interpretation
Scenario A: A man deliberately, and not in self-defense, runs into another man he does not know and injures the second man so severely that man must be in the hospital several weeks.	football	street fight
Scenario B: A man picks up a stick and proceeds to hit another man whom he does not know with the stick.	hockey	aggravated assault
Scenario C: A mother deliberately and not in self-defense strikes her daughter, causing pain.	parent spanking own child for disciplinary reasons	child abuse

It is not possible to give a precise figure on how many children experience child abuse, or assault, but it has been estimated that nearly 2 million children a year in the United States are subject to abusive behavior from family members or others (Gelles and Straus, 1979; Gelles, 1979). One observer estimated that over 300,000 children a year are the subject of sexual abuse (Sarafino, 1979). A national estimate figured that as many as one child in 100 have been physically maltreated, and even that may be an underestimate (Garbarino, 1989: 224). Furthermore, batterings and instances of sexual abuse are only rarely one-time events. Children are frequently assaulted many times during a single year. The continuation of assault most likely has to do with the fact that the victims are relatively powerless to terminate the abuse. Some observers are concerned that the rate of sexual abuse of children appears to be increasing, not only in the United States but in other countries as well (Finkelhor, 1982). Whether the increase in reports of child sexual abuse during the past decade are due to the increasing frequency of this crime or to increased awareness that has led to more reports is impossible to say.

Concern over child abuse is not confined to the immediate physical effects of this behavior on child victims. There is some evidence to suggest that children who have been physically abused are more likely than those who have not to abuse their own children later in life (Straus, Gelles, and Steinmetz, 1980). Indeed, the intergenerational cycle of violence has been documented in a number of studies of family violence (Gelles, 1985). Violence may come to be a normative pattern in these families and an accepted manner of child rearing. At the very least, violence may be seen as an acceptable last resort to obtain compliance from children. It is conceivable that the effects of abuse can contribute to subsequent violence by the person in some way, such as the learning of violence values, the alienation of the child from authority figures, or a weakened parent-child relationship. The violence some children receive may actually model for them behavior to be engaged in later in life. Consider the following report of a young man who was beaten as a child:

> The beatings my stepfather laid on me, the terrible beatings he laid on my mother, and all the violent rhetoric took their toll on my mind. It inflamed me and made me want to go to bad. I was tired of always being messed with by people. I was ashamed of being weak and lame and letting people mess with me all the time. I didn't want to be messed with by people any more. People had messed with me long enough. If anybody ever messed with me again, I was going to go up against them. I was going to stop them from messing bad with me. If I had to, I would use a gun, knife, or anything. I didn't mess with other people, and I wasn't letting them mess with me any more. My days of being a chump who was too frightened and scared to hurt people for messing with him were over. (Athens, 1989: 60–61)

Nevertheless, a rather exhaustive review of the relevant research literature has concluded that there is little direct evidence that "abuse leads to abuse" later in life (Widom, 1989). This topic represents an important research topic for the 1990s, and more carefully conducted studies are needed to examine the possible intergenerational transmission of violent values.

A number of factors are associated with child abuse in the home. Stress, low income, low levels of parental education, and family problems (such as divorce or family emotional conflict) are all associated with child abuse (Gelles, 1985). While child abuse can take place among all social classes, it appears to be more frequent—like other forms of violence—among lower-class persons. It also seems that abusing parents are more likely to be socially isolated from others who might form a support group for the parent, particularly in times of emotional crisis. The absence of close friends, or isolation from one's family or neighbors, means that the parent must handle family, economic, and social stress alone. The development of parent support groups in some communities is a result of the awareness that some parents may have fewer resources than others and that interpersonal support can be obtained from other parents who face similar problems.

A related crime is child sexual abuse, or incest in the family. This crime usually takes place between fathers and daughters, but it can occur between mothers and sons as well (Finkelhor, 1984). Child sexual abuse involves forms of exhibitionism, fondling of genitals, and mutual masturbation, as well as intercourse. Until recently, it was thought that incest was extraordinarily rare, perhaps occurring no more frequently than once in a million families (Weinberg, 1955). Because of a renewed recent interest in family violence, incest has been studied more, and it is now thought to occur as frequently as once in every hundred families (Renvoize, 1982). It is probably safe to say that these conflicting estimates do not reflect changes in the occurrence of this behavior but, rather, greater attention and sensitivity to this problem. We are detecting more incest cases.

Spouse Abuse

As with child abuse, it is also difficult to estimate the extent of spouse abuse in this country because definitions of spouse abuse differ. While early definitions of wife abuse, for example, concentrated on damaging physical violence, subsequent conceptions after wife abuse was recognized as a national problem broadened the definition to include sexual abuse, marital rape, and even pornography (Gelles, 1985). Spouse abuse typically conjures images of strictly wife abuse, but wives assaulting husbands does occur. In fact, in a national incidence survey (Straus, et al., 1980), one-fourth of the homes where there was couple violence, men were victims. Another one-fourth involved women as victims but not offenders, and one half of the violent homes involved both men and women as offenders. In fact, in this survey, the percentage of husbands victimized was greater than the percentage of wives victimized. It is possible, however, that women are more apt to use violence in self-defense.

Schwartz (1987) has concluded from National Crime Survey victimization data that men are more likely to be struck by women than women are to be hit by men, and that men are more likely to call the police. Such findings run contrary to the expectations of most people. However, women make up about 95 percent of the victimizations from spouse abuse because they are more likely to be more seriously injured in incidences of domestic violence. Thus, the real problem of spouse abuse is a problem of the physical victimization of wives, not husbands.

Historically, it has not been considered deviant for husbands to beat their wives on occasion and within certain limits (stopping short, for example, of death or disfigurement). Ideologies that permitted husbands to be the "head of the household" and to manage their wives' affairs, extended this authoritative control to physical realms as well. Thus, husbands, in some historical sense, were obligated to keep their wives in line behaviorally and to exercise certain physical control over them. It is a far cry, of course, from these historical roots to modern physical assault of spouses, most usually wives.

Spouse abuse is associated with certain factors, such as the presence of alcohol (Secretary of Health and Human Services, 1990: 172–174), husbands who are under economic stress, and families marked by interpersonal conflict. Family violence can become a behavior pattern in some homes. Spouse abuse, like other forms of violence, is largely, but not exclusively, found among lower-class families. Spouse abuse can also be located in middle- and upper-class families, although there appear to be fewer of these cases. However, it is still not clear whether the fewer reports of spouse abuse outside of working and lower class communities reflects genuine differences in behavior, or the greater invisibility of spouse abuse in middle- and upper-class homes. Clearly, middle- and upper-class spouses may have alternatives not available to lower-class spouses. For example, having enough money to go to a motel room for the night would be a luxury in many lower-class homes and this forces the spouse to maintain physical contact with the abusing spouse. Other alternatives in the form of counselors, friends with resources, and greater awareness of community services might also distinguish lower- from middle- and upper-class victims.

Cases of spouse abuse are handled as instances of assault and aggravated assault by criminal justice officials, if the situation warrants. Persons convicted of these charges by the courts are subject to terms of incarceration, probation, and fines. Generally, it has been common for the police to adopt some informal handling of domestic disputes, such as physically separating the conflicting parties for a time or referring them to a social service agency. Critics charges that this offered no protection to the victim of the abuse and such as response amounted to no response at all. As a result, some states have now implemented mandatory arrest laws with respect to domestic disputes. That is, the police are required to make an arrest every time they respond to a domestic dispute; the police have no discretion in the matter. One evaluation of such a law in Minnesota suggested that one effect of the law has been to deter offenders from subsequent violations (Sherman and Berk, 1984), but a replication of that study in Nebraska failed to detect any deterrent effects from mandatory arrest (Dunford, Huizinga, and Elliott, 1990).

Criminal sanctions are usually brought against persons convicted of spouse abuse, but, like most assault cases, prosecution of spouse abuse cases relies on the assistance of the victim. Successful prosecution of these cases without victim cooperation, like most other criminal cases, is rare. Unfortunately, this assistance is sometimes difficult to obtain for prosecutors because the spouse, say the wife, may be reluctant to prosecute the husband, particularly if she fears retaliation or plans on returning to him. As a result, it is difficult for criminal justice officials who have to deal with these cases to insure justice is done to all parties.

A similar type of problem involves the abuse of elderly persons, such as individuals in nursing homes or elderly persons who live with younger family members (Pillemer and Wolf, 1986). There are no reliable estimates of the extent of this problem, although there is reason to believe that the problem is increasing as the number of elderly increase (Pagelow, 1989). Cases of the abuse of elderly persons are complicated by the victims sometimes refusing to report instances of abuse because they fear retaliation or, if the victims are at home, they may fear being sent to a nursing home. Even when reported, cases involving the abuse of an elderly person are hard to prove. Elderly persons bruise easily and fall often, evidence of some physical injury is not sufficient to suggest abuse. Alternative explanations for physical injuries can always be constructed by others. Consequently, many instances of elderly abuse do not come to the attention of authorities. Even when reported, the elderly victim, because of other problems associated with aging, may not be a credible witness. Clearly, there are many characteristics shared by children and the elderly as potential victims for abuse, including their general powerless statuses.

FORCIBLE RAPE

Forcible rape, or unlawful sexual intercourse with a female against her will, is distinguished from statutory rape or sexual intercourse with a female with her consent but under the legal age of consent, usually 18 in the United States and 15 in Europe. Most rape statistics also include sexual assaults or attempts to commit rape by force or threat of force. Statutory rape, however, is excluded because it involves no force. The Federal Bureau of Investigation (1990: 15) recorded 94,505 rapes in 1989, but forcible rape, as will be discussed shortly, is notoriously underreported to the police and these official estimates are therefore underestimates of the total number of rapes.

Patterns of Rape

Rape is of great concern to a large number of women. Many women are aware of the possibility of sexual assault in many social situations and this awareness leads them to take precautions at times when crime prevention would not occur to men. College women, for example, often think ahead about times when precautions are appropriate, such as walking on campus at night or from class to class, returning to a dormitory from the library, and attending a social event without a male escort. Fear of rape by women is a very real and powerful motivation in their behavior (Warr, 1985). The kind of rape that often elicits the most fear is an attack by a stranger.

Jack Doe (not his real name) was arrested in Spokane, Washington for several rapes in 1987 (Spokane *Spokesman-Review*, November 13, 1987, pp. 1 and 6.) He admitted to raping at least 8 women who ranged in age from

5 to 60. Doe, who raped his victims in their homes, described his methods: "It's like doing a burglary. There is nothing to it. All you have to do is go down a street, see a house, see a light on and go up to the house and look in." Doe would keep a periodic watch on the house, keeping track of the comings and goings of the residents, waiting for a time when the victim was home alone. When he was ready, Doe would approach the house by bicycle, cut the phone lines, and enter the house, usually through a window. After the assault, he would leave the house and pedal away. He used no weapon and used force only once when a women actively resisted. Doe would not return home immediately. He would ride around, sometimes for hours. Why did he do it? Doe asserts it was a cry for "help." "The only way I thought I could get help was to do the rapes," he said. Doe's biography included instances of sexual abuse and considerable institutionalization in foster homes and detention centers.

Jack Doe is not a "typical" rapist because there is no "typical" rapist. Rape can occur in many different situations. The Bureau of Justice Statistics has compiled information from several years of surveys with rape victims regarding the circumstances under which that crime was committed. Those studies conclude that rape is a crime that often involves persons known to the victim and crimes that occur in a place supposedly controlled by persons known to the victim. Many rapes, for example, occurred either in the victim's home (26%) or the home of a relative, neighbor, or friend (19%) (Bureau of Justice Statistics, 1987: Table 52). One-half of the rapes occurred in a public place, near a street, in a playground or park, or some public building. Fifty-seven percent of the rapes involved strangers (Bureau of Justice Statistics, 1987: Table 28).

Rape victims frequently know their attackers, however casually and for a short time, but many rapes involve strangers (Newman, 1979: Chapter 9). There is little to support the notion that most rapists are "seduced" by victims. Most rapists are young (between 15 and 25 years of age) and alcohol does not play a major role in the commission of the crime (Amir, 1971). In any case, the statistical evidence shows that most rapes are planned crimes rather than being an explosive event (although some spontaneous rapes do occur). Most convicted rapists do not have prior sex records, although a substantial number have had previous convictions for other crimes (Deming and Eppy, 1981).

Like homicide, forcible rape is predominantly an *intraracial* crime involving offenders and victims of the same race (Randall and Rose, 1984). In those rapes that involve persons of different races, the rapes involving black offenders and white victims are more numerous than those involving white offenders and black victims.

A form of rape about which most people are becoming more aware is "date rape." Date rape takes place between a male and female who know one another. The rape takes place during the course of, and usually at the end of, a social occasion involving the couple, a date. During the course of the evening, or even before, some males may have generated expectations that the evening will include sexual relations, expectations that are not shared by the female. The rape does not usually involve extreme force but expectations about sexual activity are communicated to the female and her resistance is then not taken seriously by the male. The male may feel that he "deserves" to have sex with the female because he may have bought her dinner or some other entertainment. The male refuses to accept her lack of consent, sometimes by interpreting this lack of consent as just the opposite: consent. Some males will indicate they the female said "no," but really meant "yes." Many date rapes are not reported to the police for the same reasons that other rapes may not be reported: the embarrassment to the victim.

The incidence of date rape is hard to estimate. In 1989, a survey for the Stanford Rape Education Project concluded that nearly one in three female students and one in ten male students had been raped, almost always by someone they knew (Jacobs, 1990). Students were asked if they had sex "when you did not want it because you were overwhelmed by continual arguments and pressure." Only 10 percent of the women and 25 percent of the men ever mentioned the rape to anyone, much less reported it. In fact, many of the women did not even think it was rape. Only 5 percent of the women indicated they had been coerced into unwanted sex at any time in the past by some degree of physical force; a similar number blamed alcohol or other drugs.

The Political Context of Rape

Information from police statistics and victim surveys do not tell the whole story of rape, and a full understanding of rape as a crime and form of interpersonal violence requires an awareness of the political dimensions of this crime. Although forcible rape has traditionally been seen as a sexual crime motivated by the offender's desire for sexual relations, actually it can not be described adequately without reference to the use of violence. Conceiving rape as a crime of violence rather than a crime of sex has done much to reorient thinking about rape and to challenge long-standing myths about rape. Perhaps one of the most persistent myths is that rape is impossible if the victim resists sufficiently. Rape is like other predatory crimes, however, in that the victim has no choice against the use or potential of physical force (Randall and Rose, 1984).

It has only been in the last decade or so that rape has come to be viewed as a violent rather than a sexual crime. The women's movement has done much to promote the violent conception of rape and, as a result, there have been a number of legal reforms and alternative theories of rape that recognize the linkage of this crime with other violent crimes. Many observers regard forcible rape as a "political act" because it is an exercise of power by one group (males) over another (females). Brownmiller (1975: 254) has said that "rape is to women what lynching is to blacks: the ultimate physical threat." Rape, it is often concluded, represents an act of overt control that insures the continued oppression of women and the perpetuation of a male-dominated society. While recognizing the violent nature of forcible rape, the expectation of sexual gratification for the offender would seem to be another dimension, however, that is not present in ordinary assault and beating of women.

There is another political context to rape. The law has adopted what has been termed a "paternalistic" approach to women: females typically have been seen as being in need of protection and shelter from some of the harsher realities of life, and rape laws appear to treat females as though they are the property of men. As such, the crime of rape is as much a crime against men (husbands and fathers) as against women. However, the development of rape laws can not be attributed solely to conceiving women as property (Schwendinger and Schwendinger, 1981).

There are a number of similarities between many instances of forcible rape and other forms of interpersonal violence, such as homicide, assault, and spouse abuse, including characteristics of offenders and victims. This suggests that the subculture of violence may be useful in explaining rape as well as other crimes of violence. One important component of the learning of sexual roles is expectations about the role of the other sex. Groups who contribute disproportionately to rape statistics as offenders are those whose conceptions of females as objects of sexual gratification are most strong.

The male role, in addition, requires physical aggression on occasion and the assertion of masculinity, whether through the use of competitive sports, displays of physical prowess, or forcible rape may constitute an important part of the everyday lives of some males in urban slums and other lower-class areas. Because lower-class persons often have difficulty in constructing their identities in terms that most middle- and upper-class persons understand—materialism, occupational success, and social mobility—lower-class males may feel forced to seek their identity in emphasizing differences between themselves and women. Physical force and strength are devices that are suitable for this purpose; they are readily available to males, they represent a biological difference between them and women, and they convey a sense of power when used (Hills, 1980: Chapter 3). It is essential to understand that rape is a crime of violence and force. The ways in which rapists and their victims interact, even in conversation, show that rapists are interested in manipulating and exercising power over their victims (Holmstrom and Burgess, 1990).

Rape Reporting

Surveys indicate that many cases of forcible rape are not reported to the police. This is understandable for several reasons. (1) Rape is an emotionally upsetting and deeply humiliating experience for the victim. (2) There is still a strong stigma, even within their own families, for rape victims, although this is now slowly changing. (3) Victims are sometimes seen as having consented to the act either by not resisting sufficiently or by "leading on" the assailant. (4) When rapes have been reported, some officials in the criminal justice system have tended not to believe the victim and to treat her as through she had been a party to the crime. (5) Rape victims have, in the past, been subjected to humiliation in the form of embarrassing questioning by the police and prosecutors to verify that a crime has occurred. (6) The handling of rape victims in the courtroom has sometimes been very unethical with questioning, in a public forum, about the victim's previous sexual history, the provocative circumstances relating to the rape, and the extent to which the victim employed physical resistance.

There is evidence to suggest that the increase in the number of officially recorded rapes is due in part to increased reporting. The FBI's *Uniform Crime Reports* indicates that the number of rapes and the rate at which it is committed (which takes into account population changes) has been increasing over the past decade (Federal Bureau of Investigation, 1990). Victimization surveys of the general population, however, have found a very constant rate of forcible rape in the community (Bureau of Justice Statistics, 1987). This means that the increase in the police statistics is most likely due to the increased willingness of victims to report the rape rather than more rape crimes in the community. Undoubtedly, rape crisis centers and other, usually private, organizations have aided in this increased reporting of this crime.

The reasons that rape is reported to the police are varied. Among rape victims in 1985, the two principle reasons included "to stop or prevent this incident from happening again" (30%) and "to punish the offender" (27.4%) (Bureau of Justice Statistics, 1987: Table 97). Other reasons included the need for help after the rape, the desire to obtain evidence or proof, and the victim feeling that it was her duty to report the crime. Victims often cited multiple reasons for reporting the rape. Reasons for nonreporting included the feeling that nothing would happen as a result of contacting the police or fear of retaliation from the offender.

The Nature of Legal Reform

A number of changes have occurred in the criminal justice system as the public has become more sensitized to the special needs of handling rape cases. Police departments in most large cities, for example, may have a special unit to handle rape complaints. A problem in the past has been that the rape victim has been grilled or interrogated by the police in an effort to substantiate the charge of rape. The presumption seemed to be that the victim was lying. Needless to say, the process of police questioning could and did produce additional trauma for the victim. Today, it is more common that the questioning of the rape victim takes place under more understanding circumstances. A woman officer might handle this part of the investigation and a more empathetic attitude is attained in the questioning.

In forcible rape, the lack of consent is legally crucial. Some state codes have been revised to emphasize the use of force or assault, and penalties have been increased accordingly. The outcome of these cases in the criminal justice system has also been influenced by the extent to which the victim resisted or actually "precipitated" the crime. A study of rape in Philadelphia reported that one in five forcible rapes were "victim precipitated" in the sense that the victim actually, or so it was interpreted by the offender, had agreed to sexual relations, but then either retracted her consent before the act, or did not resist strongly enough (Amir, 1971: 266–270). This, of course, is not to say that forcible rape is to be condoned in those situations in which the victim somehow "invited" sexual intercourse and then refused. The idea of victim precipitation as applied to rape is simply not precise enough to describe adequately some aspects of the relationship between the victim and offender. Other studies have not found the same role for victim precipitation. Curtis (1974), for example, compared degrees of victim precipitation in four crimes against the person: murder, aggravated assault, robbery, and forcible rape. The findings pointed to the conclusion that some sort of provocation on the part of the victim was common with murder and aggravated assault, less frequent but still noteworthy with robbery, and of the least relevance with rape. Greater explication of the nature of victim precipitation would have tremendous influence in reforms regarding the nature of some rapists' responsibility for their actions.

Theories of Rape

Psychiatric and psychological approaches to rape stress such factors as hidden aggression on the part of rapists and the ability of psychiatrists to classify rapists into such categories as "power rapists" (Groth, 1979) and "sexual rapists" (Cohen, et al., 1975). It is impossible to deny the element of aggression in rape, although its psychological meaning to rapists remains unclear. Still other theorists have attempted to combine approaches that emphasize learned attitudes toward the use of violence with situational inducements to produce a rape (Gibson, et al., 1980). These theories have not received systematic test, and debates among theorists will continue until such ideas can be verified empirically.

The importance of changing sexual roles, the effects of family background, and expectations regarding sexual relationships all pose important questions with respect to rape. The connection between using sexual means to communicate aggression and other means has yet to be fully explored. Testing such theories in offender populations, however, is difficult since incarcerated rapists may be quite different from noninstitutionalized rapists. Thus, it is not clear that results from the institutionalized sample can be generalized to noninstitutionalized persons (Deming and Eppy, 1981: 365–366).

Sociological theories of rape include: (1) the idea that rape represents an extension of legitimate violence in society; (2) the relationship between rape and the degree of gender inequality (the more equal the sexes, the less likely rape is to occur); (3) rape results from the effects of pornography which often depicts women as sexual objects; and (4) rape results from value conflicts in our society. A test of these theories found support for each, except the theory that rape represents some "spill over" from the degree of normal or legitimate violence in society (Baron and Straus, 1989). Baron and Straus suggest that the relationship of rape and causal factors is exceedingly complicated and it is difficult to interpret empirical tests unambiguously. The relationship between rape and the rates of readership of pornography, for example, might suggest that readers are "learning rape," but it might also suggest that pornography readership and rape are related because of some third factor, such as a masculine ethic or culture.

Another sociological theory of rape may be the subculture of violence theory (Amir, 1971). This theory, introduced primarily to explain the conduct of murderers and assaulters, is also applicable to rape. This is particularly evident when one examines the characteristics of those persons who commit rape. The profile of rape offenders resembles that of murderers and assaulters: young, black, lower-class males in inner-city areas of large metropolitan communities. These are the groups identified by Wolfgang and Ferracuti (1982) as subscribing to the subculture of violence. Persons who subscribe to the subculture come to view women as sexual objects and possibly as accepted targets for other forms of aggression as well. As with other theories of rape, however, it is necessary that the subculture of violence theory be tested before it can be accepted.

SOCIETAL REACTION TO CRIMES OF PERSONAL VIOLENCE

As expressed in law, the societal reaction to murder, aggravated assault, and forcible rape is extremely severe. Penalties include lengthy prison terms, and, under some circumstances, capital punishment. Sociologically, however, most murderers are among the least "criminal" of all offenders. Persons who commit murder in a personal dispute do not conceive of themselves as criminals, and rarely are they recidivists. Such offenders do not engage in crime as a career, as defined here, nor do they progress to subsequent, more serious, criminality. Their criminal careers usually terminate with their apprehension for that crime. Most murderers are not eligible for probation and, therefore, they spend much more time in prison compared to other offenders. Assaultists typically have a longer record of offenses, most of these being crimes against persons. The use of violence, however, does carry the risk of death of the other party, and persons with a history of assaultive behavior are likely to be reinvolved in this kind of crime. Persons convicted of forcible rape are also subject to long prison terms, with longer terms given to those rapists who inflict serious physical injury on their victims. The full extent of victimization, however, can not be assessed strictly in terms of the degree of physical injury. Rape victims experience considerable psychological injury as a result of this crime.

SUMMARY

Crimes of violence are among the most feared crimes by the general public. Murder, aggravated assault, and forcible rape are all crimes of violence. They involve attacks, or threatened attacks, on a victim's person and they are regarded by the legal system to be extremely serious crimes requiring severe responses. Official police statistics have generally reported increases for these crime categories throughout the past decade, although victimization surveys have indicated that rates of homicide and rape have generally remained reasonably stable. Patterns of offending for these crimes are similar although not identical. Offenders are likely to be young, lower-class minority males who live in inner city communities. Victims are also likely to be from such groups. Murder and aggravated assault are statistically likely to be unplanned crimes tied to particularly emotional circumstances and chemical substances such as alcohol and drugs.

In contrast, rape is more likely to be a planned crime. Rape in particular continues to undergo a social redefinition process. Rape is now generally regarded as a violent crime, but there is still confusion among many people about rape situations, such as date rape and spouse rape. Some rape situations are not perceived to be instances of rape because the element of consent is ambiguous. Family relationships represent another area in which interpersonal violence occurs. Family violence has generated much national attention in recent years. Although husbands can be victims of family violence, most concern and policy efforts have been directed toward preventing the victimization of wives, children, and the elderly.

There is no generally accepted theory of interpersonal violence, although the similar patterns in offending suggests that a general theory that would explain murder, rape, instances of family violence, and assault is possible. Sociological approaches, such as the subculture of violence theory, have emphasized the learning of violent values (for homicide and aggravated assault) or values that permit deindividualizing women and conceiving of them as sexual objects and objects to be dominated. Most generally, crimes of interpersonal violence can be seen as the result of unequal power relations among the participants, and violence as an attempt to restore a previous power relationship. Where relationships are marked by asymmetry and inequality, violence is more likely to maintain or reestablish power relationships.

Violent offenders receive severe penalties from the criminal justice system. While most instances of homicide are eventually reported to the police, reports of some violent crimes such as forcible rape are troublesome, since many victims are reluctant to report these crimes. Increased awareness brought on by the women's movement has generated more reporting behavior, thereby increasing rape crime rates and more official involvement from criminal justice officials.

SELECTED REFERENCES

Brownmiller, Susan. 1975. *Against Our Will: Men, Women, and Rape.* New York: Simon and Schuster.

> A feminist perspective on rape. Brownmiller conceives of rape as only one aspect—but perhaps the most violent one—of the "battle of the sexes." Rape is seen as a means by which men control women and maintain their privileged positions.

Luckenbill, David F. 1984. "Murder and Assault." Pp. 19–45 in *Major Forms of Crime.* Edited by Robert F. Meier. Beverly Hills, CA.: Sage Publications.

> An excellent summary of the literature on homicide and assault, particularly serious forms of assault. Both empirical regularities in offending and theories of murder and assault are discussed.

Ohlin, Lloyd E. and Michael Tonry, eds., 1989. *Family Violence.* Chicago: University of Chicago Press.

> An invaluable collection of papers on all aspects of family violence.

Randall, Susan and Vicki McNickle Rose. 1984. "Forcible Rape." Pp. 47–72 in *Major Forms of Crime.* Edited by Robert F. Meier. Beverly Hills, CA.: Sage.

> Another summary, this one of forcible rape theory and research. The authors describe general patterns of rape and also spend time discussing implications of this crime for the criminal justice system and handling of rape cases.

Weiner, Neil Alan, Margaret A. Zahn, and Rita J. Sagi, eds., 1990. *Violence: Patterns, Causes and Public Policy.* New York: Harcourt Brace Jovanovich.

> A useful collection of papers dealing with, among other topics, individual violence, rape, violence in the workplace, mass media effects, causes, and treatment/policy approaches.

Weiner, Neil Alan and Marvin E. Wolfgang, eds., 1989. *Violent Crime, Violent Criminals* and *Pathways to Criminal Violence.* Newbury Park, CA: Sage.

> These two edited volumes contain papers on different aspects of violence, such as the relationship between violence and drugs, mental disorder, race, and gender.

Economic and Political Criminality

Crime is a fact of modern life. Most persons have committed at least one criminal act in their lifetimes, even if the act was relatively minor, and many persons have had the unpleasant experience of being victimized by crime. In recent years, for example, about one-quarter of all households in the United States were victimized by a crime of violence or theft (Bureau of Justice Statistics, 1990). Most of these victimizations were property offenses and most of these involved small dollar amounts of loss or damage. Clearly, crime is a major form of deviance.

Criminal behavior represents a heterogeneous collection of acts. Some criminal acts, such as taking an item of little value, are of little consequence in themselves, while other criminal acts, such as spying for a foreign government, can have enormous consequences for many people. Criminality reflects the diverse behavior of diverse people. Criminals differ in the extent to which they identify with crime and other criminals, the degree to which they are committed to crime as a behavior, and the extent to which they progress in the acquisition of more sophisticated criminal norms and techniques. An important theme of this chapter is to explore this diversity. Here we discuss a variety of criminal behavior systems, including occasional property offenders, conventional criminality, white-collar and corporate criminality, political crime, organized crime, and professional criminality.

OCCASIONAL PROPERTY OFFENDERS

Many offenders have criminal records consisting of little more than an occasional or infrequent property offense, such as illegal auto "joy riding," simple check forgery, misuse of credit cards, shoplifting, employee theft, or vandalism. Such crimes are largely incidental to the way of life of the occasional offender and in no way do such offenders make a living from crime, nor do they play a criminal role. This type of criminal behavior is usually situational; the offense is often committed alone, and the offender seldom has had prior criminal contacts (Hepburn, 1984). With some exceptions, the offender has little group support for the behavior. Usually, few skills are required to commit these crimes. For example, inadequate supervision of mass-displayed merchandise in stores presents almost limitless opportunities for shoplifting and training in sophisticated shoplifting techniques is unnecessary.

Occasional offenders do not conceive of themselves as criminals, and most of them are able to rationalize their offenses in such a way as to explain it to themselves as a noncriminal act. A shoplifter, for example, might justify the behavior by thinking that a large store can afford shoplifting or that there had been no intention to steal the car, only to borrow it. There is no evidence that occasional offenders make any effort to progress to types of crime requiring greater knowledge and skills. Only a few kinds of occasional crime can be discussed here, and these include auto theft, check forgeries, shoplifting and employee theft, and vandalism.

Auto Theft

Auto theft in the form of "joy riding" is not a career type of offense. Auto theft, strictly speaking, involves stealing a car with the intent to keep the car or parts of the car, while joy riding involves taking a car without the owner's permission but with no intent to keep it. Joy riding is mainly done only by youth and the offenses are usually sporadic (Wattenberg and Balistrieri, 1952). Cars may be taken, driven for a time, and then abandoned. It involves no techniques usually associated with the conventional career types in which "stripping" is learned, as well as the selection of special kinds of cars, and "fences" are found for the sale of the car or its parts (Steffensmeier, 1986).

Not all auto theft is of the joy riding variety, and not all offenders are youth. Offenders may steal an automobile not only for the short-term transportation of joy riding, but for long-term transportation, to use in the commission of another crime, or to sell (McCaghy, Giordano, and Hensen, 1977). Age and sophistication in stealing enter into these types of auto theft.

Check Forgeries

It has been estimated that three-fourths of all check forgeries are committed by persons who have no previous pattern in such behavior. Lemert, who studied a sample of nonprofessional forgers, concluded that such persons generally do not come from a delinquency area, have no previous criminal record, and have had no contact with delinquents and criminals. They do not conceive of themselves as criminals, nor do they progress to more serious forms of criminality. Lemert suggests that this offense is a product of certain difficult social situations in which persons find themselves, a certain degree of social isolation, and a process of "closure" or "constriction of behavioral alternatives subjectively held as available to the forger" (Lemert, 1972: 139). Check usage, while still high, is declining, increasingly being replaced with credit cards for all types of purchases and services. These cards may be unsupported by sufficient bank funds as checks are required to be, and when they are stolen, they may be used fraudulently (Greenberg, 1982).

Shoplifting and Employee Theft

These crimes are closely related and sometimes combined into the broader category of "inventory shrinkage," a term used to denote the loss of merchandise from illegal activities (such as shoplifting and employee theft) as well as honest, unintentional mistakes (such as bookkeeping errors). For this reason, it is impossible to estimate accurately the total amount lost annually by merchants to illegal activities. Certainly, however, the total is in the billions of dollars a year (Baumer and Rosenbaum, 1984; Meier, 1983). The most extensive study of employee theft involved an examination of nearly 50 business corporations in three metropolitan areas that included retail stores, general hospitals, and electronics manufacturing firms (Clark and Hollinger, 1983). In retail stores, the most common form of theft was in the form of abusing employee discount privileges for others, while in the hospitals taking medical supplies was the most often reported theft activity. In the manufacturing firms, the taking of raw materials was the most frequent kind of theft. Generally, young, new, and never married employees were more likely than others to engage in employee theft; these employees were also most likely to be the most dissatisfied with their jobs. These violations are in addition to other counterproductive but not illegal behavior of employees, such as taking excessively long lunch and coffee breaks, purposively slow or sloppy workmanship, and the misuse of sick leave.

Shoplifters may be found among all groups of persons. Generally, however, shoplifters appear to be either youth, "respectable" employed persons, or even housewives, largely of the middle class (Klemke, 1982; Cameron, 1964: 110). The latter can generally afford the things they steal, and the former many times do not have a need for the items they take. Most shoplifting by nonprofessionals involves small and inexpensive items. Current methods of large-scale merchandising found in supermarkets, discount stores, and merchandise marts seem to invite shoplifting by leaving goods unsupervised so that customers can touch the items. Just how small the risk of apprehension actually is was demonstrated in a study in which researchers, with permission of the management but without the knowledge of employees, deliberately stole items from department and grocery stores (Blankenburg, 1976). Less than 10 percent of all shoplifting activities were detected; most customers were unwilling to report even the most flagrant cases they observed.

Occasional shoplifters, as opposed to professionals, generally do not conceive of themselves as criminals nor do they continue their crimes after apprehension in many instances. Most offenders rationalize their offenses on the basis of beliefs such as "the store can afford the loss," "they 'owe' me this item because of all the business I have given them in the past," and so on. Employee theft is rationalized in a similar manner (Robin, 1974). There is reason to believe that employer sanctions can diminish employee theft, although informal sanctions from one's fellow workers are more effective deterrents (Hollinger and Clark, 1982).

Vandalism

It is virtually impossible to estimate the amount of property loss due to vandalism, although an estimate in 1975 reported costs of around $500

Anti-Shoplifting Statutes

By the end of 1990, 39 states had enacted legislation to help prevent (not control after it occurs) shoplifting. The legislation permits retailers to send "civil demand" letters to people accused of shoplifting. The letters ask the person to pay a penalty of $100 to $200 in addition to returning the merchandise. In exchange, the retailer agrees not to sue for civil damages.

One security officer credits the law with a reduction in shoplifting:

It's an extremely significant loss prevention issue. We saw that tacking on an additional $200 penalty impacted the parents' wallets.

The reduction in juvenile shoplifting alone may be huge. To date, there have been no systematic studies of the law's effectiveness.

Source: *Wall Street Journal*, October 15, 1990, p. 21.

million for school vandalism alone (Bayh, 1975). Vandalism is almost exclusively a crime of juvenile offenders, although it is not unheard of that some youthful adults may commit this crime. Moreover, vandalism is a worldwide phenomenon, and vandals' targets are much the same regardless of country: schools and their contents; public property in the form of park equipment, road signs, and fountains; cars; vacant houses and other buildings; and public necessities such as toilets and telephones. Likewise, graffiti on walls and public places causes much expense in its removal.

Most vandals have no criminal orientation and they conceive of their acts more in terms of "pranks" or "raising hell" (Wade, 1967). The fact that often nothing is stolen reinforces this self-conception as prankster, not delinquent. In spite of this conception, one study has found that peer relationships and adult-child conflict were the best predictors of vandalism among a sample of middle-class adolescents (Richards, 1979). Groups commit acts of vandalism, but the acts do not derive from any subculture. Acts of vandalism seldom utilize or even require prior sophisticated knowledge; they grow out of collective interaction. Few are planned in advance; it is essentially spontaneous behavior.

Societal Reaction to Occasional Offenders

In most cases, societal reaction to occasional offenders is not severe unless the theft or damage is particularly large. Occasional offenders frequently do not have prior records and they are often dismissed or placed on probation by courts. Many occasional offenders terminate law-breaking behavior upon apprehension, as in the case of many nonprofessional shoplifters (Cameron, 1964: 165), although some offenders do continue their crimes, particularly in the absence of social controls, as in the case of employee thieves (Hollinger and Clark, 1982). Certainly, occasional property crime provides the entry into crime for virtually all career offenders. The great majority of persons who engage in occasional property crime, however, do not progress to subsequent criminality.

Occasional offenders, because they are seldom committed to delinquency and criminality, are often seen as good candidates for programs that operate outside the justice system. In the Netherlands, for example, juvenile vandals might participate in a diversion program that requires them to repair or replace the property damaged by their crime (Kruissink, 1990). This might take the form of repainting over some graffiti, paying for property that has been damaged by the vandal, or replacing property destroyed by the vandal.

CONVENTIONAL CRIMINAL CAREERS

Conventional criminal offenders provide the stereotype of a serious "criminal" to most persons. They move progressively from participation in a youth gang, where they engage mainly in theft, to adult criminal behavior of a more serious and more frequent type, chiefly burglary and robbery. Their experience with crime, however, does not lead them to commit only one kind of offense since specialization in certain offenses is rare among conventional offenders, although property offenders tend to commit subsequent property offenses and violent offenders to commit subsequent violent offenses (Kempf, 1987). Conventional offenders continuously acquire more sophisticated criminal techniques and develop rationalizations to explain their crimes as they move from petty to more serious offenses. During this progression, they have many contacts with official agencies, such as the police, courts, juvenile authorities, institutions, reformatories, and, finally, prison. Their careers continue to progress, but they never become as sophisticated as professional offenders. Their careers in crime usually terminate in middle-age.

The career patterns of conventional offenders can be illustrated in a comparison of persons convicted of armed robbery with a group of other property offenders (Roebuck and Cadwallader, 1961). As juvenile delinquents, the former frequently carried and used weapons and their arrest histories showed an average of 18 arrests. The armed robbers showed early patterns of stealing from their parents, school and on the streets, truancy, street fighting, association with older offenders, and membership in gangs. Compared with other offenders, the armed robbers also had more extensive records of previous acts of violence and based their claims to leadership in gangs on superior strength and skill.

Studies of offenders who have been arrested offer further insights into conventional criminal careers. One such study followed over 4,000 persons who had been arrested for robbery or burglary through the criminal justice system in Washington, D.C. between 1972 and 1975 (Williams and Lucianovic, 1979). These persons differed from other property offenders by being younger, more often male, more often African American, and less likely to be employed than other defendants. The median age of the robbers was 22 and the median age of the burglars was 24. Two-thirds of the offenders had been arrested previously as an adult (juvenile records were not available), and most of these offenders were likely to be recidivists.

Another study examined about 30,000 incarcerated property offenders (persons who had been imprisoned for robbery, burglary, or both) to discern the relationship between their crimes and their participation in conventional occupations. (Holzman, 1982). It was discovered that about 80 percent of the offenders had been employed prior to the offense for which they had been imprisoned, and, of these, 95 percent had been employed full-time. This suggests that much repeated property crime is committed by persons who hold legitimate occupations. Even keeping in mind that institutionalized offenders are not representative of all offenders, it may be that robbers and burglars who are institutionalized are those who are near the end of their criminal careers and making the transition to conventional society. In any case, the stereotype of the conventional criminal as unemployed or, as reported above in the study in the Washington, D.C. criminal justice system, as less likely to be employed than other offenders, may require revision.

Gibbons distinguishes between conventional and professional property offenders but suggests that the difference is only a matter of degree (Gibbons, 1965: 102–106). Professional "heavy" criminals (that is, those who commit crimes with the threat and availability of violence) are more sophisticated than the semi-professional property offender. These latter offenders tend to be relatively unskilled in committing crimes, they do not make much money as a result of their crimes, and they work at crime on a part-time basis, being employed at least to some extent at other, legitimate, occupations. Semi-professional offenders tend to develop a lengthy arrest record and a record of imprisonments for their crimes.

Self-Conception of Conventional Offenders

As conventional offenders continue to associate with other youth of similar backgrounds in juvenile gangs, and as they progress in their offense careers, they also develop a criminal self-conception. As they identify more and more with crime, their criminal self-concepts become more concrete. Offenders who pursue criminal activities sporadically tend to vacillate in the conceptions they have for themselves, but for those who commit offenses regularly and are continuously isolated from law-abiding society, a criminal self-conception is almost inescapable. They are also more and more severely dealt with by the law, further cementing their regard of themselves as criminals.

Those offenders who maintain conventional occupations are better able to maintain a noncriminal self-concept, although it is difficult for these criminals to reconcile their legitimate and illegitimate income. Persons who commit crimes as a form of "moonlighting" do acquire a substantial record by the time they are in their mid-twenties and, in the study reported earlier, the average number of incarcerations—not convictions—was four and the average age of the offenders studied was about 27 (Holzman, 1982: 1791). It is hard to maintain a noncriminal self-concept under those circumstances, although some moonlighters undoubtedly do just that.

Societal Reaction to Conventional Offenders

The severe punishments given those persons who have committed robberies and burglaries indicate the strong societal reaction to conventional offenders. Such penalties in part represent society's desire to protect property and to punish harshly whenever violence is used to obtain it. They also represent a difference in orientation toward this type of lower-class crime as compared with occupational and corporate crime. The processing of this type of offender often results in a long arrest record and a series of incarcerations for this type of crime. The offenses these persons commit, often with little skill, may lead to a considerable risk of apprehension and a high risk of conviction and incarceration in prisons. As a result, studies of conventional offenders usually indicate that the chances of recidivism are quite high for this type of offender (see Vera Institute of Justice, 1981). In spite of this, many spend much time in correctional institutions, and since society generally holds the fact of imprisonment against a person as much as what the person did, these offenders are among the most stigmatized.

ABUSES OF POWER: OCCUPATIONAL AND ORGANIZATIONAL CRIME

Some persons commit crimes in connection with their occupations. These offenses occur in all types of occupations and they take many different forms (Geis, 1984). *Occupational* crime is a term to refer to offenses committed by persons in the course of their occupation (Sutherland, 1949 and 1983). When offenses involve the entire firm or industry, the term *organizational*, or corporate crime is used.

Occupational Crime

Crime is extensive among persons in many occupations and in business. *Repairpersons*, for example, who service automobiles, television sets, and other appliances make many fraudulent repairs (Vaughan and Carlo, 1975). Auto mechanics may fail to replace parts, yet charge customers for them, or simply paint old parts to make them look new. Occupational crime may include income tax violations, illegal financial manipulations such as embez-

Common Terms for Criminal Abuses of Power

Type	Short Definition
Occupational Crime	A crime committed in the context of one's occupation.
Corporate Crime	A crime committed by an organization in the course of its regular business.
White-Collar Crime	A crime committed during the course of middle- or upper-status occupations.

Sources: Edwin H. Sutherland, *White Collar Crime* (New York: Dryden, 1949) and Richard Quinney, "The Study of White Collar Crime: Toward a Reorientation in Theory and Practice." *Journal of Criminal Law, Criminology, and Police Science*, 55 (1964): 208–214.

zlement and various types of fraud, misrepresentation in advertising, expense account misuse, and bribery of public officials, to name only a few examples. In the investment business, fraudulent securities are sold, asset statements are misrepresented, and customer assets may be used illegally by brokers. Stock brokers themselves may engage in "insider trading" where they gain illegal profit based upon information they have concerning possible mergers or corporate takeovers before that information is made public. A series of insider trading scandals took place in the late 1980s and several prominent brokers and investment bankers were incarcerated as a result. In each instance, the offender had used a position of trust to commit the crime (Shapiro, 1990).

A relatively recent type of occupational criminal behavior has developed in the area of computer technology, and usually it involves offenses connected with businesses. Because more and more businesses and individuals are depending on computers in everyday life, we have only recently become aware of a number of ethical issues surrounding their use (Ermann, Williams, and Gutierrez, 1990). It is increasingly clear that computers can be a "weapon" with which to commit a crime as well as the object of criminality. Computers can be used to perform unauthorized functions, such as monetary transactions and transfers. In a large bank, for example, it might be possible to reduce all account balances by a very small amount (such as 1 percent) and place the funds into another new account. Depending on the balances, this could result in a great deal of money (Tien, Rich, and Cahn, 1986). In other computer crimes, unauthorized information could be obtained from computers, or information could be changed, such as account balances in a bank or university grades in the registrar's office. Attempts to prevent this type of crime include new code systems to prevent unauthorized access to computer files. Some states have also written legislation to deal with such crimes in the criminal justice system, although traditional police methods are not adequate to deal with this crime. The police are not sufficiently well trained to detect or solve these kinds of crime. Computer crimes are often brought to light by "whistle-blowers" (offenders who decide to turn in the other offenders) or persons with technical training who notice that things are not as they should be.

Politicians and government officials commit crimes for personal gain, including personal misappropriation of public funds or the illegal acquisition of these funds through padded payrolls, through the illegal placement of relatives on government payrolls, or through monetary "kickbacks" from appointees. Even local government officials may violate the law, as in a recent Idaho case where city officials in a small town had manipulated utility bills to save themselves thousands of dollars in charges. *Labor union officials* may engage in such criminal activities as misappropriation of union funds, defiance of government orders to enforce rules within the union, collusion with employers to the disadvantage of their own union members, and the use of fraudulent means to maintain control over unions.

Certain activities in the *medical profession* are not only unethical but illegal, including giving illegal prescriptions for narcotics, performing illegal abortions, making fraudulent reports such as in Medicare payments and giving false testimony in accident cases, and fee splitting. Fee splitting, in which a doctor shares or splits the fee charged with the referring physician, is against the law in many states because of the danger that such referrals will be based on the size of the fee rather than on the proficiency of the practitioner. Patients may also suffer from unnecessary surgery performed by doctors (Lanza-Kaduce, 1980), and from increased medical bills when they are charged in excess of what a given service is normally worth. Individual physicians may also bill parties other than patients for services that were not rendered or for services that are billed more than once (Pontell, et al., 1982). No less crucial are cases of *pharmacists* who violate federal and state regulations pertaining to the content of prescriptions and the control of narcotics. In one study, it was concluded that deviance is tolerated and even condoned by the "norm groups" because of the nature and press of modern pharmaceutical practices (Wertheimer and Manasse, 1976).

Other professional groups, such as *lawyers*, may violate the law as well. Lawyers may engage in such illegalities as misappropriating funds in receivership, securing perjured testimony from witnesses, and collecting fraudulent damage claims arising from accidents. Such instances, usually called "ambulance chasing," involve attorneys having clients sign a contingent-fee contract entitling the lawyer to a fixed percentage of any settlement out of an accident where the client has been injured (Carlin, 1966). When such cases are discovered, the offender is more apt to the disbarred than prosecuted criminally.

Self-Conception of Occupational Offenders.

A major difference between occupational crime and many other forms of crime lies in the offenders' conception of themselves. Since the offense takes place in connection with a legitimate occupation, the occupational offender generally regards him or herself as a respectable citizen, not a criminal. Because occupational offenders are likely to regard their actions as "lawbreaking" but not criminal, they are similar to those convicted for such crimes as statutory rape, nonsupport, or drunken driving. The higher social standing of most occupational offenders compared to conventional criminals makes it difficult for the general public to conceive of occupational offenders as being involved in "real" criminality; this also influences the noncriminal self-conception of occupational criminals (Coleman, 1989).

One of the strongest factors helping occupational offenders maintain a noncriminal self-concept is the fact that such offenders are otherwise "respectable." This maintenance of a noncriminal self-concept is an essential element in occupational crime. Embezzlement is generally committed by persons in positions of financial trust, such as office employees, bookkeepers, or accountants, who have access to financial records. In a study of embezzlers, Cressey (1971: 30) found the offenders to be trusted persons who stole funds under three conditions: (1) a nonshareable financial problem; (2) knowledge of how to violate the law and the violators' awareness that the financial problem could be resolved by violating a position of financial trust; and (3) suitable rationalizations for the embezzlement of funds to resolve the individual's self-conception as a trusted person. The trust violators defined the situation as noncriminal in nature; for example, they justified their stealing as merely "borrowing" or an act over which they were not completely responsible. While researchers in the pharmaceutical industry may deliberately falsify research findings on the side effects of drugs, they are also likely to feel they are not criminals because they are also involved in the manufacture of other medicine and drugs that clearly save lives and help people (Braithwaite, 1984).

Learning Occupational Crime. Persons may learn to violate the law in any legitimate occupation and from virtually any source. In some instances, bosses can teach their employees about theft by their example or failure to take action when they are aware of violations.

> On one of many occasions, the manager [of the mall book store] went round with a plastic bag which she filled with about 25 romance novels [which] were worth about $2.50 each. These were taken home to her apartment but they were never returned. On another occasion, she filled a cardboard box with hardback books which came to roughly $275 worth. These were being taken home to be given to her family as Christmas gifts. They were never paid for. (Adams, 1989: 32)

The learning of occupational crime involves not only techniques, but rationalizations, norms, and attitudes as well. Violations may be increasingly seen as "normal" in business situations, and expected behavior in other situations. Occupational offenders are part of a larger social group from which they derive social and other support and even encouragement, as is the case with many other forms of crime. What distinguishes occupational offenders and their subcultures from other deviants and their subcultures is the amount of power and social standing held by the occupational offenders by virtue of their other social roles (see Gandossey, 1985). The occupational offender's subculture provides support for legal as well as illegal activities by blurring the line between legal and illegal practices. Violations are further facilitated because of the inherently complex nature of many business transactions and the importance of trust in negotiating business deals (Shapiro, 1990). There is much opportunity to commit occupational crimes in these settings.

A number of factors may isolate persons in various occupations, such as business, from unfavorable definitions of illegal activity (Vaughan, 1982). First, the mass media decry conventional crime while often treating occupational crime, unless it is particularly sensational, much more leniently.

Second, persons in higher status occupations are often shielded from severe criticism by government officials, many of whom either were formerly in business, have accepted contributions from business sources, or associate socially with business people in clubs and other organizations. Finally, business people tend to associate chiefly with one another, both professionally and socially, so that the implications of white-collar crime are not objectively scrutinized. In other words, a combination of learning and opportunity may best explain such acts (Coleman, 1989).

Societal Reaction. Occupational offenders are seldom apprehended, partly because of limited government agency enforcement staffs, and partly because of the complexity of these crimes. When they are used, legal actions most usually include the use of administrative sanctions, such as fines and license withdrawal. Sometimes, control is left to a professional organization, such as the bar associations or medical society; but these bodies seldom employ severe sanctions against their members, and even when they do, there is no assurance that occupational offenders will be seriously stigmatized. One study reported that malpractice suits against physicians had no negative effect on their practices and one physician had his business actually *increase* as a result of the suit. It seemed that other physicians felt sorry for the offenders and increased referrals to them because they thought they would not otherwise get patients (Schwartz and Skolnick, 1964: 111). A study of pharmacist illegalities found that offender pharmacists were not socially rejected by other pharmacists (Wertheimer and Manasse, 1976).

This is not to say that all occupational offenders are immune from legal action. Some white-collar criminals are convicted in criminal court and are incarcerated for their offenses. It is often claimed, however, that in general even if occupational offenders are apprehended, they are given preferential treatment at the time of sentencing. Two studies have explored this issue, examining a number of white-collar crime cases in federal criminal courts. Nagel and Hagan (1982) were unable to find any evidence of sentencing disparity among the cases of 6,518 offenders sentenced between 1974 and 1977. Wheeler (et al., 1982) not only found little evidence for sentencing disparity, but actually found that the severity of sentence increased with the offender's social status, with higher status offenders receiving more severe sentences. These findings may reflect an increased awareness of occupational crime and its harmful consequences on society.

Corporate Criminal Behavior

Although the law treats corporations as tangible "persons," corporate behavior can not be explained within the same framework of occupational crime. The immense size, the diffusion of responsibility, and the hierarchical structure of large corporations all foster conditions conducive to organizational deviance. For these reasons, corporate crime must be viewed as organizational behavior designed to achieve organizational goals (Vaughan, 1982). Large corporations have a complicated organizational structure. It consists, on the one hand, of those with a great deal of power to make decisions; the board of directors, the top executives like the chairperson of the board, the president, the chief executive officer (CEO), and the vice-presidents. On the other hand, there are those with much less power: the middle managers, supervisors, and the workers. As a legal entity, a corporation uses capital provided by the shareholders or stockholders who, technically, own the corporation. It is the top managers, however, who largely control the corporation.

Large corporations are huge economic conglomerates with assets often totaling in the billions of dollars. In 1988, the annual combined sales of the largest U.S. corporations, the *Fortune 500*, totaled nearly $2 trillion, with profits of $115 billion. General Motors' sales of $121 billion made it the world's largest corporation. Along with the large corporations' greatly increased productive power, an equally significant potential for social harm and a lack of social responsibility has evolved. This potential is made all the more real, given the tendency for corporations to enter into relationships with one another to ensure continued access to resources, such as customers or raw materials (Mizruchi, 1987).

The Nature of Corporate Violations. Corporate violations are extensive. Over one two-year period, the federal government charged nearly two-thirds of the *Fortune 500* corporations with law violations, half of them of a serious nature (Clinard and Yeager, 1980: 118). Violations are not confined only to the largest corporations; corporations of virtually any size can commit many of the kinds of violations that constitute corporate criminality. Estimates of the cost of crimes range as high as $200 billion a year as compared to the estimated annual losses from street crimes of $3 or $4 billion. The cost of a single incident of corporate crime may run into the millions or even billions of dollars, as in Exxon's $2 billion illegal gasoline overcharges during 1974–1981. These violations include restraint of trade (price-fixing and monopoly control); fraudulent sales; illegal financial manipulation; misrepresentation in advertising; issuance of fraudulent securities; income tax violations; misuse of patents, trademarks, and copyrights; manufacture of unsafe foods and drugs; illegal rebates; unfair labor practices; maintaining unsafe working conditions; and environmental pollution and political bribery.

Government investigations have shown that numerous corporations and their executives have knowingly concealed the fact that their unsafe products have caused death and injury to tens of thousands of consumers. Among the most notorious have been Firestone's highly defective steel-belted radial "500" tires, Ford's Pinto automobile whose design flaw caused scores of serious injuries and deaths (Cullen, Maakestad, and Cavender, 1984), and the Dalkon Shield, a defective contraceptive device that resulted in hundreds of deaths and injuries to women (Mintz, 1985). Far too often corporate plan workers are injured due to unsafe working conditions or become seriously ill from a work-related disease, such as cancers caused by a chemical agent or lung diseases from asbestos. Many large corporations have shown contempt for laws to protect the environment. So serious is the corporate abuse of the environment that in 1989 a large coalition of environment, religious, and investment groups joined to issue a corporate code of conduct called the "Valdez Principles" after the disastrous Alaskan oil spill caused by the Exxon tanker Valdez.

Explaining Corporate Criminal Behavior. Many corporate objectives are economic, so it is unsurprising that such crimes as antitrust violations are influenced by a deterioration of business conditions and decreased profits (Simpson, 1987). But economic factors alone fail to predict the extent of a firm's illegal behavior (Clinard and Yeager, 1980: 127–132). From an organizational standpoint, corporate unethical practices and law violations can be attributable to the internal structure of a corporation. One might best explain a corporation's law violations as a product of (1) a unethical or illegal corporate culture, or (2) by the unethical or illegal conduct of the corporation's top executives. All corporations eventually build up a distinctive cultural pattern permeated with the firm's position on ethical standards and law obedience. These factors include an emphasis on maintaining a good corporate reputation, attitudes toward corporate expansion and power, its sense of social responsibility, and the degree of its concern for its employees, consumers, and the environment. In pursuing its objectives, the corporation may proceed ethically or unethically, in compliance with the laws or toward violation of them.

Corporations that take primarily the unethical road tend to socialize its members to accept a climate of unethical behavior conducive to criminality. A former Security and Exchange Commission enforcement chief once said that "our largest corporations have trained some of our brightest young people to be dishonest." In a large-scale illegal price fixing conspiracy in the large folding carton industry, a corporate executive testified that "each was introduced to price fixing practices by his superiors as he came to the point in his career when he had price fixing responsibilities" (Clinard and Yeager, 1980: 64–65).

Widespread prevalence of unethical and illegal practices characterizes the milieu of certain industries (Sutherland, 1949: 217–220). This particularly characterizes four industries—auto, oil, pharmaceuticals, and defense (Clinard and Yeager, 1980: 119–122; Clinard, 1990: 21–90; Braithwaite, 1984). The auto industry has for a long time been tarnished by widespread disregard for laws designed to protect the safety of consumers, reduce consumer fraud, and protect the environment. The oil industry has a long history of industry-wide violations that include price fixing, illegal overcharges, illegal campaign contributions, and environmental pollution. Pharmaceutical corporations have frequently been involved in the production and distribution of unsafe medications and drugs. Most of the giant defense corporations have habitually engaged in fraudulent cost overruns, fraudulent charges, and bribing government officials.

The other internal source of corporate misbehavior lies in the role played by top management, particularly that of the chief executive officer (CEO). Admittedly, the complex structural relationships within large corporations sometimes make it difficult to disentangle delegated authority, managerial discretion, and the ultimate responsibility of top management. Top management communicates the corporation's goals to middle managers who, in turn, are responsible for achieving those goals. The pressure on middle management to achieve goals, using either legal or illegal means, can be intense since prestige, promotions, and salary bonuses often rest on the outcome. When violations are subsequently discovered top management can claim they were unaware of, or had no part in, middle management's decisions to break the law. Clearly, top management can set the ethical tone in the corporation. One middle manager furnished an example:

> Ethics comes and goes in a corporation, according to who is in top management. I worked under four corporation presidents and each differed. The first was honest, the next one was a "wheeler-dealer." The third was somewhat better, and the last one was bad. According to their ethical views, varying ethical pressure was put on middle management all the way down. (Clinard, 1990: 172)

Some top executives consider profits to be the only "bottom line." The ultimate test of good and effective management is how profitable the corporation is, not how moral it is. Corporate executives do not consider themselves criminal offenders.

Societal Reaction and Control of Corporate Crime.

Public opinion polls have shown that large segments of the public have grave doubts about the honesty and integrity of major American corporations (Harris, 1987). Yet, the corporations are so large, and the violations so complex and so diffused among great numbers of persons that it is difficult to generate the same degree of public reaction to the crimes committed as for conventional or occupational crimes. This is not to say that the public is indifferent to these crimes. On the contrary, it is clear that the public condemns corporate illegality to a great extent. Corporate crimes that result in physical injury or death are seen as more serious than those involving property loss or damage (Meier and Short, 1982). Still, the nature of these violations often precludes a strong negative public reaction to each offense. Many times, for example, persons do not know that they have been victimized by corporate crimes. A price conspiracy between breakfast food manufacturers may produce a two-cent increase in the price of breakfast cereal, an increase that most consumers would not notice or would be attributed to inflation, not crime. The total cost to consumers could well be in the millions of dollars.

Government regulations have been effective in controlling some corporate violations, but often enforcement of laws is negligible and probably a large proportion of all violations go undetected because of a lack of adequate enforcement staff and prosecutorial zeal. It also appears not to be feasible to rely upon corporate self-regulation, such as corporate or industry codes of ethical conduct, to control corporate misconduct (Mathews, 1988). Such codes have little or no possibility of being followed by severe enforcement measures. There is reason to believe, however, that aggressive enforcement activity might be a deterrent to corporate criminality (Braithwaite and Geis, 1982). Corporate offenses are often the result of planning and other rational processes rather than emotional, spontaneous acts. As such, the effects of legal sanctions might be to prevent such crimes.

As a result of consumer pressures, stronger actions in cases of corporate violations (i.e., court decisions making corporate officers personally liable, monetary penalties heavier), in the future seem certain. Corporate fines and other monetary penalties are generally small in terms of the immense corporate assets and sales. Proposals have been made for making corporate fines much larger by making them exceed illegal profits or making the fines correspond to the degree of harm to the public. The use of the criminal sanction with respect to those convicted of the insider trading scandals of the 1980s is an unmistakable signal to occupational violators. Other sanctions might also be effective in reducing corporate criminality, such as the use of adverse publicity (Fisse and Braithwaite, 1983). Corporations fear widespread adverse publicity for their illegal conduct, and court decisions that include a provision for publicizing the violation might have a significant impact. In 1990, for example, that proposal was made to the California legislature. Such innovative sanctions, along with a greater use of criminal sanctions when appropriate would, some think, do much to help control this kind of criminality.

POLITICAL CRIMINAL OFFENDERS

Political crimes can be divided into two types: crimes against the government and crimes by the government. Each of these types involves different offenses and the societal reaction to each differs.

Crimes Against Government

Crimes against the government include attempts to protest, to express beliefs about or alter in some way the existing social and political structure. Persons who engage in these acts may be employed by a foreign government, or they may be expressing some deeply-felt personal political conviction. Examples of these acts would be treason, sedition, sabotage, assassination, hijacking, violation of military draft laws, civil rights violations, and actions that arise from a perceived conflict of state and religious tenets. Most such acts are motivated by a desire to improve the world or the existing political system, and political motivation is considered by some to be one of the main definitional criteria of political criminals (Minor, 1975).

Characteristics like age, sex, ethnicity, and social class do not differentiate most political offenders as a whole from the general population. Despite this, Turk (1982) has found that persons who express some dissent with the government and push for change are more likely to derive from middle-class backgrounds and to be more politically conscious. In most cases, political offenders do not conceive of themselves as criminals and, in fact, they may contend that their actions are in no way criminal. For example, draft resisters during the Vietnam war frequently saw their behavior as more "moral" than the government who labeled them as criminals.

Political offenders may perceive themselves as political revolutionaries. Their goals are usually ideological rather than personal. The political offender is usually committed to some form of political and social order, although it is not usually the existing one. Schafer (1974: 146) has referred to the political offender as a "convictional" criminal "because he is convinced of the truth and justification of his own beliefs, and this conviction in him is strong enough to cause him to give up egoistic aspirations as well as peaceful efforts to attain his altruistic goals." Such persons are more likely to recognize that their actions are illegal, although they may not feel bound to the laws that define those acts as illegal.

Persons who resist a political system, whether through dissent, evasion, disobedience, or violence, are often recruited from the more politically sensitive groups in the community. Turk (1982: Chapter 3) suggests that this is why both upper- and lower-class persons may be political resisters and why upper-class persons may be even more likely to become political resisters. Stereotypes about conventional or street criminals simply do not apply to political criminals. In a general sense, however, just as many conventional offenders come to learn norms that permit or condone criminal activity, the political offender may learn political values that permit political crimes.

Political criminals may receive group support from others who share the views of the political criminal. Such support may come from sympathetic political groups or simply from interested individuals. Sometimes these offenders also receive support, both social and material, from other segments of society who are not as committed to overt social action. Such groups serve the same purpose as subcultures do for other kinds of deviants: they lessen the stigma from the outside society by offering solidarity and means to interact with persons of similar ideas, and they facilitate other contacts in a social network committed to political resistance.

Crimes By Government

Violations of law by governments have been extensive in many countries over a long period of time. Just in the United States alone, corruption during the presidencies of Eisenhower and Johnson affected their closest advisors, both of whom were required to resign. In the 1970s, violations by President Nixon and his associates resulted in the imprisonment of 25 high-ranking officials, including the attorney general and two top presidential aides (Jaworski, 1977; Douglas, 1974). The President himself was spared from the possibility of criminal action by a presidential pardon. The violations of law by these men included obstruction of justice, conspiracy to obstruct justice, perjury, accepting contributions or bribes from corporations, bribing persons to prevent testimony, illegal tactics or "dirty tricks" in conducting election campaigns, and misuse for personal purposes of the FBI, the CIA, and the Internal Revenue Service.

In the summer of 1987, the nation witnessed a similar set of hearings on television concerning President Reagan's authorization to send arms illegally to Iran in exchange for political hostages. High-placed officials in the administration were actually involved in two operations, one that would exchange arms for hostages, the other which would funnel the profits from the arms sales to aid political guerillas in Nicaragua. While the former operation may have been a misjudgment (no hostages were ever returned in exchange for arms), the sending of money to the Nicaraguan rebels (called "Contras") was expressly forbidden at the time by the U.S. Congress.

Other kinds of crimes by the government can be found at a local level of government. Police officers may be involved in misconduct and brutality, a well as corruption, illegal use of force, harassment, and illegal entry and seizures. Other public officials may engage in deliberate failure to perform their duties or take bribes to perform an action other than that prescribed by law. In other countries, there has been extensive use of imprisonment, torture, and murder by governments of some countries, such as Uganda, South Africa, Chile, Argentina, and El Salvador (see Amaya, et al., 1987).

Violations of law can, and do, occur at all levels of government from local police and prosecutors to the federal authorities. Such violations may also include deliberate neglect of duties and abuse of privileges. Agents of regulatory agencies, such as building inspectors, may permit contractors to build without necessary permits if the contractors pay the inspectors money (Knapp Commission, 1977). Politicians may do "favors" for persons who supply substantial amounts of campaign money. Police officials may suppress or manufacture evidence to secure a conviction, and prosecutors may engage in conduct that is prohibited by procedural law.

Societal Reaction to Political Offenders

The degree of societal reaction to political offenses against government depends on the extent to which the government's authority is accepted.

Public officials usually do not recognize the moral character of political offenders; rather they react severely toward such offenders because they threaten the existing political structure of which they are a part. Official reaction may, therefore, be severe.

Offenses committed by the government, on the other hand, are seldom reacted to strongly unless the situation becomes particularly flagrant, as in the Watergate scandal, or when there is a change in government. Individual acts of politicians and government officials are more frequently punished, but not with the severity accorded conventional criminal offenders.

ORGANIZED CRIME AND CRIMINALS

Members of organized criminal syndicates earn their living from a number of criminal activities including prostitution, pornographic establishments and films, usurious financial loans, illegal gambling, illegal narcotics, racketeering, and the sale of stolen goods to voluntary customers. Some of the profits from these illegal operations are also used to fund legitimate business concerns (Ianni and Reuss-Ianni, 1983). It is the relationship of illegitimate businesses to legitimate businesses, as well as the public demands for the services provided and the relationship of syndicates with local political structures, that make organized crime and the organized criminal an incredibly complex and difficult problem.

Some observers claim there is a feudal basis to organized crime that appears to rest on the "family" as associated with the terms "Mafia" or "Cosa Nostra." The general view of most criminologists, and crime control specialists, is that there exists about 24 "families" of Cosa Nostra operating in large cities and directed by a council which governs the activities of these families (Cressey, 1969). This council divides operational territories and settles jurisdictional disputes among the families. Most cities have only one such family, if any at all, which may contain as many as 700 persons, but New York City is reported to have five. The wealthiest and most influential families are said to reside in New York, New Jersey, Illinois, Florida, Louisiana, Nevada, Michigan, and Rhode Island, and there are syndicates in other places as well (Rowan, 1986). The claims for the existence of such groups and these organizations are questioned by some who feel that the belief in the Mafia in the United States is the result of a combination of ulterior motives and sensational reporting (Smith, 1975). Others have stated that the existence of a national organization of criminal syndicates has never been demonstrated (Reuter, 1983; Morris and Hawkins, 1971).

Because there are multiple conceptions of the organizational structure of syndicated crime (see Abadinsky, 1981), working definitions that avoid reference to that structure are preferred. Rhodes' definition is illustrative:

> Organized crime consists of a series of illegal transactions between
> multiple offenders, some of whom employ specialized skills, over a continuous

An Organized Crime Family

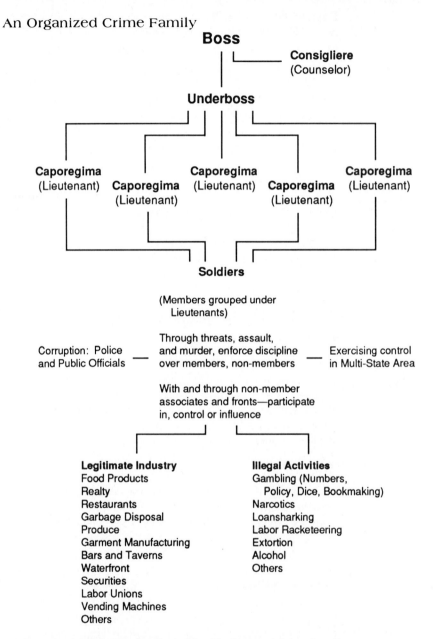

Boss

Consigliere
(Counselor)

Underboss

Caporegima
(Lieutenant)

Caporegima
(Lieutenant)

Caporegima
(Lieutenant)

Caporegima
(Lieutenant)

Caporegima
(Lieutenant)

Soldiers

(Members grouped under
Lieutenants)

Corruption: Police
and Public Officials

Through threats, assault,
and murder, enforce discipline
over members, non-members

Exercising control
in Multi-State Area

With and through non-member
associates and fronts—participate
in, control or influence

Legitimate Industry
Food Products
Realty
Restaurants
Garbage Disposal
Produce
Garment Manufacturing
Bars and Taverns
Waterfront
Securities
Labor Unions
Vending Machines
Others

Illegal Activities
Gambling (Numbers,
    Policy, Dice, Bookmaking)
Narcotics
Loansharking
Labor Racketeering
Extortion
Alcohol
Others

An Organized Crime Family. From *The Challenge of Crime in a Free
Society*, Report of the President's Commission on Law Enforcement and Administration of Justice (Washington, D.C.: Government Printing Office, 1967), p. 194.

period of time, for purposes of economic advantage, and political power when necessary to gain economic advantage. (Rhodes, 1984: 4)

The structure of organized crime might best be seen as an organization that facilitates business dealings among members (Haller, 1990). Much like the Rotary Club serves the economic interests of legitimate business people by facilitating the building of business contacts, so too do illicit enterprises like organized criminal syndicates. These syndicates also provide the means to settle disputes among members when there are difficulties in these illicit dealings, regardless of whether the issue concerns bootleg liquor, illegal gambling, or prostitution.

In addition to facilitating the commission of illegal acts, syndicates are also organized to help keep members out of legal entanglements. Connections with political machines or with branches of the legal system, such as the police or courts, insure some immunity from arrest, or, if there is an arrest, make possible a "fix" to have the charges dropped. Syndicates maintain close relationship with members of conventional society through politicians, police officers, and civil servants by making direct payoffs or delivering votes, honest or fraudulent. One study of the relationship between an organized criminal syndicate and city officials indicates that organized crime probably could not flourish without the active support of those in legitimate power in city government or public affairs (Chambliss, 1978).

Organized Criminals

There has been practically no first-hand research on characteristics of organized criminals, and the large-scale organization of criminal syndicates makes generalizations about backgrounds of members difficult (Cohen, 1977). Most syndicate members came originally, however, from inner-city areas and most have had a record of youth crime (Ianni, 1975; Anderson, 1965). Many of their histories resemble a conventional criminal career in which there is a progression in a long series of delinquencies and crime, as well as association with a tough gang of young offenders. Instead of ending their criminal careers as delinquents, offenders have continued their activities in association with some organized criminal syndicate. This is largely a matter of opportunity and not all youthful offenders who wish to continue are able to do so.

Organized Criminal Activities

For the most part, seven areas predominate in organized crime: (1) illegal gambling; (2) racketeering; (3) illegal drugs; (4) usury or loan sharking; (5) illicit sex; (6) sale of stolen or hijacked goods; and (7) control of legitimate business. While some syndicates may be involved in related illegal activities, these appear to dominate the interests of syndicates. One estimate placed the annual gross income from the activities of organized crime to be in excess of $50 billion, a figure that is greater than all U.S. steel, iron, copper, and aluminum manufacturing combined—or about 1.1 percent of the GNP (Rowan, 1986).

Illegal Gambling.   The lucrative returns from illegal gambling make it most attractive to organized crime. In most cities, there are illegal book-makers associated with crime syndicates who specialize in bets on sporting events, such as horse races and football, basketball, and baseball games. Formerly, nationwide betting was concentrated on professional events only, but college football and basketball have also drawn a great deal of attention. "Numbers" is a very popular form of gambling on the east coast in partic-ular. It involves placing a bet on a sequence of three numbers, such as the last three digits of the daily U.S. Treasury balance (Light, 1977). A com-plicated organization is required to secure the bets, record them, and pay off the winners. Numbers runners (who record the bets and collect the money from betters) are usually small-time criminals who are part of the larger syndicate. Generally they receive between 15 and 25 percent of all bets they take in, from a third to a half of the losses of a new customer, and a 10 percent tip from the customer who wins (Plate, 1975: 75).

Racketeering.   A major source of income for criminal syndicates is from racketeering, the systematic extortion of money from persons and orga-nizations to force some service on them, to obtain funds, or to maintain control (Block and Chambliss, 1981: Chapter 4). Most racketeering is con-centrated in organizations that deal with services or commodities, such as the wholesaling of perishable goods and the laundry and cleaning businesses. Concerns are forced to pay tribute to "protect" themselves from violence and damage. Businesses that deal in cash (such as vending machines and pinball and video games) are favorite targets since the extortion payments can not be as easily traced. Racketeering has been particularly prevalent in the motion-picture business, the building trades, liquor stores, laundry and cleaning establishments, and the waterfront, trucking and loading business (Ianni and Reuss-Ianni, 1983).

Illegal Drugs.   A major role of organized crime is in importing and distributing illegal drugs. This topic will be discussed in more detail in Chapter 8.

Usury.   Usury, or loan sharking, is an important source of organized crime profits. Money is made available at rates far above the legal limits to persons who are in desperate need of cash but who have neither the collateral nor the financial reputation to secure it through legitimate sources.

Illicit Sex. Organized crime plays a part in shakedowns in prostitution, in the ownership of topless and bottomless bars, and in the distribution of pornographic literature and films. Recent concern over pornography has opened the question of the extent to which the production and distribution of pornographic materials is controlled by criminal syndicates (Reuter, 1983: 95–96; Attorney General's Commission on Pornography, 1986).

Sale of Stolen or Hijacked Goods. Organized criminals have engaged in the sale, often in interstate commerce and sometimes the actual theft, of valuable goods. Some of these goods are stolen from airports and similar loading or storage facilities. More recently, they have engaged in the sale of stolen credit cards (Tyler, 1981).

Legitimate Business. In addition to the control of illegal activities, organized crime has infiltrated legitimate businesses. This has been accomplished through illegal means or the investment of large sums of money from illegitimate sources. Two recent accounts have identified the relationship between legitimate sources of income and illegitimate sources for criminal syndicates. Ianni (1972) reports that the "Lupollo" family (the fictitious name for the family he studied) operated several legitimate businesses, including a realty company, food processing companies, an ice-and-coal delivery business, and garbage disposal enterprises. Anderson (1979) reports that the "Benguerra" family (the fictitious name for the family she studied) were involved in 144 legitimate businesses, including eating and drinking places, retail trade businesses, firms for manufacturing and processing food, construction and building services, casinos and travel services, and vending machines. While legitimate businesses may be used as fronts, they also provide major sources of revenue.

Societal Reaction to Organized Crime

Although societal reaction is strong against organized crime, in actuality the public feels ambivalent about organized crime (Morash, 1984). Many persons are intrigued by the "Mafia" and their purported power. Organized crime has been the object of relatively romantic treatments in several films and often in other media where organized criminals are presented as being sympathetic and even kindly persons. In any case, most observers feel that in spite of strong legal measures, organized crime has not been controlled very much. Many reasons may be cited for this failure: (1) The very nature of organized crime creates a major problem. As "illegitimate" business, it usually consists of different types of individual crimes rather than the "organized" element being a crime itself.

> It is not against the criminal law for an individual or group of individuals rationally to plan, establish, develop, and administer an organization

designed for the perpetration of crime. Neither is it against the law for a person to participate in such an organization. What is against the law is bet taking, usury, smuggling, selling narcotics and untaxed liquor, extortion, murder, and conspiracy to commit these and other specific crimes. (Cressey, 1969: 299; also see Homer, 1974: 139–168)

(2) The public demands many of the services provided, such as gambling and usury. (3) Many difficulties are encountered in obtaining proof of criminal activities, particularly because of the intimidation of witnesses. (4) The corruption of public officials prevents possible prosecution. (5) Greatly lacking are effective resources to deal with the possible nationwide operations of some crime syndicates. (6) Local, state, and federal organized crime agencies are poorly coordinated. (7) Even the sanctions that are available are ineffectively used, sentences seldom being commensurate with the financial rewards obtained from the illegal activities. (8) Finally, the image of the organized criminal as a great success and even a "hero" to certain slum dwellers makes it difficult to deal with this problem by soliciting community support (Ianni and Reuss-Ianni, 1983).

Unfortunately, organized crime has become an integral part of American society. Organized crime is firmly planted in the values and desires of large segments of the population. It offers goods and services that are demanded by many persons: gambling, illicit sex, loans difficult to obtain, illegal drugs, and, before the repeal of Prohibition, illegal alcohol. Organized crime could not exist without the demand for these goods and services, although many are able to satisfy some desires legally, such as going to Las Vegas or Atlantic City for gambling, and using legal drugs. In addition to satisfying these demands, criminal syndicates have attempted to insure high demands by promoting their services.

The most logical method of control over organized criminal activities is legal in the form of new, more far-reaching statutes and in the form of more severe penalties for conviction of such statutes. Other solutions include the decriminalization of the activities of organized crime. If gambling and the use of presently prohibited drugs were made legal, two possible consequences would include: (1) organized criminals would find these areas not lucrative and get out of the business of supplying these services, thereby depriving them of an important source of revenue; and (2) the state could benefit from increased revenue that would otherwise be going to organized crime. A number of states now offer forms of gambling to its citizens in the form of lotteries, although it is difficult to say whether this has been much competition for the illegal forms of gambling presently available. There is no evidence that legalization of gambling would eliminate the profits to organized crime since the members of criminal syndicates might well shift to another area, such as when they turned to gambling after the repeal of Prohibition.

In any case, it seems clear that the more fruitful approach is to attempt to deal with organized crime rather than organized criminals. The apprehension of individual offenders does little to alter the basic structure of public demands for presently illegal goods and services. And the demand for such goods and services is enormous. One estimate has placed the potential revenue income from the "hidden economy" (a term that denotes economic exchanges, such as for drugs and sex, that are not recorded as "legitimate" through reporting on taxes or other normal accounting forms) as being over $100 billion in 1974, a figure that must be adjusted upward since that time to account for inflation (Simon and Witte, 1982: xiv). Most of this exchange reflects organized crime, and there is no known way to alter consumer preferences for illegal goods.

THE PROFESSIONAL OFFENDER

Of all offenders, "professionals" have the most highly developed criminal careers, social status among criminals, and skills (see generally Roebuck, 1983). Probably more than any other form of deviance, this behavior can best be understood by the application of sociological concepts. They probably have never been very numerous, and there is an indication that their numbers have declined to such an extent that Cressey has referred to them as "old-fashioned criminals" (Cressey, 1972: 45). Various preventive devices have made professional forgery, for example, more difficult, and there is a possibility that it, as well as other types of professional crime, is declining. Professional offenders are characterized by skill (a complex of techniques for committing crimes), status (their positions in the world of crime are high), consensus (they share values, attitudes, and beliefs with other professional criminals), criminal association (they associate primarily with other criminals), and organization (not only do they frequently commit crimes as part of some larger group, but that group provides social support and assistance as well) (Sutherland, 1937).

Skill

Professional criminals as a group engage in a variety of highly specialized crimes, although the type of crime committed does not necessarily denote whether the offender was a professional or an amateur. Both amateurs and professionals can commit such crimes as picking pockets, shoplifting, and forgery. The degree of skill possessed by an offender, however, usually does indicate whether the person is a professional or not. Professional offenders acquire substantial skill in committing one particular crime, such as pickpocketing, shoplifting, confidence games, stealing from hotel rooms, passing forged checks and securities, counterfeiting, and various forms of sneak-thievery from offices, stores, and banks. The specialized skills of the professional check forger are quite different from those of the amateur or naive forger (Lemert, 1958; Inciardi, 1975).

Professional pickpockets, or "class cannons" as they call themselves, ply their trade at airports, race tracks, amusement parks, and other areas frequented by tourists (Maurer, 1964). Working in a "mob" of from three or four to up to ten, each with a specific role to play, they select a "mark" (victim) on the basis of their guesses, as well as the mark's dress and demeanor, as to whether or not the person has money. The actual snatch is quick, sometimes with a jostle or bump, the cannon quickly passes the loot to another member of the mob traveling in the opposite direction as the mark and the cannon who, in turn, passes the loot quickly to another member of the mob. This may continue until the mob leaves the area. Although the amount varies, it has been estimated from their own reports that average incomes of about $25,000 a year are not uncommon for class cannons, a figure, of course, that is not taxed. A hard-working mob will "hit" 10 to 15 marks a day with an average take of $10 to $25. When detected, which is infrequent, pickpockets frequently talk their way out of trouble by offering the victim's money back, often with a bonus (Inciardi, 1984).

The *professional shoplifter* sells the merchandise rather than using it as does the amateur. Professional shoplifters usually work in a small group, touring the country, usually staying only long enough in a given place to "clout" (shoplift) and dispose of the stolen merchandise. Professionals are skilled in various techniques, and usually use such devices as "booster" bags (with false bottoms), specially prepared coats to receive stolen items, and special boxes that appear to have been purchased elsewhere in which to hide merchandise (Cameron, 1964: 42–50; Meier, 1983).

Confidence games are divided into "short cons" and "big cons." In the former, money is secured illegally from an individual directly and in a brief period of time through the sale, for example, of false jewelry (Roebuck and Johnson, 1963; Gasser, 1963). Sometimes short cons take the form of fixed gambling events, such as card or dice games (Prus and Sharper, 1979). The big con requires more time and involves a larger sum of money since the "mark is put on the send" (victims have to go to another place—a bank or their place of business—to get the money). A great asset is that most victims are also out to violate the law, either in accepting an illegal proposition or in engaging in crime to raise the money for the confidence game. A special kind of con man is the professional *hustler* who makes his living betting against his opponents—often with a backer—in various types of pool or billiard games by hiding his own well-developed skills in those areas. His conning involves an "extraordinary manipulation of other people's impressions of reality and especially of one's self, creating 'false impressions' " (Polsky, 1964: 14; see also Hayno, 1977; Walker, 1981).

Status

The high status of professional criminals is reflected in the attitudes of other criminals and by the special treatment accorded them by the police, court officials, and others. Nonprofessionals tend to look up to the professionals and to aspire to their status. Desire alone, however, is not sufficient. To become a professional thief, one must have the opportunity to become one, to learn the techniques of crime, as well as the attitudes that permit their use. The status of the professional criminal is partly a function of the fact that there are so few of them; as with gold, rarity creates value.

Criminal Association

As in other professions, professional criminals are trained in their occupation. A newcomer learns from other professionals the techniques that have been used for generations to commit crimes (Inciardi, 1975: 5–13; Inciardi, 1984). Persons may be recruited into professional crime in their twenties after they have demonstrated their potential, or they may be recruited after showing some specialized skill, as in the case of professional counterfeiters who recruit an engraver. Initially, the "apprentice" criminal will be given small jobs to improve his or her skills. During this apprenticeship, the novice progresses in skills and criminal attitudes, learning the code of his profession and how to dispose of stolen goods (Klockars, 1974; Steffensmeier, 1986).

Associations with other professional criminals is also important in learning about criminal opportunities and keeping up on news in the profession. Professional criminals will learn, for example, that the "heat is on in Milwaukee" and to avoid such places at least temporarily, or that "Des Moines is a good town at the moment" and to gravitate toward such places. For some crimes, it is important to know which other professional criminals are in town to help in committing certain crimes that require more than one person.

It is through association with other criminals that the specialized language, or argot, of professional crime is learned. The language is not employed to hide anything, or its use in public would attract attention among laypersons. It is handed down from generation to generation; hence, many terms used by professional criminals, like some terms used by doctors, can be traced back several hundred years (Maurer, 1949: 282–283; Sutherland, 1937: 235–243; and Maurer, 1964: 200–216). This gives unity to the group, a sense of history and some measure of pride to professional criminals to be part of something that has such a tradition.

Organization

Professional criminals are generally loosely organized, although this depends somewhat on the nature of the crime they commit. Some criminals such as pickpockets work in groups over a long period of time, although there is usually a turnover in personnel. Others, such as professional "heavy" criminals (robbers and burglars) form groups for specific jobs and then disband (Eisenstadter, 1969; Chambliss, 1972).

Societal Reaction to Professional Crime

Generally, the public is not as aware of professional crime as it is of organized or conventional criminality. Occasionally, a well-publicized case or a popular motion picture, such as *The Sting*, brings this type of crime to the public's attention. But even then, this form of crime is sometimes romanticized and made to seem innocuous. The apprehension of the professional criminal does not mean he or she will be convicted or punished. Professional criminals are often given preferential treatment by the police who, because they often see themselves as craftsmen in the same profession, recognize and respect the professional criminal's ability and skills. Those professionals who are prosecuted may be released if their case is "fixed" through the bribery of a police officer, judge, or prosecuting attorney. Moreover, the fact that such criminals usually do not have a lengthy prior record of previous arrests gives the professional the appearance of being a noncriminal and one to whom a light sentence might be appropriate.

SUMMARY

Criminal behavior is very heterogeneous. It is sometimes committed by groups, sometimes by individual offenders. Some crime involves violence, some does not, such as in instances of theft. Crime can be committed by middle- and upper-status persons in the course of their occupations or by groups of individuals, as corporations, to achieve organizational goals. Criminals differ in the extent to which they identify with crime and other criminals, the degree to which they are committed to crime as a behavior, and the extent to which they progress in the acquisition of more sophisticated criminal norms and techniques. An important theme of this chapter is this diversity.

Occasional property crime is crime that is tied to specific situations. Occasional property offenders do not usually conceive of themselves as criminals, and they are able to rationalize their criminality. These offenders usually uphold the general goals of society and find little support for their behavior in subcultural norms. Most of these offenders do not progress to more serious forms of criminality and the societal and legal reactions is usually mild when the offender has no previous record of crime.

Conventional criminal behavior is sometimes called "street crime." Offenders begin their careers early in life, often in the context of a gang. They frequently vacillate between the values of conventional society and those of a criminal subculture. Some continue their association with other crimes while others leave crime after adolescence. These offenders usually accumulate many arrests and convictions for their crimes and, often, relatively severe legal penalties.

Governments create laws to protect their own interests and existence. Criminal behavior that violates these laws is considered *political criminal behavior*. Specific criminal laws, such as conspiracy laws, as well as traditional laws are made to control and punish those who threaten the state. Political offenders, acting usually out of conscience, do not usually conceive of themselves as criminals and often regard the government they are trying to change as the "real" criminal. Governments can also act illegally through its agents. Crimes by the government usually receive less condemnation than crimes against the government.

Occupational criminal behavior is crime that is tied to a specific occupational context. These offenders do not consider themselves criminals and are able to rationalize their conduct as being part of "normal" occupational duties. Some occupations, or groups within occupations, tolerate or even support the offenses. This may also be the case in particular industries that have high violation rates. Because they are committed by "respectable" people, societal reaction has historically been mild but it appears that the public is becoming less tolerant of this kind of crime.

Corporate crime is collective law breaking in order to advance organizational goals such as the maximization of profits. Law violations are a product of either an illegal or unethical corporate culture or the conduct of corporate top executives. They do not consider themselves to be "criminals" even if they are convicted of breaking the law.

Organized or syndicated criminals pursue crime as a livelihood. At the lower levels of these syndicates, these offenders conceive of themselves as criminals and are isolated from the larger society. At the upper levels, syndicated offenders associate with other members in society, such as politicians and lawyers. Syndicates provide the illegal goods and services that are demanded by legitimate members of society. There is no agreement on whether these syndicates are coordinated at the national level. The public tolerates this form of crime, partly because of the services it provides, and partly because of the complexity of dealing with it.

Professional criminals pursue crime as a means of making a living and as a way of life. They conceive of themselves as criminals and are often proud of their skills and criminal accomplishments. They associate with other criminals and they enjoy high status among other offenders. Professionals and amateurs may commit the same kind of crime, but professionals are more sophisticated. Few professionals accumulate a lengthy criminal record, not only because the offenders are good at what they do and are often successful in eluding police, but also because many cases are "fixed" in the course of legal processing.

SELECTED REFERENCES

Braithwaite, John. 1984. *Corporate Crime in the Pharmaceutical Industry*. London: Routledge and Kegan Paul.

An examination of corporate violations within one industry. The book includes many examples of specific violations and the response to those violations by the law and within the industry. An interesting discussion of the application of the concept of rehabilitation to corporations.

Coleman, James W. 1989. *The Criminal Elite: The Sociology of White Collar Crime,* 2nd edition. New York: St. Martin's Press.

A textbook account of white-collar and corporate crime. It contains a discussion of much of the relevant literature on white-collar crime, as well as a discussion of some of the conditions under which these violations take place.

Morash, Merry. 1984. "Organized Crime." Pp. 191–220 in *Major Forms of Crime.* Edited by Robert F. Meier. Beverly Hills, CA: Sage.

A summary of the literature on syndicated or organized crime. The book discusses the role of law enforcement, both federal and local, in the control of this behavior.

Turk, Austin T. 1982. *Political Criminality: The Defiance and Defense of Authority.* Beverly Hills, CA: Sage Publications.

A valuable discussion of the concept of political crime, its definition, and review of the research on the topic. The book goes beyond a mere review, however, by offering new directions for research and theory in political criminality.

Steffensmeier, Darrell J. 1986. *The Fence: In the Shadow of Two Worlds.* Totowa, NJ: Rowan and Littlefield.

A readable account about a type of property offender that makes possible other kinds of offenses and offenders. Without fences, many property offenders would not be able to dispose of their illegally obtained goods.

Drug Use and Addiction

No terms are more confusing, or misleading, than "drugs," "drug users," or "under the influence of drugs." Although widely used in the media and in public discussion, these terms actually refer to a tremendous variety of often disparate things. They differ in the substance used, the expectations of what effect they will produce, and the immediate environment in which they are taken. The range of substances used is wide: alcohol, nicotine, hallucinogenic drugs such as marijuana, stimulants like cocaine, mind-altering drugs, mood modifiers, and psychoactive narcotics like heroin that affect the central nervous system and influence mood behavior and perception through action on the brain.

Many people regard drug use as one of the major problems facing society today. Yet, those same people would characterize the use of some drugs under certain circumstances as benign and that of others as beneficial. Still other drugs are connected to social and individual situations. Clearly, there are many different reasons for drug use. Drugs are used in medicine to treat disease, to ease pain, or to control emotions; drugs are used for sedation, for social relaxation, and for relief from tensions or boredom; they are used for pleasure, to satisfy curiosity, and to open the mind to new feelings of sensitivity or spirituality; they are used to create a bond of fellowship and to increase sexual performance. For whatever reason drugs are used, it is important to understand the nature of drugs, their use and misuse (and how misuse is judged), as well as the social meanings they have for both those who use them and those who do not.

DRUG TAKING AS DEVIANCE

The deviant nature of drugs is not related to a particular drug or characteristics of particular drugs. The purpose of the drug use may be important. The same drug when used for medical purposes, such as when an opiate is used as a painkiller after surgery, is not considered deviant as opposed to using the drug merely to avoid withdrawal symptoms from addiction. The deviant nature of drug use is not related to the physical properties of the drug. For example, some drugs such as heroin that produce physical dependency are considered deviant, but other drugs that also produce physical dependency, such as caffeine, are not as deviant.

The concept of "drug" refers to something that has a chemical basis, but beyond this generality, there is nothing to distinguish nondeviant from deviant drug use. One definition of drug is "any substance, other than food, that by its chemical or physical nature alters the structure or function in the living organism" (Ray, 1983: 94). But such a definition of a drug would include car exhaust fumes, a bullet, perfume, a cold shower, penicillin, and ammonia. As a result, Goode (1984: 15) argues that "a drug is something that has been arbitrarily defined by certain segments of society as a drug." In fact, the concept of drug is a socially created one as is our attitudes toward drugs. The term "drugs" has certain connotations that are socially determined and usually negative.

Whether the use of a particular drug is deviant depends on norms, which are also socially created. Norms may define the nonmedical use of certain drugs as deviant. The use of opiates by a physician to control pain in a patient is far different from the use of opiates by recreational users who wish to "get high." And, because norms vary, so too do conceptions of what drugs are considered deviant. In India, for example, a country in which there is a strong, religiously associated aversion to alcohol in the higher castes, the use of bhang, a liquid form of marijuana mixed with milk or fruit juice, is not only tolerated but actually prescribed by custom and religious usage. It is often expected that bhang be served at weddings, and if certain priests who use marijuana and opium become addicted to the opium it is not considered particularly reprehensible or unusual.

Norms can change over time. This is why the use of drugs that are not considered deviant at one time may be so considered at another time.

Two important drugs were derived from opium—morphine, a potent drug, in 1804 and heroin, about three times as powerful as morphine, in 1898. These drugs, as well as opium, became widely used in the nineteenth century in the United States, particularly among women who took them in patent medicines for "female disorders." At that time, many of these drugs could be easily and legally purchased. In fact, heroin was first made by pharmaceutical chemists to be sold over the drug store counter as a cough remedy. Cocaine was first isolated in the late 1850s but it did not become popular in the United States until the 1880s when it was proclaimed a wonder drug and sold in wine products as a stimulant (Morgan, 1981: 16).

To determine whether the use of drugs is deviant one must identify the norms that govern the use of that drug and the situations where use is prescribed. It is also important to determine whose norm it is, since the norms of some persons, such as those who abstain completely from drugs, will be different from others, such as those who use drugs occasionally in social situations. These norms, in turn, will be different from those of addicts. Groups differ in their conceptions of what drug use is "deviant." Given such differences, it should be clear that deviant drug use can be "created" by persuading others to adopt a group's norms. This process of promoting one's own norms is reflected in changing public attitudes about different drugs and their use.

Social Attitudes About Drugs

Most attitudes about the deviant nature of drug use were shaped during the twentieth century. Prior to this time, drug use in many forms was widely tolerated in the United States. During the nineteenth century, drug addiction was regarded as a personal problem, and addicts were generally pitied rather than condemned. It was not until later that addicts become regarded as disreputable characters, and that addiction was associated with criminal behavior. In fact, it was only during this century that laws were passed prohibiting the use of drugs such as marijuana, heroin, and cocaine.

Changes in public opinion, and subsequent changes in the legal status of drug use, seem to occur when usage becomes associated with persons or life-styles that are in some sense disvalued. Around the turn of the century, opium smoking was prevalent among certain criminal elements. Marijuana and cocaine use became associated with inner-city life and use was concentrated more in urban areas especially among immigrant groups. Cigarette smoking, once considered acceptable, moved between tolerated and disvalued when it was associated with certain groups. In the 1870s, for example, cigarette smoking was strongly condemned by many groups and individuals because it was most common among urban immigrants who had low social status and who drank heavily (Troyer and Markle, 1983). As was discussed in Chapter 1, the recent changes in legally prohibiting smoking from public places did not come about because of the widespread agreement of the health risks of smoking. Rather, groups of persons have lobbied legislatures and city councils on the basis of individual rights not to breathe the smoke of others. Generally, the drug use that is associated with socially marginal persons is more likely to be considered deviant than the drug use among the well-to-do.

Negative public opinion can result from information about the health hazards of drug taking; such hazards have been known for some time. The ability to document health risks from tobacco does not explain why smoking has become deviant, since creating deviance is a social process. The consumption of certain drugs becomes deviant because individuals and groups are able to define those drugs as deviant.

Two Examples

The process of creating deviant drug taking involves the actions of individuals and groups who believe, for whatever reason, that taking the drug is wrong. Such a judgment, of course, involves a definition of deviance. The specific processes whereby "disreputable" persons come to be associated with "disreputable" drugs and the importance of groups in creating deviance can be illustrated with marijuana and cocaine.

The Case of Marijuana.

The first major piece of legislation that stemmed from these attitudes was the Harrison Act. Passed in 1914, the Harrison Act strictly regulated the sale and use of opiates and cocaine in the United States. This legislation, as well as subsequent statutes, made the selling and use of these drugs, as well as marijuana, illegal without a doctor's prescription (Musto, 1973). Consequences of the legislation included making addicts criminals and recasting drugs as mysterious and evil substances. The Marijuana Tax Act, passed by Congress in 1937, was designed to stamp out its use by criminal law proceedings. This law, which was brought about through the actions of special interest groups (Becker, 1973: 138–139; Galliher and Walker, 1977), influenced the states to pass similar criminal legislation. The Marijuana Tax Act clearly influenced not only public opinion regarding marijuana use, but also the subsequent legislative climate about marijuana use, possession, and sale. Over the next three decades, marijuana came to be seen as a major national problem.

Three interrelated factors had fostered the definition of marijuana as a national problem (President's Commission on Marijuana and Drug Abuse, 1972: 6–8). First, the illegal behavior was highly visible to all segments of society; some users did not attempt to hide their behavior. Second, such drug use was perceived as a threat to personal health and morality as well as to that of the society. Third, it grew out of broader issues affecting changes in the status of youth and wider social conflicts and issues. The symbolic aspects of marijuana were also important factors. Marijuana came to be associated in the 1960s and 1970s with a youth "movement," defiance of established authority, the adoption of new life-styles, the emergence of "street people," campus unrest, general drug use, communal living, and protest politics. In short, marijuana came to symbolize the cleavage between youthful protest and mature conservatism, between the status quo and change.

As such, marijuana itself was not the object of all the legislation, but rather it is the style of life that goes with marijuana smoking. This is also the case with other drugs since the public's attitudes toward users of other drugs, particularly heroin, seems to be informed by a stereotype of what has been termed the "dope fiend" myth, a stereotype that views all addicts as virtually criminal and unproductive. This stereotype has produced a highly negative reaction on the part of most persons to drug addicts, regardless of the drug and other circumstances.

The Case of Cocaine.

Cocaine use has traditionally been reserved for persons of means. The harvest of the coca leaves, the shipment of the leaves through rugged South American jungles and mountains, and the later refinement of the drug to its usual crystalline form all took a great deal of time and effort. Smuggling the drug into the United States had its own risks. Cocaine that made it to consumers was therefore expensive and relatively rare, compared to some other drugs. Cocaine use was found exclusively among wealthy people and use was confined to occasional periods; chronic use appears to have been rare (Grinspoon and Bakalar, 1979).

Changes occurred in the last decade. Better and easier transportation as well as a ready market in the United States provided smugglers with greater incentives than previously to ship cocaine. Freed from slow burro traffic and the dangers of narrow mountain passages, cocaine harvesters could bring more product and smugglers could import more of the drug to the United States. The price began to drop. Patterns of use began to change and cocaine could be obtained by working persons, students, and others. Patterns of cocaine use are changing rapidly, not only in the United States, but also in Europe, South America, and cities in the Far East (Cohen, 1987).

Public concern over cocaine peaked in the mid-1980s. While public attention of marijuana subsided, concern over heroin and cocaine increased. In 1986, for the first time since the Gallup poll on education began (in 1968), the public identified drugs as the biggest single problem confronting schools (*Lewiston* (Idaho) *Tribune*, August 24, 1986, p. 6A). That same year, President Reagan and Vice-President Bush produced urine specimens to prove they were not drug users and they advocated a systematic drug testing program in schools, workplaces, and government. This concern was fueled by the death of two well-known athletes, Don Rogers and Len Bias, as a result of complications from cocaine use. Drug use among the young, always a concern, took on special urgency with the discovery of cheaper form of cocaine, called "crack," that flooded large urban areas.

It really did appear, as President Reagan proclaimed, that there was a "war on drugs." The media publicized the "war" with documentaries and special reports on television, and special issues of news magazines. Publicity about the war on drugs was accompanied by reports and editorials about the need for treatment for users. On September 2, 1986, CBS broadcast "48 Hours on Crack Street," a show that, according to the Nielsen ratings, achieved the highest rating of any news documentary on any network in over five and a half years (Trebach, 1987: 13). Critics charged that the so-called war neglected alcohol, which destroys more lives than all other drugs combined, and nicotine in tobacco which counts more victims than cocaine and heroin (Trebach, 1987). It also appeared that cocaine deaths had actually declined in the year prior to the declaration of war. No matter, cocaine and heroin were the most feared drugs and public attitudes reflected a "get tough" attitude toward those buying or selling them.

Changing attitudes toward the use of various drugs is reflected in national surveys of high school students concerning the risk to users of such drugs. These surveys, sponsored by the National Institute of Drug Abuse (NIDA), indicate that the proportion of high school seniors who perceive some harm from the use of drugs, including marijuana, amphetamines, opiates, barbiturates, LSD, cocaine, and tranquilizers, has increased throughout the 1980s for most types of drugs (Johnston, et al., 1987: 120). Perceptions of harm are not restricted only to illicit drug use, but extend also to such drugs as alcohol and cigarettes as well. Perceptions of harm, however, are tempered in these younger users. For example, only two-thirds of the high school seniors in 1986 reported that regular cigarette use was harmful to the smoker, despite all that is now known about the health risks of tobacco use. Rates of disapproval among high school seniors has also increased during the 1980s for the use of all drugs. The disapproval of cocaine use is strong (80 percent of the high school students say they disapprove of even trying it) but an appreciable minority of students are willing to take the chance of using it.

Public Policy and the War on Drugs

The current concern over drugs is widespread and has taken on many aspects of public hysteria. The present "War on Drugs" has failed to take into account two points having to do with the overall context in which drug taking is found. First, public concern over drugs is faddish. It appears that some drugs are "in" at certain times and "out" at others. That is, public concern about drugs focuses for a short time on a particular drug and then moves on to another, independent of any characteristic of the drug or the result of the public attention. In the 1960s, marijuana was of great concern. In the 1970s, there was much national discussion about methaqualone or "quaaludes" (especially after actor Freddie Prinz's death) and "Angel Dust."

Heroin is always in vogue in drug discussions, but it appeared to be an epidemic in the 1970s. Through the mid-1980s, cocaine and heroin were the most talked about drugs.

Second, drug taking behavior, in general, is so much a part of the behavior patterns of people in the United States that it is inconceivable that all segments of the general public will abstain from all drug use. Drug taking is learned initially by most persons in the context of using legally available drugs. These are obtained readily, George Carlin reminds us, from places called "Drug Stores." Taking these legal drugs is the first time most people learn about the connection between chemical substances and desired end results, such as reducing headaches, increasing bowel regularity, suppressing appetite to help us look thinner, clearing stuffed noses, or keeping us awake for night driving or studying. The use of alcohol and tobacco have a long history in this country and they are often tied into specific social situations. The ultimate roots of the desires to take illegal drugs reside in our desires to use legal drugs.

Drug taking will undoubtedly continue to be widespread through the population. What varies is the kind of drugs that are used. Those drugs that are considered the most deviant are likely to be those most used among less powerful groups in this country, including lower-class individuals, those in socially marginal occupations, students, and those not fully assimilated to the United States.

LEGAL DRUG USE

There are many drugs that are legal and whose use is approved, such as alcohol, tobacco, tranquilizers for relaxation, barbiturates for sleeping, and other minor pain killing drugs, such as aspirin. Most of these drugs are sold "over the counter" (OTC), a phrase that means that a physician's prescription is not necessary for their purchase. It has been estimated that there are over 300,000 different drugs that are available for purchase OTC (Schlaadt and Shannon, 1990: 270). Americans spend more than $6 million a year on such products, and this figure increases about 10 percent a year (Ray, 1983). Much more is spent on caffeine in the form of coffee and nicotine in the form of cigarettes and other tobacco products. Coffee is an important element in international trade (it is number two behind oil), and there are powerful economic lobbies that stand ready to protect trade in coffee and cigarettes. The proportion of persons who report drinking coffee in the United States has declined over the past two decades, as has the percentage of cigarette smokers, but cigarette sales have increased at the same time. There may be fewer smokers, but they seem to be smoking more. Worldwide, about four and one-half trillion cigarettes were sold in 1981. The biggest sellers in the United States are Marlboro, followed in order by Winston, Salem, and Kool. It has been estimated that there were 53.6 million cigarette smokers in the United States in 1990, a decline of 11 percent from the number in 1985 (*Des Moines Register*, December 20, 1990, p. 6A).

In addition to OTC drugs, there are many more drugs that are sold with a doctor's prescription. These drugs are for purposes of helping induce sleep or helping to stay awake, to relax or to stimulate. Regardless of purpose, prescription drugs are usually considered more dangerous in terms of addiction potential than OTC drugs. Prescription drugs may be the result of a response to stress. If so, there are clear sex differences in how stress is handled with drugs. Women are more likely than men to consume prescription drugs, such as tranquilizers (Siegal, 1987: 111). Women are also more likely than men to be consumers of medical services. Men, on the other hand, are more likely than women to use alcohol to cope with stress.

For many persons, legal drug use is a daily event. Many people start their day with a morning cup of coffee and a cigarette; they may take some aspirin for a mid-morning headache; as the day wears on, they may think they require a tranquilizer; cocktails may be consumed before dinner (and maybe at lunch); and, finally, some pills may be taken to get to sleep (or to stay awake) at the end of the day.

Decaffeinated coffee has become popular among some coffee drinkers, although this lowers but does not eliminate caffeine. Similarly, low tar and nicotine cigarettes are popular, but, again, these products reduce but do not eliminate nicotine. A high caffeine cola soft drink, called *Jolt*, was popular on many college campuses in 1987 for late-night studying. Coca Cola, which once contained a small amount of active cocaine, has decocainized its product since 1903 (Poundstone, 1983). But even if the company were successful in removing 99 percent of the drug from its product, which is pretty good, there are millions of cocaine molecules left, just as there are caffeine molecules from decaffeinated coffee. The amount left is so insignificant that it has absolutely no effect on the body, but as long as Coca Cola is made from coca leaves, it can not avoid containing some cocaine.

Over 70 million prescriptions are written each year for minor tranquilizers, such as Librium, Valium, and Verstran. The use of amphetamines, which include all types of stimulant drug products and pep and diet pills, now reaches extensive proportions. The United States may be the most weight- and beauty-conscious society ever to exist and the money spent on diet and beauty aids (books, pills, tapes, exercise equipment, and magazines) rivals only that spent on cosmetics by both men and women. Various appearance norms place some individuals under enormous stress to look a certain way, to have a certain type of body and clothing (Schur, 1983). This kind of stress may result in eating disorders, such as anorexia nervosa and bulimia, and the consumption of amphetamines to manipulate one's body metabolism. Through the 1960s, most large cities contained large numbers of intravenous amphetamine users (Kramer, et al., 1967). Amphetamine users experience no physical withdrawal symptoms from amphetamines, although the body can build up a tolerance that requires higher dosages to achieve the desired effect. Barbiturates, a drug that acts as a depressant on the central nervous system, are sometimes used in connection with amphetamines to achieve weight loss.

Virtually all of the barbiturates and amphetamines in use are manufactured by legitimate drug companies. These drugs are then diverted into the illicit market via hijackings and theft, spurious orders from nonexistent firms, forged prescriptions, and numerous but small-scale diversions from family medicine chests and legitimate prescriptions. The number of pills from such sources can add up rapidly when you consider that there are over 60,000 pharmacies staffed by over 150,000 pharmacists dispensing prescriptions written by over 300,000 physicians (Ray, 1983: 59). Amphetamine and barbiturate use is spurred by media advertisements for these kinds of products. Nearly a billion dollars a year is spent on advertising by drug manufacturers to reach and convince physicians to use their products in making prescriptions. Physicians are constantly bombarded by a barrage of "drug literature"—by direct mail and in medical journal advertisements—that reflects some of the best marketing techniques of Madison Avenue (Seidenberg, 1976). The use of photographs of attractive women (frequently wearing little clothing in the case of dermatology-related drugs), sensational situations, and slogans permeate the advertisements, on the basis of which physicians chose one drug over another. Physicians are subject to such advertising because most physicians, even ones who prescribe drugs, are not fully expert on the many different kinds and side effects (Schlaadt and Shannon, 1990: 326–327).

Physicians are not the only ones subjected to drug advertisements. Drug manufacturers are keenly interested in the public being aware of the over the counter drugs they make. The makers of Anacin have spent $27.5 million a year to persuade people that their product is better than aspirin in spite of the fact that there is only one effective active ingredient in Anacin for treating pain—aspirin (Kaufman, Wolfe, and the Public Action Citizens Health Group, 1983). Millions of Americans turn to nonprescription drugs in the form of tablets, capsules, and syrups to relieve the symptoms of everyday common ailments. These shoppers can choose from an estimated 300,000 "different" products which are made from only 1,000 or so different ingredients. Some persons select drugs that work differently than they may suspect they do. Take, for example, a common cold and a person trying to relieve the symptoms of that cold. He might choose a substance advertised to be that "night time sniffling, sneezing, coughing, aching, stuffed head, fever so-you-can-rest medicine." What such ads neglect to say is that Nyquil costs about $10.60 a fifth, is 50 proof, and is usually taken with a shot glass that comes with the "medicine."

TYPES OF ILLEGAL DRUGS

Drugs can be classified into two categories depending on the drug's general effect on the body—depressants and stimulants. As the names imply, depressants decrease mental and physical activity in varying degrees, depending on the dosage, and stimulants excite and sustain activity and diminish symptoms of fatigue. The most talked-about depressant drugs whose use is considered deviant are morphine, heroin, methadone, and marijuana. *Opiates* account for the greater proportion of drug addiction in the United States. Heroin and morphine, which, together with semisynthetics and synthetics, such as methadone and ineperidine, with qualities similar to real opiates, are the most well known opiates.

The opiates heroin and morphine, white powdered substances derived from opium, are most frequently taken by injection, either subcutaneously or directly into the vein. Heroin and morphine are both highly toxic, or poisonous. Great care must be exercised in their use and they must be diluted before use. Almost immediately after the injection of either drug, the person becomes flushed and experiences a mild itching and tingling sensation. Gradually, the user becomes drowsy and relaxed and enters a state of reverie. Soon this state of euphoria is reached only with higher doses of the drug. Thus, the addict builds up *tolerance* for the drug as well as physical dependence upon it. As this tolerance increases, the addict becomes comparatively immune to the toxic manifestations of the drug. With morphine, for example, the tolerance may be as high as 78 grains in 16 hours, a dosage strong enough to kill 12 or more unaddicted persons. The safe therapeutic dosage of morphine given in hospitals is usually considered to be about one grain in the same period of time.

The heroin or morphine addict becomes physically dependent upon the drug over a varying length of time, usually quite short, the addiction increasing slowly in intensity thereafter. Authorities generally agree that this dependence is more the result of regularity of administration than the amount of the drug or the method of administration. That is, physical dependence is the consequence of regular use. If addicts are receiving their usual daily supply they are not readily recognized as addicts. Even intimate friends and family may not know of the addiction. If addicts do not receive their daily supply, however, clear characteristic symptoms, referred to as *withdrawal distress* or the abstinence syndrome, will appear within approximately 10–12 hours. Addicts may become nervous and restless. They may develop acute stomach cramps, their eyes may water and their nose run. Later, they stop eating; they may vomit frequently, develop diarrhea, lose weight, and suffer muscular pains in the back and legs. The "shakes" may develop during this period and if the addict can not get relief by obtaining drugs, considerable mental and physical distress results. Consequently, an addict will go to almost any lengths to obtain a supply of drugs to relieve the suffering of withdrawal distress. Once the drugs are obtained, the addict will appear normal again within about 30 minutes after administration.

Methadone (dolophine) is a synthetic narcotic analgesic originally developed in Germany during World War II. It is a potent, long-lasting narcotic that comes in pill, injectable liquid, or oral liquid form. It has been

Some Current Drug Jargon or Argot

Crack: A purified, smokeable form of cocaine, usually mixed with baking soda and sometimes other chemicals, and cooked. Crack is usually sold in chunks (about the size of peas) packaged in vials less than one inch long, originally made to hold perfume samples. The drug produces one to five minutes of euphoria marked by hyperactivity.

Pancakes: When the cocaine and baking soda are cooked to make crack, usually in a frying pan, the finished product comes out resembling a pale, thick pancake, which is then broken into small chunks to be sold.

Nickels: $5 worth of narcotics, sold in bags, or in the case of crack, a three-quarter inch vial.

Trays: $3 worth of narcotics, sold in bags or, in the case of crack, a half-inch vial.

Jumbos: A $10 vial of crack, usually sold in a one-inch vial.

Deck: A $10 bag of heroin.

Bundle: Ten decks equal one bundle of heroin, which sells for about $100.

Stem: Glass pipes of various shapes, with a hole at one end for inhaling crack, which is melted with a propane or butane lighter outside the stem. Some users punch a hole in the side of a juice bottle, where they insert a straw. Then they heat the crack in a piece of foil placed over the bottle opening and inhale the fumes through the straw.

Crank: Amphetamines, also known as speed, a once-popular drug showing signs of resurgence.

Ice: A smokeable form of amphetamine that originated in Asia and spread to Hawaii and the west coast of the U.S. A few puffs of ice can produce a high similar to that of crack but which lasts all day.

Speedball: A mixture of heroin and cocaine, usually injected as a liquid. It brings an instantaneous euphoria like cocaine, without the accompanying irritability, followed later by a calmer heroin high. The same mixture, but using crack cocaine and smoked rather than injected, produces similar effects and is called "parachute."

Spikes: Hypodermic needles used to inject heroin, cocaine, or speedballs.

Skin Poppers: Addicts who inject narcotics into a pinched fold of skin, which gives them a slower high, because it takes a longer time to reach the bloodstream and the brain.

Mainliners: Addicts who inject narcotics directly into their veins, which produces a faster high because the drug goes directly into the bloodstream and reaches the brain faster.

Dust, Angel Dust: An animal tranquilizer known as PCP, which is usually used to lace marijuana. A bag of dust, which sells for about $8, usually contains enough for about two small cigarettes.

Adapted from: *The Daily News*, (Pullman, Washington), September 9, 1989, p. 4B.

used for some years to treat heroin addicts who are put on a methadone-maintenance program with one dose that may last up to a day, as contrasted with about three minimum daily dosages taken by most heroin addicts. The idea behind methadone maintenance is to increase the average dosage until the addict develops enough "cross tolerance" so that the euphoria from other drugs, particularly the less potent heroin, is impossible. Methadone is physically addicting. It does not always block the euphoric effects of other opiates and it has increasingly presented problems as it has been used in place of other drugs, although such problems are less than those caused by chronic consumption of heroin (Kreek, 1979).

Cocaine, the best known stimulant drug, is found as a white powder and most commonly inhaled or "snorted" through the nose. Some users prefer to "freebase" the cocaine by combining it with volatile chemicals. This increases the potency of the drug and permits oral or intravenous injection. Cocaine has traditionally been considered a recreational drug and one that facilitates social interaction. Cocaine produces a euphoria, a sense of intense stimulation, and a sense of psychic and physical well-being accompanied by reduced fatigue. Cocaine in its crystalline structure used to be very expensive. The combination of the high price and the exotic properties attributed to it has contributed to cocaine's street reputation as a high status drug. In recent years, cocaine has become more plentiful in the United States and its price has declined. American use patterns are characterized by infrequent use of small quantities of cocaine (Grinspoon and Bakalar, 1979). Cocaine does not appear to produce physical dependency, but some observers claim that it is nevertheless addicting because of the chemical changes it produces in the brain (Rosecran and Spitz, 1987: 12–13). At present, cocaine use appears to be widespread, involving large numbers of people, although chronic use may be a feature of only some subgroups in society.

Crack is a drug made from cocaine mixed with water and baking soda or ammonia. The result looks like small ball-shaped bits about the size of large peas. In some communities, it may be combined with amphetamines. It is most often smoked in a pipe and makes a crackling sound when ignited. Cocaine can also be combined with amphetamines into a substance similar to crack. While cocaine may cost more than $100 per gram (less than a teaspoon), crack has been sold for as little as $10 each. Because of this, crack is more accessible to children and young users. Crack has been likened to heroin in terms of its physical addiction potential and physical harm to the user.

Barbiturates are sedatives and hypnotics that exert a calming action on the central nervous system. These synthetic drugs, when properly prescribed and taken, have no lasting adverse effect. The patient's system will absorb the drug and render it harmless by liver and kidney action. Eventually, the body will pass whatever residue may be left. If carelessly used, however, barbiturates often lead to psychological dependence and to physiological addiction. In their direct action on the body, they are potentially more harmful and dangerous than opiates, and an overdose may well lead to death because the drug can depress the brain's respiratory control to the point where breathing ceases (Smith, et al., 1979).

Methamphetamine (sometimes called "crank") is a derivative of amphetamine. Made in a laboratory, "meth" is a drug that produces a cocaine-like high. Crank is similar to cocaine in some respects—both are white powders and both are usually snorted or injected—but whereas a cocaine high might last for half an hour, a crank euphoria may last all day. Crank gives users long-term energy and some users might stay awake for days at a time, always feeling full of energy. There is little evidence that crank is a major drug in the United States, although some speculate that if legal efforts to control the importation of cocaine into the United States are successful, there may be more users of crank for simple economics: cocaine will be too expensive and crank can be made locally.

The *hallucinogens* include marijuana and hashish and the "consciousness expanders," such as mescaline and peyote, produced from certain mushrooms, morning glory seeds, and other plants. They also include LSD, a chemical synthetic made largely from lysergic acid. It is not clear exactly in what physiological manner the hallucinogens work; while it is obvious they have a chemical effect on the brain, the process is not known exactly. These drugs are not thought to be addicting or physically habituating, but the sensations they produce, startling and sometimes pleasurable, may lead to repeated usage. The effects of such natural hallucinogens as peyote are not as great as those of LSD if not taken in prolonged dosages, but LSD is a different matter. A tiny amount (1/300,000 of an ounce) of it causes delusions or hallucinations, some pleasant while others are terrifying. It tends to heighten sensory perceptions, often to the point where they are wildly distorted. Although the experience usually lasts from 4 to 12 hours, it may continue for days.

Marijuana (marihuana)—bhang, hashish in stronger cake form, cannabis, or popularly "grass" or "pot"—is derived from the leaves and tender stems of the hemp plant, often known as Indian hemp, and it is usually inhaled by smoking specially prepared cigarettes called "reefers" or "joints." The general technical term for this drug is *cannabis*, an annual herb native to Asia. The usual effect produced by smoking these products is euphoria, an intensification of feelings, and a distorted sense of time and space, all with few unpleasant after-effects. In spite of some controversy about the effects of using marijuana for prolonged periods of time, use is usually not considered physiologically addicting, although it may, to some extent, be psychologically addicting. Research on long-term effects from using marijuana has been hampered by the fact that chronic users of marijuana also frequently use other drugs, making it difficult to identify effects from marijuana alone (Petersen, 1984). It appears that marijuana use affects motor skill performance, for example, in auto driving, but claims that use causes psychotic episodes and bodily and brain damage are disputable.

MARIJUANA USE

Marijuana is the most widely used illicit drug in the United States based on the percentage of the population that has ever used the drug. The manufacture, sale, and use of marijuana is a crime in the United States, although some jurisdictions have reduced the penalties for possession of a small amount of marijuana to a misdemeanor. Marijuana use is considered deviant by many groups, although its use is condoned and even encouraged in other groups. As one marijuana smoker put it: "Even though it is not the norm of society, enough people do it to make it acceptable" (Wilson, 1989: 58).

Extent of Marijuana Use

There is little doubt about the extensive use of marijuana. Estimates of the proportion of the population who have ever used marijuana (prevalence) indicate that a high proportion of all persons have tried marijuana at least once. In 1972, the National Commission on Marijuana and Drug Abuse reported from surveys in the United States that an estimated 24 million persons had tried marijuana. Of this number, 8,300,000 generally used it less than once a week, and there were about 500,000 "heavy" users, those who used it more than once a day. The Commission found that the use of marijuana had tripled in the two and a half years before the report was issued (National Commission on Marijuana and Drug Abuse, 1972).

The National Institute of Drug Abuse annual surveys of high school seniors documented an increase in lifetime prevalence (i.e., the proportion of those who had ever used marijuana) throughout the 1970s. Beginning in 1979, however, the proportion of high school seniors who had ever used marijuana began to decline slightly (Johnston, et al., 1987: 46). By 1979, slightly over 60 percent of all high school seniors reported that they had tried marijuana at least once. By 1986, this figure had declined to 51 percent (Johnston, et al., 1986: 47). For persons in their mid-twenties, the annual prevalence of marijuana (i.e., the proportion of young people not in college who have used marijuana in the last year) is 37 percent (Johnston, et al., 1987: 162). Among college students, the annual prevalence has dropped from 51 percent in 1980 to 41 percent in 1986 (Johnston, et al., 1987: 206). Daily marijuana use among college students fell by more than half from 7.2 percent to 2.1 percent.

The decline in reported daily use among high school seniors is even greater than the decline in the percentage of high school students who have ever used marijuana. The percentage who reported daily use reached a peak of nearly 11 percent in 1978 but had declined to under 4 percent by 1986 (Johnston, et al., 1987: 50). Undoubtedly, part of the reason for this reduced usage is the fact that high school seniors are likely to perceive more risk—physical, medical, legal—to using marijuana than ever before. The percentage who reported they believed that regular use of marijuana posed physical risks to the user increased from 43 percent in 1975 to over 71 percent in 1986 (Johnston, et al., 1987: 120). The survey of the drug usage among the class of 1990 high school graduates showed that marijuana use continued to decline and that most of this decline was due to reduced demand rather than reduced supply (*Des Moines Register*, January 25, 1991, p. 9A). Probably a third of the high-school students in the United States has tried marijuana at least once.

Marijuana use patterns show that it is still confined mainly to younger persons in their twenties or younger. Experimentation and continued use among persons over 35 is rare. Marijuana use is not found to be more prevalent in one social class than another. At the present time, about a quarter of the total American population has used the drug at least once.

Using Marijuana and Group Support

Marijuana is essentially a social drug. Its use is linked to social situations and use is a group activity. There is no marijuana equivalent to the "secret drinker" or solitary opiate addict. One becomes a marijuana user, either on a regular basis or irregular basis, through a learning process, bolstered by subcultural support for its continued use. When one wishes to use marijuana, others who have had experience with the drug must show the initiate how to do it. Becker (1973: 235–242), in a study of how persons come to be marijuana users, found that the new user must learn three things: (1) how to smoke the drug in a way that will produce certain effects; (2) how to recognize these effects and connect the drug with them; and finally, (3) to interpret the sensations as pleasurable. First-time users do not ordinarily "get high" because they do not know the proper technique of drawing the cigarette or pipe and holding in the smoke. Even though there may be pleasurable sensations, the new marijuana users may not feel that the pleasures are enough, or they may not be sufficiently aware of their specific nature to become regular users. Feeling dizzy, being thirsty, misjudging distances, or a tingling scalp may not of themselves be pleasurable experiences. Association with other marijuana users helps to define sensations that may have been annoying into those that are pleasurable and to be looked forward to.

The use of marijuana is essentially a group activity, lending itself to friendships and participation in a group setting. It is smoked in intimate groups and marijuana use may be very functional for the group. Here is one user recounting his initial experiences:

> The people in my group of friends that smoke marijuana began as a result of pressure from peers combined with a general curiosity. When their friends started smoking marijuana, it became "the thing to do" and it made them wonder what it was like. (Jones, 1989: 59–60)

Marijuana can be part of the pattern of social relations in some groups and contribute to the long-term continuing social relations within the group, a certain degree of value consensus within the group, a convergence of values as a result of progressive group involvement, and the maintenance of the circle's cohesive nature, among other things (Goode, 1970). Group contacts with respect to marijuana are needed in order to get a supply, to learn the special technique of smoking to gain maximum effect, and to furnish psychological support for engaging in an illicit activity.

Continued marijuana use also requires group support. Association with others for purposes of marijuana use may also lead to the use of other drugs. As such, it is not the use of marijuana, per se, that can lead to other, possibly more dangerous drugs, but association with and membership in a group that may condone experimentation with drugs. There are, of course, subcultures with other drugs such as heroin, but the nature of these groups is different, just as the drugs themselves are different. A major difference between subcultures for marijuana and those for heroin is that marijuana use is overwhelmingly recreational in nature, but heroin is not. Heroin is more likely to be used alone or with another person in order to share resources, such as money, "works" (or drug paraphernalia), and the like. Marijuana is much less likely to be used alone, but there are some chronic (daily) users who engage in mostly solitary use (Haas and Hendin, 1987). These are older, more experienced users for whom marijuana has become an important part of their lives. As such, their use differs substantially from that usually found.

Marijuana and Heroin

Many people believe that heroin addiction is facilitated by marijuana use; in fact, may people think that marijuana smoking causes people to want to use stronger drugs, like heroin and cocaine. This idea can be called the "gateway theory" where marijuana serves as a gateway to the use of other, more serious drugs. So far, the relation between using marijuana and that use leading to the use of other drugs has not been substantiated. The National Commission on Marijuana and Drug Abuse (1972: 88–89) concluded that the overwhelming majority of marijuana users do not progress to other drugs, either remaining with marijuana or changing to alcohol: "Marijuana use per se does not dictate whether other drugs will be used; nor does it determine the rate of progression, if and when it occurs, or which drugs will be used. The user's social group seems to be the strongest influence of whether other drugs will be used; and, if so, which drugs will be used."

It does seem to be the case that many heroin users had used marijuana prior to their use of heroin, but this in no way establishes a causal connection between taking marijuana and subsequent heroin use. The fact remains that most persons who use marijuana do not use heroin. National surveys of current use have disclosed that about 62 million people have tried marijuana and that there are more than 18 million current users of marijuana compared with less than 6 million current users of cocaine (the number of current heroin users was so small that reliable estimates were not possible) (Trebach,

1987: 83). If marijuana were the "gateway drug" that some claim, many more of the 62 million users would have gone on to these other drugs. ⚹

THE USE OF OPIATES AND ADDICTION

The Meaning of Addiction

The term "addiction" refers to physical dependence, "an adaptive state of the body that is manifested by physical disturbances when drug use stops" (Milby, 1981: 3). As clear as that term seems to be, there seems to be a tendency to apply the term "addiction" to anything that people do repeatedly. This would be a mistake (and the term would lose its meaning) and a distinction must be made between the behavior representing an addiction and behavior that is engaged in for other reasons (Levison, et al., 1983).

Ann Wilson Schaef (1986), who regards an addiction as "any process over which we are powerless," estimates that as many as 96 percent of all Americans either live with addicts, by this definition, or grew up with one affecting their lives. Furthermore, a variety of activities (making money, work, worry, religion, sex), substances (food, nicotine), and relationships (marriage, love affairs) can be addictive. Thus, we can be addicted to our spouses, to a particular food, to school or jobs, and to the color of the paint

Are We All Addicts of Some Kind?

Anne Wilson Schaef believes we are all addicts. Some of us, she says, are chemically dependent, others suffer from "process" addictions which are addictions to things or people rather than drugs. We can become addicted to alcohol and other drugs, but we can become addicted, Schaef says, to things like:

- television
- gambling
- work
- caffeine
- a relationship
- sex
- shopping

Schaef even asserts that Pope John Paul II is a "sex addict" because one doesn't have to act out sexual behavior to qualify, but merely be obsessed with the subject or with controlling other people's sex lives.

Adapted from *Des Moines Register*, October 23, 1990, p. 3T.

in our bedrooms. Surely, with such an expansive definition, the meaning of the term "addiction"—or any other term for that manner—is lost.

The broadening of the meaning of addiction appears to be part of a larger fad that conceives of many behaviors as diseases in the absence of a defensible medical rationale (see Peele, 1990).

> The broadening of the meaning of addiction is part of a growing fad to medicalize many different behaviors. Drink too much? It's a disease. The child of an alcoholic? You have a disease. Overeat or gamble? Both diseases. Sex-obsessed? Definitely a disease. Workaholics, compulsive shoppers, fitness freaks, drug users, whatever the behavior: The growing trend is to call it a disease (Sullivan, 1990: 1T).

In many instances, persons who have been said to have these behavioral "diseases" have participated in a 12-step program, similar to that of Alcoholics Anonymous (see Chapter 9). These programs begin with the assertion that one is powerless over the addiction and one must therefore put faith in a larger power to overcome the addiction. There are many such "anonymous" groups, including not only the well-known Alcoholics Anonymous and Narcotics Anonymous, but such organizations as Sexaholics Anonymous, Emotions Anonymous, Debtors Anonymous, and Workaholics Anonymous. A social worker has noted that, "This rise in AA [style] groups has become a kind of secular religion. It's like witnessing. People come in and say 'I have a disease.' It's similar to people standing up in church and saying 'I am a sinner.'" (quoted in Sullivan, 1990: 1T).

For purposes of discussion here, the following distinctions are necessary. Drugs that result in physiological consequences once the drug is withdrawn from the user are said to be physically dependence-producing, or addicting. An addict is one who experiences distress as a result of not having a drug, whether that drug is alcohol, heroin, or an amphetamine. Persons who fail to experience distress from not taking a drug, such as alcohol or heroin, are said here not to be addicted.

There are, however, differences in addictive behavior that is not captured in that term. As a result, many professionals prefer to talk about "tolerance," "dependence," and "abstinence syndrome." Persons who use certain drugs may become physically dependent upon the drug and abstinence from the drug causes withdrawal distress. Most of these persons have built up their tolerance of the drug so that larger and larger doses are required to produce the desired effect. This is the case with opiate addiction where the addict consumes ever greater amounts to achieve the same effect. When the needed dose, usually in the form of heroin, is not obtained, withdrawal symptoms occur after eight hours. The eyes and nose water, perspiration becomes profuse, the body aches, and nausea and diarrhea may occur.

Patterns of Heroin Use

People have used heroin for thousands of years; in fact, opium use was legal in the United States until early in the twentieth century. One can experience the effects of heroin by inhaling, smoking, or, more commonly, injecting it into the veins. Patterns of heroin use have changed over time. In the Nineteenth Century, for example, about two-thirds of the users, according to early surveys, were women, and addiction in the medical profession was noted (Morgan, 1981). Most observers felt heroin use was less prevalent in the lower as opposed to the middle and upper classes at the time. The average age of addicts was then found to be between 40 and 50, and some investigators believed that addiction was a problem of middle age, since most addicts took up the habit after the age of 30 (Lindesmith and Gagnon, 1964: 164–165).

The first widespread use of heroin in large cities was reported after 1945 (Hunt and Chambers, 1976: 53). Use remained at low levels through the 1950s and then increased rapidly during the 1960s. Levels of use reached peaks in most cities during 1968 and 1969. By 1971, all large cities (over one million) had experienced their peak use, although smaller city use continued to increase during the 1970s. Information on patterns of use throughout the 1980s suggests that the levels of addiction present in the late 1970s may have leveled off with estimates of the number of heroin addicts in the early 1980s being close to those estimated to exist in the late 1970s (Trebach, 1982; Trebach, 1987). Although heroin addiction is not a growing problem, it is still the case that there are more addicts today than two or three decades ago.

Selected Glossary of Terms Used by Addicts

Burn (*n, v.t.*): less heroin than an amount of money should purchase.
Cooker (*n*): small receptacle in which to dissolve heroin.
Dynamite (*adj.*): high-quality heroin.
Fix (*n, v.t.*): inject heroin intravenously.
Flash (*n.*): physical sensation of heroin entering system.
Goofballs (*n.*): barbiturates.
Heart (*n.*): courage.
House Connection (*n.*): dealer who operates in a private home.
Hype (*n.*): heroin addict.
The Life (*n.*): life of a heroin addict.
Panic (*n.*): shortage of heroin.
Ropes (*n.*): veins.
Smack (*n.*): heroin.
Strung Out (*adj.*): physically addicted to heroin.
Tracks (*n.*): scars along a vein resulting from frequent injection.
Works (*n.*): paraphernalia for injecting heroin.

Adapted from Agar, Michael. 1973. *Ripping and Running: A Formal Ethnography of Urban Heroin Addicts*. New York: Seminar Press, pp. 157–166.

The linking of the disease AIDS to the use of dirty heroin needle use has generated a new fear among addicts. The extent of AIDS among addict populations is not now known with precision, but the percentage may be high. Dr. Stephen Joseph, New York City health commissioner, estimated in October 1987 that half of the city's 200,000 intravenous drug users (many of whom are opiate users) have AIDS (Spokane *Spokesman-Review*, October 23, 1987, p. 6). Other heterosexuals who contact AIDS clearly run a very high risk by sexual contact with intravenous drug users. The percentage of addicts who have AIDS in Europe has also been increasing. AIDS victims who are also heterosexual intravenous drug users increased from 6 percent in 1985 to 12 percent in 1986, a 100 percent increase (National Institute of Drug Abuse, 1987: 2). The influence of the AIDS virus on patterns of heroin use is still unknown, but such figures suggest that there may be changes.

While the concern over AIDS may have depressed heroin sales and use, a new, more potent form of heroin that began to be seen on the streets in 1991 poses new problems (*New York Times*, April 28, 1991, p. 4E). For years, a bag of heroin (which is about the size of a pencil erasure ground up) contained less than 10 percent of the drug mixed with adulterants. In recent years, some bags have been found that contain 40 to 45 percent heroin. With a concentration this pure, the drug can be snorted like cocaine, thereby eliminating needles and the risks of sharing needles with other users. The supply of heroin worldwide may have increased in the early 1990s and this has led to a lower selling price, thereby making the drug more available.

Number of Heroin Users

The estimated number of drug addicts in the United States has increased over time. For example, with respect to heroin addiction, in the late 1960s, two experts estimated for 1967 that there were 108,500 addicts in the United States (Ball and Chambers, 1970: 71–73). The Federal Bureau of Narcotics and Dangerous Drugs, however, estimated there were 64,000 addicts in 1968 (Milby, 1981: 75). The differences in these estimates are due to differences in data sources. Estimates in the 1970s, however, represented drastic increases from those estimates of addicts in the 1960s. One estimate put the number of active heroin users in the United States in 1975 at 660,000 (Hunt and Chambers, 1976: 73). More recent estimates suggest that there may be about 500,000 heroin addicts, persons who use heroin every day, with an additional 3.5 million "chippers," or occasional users, in this country (Trebach, 1982: 3–4). The existence of so many persons who are occasional users of heroin suggests that the addiction process is not simply a result of taking heroin. The addiction process involves social factors usually of a group nature.

The United States may lead all countries in the number of addicts. An international survey of research scientists and physicians in 25 countries published in 1977 reported that the largest number of opiate addicts in the general population was the United States with 620,000, followed by Iran with 400,000, Thailand with 350,000, Hong Kong with 80,000, Canada with 18,000, Singapore with 13,000, Australia with 12,500, Italy with 10,000 and the United Kingdom with 6,000 (Trebach, 1982: 6–7). As is usual in these matters, there are differences in the estimates from one country to another regarding the precise number of opiate addicts, but the data is instructive with respect to broad trends.

Who Uses Heroin?

Researchers have identified at least five types of heroin addict roles: (1) "expressive students" who try drugs, including heroin, in attempts to explore the cultural and social world; (2) "social world alternators" who use heroin in connection with exploiting diverse social opportunities in the pursuit of hedonism; (3) "low riders" who use heroin symbolically to express cohesiveness in motorcycle groups; (4) "barrio addicts" who use heroin as a ritual of sociability; and (5) "ghetto hustlers" who use drugs as an aspect of the adventure of "living by one's wits" (Lewis and Glaser, 1974). In addition to these types of addicts, there are also "chippers," or persons who use heroin only infrequently and without developing a physical dependency. Maintaining many of these roles requires involvement in an addict subculture in order to receive social support from others and to guarantee a supply of heroin (Hanson, et al., 1985).

Heroin use among persons under 20 years of age is not common. The NIDA national survey of high school youth in the United States in 1986 found that the percentage of students who reported ever having tried heroin to be less than one percent (Johnston, et al., 1986: 25). This figure represents a 50 percent reduction from the comparable figure in 1975, but the number of users in each case is very small. Figures of reported heroin use among college students and young adults are also very low (i.e., one percent or less).

Addiction to opiates is now heavily concentrated among young, urban, lower-class males from large cities, particularly among African Americans and Puerto Ricans. The relationship between heroin use, low socioeconomic status and minority status has been documented in a number of studies (cf. Redlinger and Michael, 1970). The high heroin addiction rates among African Americans is partially a product of the fact that the concentration of traffic in black areas in many cities makes the drug particularly available there and this facilitates the development of a street addict subculture. The proportion of all addicts who are black is nearly 30 percent, and most of these are concentrated in the Northeastern part of the United States (Chambers and Harter, 1987); roughly 200,000 of the estimated 500,000 regular users of heoin in the United States live in New York City.

Although most addicts are of the street addict variety, certain occupations are known to contain more risks toward heroin use. The medical profession, especially physicians and nurses, also has an excessive number of addicts, as does the entertainment industry (Winick, 1961). Goode (1984: 230) estimated that there were three to four thousand physician narcotic addicts in the United States or about one in every 100 physicians (based on a total of about 350,000 physicians), a figure that is far higher than for the population as a whole. In fact, the level of general drug use among physicians is much higher than among nonphysicians (Vaillant, et al., 1970). Doctors can prescribe drugs, which means they can obtain them easily and rather inexpensively. Moreover, physicians have knowledge of what drugs can do for someone who is tense or tired, which is an important factor in their coming to use drugs. Many of these physicians do not come to the attention of authorities because they can often maintain their addiction without detection.

Many of these physician addicts become users as medical students. The pressures on students are great: the hours are long, and the stress substantial, in part because students work in situations they do not control.

> There is no release from the constant bombardment of the work, no one to off-load it onto. Some want to talk but don't know how. Learning to be emotionally mature to cope with this kind of stress is not part of the medical student's education. Students come up with different ways out. Drugs provide one conscious form of release. As Frank explains, 'It was never a conscious thing on my part. I never said, 'OK enough, I'm going to get snowed.' It was gradual. (Hart, 1989: 80)

Students learn that the use of drugs can help relieve stress and help maintain an even disposition during a hard work day that may last as long as 36 hours. They also learn that drugs are preferable to alcohol because it is necessary to conceal the fact of their use while on the job. As one student pointed out: "If you drink, people can smell it. But if you're loaded they can't tell. You can explain the dilated pupils, the staggering by saying you hadn't slept or you've got the flu" (Hart, 1989: 80).

Performers in the entertainment world, such as jazz musicians, sometimes become marijuana users and may use other drugs, largely because such drug use appears to be much less disapproved by their associates than by the general population. Drugs may be very functional to musicians: it may help a musician over periods of unemployment, long periods of time away from home while "on the road," and, depending on the circumstances, to perform certain kinds of music. A study of jazz band musicians concluded that drugs were a "normal" part of the life of these persons (Winick, 1959–60). Heroin had been used at least once by 53 percent of the musicians. Many of the users attributed drug use to "the road."

> I was traveling on the road in 1952. We had terrible travel arrangements and traveled by special bus. We were so tired and beat that we didn't even have time to brush our teeth when we arrived in a town. We'd get up on the bandstand looking awful. The audience would say, 'Why don't they smile?

They look like they can't smile.' I found I could pep myself up more quickly with heroin than with liquor. If you drank feeling that tired, you'd fall on your face. (Winick, 1959–60: 246)

BECOMING AN OPIATE USER

As has been pointed out, there is more to becoming an opiate addict than simply using drugs. Sociological and social psychological processes in becoming an opiate addict are very important and are reflected in information pertaining to (1) initial use and the length of addiction, (2) types of addicts and the role of addict subcultures, and (3) various theories of opiate addiction.

Opiate addiction is learned just as any other behavior is learned—primarily in association and communication with others who are addicts. Some indication of this is the fact that although drug addiction used to be common among Chinese in the United States, by the late 1960s this group of addicts had almost ceased to exist (Ball and Lau, 1966). The usual pattern is that of association for other reasons rather than one person seeking another simply because the other person is an addict. Opium addiction is taken over by an individual in much the same way as other cultural patterns are transmitted. Addicts must first learn how to use drugs. They must be aware of the drug, know how to administer it, and be able to recognize its effects. Beyond this, there must be some motive for trying the drug—to relieve pain, please someone, achieve acceptance in a group, produce euphoria, for kicks, or achieve some other goal. This goal need have little to do with the narcotic's specific effects.

Most drug addicts are knowingly initiated into drug usage, usually in their teens, by a friend, acquaintance, or marital partner. Only a few of the addicts interviewed by Bennett (1986) reported that they were pressured into taking the heroin. Fewer than 10 percent of the addicts were introduced to the drug by a stranger. Most also reported their decision to try heroin was made some time prior to their actually trying the drug. One said: "I'd heard about heroin for as long as I could remember. I wanted to try it. I knew I'd take it, but I didn't know when. I took it at the first opportunity" (Bennett, 1986: 95). Another said: "Drugs fascinated me from an early age, especially the junkie culture. Let's put it this way: I wasn't worried about becoming an addict, I wanted to become one" (Bennett, 1986: 95).

Rarely does the use of drugs during illness lead to addiction. A large number of heroin users first took the drug out of a curiosity they developed from persons who were already addicted. Others take their initial dose of heroin out of a desire to "try anything once." Some adolescents take drugs for the "kick" as something tabooed by "squares" and to heighten and intensify the present moment of experience and differentiate it from the routine of daily life (Finestone, 1957). The chain reaction process of addiction has often been called "sordid and tragic pyramid game" in which the average addict introduces several friends into the habit, often as a means of solving his own supply problem. Persons are sometimes initiated at parties where the first "shots" are "on the house" (Chein, et al., 1964: Chapter 6). Most frequently, this initial use is done with drugs made available by friends, family members, or acquaintances. Initial use among women seems to be highly influenced by a man, especially a sexual partner, who used heroin daily (Hser, et al., 1987). It is unsurprising that others are involved in the initiation process. Techniques for proper injection and supply must be learned from others, as well as rationalizations for continued use. Social and interpersonal support is also necessary. Far from representing retreatist or withdrawal behavior, the use of narcotics requires the help and support of others. Addicts need one another.

For this reason, there is an unmistakable group nature to heroin addiction. The processes of initial and continued use point to the importance of a social learning interpretation of heroin addiction. Waldorf (1973) found that the initial heroin use among a New York sample of addicts was *not* a solitary activity. "Persons are initiated in a group situation among friends and acquaintances. Only 17 (4 percent) of our sample of 417 males reported that they were alone the first time they used heroin; by far the majority (96 percent) reported that they used heroin the first time with one or more persons" (Waldorf, 1973: 31). The other persons were almost always friends, usually of the same sex.

The Process of Addiction

Addiction may follow a "natural history" with a progression from preliminary, initiate stages to a stable habit to termination of the addiction. Waldorf interviewed over 200 ex-addicts to inquire about the processes of their addiction and subsequent abstinence. He identifies the following phases of an addiction career:

1. Experimentation or Initiation—usually with peers; most users terminate opiate use after this stage, but some continue.

2. Escalation—this follows a pattern of frequent use over a number of months and results in daily use, physical addiction, and increased tolerance; some users "chip" (use heroin infrequently and without developing a physical dependency) at this stage.

3. Maintaining or "Taking Care of Business"—heroin use is relatively stable in this stage and the addict is still able to "get high"; psychologically, the addict is confident that although there is little question about whether the person is addicted, the addict is optimistic about keeping his or her job and maintaining the habit at the same time.

4. Dysfunctional or "Going Through Changes"—in this stage, the addict may experience jail or treatment for the first time, and maybe other negative aspects of the habit as well; the addict may try to quit the habit, either with others or alone, but fails.

5. Recovery or "Getting Out of the Life"—in this stage, the recovered addict develops a successful attitude to quit drugs; the addict may be in treatment or not; this involves major life changes

6. Ex-Addict—this final stage involves the acquisition of a new social identity: an ex-addict; this role is adopted by successfully treated addicts working in treatment programs (and it is very seldom adopted by untreated ex-addicts). (Waldorf, 1983)

Persons become addicted more or less quickly. Some persons can use heroin infrequently for long periods of time without becoming addicted. Other will move to a full addiction only after a year or more of heroin use, while still others who progress from occasional to regular use do not use drugs all the time (Bennett, 1986). There is some evidence that women may become addicted more quickly than men (Anglin, et al., 1987), but there are great individual differences among heroin users in their addiction careers.

After initial use and progression to a full-blown habit, addicts must learn to maintain their habits within the realities of scarce resources (money) and legal and social rejection from others. One user described the following "typical" day:

> You have to be on your toes 24 hours a day. The only time you can afford to sit back is when you're dealing and have a supply of drugs. All the rest of the time, it's a never-ending process of scoring drugs, fixing, maybe getting high—maybe not—for an hour or so, and then starting all over again.
> You have to figure out where you are going to get your next money, who you know who can help you out, who can front you some drugs. It's an everlasting process that might go on through the night; if you don't have drugs, you can't sleep anyway. So you work through the day, through the night, and into the next morning. You've got to be on call 24 hours a day. (Coombs, 1981).

All studies evaluating treatment of narcotic addiction point to a high rate of relapse among opiate addicts. Those leaving treatment programs seem to return to heroin at a high rate, with figures varying from 70 to 90 percent. Yet, many persons do eventually leave addiction and refrain from opiates for long periods of time, often without the benefit of treatment. The reasons for leaving addiction are not clear, but it appears that it is a function of some life cycle of addiction and that addicts leave the more youthful drug subculture as they grow older. The role demands of coping and hustling become too difficult or are in major conflict with new roles. All this may lead the addict to abandon, for example, the street addict role and seek abstinence through treatment or other means. Some may simply "burn out" from playing the addict role. People who work with addicts frequently observe this "burn out" from addiction.

Other addicts may make a conscious decision to stop using heroin. Such decisions might be based on the negative features of being an addict, such as the wanting to avoid the expense of a heroin habit, the lack of meaningful goals, or avoiding the rejection from nonaddicts. A 45-year-old man who had been addicted for six years described his decision as follows:

> The one thing I did know was that I didn't want to be known as a tramp and I didn't want to feel like a nobody, a nothing. I was a very proud person, so I had to do something about those drugs. Either it was going to control me, or control my life, or I was going to stop using. (Biernacki, 1986: 51)

Such addicts might shun conventional treatment programs and stop using heroin the same way some cigarette smokers terminate their drug: "cold turkey"—that is, by simply not using the drug and enduring the physical discomfort that follows termination of the drug. Some addicts, like some cigarette smokers, are more successful in this effort than others. Factors associated with successful termination of heroin include high resolve, a changed or changing personal identity, and a desire to reestablish conventional social relationships (Biernacki, 1986).

The Addict Subculture

As suggested above, the idea that addicts are escaping or retreating from social life has not been supported in the research literature. The career type of addict, or the "street addict," is a member of a drug subculture largely made up of urban slum-dwelling male members of minority groups who adhere to a deviant drug-related set of norms (Stephens, 1991). They favor the use of heroin and cocaine, use the intravenous method of administration, have frequent withdrawals from drug use, and engage in many hustles to support their habit. Heroin use is not "escape" in a psychological sense, nor is it "retreatist" in terms of Robert Merton's anomie theory (see Chapter 4). Rather, heroin use can represent new membership in a close-knit society as in the case of the street addict. Their lives revolve around (1) the "hustle" (obtaining money illegally to purchase drugs), (2) "copping" (buying heroin), and (3) "getting off" (injecting heroin and experiencing the drug) (Agar, 1973: 21). Conventional activities occupy little time in the life of street addicts. Maintaining the role of street addict requires involvement in an addict subculture in order to receive social support from others and to guarantee a supply of heroin (Hanson, et al., 1985).

Drugs must be imported illegally into the country and then distributed through suppliers and peddlers. Those who use drugs are, to a large extent, also part of this subculture, since drug addicts must generally associated with "pushers," usually other addicts, in order to secure their supply. The structure of the distribution system whereby heroin is imported and made available to users is like most other distribution systems dealing with the importation and sale of products, except that there is more concern for reducing legal risks (McBride, 1983). For this reason, criminal syndicates have been associated with narcotics.

Just as one must learn to become a member of a drug using community, so too must one learn to become a drug dealer or smuggler. The individuals in the drug dealing and smuggling community studied by Adler (1985: 3) became committed to drug trafficking because of the uninhibited life-style it provides them. In many respects, in fact, for these individuals, drugs are the principal means by which to express a way of life. The profits from selling marijuana, amphetamines, and cocaine lend themselves to a hedonistic life-style with plenty of free time and cash. The knowledge to participate in a drug dealing subculture is acquired either by on-the-job training or having someone take special interest in the new recruit, a kind of sponsorship arrangement. There are some skills these individuals must bring with them. One smuggler described the criteria he used in recruiting new members to his crew:

> Pilots are really at a premium. They burn out so fast that I have to replace them every six months to a year. But I'm also looking for people who are cool: people who will carry out their jobs according to the plan, who won't panic if the load arrives late or something goes wrong, 'cause this happens a lot . . . and I try not to get people who've been to prison before, because if they haven't they'll be more likely to take foolish risks, the kind that I don't want to have to. (Adler, 1985: 126)

Norms in an addict subculture are organized around the supply and support of heroin use. There are a series of rationalizations that permit use as well as justifying the recruitment of new members so that continued supply can be assured. There is also defensive communication with its own argot for drugs, suppliers, and drug users which must be learned by initiates. The support of the habit requires a complex distribution system of the illegal drugs. Such information as the coming or the power or the kind of heroin for sale and where, is said to pass rapidly and accurately, with greater safety than provided by the telephone, and information is sifted out according to the reliability consensus of different persons.

A Rand Corporation study on the economics of drug dealing found that the typical daily drug dealer between 18 and 40 years old nets $24,000 per year, tax free (*Des Moines Register*, July 11, 1990, p. 1A.) Data were obtained from interviews with persons charged with crimes, on probation, and from other studies. It was discovered that many of the dealers need the extra money to support their drug habits. The Rand study reported that two-thirds of a sample of probationers earned an average of $7/hour on legitimate jobs and $30/hour as "moonlighting" drug dealers working a few hours or days each week when the demand is the highest.

Suppliers and most addicts live in a world that often has its own meeting places, values, and argot. Possibly nothing more clearly demonstrates the fact that addiction has cultural components than the argot which is used. It includes special terms for the drugs, for those who supply the drugs, and for addiction itself. It also includes special descriptive terms for those who use the drugs.

In summary, through the addicts' subculture, they are able to connect with dealers, employ a number of "hustles" to secure money for drugs, and to protect themselves from outside interference, particularly from the police. The addict subculture thus performs a number of important functions for the addict, not the least of which is the opportunity to meet others like themselves and benefit from mutual association. The drug subculture, however, has its negative aspects as well since it isolates the addict from conventional society. The only persons the addict is likely to know are other addicts and these people tend to reinforce the addiction process rather than providing positive social support to remain off drugs. Like other deviant subcultures, the drug subculture does not prepare its members to reenter the conventional world, and, in fact, it inhibits this reentry.

A SOCIOLOGICAL THEORY OF ADDICTION

Lindesmith (1968), one of the leading sociologists in the area of opiate addiction, has explained addiction to opiates on the basis of the addict's association of the drug with the distress accompanying sudden cessation of its use. Lindesmith's is a social psychological theory. It explains addiction in terms of social psychological processes that link drugs, their effects, and the desire to eliminate withdrawal symptoms. Drug addiction is a condition where drugs are used because of the fear of the pain or discomfort associated with withdrawal. Addiction is *not* simply a physical process whereby the consumption of a drug automatically produces dependency. Users who fail to realize the connection between the distress and the opiate seem to escape addiction, whereas, Lindesmith claims, those who link the distress with the opiate and, thereafter, use it to alleviate the distress symptoms invariably become addicted.

In becoming an addict, the individual changes his conception of himself and of the behavior he must perform as a "drug addict." Addicts must learn the proper use of drugs and sources of supply from others; they must also learn to connect withdrawal symptoms with the continued use of heroin. Addicts, in other words, must be socialized to addiction. The more addicts associate with others who are "hooked," the less likely they are to free themselves from dependence on drugs. In short, addicts come to play the role of "addict" (Lindesmith, 1968: 194). After a while, the primary reason for continued use becomes the relief of withdrawal symptoms rather than the euphoria produced by the opiates.

One problem with Lindesmith's theory is that it dismisses completely the motivation of euphoria—the desire to "get high"—in explaining continued heroin use. In a study of long-term addicts, McAuliffe and Gordon (1974) found that these addicts do indeed experience euphoria, although not to the extent that newer addicts do. The reason fewer long-term addicts get high or obtain a rush is that they usually do not have enough money to buy the amount of drugs needed to produce a desired level of euphoria. Such addicts, however, do appear to orient their behavior around the achievement of euphoria as produced by heroin. Richard Stephens (1991) found that street addicts use heroin not necessarily to counter the pain suffered by withdrawal, as Lindesmith has maintained, but in order to experience the euphoria of drug use "highs" in themselves. The group or subcultural aspects of the addiction process are paramount. To the street addicts, their self-concept, their sense of personal worth, and their status all revolve around the role of taking heroin to experience a drug high.

In response to such criticisms, Lindesmith (1975) has replied that he never claimed that drug addicts used withdrawal symptoms as the only motivation for continued addiction. Physical dependency may generate its own motive for continued use, but an addict's craving for drugs is produced or caused not by a motive but by the repetition of a particular experience. Physical dependency can produce the nature of the withdrawal distress, but the interpretation of those symptoms and the subsequent continued use of heroin are cognitive, not physical, dimensions of the addiction process. Lindesmith also concedes that euphoria may be the "bait" to lure persons to continued use, even though euphoria is not a frequent concomitant of continued use thereafter. But euphoria does not explain the behavior of persons who continue use in its absence, only the cognitive element can do this. In any case, Lindesmith's theory is not phrased in sufficiently precise language to permit empirical test (e.g., Platt, 1986). So, while Lindesmith claims that the addict must "learn" to be an addict, he does not identify this learning process in terms sufficiently clear to be tested.

COCAINE USE

When cocaine was first discovered in the 1800s, it was thought by some to be a medical wonder drug. Among the first noticeable effects were the ability of the drug to suppress fatigue, an effect that did not escape the attention of Sigmund Freud. By the end of the 1880s, Freud and others had given up hope of cocaine's medical uses, but because of its unregulated status, cocaine quickly became a staple in the patent medicine industry. It was not until the Pure Food and Drug Act of 1906 that cocaine moved underground. At the time, users included such socially marginal groups as criminals, jazz musicians, prostitutes, and African Americans. Later, cocaine would become associated with beatniks and, still later, movie stars and professional athletes. By mid-century, cocaine was a rare and exotic commodity. It had become a rich person's drug, the "champagne of the street."

The Cocaine Highway

By the late 1960s and early 1970s, cocaine use in the United States was more prevalent as a result of two particularly important events (Inciardi, 1986: 73–74). First, The U.S. Senate passed legislation that reduced the legal production of amphetamines and placed strict controls on depressants. Second, a decision by the World Bank allocated funds for a new highway in the high jungles of Peru. The growing of coca leaves had always been popular around the Peruvian Andes but only small amounts of leaves were available for processing into cocaine. Travel from the slopes to population centers

Making Cocaine

What You Need	Why You Need It: The Six "C"s
1. A jungle location	*Coca leaves:* coca grows best at altitudes over 1,000 feet in a hot, humid climate with heavy rainfall. With proper care and fertilization, leaves can be harvested every 35 days or so.
2. A processing plant	*Coca Paste:* dried leaves are put in a plastic-lined pit with water and sulfuric acid. Someone wades through the mixture periodically to stir it up. After several days, the liquid is removed, leaving a grayish paste.
3. Some chemicals	*Cocaine base:* water, gasoline, acid, potassium permanganate, and ammonia are added to coca paste to form a reddish brown liquid which is then filtered. Drops of ammonia produce a milky solid.
4. More chemicals	*Cocaine hydrochloride:* filtered and dried cocaine base is dissolved in a solution of hydrochloric acid and acetone or ethanol. A white solid forms and settles.
5. Electrical capacity	*Cocaine:* cocaine hydrochloride is filtered and dried, under heating lights, to form a white, crystalline powder. Cocaine is now ready for distribution, usually in one kilogram packages for $11,000 to $34,000 each.
6. Chemicals, once again	*Cutting:* before reaching the street, cocaine is diluted, or cut, with sugars, such as mannitol (a baby laxative), or local anesthetics, such as lidocaine. Usually sold in 1-gram packages for $50 to $120.

where the drug could be refined and exported was difficult, requiring many hours with pack mules and some danger from the terrain. The World Bank's construction of a paved highway through the Huallaga Valley opened up transportation routes for the shipping of coca. The reduced supply of amphetamines and sedatives in the United States produced a ready market for this increased cocaine supply.

Inciardi (1986: 73–74) describes the processing.

> At [secret, near-by] jungle refineries, the leaves are sold for $8 to $12 a kilo. The leaves are then pulverized, soaked in alcohol mixed with benzol . . . and shaken. The alcohol-benzol mixture is then drained, sulfuric acid is added, and the solution is shaken again. Next, a precipitate is formed when sodium carbonate is added to the solution. When this is washed with kerosene and chilled, crystals of crude cocaine are left behind. These crystals are known as coca paste. The cocaine content of the leaves is relatively low—0.5% to 1.0% by weight as opposed to the paste, which has a cocaine concentration ranging up to 90% but more commonly only about 40%.

The cocaine highway leads from the jungle refineries to the Amazon river and, then, to the Atlantic Ocean. A number of cities are used as transportation centers. Santa Cruz, Bolivia is a major meeting point for Columbian and American buyers of cocaine, and the city is a major point of departure for smugglers. Other important cities include Tingo Maria and Iquitos, both in Peru, the latter of which is important because it can be serviced by commercial ships. Leticia, Columbia is a town of only a few thousand but its marina boasts a number of overpowered outboards and other high-performance racing boats, as well as a number of small seaplanes (Inciardi, 1986: 75).

From these beginnings, the movement of cocaine spreads to the United States and other points. Deserted airstrips and obscure combinations of air-sea routes chosen because they are hard to patrol are used. By the time cocaine has been diluted several times, with among other things baking soda, caffeine, powdered laxatives, it has an average purity of 12 percent and sells on the street for near $100 a gram. With this process, 500 kilograms of coca leaves, worth $4,000 to the grower, results in 8 kilos of street cocaine worth $500,000.

Extent of Use

Cocaine has become the illicit drug of choice for many persons in the 1980s and its increased availability was helpful in introducing many new users to the drug. The Chief of Staff of the House Select Committee on Narcotics Abuse and Control was quoted in 1985 (Trebach, 1987: 178) as saying that "It [cocaine] is dropping out of the skies. Literally. In Florida, people are finding packages in their driveways that have fallen out of planes." The increased availability of cocaine was responsible for new use patterns. Previously, use was confined to a small group of relatively wealthy persons because the drug was expensive and scarce. Cocaine is now more commonly available and some forms of the drug are cheaply obtained. As a result, cocaine is found among all segments of the population and in all areas of the country.

The use of cocaine is considered by many to be a major health problem with regular users estimated in the mid-1980s to number about 10 million with perhaps 5,000 persons each day trying the drug for the first time (*Newsweek*, February, 25, 1985, p. 23). An earlier national survey conducted in 1982 reported that slightly more than 21 million persons total had ever tried cocaine (reported in Stephens, 1987: 45).

A survey by the National Institute of Drugs estimated in 1990 that although daily or near-daily use of cocaine increased, the casual use of cocaine declined over 70 percent from 1985 to 1990 (*Des Moines Register*, December 20, 1990, p. 1A). The survey also reported that the number of crack cocaine users remained constant over the same time period. The survey put the figure of current cocaine users at 1.6 million, down from an estimated 5.8 million in 1985. Differences in the definition of "current user" and estimation procedures, and the problems of high nonresponse in the survey makes for differences in the number of users.

There are many reasons to use drugs, including cocaine, and these may differ between men and women. One study found that women were more likely to give specific reasons for using cocaine than were men, who tended to use cocaine as part of some overall pattern of deviant behavior (Griffin, et al., 1989). This is similar to parallel findings in research on alcoholism where women, there too, were able more often than men to cite specific reasons for their excessive drinking, such as death of a spouse.

Youthful cocaine use is of major concern in the United States. The NIDA survey of drug use found that 18 percent of high school seniors in 1986 reported that they had tried cocaine (Johnston, et al., 1987: 27). Current use (i.e., use in the prior 30 days) among these seniors has risen in recent years from 4.9 percent in 1983 to 5.8 percent in 1984 to 6.7 percent in 1985. This percentage dropped to 6.2 in 1986. Reported use among high school seniors in 1987 was even lower (Johnston, 1988). High school students have only recently, however, appreciably changed their beliefs regarding the harm from experimenting with cocaine. Only 34 percent of high school seniors believe that there is great risk in trying cocaine once or twice, a figure that has not changed since 1978. The rates of disapproval of cocaine use are higher, with 87 percent of seniors disapproving of even trying cocaine, and about 97 percent in 1987 disapproving of regular cocaine use (Johnston, 1988). The survey also disclosed that cocaine seemed to be readily available on college campuses and that a number of college students have experimented with the drug. By the end of their fourth year of college, about 30 percent of all college students will have tried cocaine. About 17 percent of the college students surveyed indicated they had used cocaine within the last year, a figure that has remained virtually unchanged since 1980 (Johnston, et al., 1987: 206). This steady use of cocaine compares with the declines in use among college students of other drugs.

There is substantial public concern over cocaine and crack and that concern is justified in terms of the extent of use and the negative consequences of using cocaine. But it appears that the antecedents to or causes of cocaine use are general rather than specific. One study, for example, reported that the processes that led users to cocaine were little different from those that led others to use heroin or other illicit drugs (Newcomb and Bentler, 1990).

Methods of Use

Cocaine has been found to be a versatile substance that can be administered in a number of ways. Cocaine users in South America had long smoked cocaine paste, which contained from 40% to 90% pure cocaine. The smoking method is now found in the United States. Among chronic users, the intravenous use of cocaine is more common this method permits combining cocaine with other drugs, such as heroin. This combination, called a "speedball," intensifies the euphoric effect although it is a dangerous practice; speedballing killed actor John Belushi.

Cocaine is most commonly administered by inhaling ("snorting") it through the nostrils. The powder is arranged ("cut") in a line, or several lines, and a small straw (a rolled up dollar bill will do) is used to take up the drug into the nose. This produces a relatively short experience of about 20 or 30 minutes in duration. The short duration of the effects of the drug is the main reason that some users use multiple doses of the drug. One source reported that some users may use up to 10 doses per day (Cox et al., 1983). The user remains alert and in full mental control and enjoys the benefits of no hangover, physical addiction, lung cancer, risk from dirty syringes, or burned-out brain cells (Inciardi, 1986: 78). The cocaine euphoria is immediate and virtually universally regarded as pleasurable. Because it seemed to have few negative side effects, cocaine developed a reputation as being an "ideal" drug.

The use of crack is much more recent. The advent of crack is usually traced to about 1985. Crack came about, according to Trebach (1987: 178), in large part because of the glut in the cocaine market. "In a sense," he says (p. 178), "crack was a packaging and marketing strategy to deal with the economic problem of an excess of cocaine supplies." Crack is most usually smoked in small pipes and its use is more common among younger drug users because it is relatively cheap. Crack is now common in the classrooms and streets of virtually all large cities. The effects of the drug are even more immediate than snorting the substance through the nostrils and crack is thought by some to be more addicting than regular cocaine and, perhaps, heroin. Yet, one observer reports: "I have searched. My assistants have searched. We have gone through many government reports. We have quizzed government statistical experts. We have yet to discover one death in which the presence of crack was a confirmed factor" (Trebach, 1987: 12).

The Consequences of Cocaine Use

Cocaine does not produce physical dependency and the user does not build up increasing tolerance of the drug, like heroin. Most users use small amounts of the drug infrequently. Some users, however, take increasing amounts of the drug and increase the frequency of their use. This has led some to claim that the idea that cocaine is nonaddicting came from limited information due to the recent scarcity of the drug (Gonzales, 1987; Cohen, 1984). There is much debate among experts on the issue of physical dependency. Stephens (1987: 35) suggests that "The current evidence favors the conclusion that physical dependency [with cocaine] may occur," while Inciardi (1986: 79) asserts that the drug is nonaddicting and habitual use is the result of "psychic dependence"—a motivation to use cocaine to avoid the feeling of depression chronic users experience when they stop using cocaine. This desire to use cocaine is not physically based, Inciardi claims, but is psychogenic, "emanating from the mind" (Inciardi, 1986: 80).

Regardless of whether it reflects physical or psychic dependence, chronic, heavy cocaine use does lead to a number of negative consequences, including a paranoid psychosis state and a general emotional and physiological state of debilitation. Short-term effects can include heart failure, respiratory collapse, fever, and sudden death (Welti, 1987). Damage to the septum of the nose can also occur after prolonged chronic use. Other long-term effects of the drug are now being investigated but seem to include depression and heart ailments in addition to the short-term effects (Estroff, 1987). These more serious consequences of cocaine use have been identified since intravenous administration and freebasing have become more common.

Perhaps the more important consideration with cocaine, as with other drugs, is not whether or not it is addicting in a conventional sense, but its effects on people's lives. This judgment can be made in terms of such things as the effect of cocaine use on the user's social relationships, employment status, school performance, and general functioning in society. Clearly, as with other drugs, many users are able to function in life with drugs being only a small portion of their lives. This is especially true for recreational drug users whose use is tied to social situations. For other users, however, drugs can become *the* most important part of their lives. They spent their time seeking and taking their drug of choice, and their other activities and relationships are fit in around this central interest.

SOCIAL CONTROL OF DRUG USE
AND ADDICTION

There are basically two strategies of control efforts with respect to illicit drugs: control the drugs themselves or the persons who use them (e.g., users, dealers, importers). Strategies directed toward the control of drugs include attempts to restrict imports of drugs and the decriminalization of drugs. Strategies directed toward persons associated with drugs include the use of criminal sanctions, specific treatment programs for drug users and addicts, addict self-help programs, and measures that attempt to prevent drug use, such as drug education programs in schools.

The Use of Criminal Sanctions

For most of this century, the United States has followed a policy of legal suppression with respect to drugs. Early legislation, such as the Pure Food and Drug Act of 1906 and the Harrison Act in 1914, was a direct attempt to use the law to restrict or eliminate the use of undesirable drugs. Supported by a number of reform groups including the American Medical Association (Courtwright, 1982), these laws, and the ones that followed, have been the principal tool of manipulating drug use in this country.

Addict Crime.

Not only is the manufacture, sale, and use of certain drugs illegal, but there are also crimes associated with drug use, such as those committed by addicts in order to secure money for drugs. Estimates of the amount of criminality by addicts are notoriously unreliable in view of the difficulties in obtaining precise figures on which crimes were committed by addicts and which by nonaddicts. In addition to the direct costs of drug addiction, there are also indirect costs, such as the costs of law enforcement, drug education programs, absenteeism, and the like.

A heroin habit is expensive. Between 1948 and 1953, heroin was available at 30 to 40 percent purity for $2 to $2.50 a packet (about 10 milligrams). In 1960, a 2 percent mixture cost $5 (Winick, 1965). The price has kept going up. The cost of a milligram of street heroin was about $1.50 in 1972, but a year later, it had increased to $5.80 (Dupont and Greene, 1973). Most addicts require between 10 and 30 milligrams per day. The cost of heroin varies from city to city, and one study reported that the price of a gram of heroin in Miami is almost twice as expensive as a gram in New York City (Brown and Silverman, 1980).

The idea that addicts will often turn to crime to support their expensive habit has been called the "enslavement theory of addiction" (Inciardi, 1986: 160–169). As tolerance to the drug is increased, and the addict requires larger and larger doses, the daily expenditure is generally more than most persons can make legitimately. Therefore, "hustles" are used to secure drugs.

Because of the high cost, most addicts support their addiction by committing crimes. In the past, crimes committed by addicts largely involved various types of stealing, such as burglary, and, for women, prostitution (Rosenbaum, 1981; Cuskey and Wathey, 1982).

The enslavement theory, however, only receives partial support. Many addicts clearly do use criminal activity to help buy heroin. For many addicts, however, crime is a well-established behavior pattern before addiction. For still other addicts, crime is a behavior pattern only at certain stages of their addition (Faupel and Klockars, 1987). It therefore represents a logical method by which to support a heroin habit, once begun. That decision is even more logical for female addicts since many of their crimes are successful, i.e., there is no arrest. A study of two groups of female addicts in Miami reported that less than one percent of all crimes committed by these groups resulted in an arrest (Inciardi and Pottieger, 1986). One reason for this low percentage is that the crimes these women committed were mostly "victimless"—drug sales and vice, especially prostitution and procuring.

Criminal Law and Addiction.   The criminal law as a mechanism of social control over drugs is used to different degrees in other countries. For example, in Malaysia and Singapore, conviction of possession of 1 kilogram of hashish or 20 grams of heroin carries an automatic death penalty. Similar offenses in Denmark bring sentences of only 60 and 90 days respectively (Rowe, 1987). Other countries have penalties in between these extremes which reflects their social attitudes in these countries.

The use of the law to attempt to suppress drug usage has been controversial in the United States. Some advocate a strategy of decriminalization which would eliminate or lessen the legal penalties for drug use while retaining criminal sanctions for persons who sell or distribute drugs. It is further believed by some that suppression has actually increased the difficulties of controlling drug traffic because it created an extensive organization for importation and sale of drugs. Addict crime is generated by attempts to obtain enough money to buy illicitly—or to steal—high-priced drugs that could be obtained legally for a fraction of the cost charged by peddlers (Johnson et al., 1985). An additional consequence is that drug users become "criminals" by virtue of using drugs. The stigma of having an arrest record makes the transition from "addict" to "ex-addict" all the more difficult.

Persons in favor of legalizing drugs also point out that in spite of huge expenditures ($10 billion at the federal level alone in 1987) and the employment of hundreds of thousands of agents and other personnel, the "war against drugs" has not been won. Increasingly large costs are required for more police, prosecutors, court trials, and prisons where it is estimated that over 10 percent of the inmates are serving time for drug law violations. Because of their illicit nature, drugs such as cocaine and heroin are extremely costly (heroin probably on the average of $30 a milligram in the United States in 1990) and this cost must be borne daily for addicts. Because of this, as has been pointed out previously, many addicts must commit crimes to support and maintain their habit. One study estimated that addicts in this country may commit as many as 50,000,000 crimes a year (Ball, et al., 1982). This figure was derived from a listing of crimes committed by a sample of 243 addicts who had committed an average of 178 crimes over the course of 11 years. Under a medical system, however, because heroin would be cheaper, there would be less need for criminal penalties; there would be less addict crime.

Because of these and other problems, many authorities and groups have recommended that drug addiction be regarded primarily as a medical or social problem, as it is in Great Britain and in many Western European countries. Trebach (1987: 383–385), for example, proposes a program of heroin maintenance where addicts would be expected to live productive lives or risk being taken off heroin. Trebach also suggests providing affordable treatment for drug users who wish it and devising more creative legal solutions to drug problems. If successful, addicts would not have to turn to crime to get their supply of drugs. Trebach also suggests placing greater controls on drugs that are currently legal, such as alcohol and tobacco, and lessening controls on currently illegal drugs. He recommends decriminalizing marijuana use and cultivation for personal use; he would medicalize heroin for addicts and pain patients by prescription, but he would not make heroin legal for casual recreational use.

Legalization assumes that persons (adults) have a right to abuse their own physical bodies without interference from law. Far more physical injury is caused by being excessively overweight than by using drugs, yet no one would propose a law on the amount of food intake. Proponents of legalization also claim that the enforcement of drug laws has often violated constitutional guarantees of personal freedom in the Bill of Rights through the abuse of search warrants, seizure of property, and protection from self-incrimination. Drug testing may result in loss of employment, while such testing does not reveal the level of alcohol use, nor is it done for alcohol use, which constitutes a potentially much graver problem. Moreover, it is claimed that it is highly inconsistent to allow two other drugs, alcohol and nicotine, to be readily available when the effects of these drugs are more devastating than perhaps all other drugs put together. Over 350,000 persons die each year from cigarettes alone.

Other observers believe that a legal approach has been more successful than critics are willing to concede. Inciardi and McBride (1990: 1) point out that although it is true that the death rates for the use of alcohol and tobacco exceed those for other drugs, "What is summarily ignored is that the death rates for alcohol and tobacco use are high because these substances are readily available and widely used, and that the death rates from heroin and cocaine use are low because these drugs are not readily available and not widely used." Legalization might increase the supply of drugs and the effects of increased drug use would fall on those for whom the effects would be most harmful. For example, we have seen that there is much drug usage in inner-city areas, or ghettos, and especially among minority group members. A large number of ghetto families are female-headed and it appears, for reasons that are not clear, that women are more likely to become physically dependent on crack than men, further increasing problems of family disruption, abuse, and child neglect.

In searching for a sound national policy, Inciardi (1986: 211) says that "it would appear that contemporary American drug-control policies, with some very needed additions and changes, would be the most appropriate approach [to control illegal drug use]." The changes Inciardi advocates include the use of the military to interdict shipments of illegal drugs into the U.S., conventional treatment efforts with greater funding and more personnel, and the enhancement of education programs aimed at prevention (Inciardi, 1986: 212–214). Inciardi points out that other nations have done little to control the manufacture of illegal drugs within their borders and that the U.S. therefore must take more aggressive measures than previously. "Drug abuse," Inciardi (1986: 215) claims, "tends not to disappear on its own" (p. 215).

In a similar vein, Kaplan (1983) has advocated a legal approach to heroin addiction where addicts who are responsible for much crime are legally coerced into treatment programs. The benefits may be small, given the limited success of heroin treatment (see below), but according to Kaplan even small benefits may be worth the legal effort. Other observers have rejected such a policy on the grounds that such actions may violate civil rights and would not, in any case, have much impact on the street addict subculture (Stephens, 1991). Yet, it remains the case that research increasingly suggests that persons who are coerced into treatment have lower relapse rates than those who are not (Inciardi and McBride, 1990: 4) perhaps because treatment success is related to length of stay in treatment and those who are coerced into treatment stay there longer than those who are not.

Treatment

This brings us to the subject of treatment for addiction. The treatment response to drug use is a major effort in the United States, as can be seen in the following figures. During 1985, it is estimated that 305,360 clients— the majority of whom were heroin-, marijuana-, or cocaine-using males aged 21–44—were admitted to about 6,000 treatment facilities (Reznikov, 1987). States reported that they spent more than $1.3 billion during that same time for alcohol and drug abuse treatment and prevention services. Of this amount, nearly 80 percent was spent on direct treatment, 12 percent on prevention programs, and the remaining money spent on training, research, and administration. Most (76 percent) of the clients of drug treatment programs were handled on an out-patient basis, 19 percent were admitted to residential facilities, and 5 percent were admitted to hospitals.

Methadone Maintenance. A person who is on drugs does not always fail to function effectively in society. Substantial numbers of addicts can become responsible and productive members of the community. For this reason, clinical programs in major urban areas have been directed at keeping thousands of addicts on *methadone maintenance* while efforts are made to bring about their rehabilitation. Methadone maintenance began in the early 1960s in New York City (Stephens, 1987: 88–92). Under this program, the addict reports daily to a clinic for a medically supervised dosage of methadone, a substitute for heroin, to prevent withdrawal symptoms. The methadone is usually given in a glass of orange juice after a urine test to determine that the addict has not used heroin prior to coming to the clinic. Methadone behaves chemically much like heroin (one can become physically dependent on methadone, just like heroin) but its effects are much more long-lasting because the drug is not metabolized into the system of the user as quickly as heroin. Therefore, one dose per day is sufficient, contrasted with the 3 to 4 doses that are required by heroin addicts. Moreover, the theory is that an addict on methadone will not return to heroin either to relieve withdrawal distress (since there is none) or to gain a euphoric reaction (because the methadone blocks this). The oral administration of methadone also eliminates problems associated with dirty needles—infection, AIDS, and risks of other diseases.

There is some evidence that methadone maintenance programs do reduce somewhat the amount of "addict crime" by providing a legal source of drugs. This particular benefit, however, is limited to those who voluntarily participate and remain in the program for a period of time (Newman, et al., 1973). Moreover, we are unsure of the long-term effects of methadone; clinical research continues on this issue (Kreek, 1979). In any case, it is clear that methadone does not deal with the causes of addiction and that the administration of methadone for heroin addicts may seem to some like the strategy of prescribing brandy for an alcoholic. Because many addicts also use other drugs, it is also clear that the use of methadone does nothing to discourage the use of nonopiates. Furthermore, there is a danger that methadone can be diverted from clinics into black market usage to be used along with rather than instead of heroin (Agar and Stephens, 1975; Inciardi, 1977).

Heroin Maintenance. Some observers, sensitive to the problems of programs like methadone maintenance, have suggested that the United States institute a "heroin-maintenance" policy. Such a program would not differ greatly from that presently found in England where addiction has been considered to be a medical problem, to be treated largely by outpatient care by physicians and social service professionals, with drugs being prescribed at low cost. Addiction has been considered a disease in England since 1920. In the 1960s, the British policy experienced difficulty since it excluded the most predominant type of drug user at the time: the recreational user. The most recent statement of the British drug policy attempts to acknowledge the recreational user by invoking the concept of the "problem drug taker," an analogous term to "problem drinker." The British approach is treatment-oriented and multidisciplinary, which broadens the handling of drug users beyond medical personnel to include social workers and community workers in addition to medical specialists. In fact, the British policy reflects a departure from the strict medical or disease model that oriented that policy for so many years (Advisory Council on Misuse of Drugs, 1982; see also Journal, 1983).

Traditionally, British physicians have been able to prescribe a minimum dosage of heroin along with their attempts to cure the addict. British officials as well as the public therefore did not regard the addict as a "criminal." Because heroin is legal and cheaper than if purchased on the black market, addicts would not have to commit crimes to secure their drugs. Nor would they therefore have to come into contact with criminals and participate in a criminal subculture. British addicts appear to be relatively noncriminal, and the addict does not have to steal or become a prostitute or peddle heroin in order to obtain heroin. The shift from this medical approach may have been motivated in part by increases in the rates of addiction in many parts of the world, including England.

Clinical Approaches with Heroin Addiction. Where hospitals are used with heroin addicts, the optimum treatment period is a few months. Newly admitted addicts are first given thorough medical examinations and treatment for any conditions aside from their addiction. Their physical condition is built up along with the removal of drugs. Currently, the most commonly used drug in treatment is methadone because it is accompanied by much milder withdrawal symptoms. The addict's dependency on heroin is replaced with that of methadone; then, that drug is slowly withdrawn. Along with methadone, the addict also receives recreational and occupational therapy, and vocational training. Upon release, the addict usually receives follow-up services as needed. There are many different therapeutic community facilities that provide in-patient care. To avoid stigma, California uses a method of "civil commitment" to hospitals rather than utilizing the criminal law. Under either alternative, however, forced treatment is involved.

The studies of the results of treatment indicate that narcotic addiction, particularly the use of heroin, is one of the most difficult forms of deviant behavior to treat as reflected in the high percentage of relapse among treated addicts. For example, one follow-up study of male addicts found that if relapse was defined as any reuse of narcotics, the observed relapse rate of 87 percent is equivalent to the 80–90 percent relapse rates reported in most other studies (Stephens and Cottrell, 1972). Results like this are not uncommon. Addiction is a difficult behavior pattern to break, and most treatment efforts, which are generally guided mainly by medical rationales, are unable to break this pattern. Addiction is a complex process involving sociological and psychological factors, such as drug associates, participation in a drug subculture, and the conception of self as an addict. For this reason, research is continuing on the use of social and psychological factors in treatment (Grabowski, et al., 1984).

In this regard, it is particularly interesting that several studies report that while heroin use is hard to terminate, many users (1) use heroin on a regular basis without becoming addicted, and (2) are able to terminate their addictions without any treatment whatsoever. A comparison of treated and untreated addicts reported that untreated addicts had smaller habits, were more likely to stop using heroin, and had more cohesive families and higher levels of self-esteem, suggesting that studies that look only at treated addicts as somehow representative of all addicts may be misleading (Graevan and Graevan, 1983). Another study reported that of a sample of 51 addicts, one third had been in drift between chronic use and abstinence, one quarter had been dependent at times but had overcome these episodes without treatment, and the remainder were dependent addicts (Blackwell, 1983). Clearly, opiate use can be situational.

Some ex-addicts terminate their dependency from heroin, not by stopping use of the drug "cold turkey," but by changing for a time to another drug, usually alcohol (Willie, 1983). In an area where progress is often difficult to define, the movement from one drug to another may strike some as no progress at all. To other observers, however, the cessation of heroin under virtually any circumstances is a welcome and successful situation. Other addicts leave addiction after reaching some turning point in their lives, usually a profoundly moving existential crisis that forces a self-realization of the need for self-change (Jorquez, 1983). Simply desiring to terminate an addiction, however, takes time since users must extricate themselves completely from "the life" and adjust to the world of "squares." Some observers feel, however, that nothing less than a total change of life circumstances, prompted by circumstances beyond the control of the addict, is sufficient for this purpose (Waldorf, 1983), while other observers report that the change to nonaddiction can be the result of rational decision-making to make a change (Biernacki, 1986; Bennett, 1986).

Since addicts are "rational actors" in their choice of the street life, they need to learn that lasting change can only take place if they take responsibility for their own behavior. The higher incidence rates of AIDS in inner-cities has been an incentive to heroin addicts to come to grips with their addiction. In this connection, it is likely that most persons who are faced with a genuine chance of dying because of their behavior will likely change that behavior (see Stephens, 1991). At the same time, the street addict subculture has isolated the addict from conventional society. The only persons addicts know well are likely to be other addicts, and they generally tend to reinforce the addiction process rather than to provide positive support to stay off drugs. Like other deviant subcultures, the street addict subculture does not prepare its members to reenter conventional society; rather, it inhibits reentry.

Clinical Approaches with Other Drugs. By the mid-1980s, the use of clinical, medical out-patient and in-patient approaches to treat marijuana and cocaine abuse were well-established. One of the best known of these is the *CareUnit*, a type of program run by Comprehensive Care Corporation in California, a for-profit corporation, which reported 1986 earnings of $192,936,000 (Gonzales, 1987: 189). CareUnits utilize unused hospital beds in local communities, so no new facilities are required. The CareUnit approach, which was endorsed by Nancy Reagan, is an inpatient program for younger users of any drug. Drug users are referred by their parents. The routine is fairly generic and drug education is stressed, as is family relationships. Beyond this, there is a program of individual therapy along with an in-patient visit, but most programs do not attempt to provide sufficiently intensive treatment to result in long-term personality changes. Even a short stay can be helpful, however, if it removes the user from a less-than-desirable environment.

Critics have found many faults with the CareUnit approach, not the least of which is that the stay is not sufficient to affect permanent change, especially when the youngster returns to the same drug-using environment. The in-patient stay is usually fixed at 28 days, a term that coincides with the maximum provided by insurance companies, and there is no systematic follow-up. The nature of the therapy in CareUnits has been called "McTreatment," to liken it to a fast-food style of behavior change. But perhaps this approach to drug problems may be effective for younger, more inexperienced users.

There are other private drug treatment programs. One such program, at Hazeldon, Minnesota stresses early intervention (Gonzales, 1987). The Hazeldon model asserts that any treatment, even involuntary treatment, is better than none and that it is never too early to begin such treatment. Intervention at Hazeldon might be the result of a "surprise party" where the drug user's spouse/parents, concerned friends, neighbors, employer or teacher, and anyone else who might have some influence attend. The "party" is supervised by a professional counselor who makes the arrangements. Each person makes a list of things the user has done recently to make life miserable for him or her and the list concludes with an ultimatum: get help or else! The point is reinforced with everyone going through a list of troubles and making the same ultimatum. In-patient treatment can begin immediately after the "party" and after the therapy, a one year's follow-up at Alcoholics Anonymous is required.

It is difficult to judge the effectiveness of many private drug treatment programs. The lack of careful follow-up information on each patient and the lack of a control group of similar drug users who did not experience the treatment program seriously inhibit generalizations. These programs usually admit only those who are early in their drug using careers and the success rate may indeed be high, but that is speculation. Perhaps the major benefit of such programs is that they continue to publicize an anti-drug message that may inhibit some use.

There are no fewer problems in assessing the effectiveness of public (or tax-supported) drug treatment programs. One major problem is that it is not possible to know beforehand what constitutes reasonable success. How many drug users must abstain before a program is a success? One researcher reported an abstinence rate of over 50 percent for the treatment of cocaine use (Tennant, 1990), but it is difficult to know whether this is a very high, very low, or very average figure.

Addict Self-Help Programs

Other methods of helping addicts can be found in self-help groups of addicts themselves, such as *Narcotics Anonymous*, Synanon, and various local groups. Narcotics Anonymous (NA) was founded in 1948 by a former drug addict and is similar to Alcoholics Anonymous in both its activities and focus. It uses an informal organizational approach to combat drug addiction. NA recognizes the difficulties faced by former addicts in refraining from drugs and attempts to provide substantial social support for abstaining from drugs. Branches exist in most large cities in the United State and Canada. New members are assigned to an older member who can be called upon to help when the newer member is having difficulties. The process in Narcotics Anonymous is similar to that of Alcoholics Anonymous in that norms and attitudes favoring the use of drugs are replaced by norms and attitudes opposed to their use. There is little firm information about the effectiveness of Narcotics Anonymous, but it does not appear that this organization has been as successful as Alcoholics Anonymous in effecting permanent change. The lack of public support, the stronger negative public attitude toward drug as opposed to alcohol addiction, and the nature of the addiction may account for this comparative ineffectiveness.

A group method that deals with drug addicts is the organization of drug addicts called *Synanon*, a word that came from an addict who was trying to say "seminar." Synanon was founded in 1958 in Santa Monica, California by a former member of Alcoholics Anonymous, Charles Dederich (Yablonsky, 1965). Typically, Synanon establishment drug addicts live voluntarily together in a number of buildings for the purpose of freeing themselves and each other from drug addiction. Synanon is a group approach to treating drug addiction that stresses interpersonal cooperation and using the group to help individual addicts with their problems. Addicts manage their own offices and carry out the physical operations of the establishment. Membership in a Synanon group can be divided into three groups, which represent stages in progress toward rehabilitation. In the first stage, they live and work in the residential center; in the second, they have jobs outside but still live in the house; in the third, persons graduate to living and working on the outside.

An important part of the program is that each evening members meet in small groups, or "synanons," of 6 to 10 members, and membership is rotated so that one does not regularly interact in the small group with the same persons. No professional persons are present and the purpose is to "trigger feelings" and to precipitate a "catharsis," or release of emotional energy. In the discussions, there is an "attack therapy" or "haircut" in which members insist on the trust and cross-examination; hostile attack and ridicule are expected. "An important goal of the 'haircut' method is to change the criminal-tough guy pose" (Yablonsky, 1965: 241). The purpose of this method is to break down defensiveness about drugs and the denial of addiction. The haircut also serves to trigger feelings and emotions about addiction and the problems of coping with the use of drugs.

A similar program on the east coast is *Daytop Village*, a residential treatment community founded by a Catholic priest in 1963 in Staten Island, New York. It is now an international organization with centers in the United States, Canada, Ireland, Brazil, Malaysia, Italy, Spain, Thailand, Sweden, Germany, and the Phillipines. Outside of Alcoholics Anonymous, this is the largest drug treatment organization in the world (Gonzales, 1987). Most centers have a waiting list of clients and treatment is relatively inexpensive (about $35 a day or $13,000 a year). The course of treatment is about two years. The program at Daytop Village views drug use in the context of family problems. New potential members are made to sit in a "prospect chair" to contemplate their need for treatment and then they must stand on the chair and beg to be admitted to the program. After admission, the new prospect is showered with hugs and encouragement and put to work. Initiates are besieged by recovering addicts who force the person to admit they are powerless over drugs and need help. Group dependency is fostered, and members are alternatively punished for bad behavior and rewarded for good behavior in the program.

It appears that the Synanon program unknowingly applies a learning or socialization theory of deviance to the treatment of drug addicts because it brings them into contact with an antidrug subculture in which they learn to play nonaddict roles. The main aspects of the program are:

1. A willingness expressed by the individual, who gives up his or her own desires and ambitions, in order to become completely assimilated with a group dedicated to "hating" drug addiction.

2. The discovery by the addicts that they belong to a group that is "anti-drug," "anti-crime," and "anti-alcohol,"as they hear over and over again each day that their stay at Synanon depends upon their staying completely free from drugs, crime, and alcohol—the group's basic purpose.

3. The maximization of a family-type cohesion wherein the members are deliberately thrown into continuous mutual activity, all designed to make each former drug addict fully realize that he or she is, in this respect, like each other member of the family-type group.

4. Explicit programs for giving to each member certain status symbols in exchange for their staying off drugs and even developing anti-drug attitudes. The entire experience is organized into a hierarchy of graded competence unrelated to the usual prison or hospital status roles of "inmate" or "patient."

5. Specific emphasis on complete dissociation from the former drug and criminal culture and the substitution of legitimate, noncriminal cultural patterns (Volkman and Cressey, 1963).

It is not possible to determine exactly the effectiveness of such residential organizations in bringing about change in addiction. The organization is secretive about its clients in an effort to protect their privacy and the nature of their treatment. Furthermore, it is not clear whether a high success rate from such programs is the result of the program itself or more a function of the program dealing only with persons who are highly motivated to terminate their addiction. Most persons, perhaps as many as two-thirds, who are admitted to programs like Synanon and Daytop Village do not complete the program. In any case, recent revelations concerning the philosophy and operations of Synanon suggest that the program may not be as effective an "anti-drug subculture" as "anti-crime subculture" as once thought (Olin, 1980).

Therapeutic communities are a very expensive treatment modality. Such expense might be worth it if the rates of success were high, but there is another problem with such programs. Programs like Synanon and Daytop Village appeal to a very small proportion of heroin addicts. So, even if exceptionally effective, such programs do not offer a practical alternative to other attempts to reduce addiction. These programs also run a risk of increasing dependency on the part of addicts. Some users have a hard time "graduating" and may stick around the fringes of the program for a long time.

Prevention of Drug Use

Two strategies underlie present attempts to prevent drug abuse: the employment of threat-like tactics to "scare" potential users into not using drugs, and the use of special education programs that alert potential users to the dangers and consequences of drugs.

Media Messages. The former strategy is difficult to evaluate because the use of scare tactics has not been formalized like programs of drug education. Scare tactics can be part of a larger program of drug information. Media messages, mostly geared toward younger users, stress the negative physical consequences of using drugs, particularly cocaine and crack. One such ad shows a hand holding an egg. "O.K.," the announcer says, "one more time. This is your brain." The egg is then shown frying in a pan. "This is your brain on drugs," the announcer says. "Any questions?" By the mid-1980s, such public service announcements on television and ads in newspapers were common.

It is virtually impossible not to become acquainted with such messages, but their effectiveness may not be great. Many users, like cigarette smokers, do not deny the potential harmfulness of drugs but some feel that drug behavior, including their own, can not be controlled. One young cocaine user expressed this sentiment:

> I think it's a good idea that the media focuses on coke. It's good to teach young kids to stay away from it. It really is bad stuff. But as you get older, you can make your own decisions. I am old enough to make my own decisions. I don't need everyone and their brother telling me what to do. The government can try to control it, but it's impossible to stop people from using drugs. Right now it is just a passing fad to say 'no' to drugs. Next year I bet no one will remember the whole campaign. (Smart, 1989: 69–70)

Drug Education Programs. Drug education programs, on the other hand, are formalized, structured attempts to provide objective information to potential users so that they might evaluate and, it is hoped, reject drug use. Drug education programs are generally confined to persons who do not have extensive drug experience or backgrounds. It has been reported that the more a person has been exposed to drugs in the past, the more the person is likely to find "drug education" among peers in the community than through a drug education program (Blum, et al., 1976). For this reason, some studies report that these programs have both retarding and enhancing effects; that is, for some persons, the drug education program will inhibit drug usage while for others it may stimulate further use. For example, a study in Canada found that there was a positive relationship between drug education programs and drug and alcohol use among adolescents (Goodstadt, et al., 1982). Clearly, these programs must select cautiously the contents of their programs as well as the population to which they are to be directed, and they must carefully assess the prior experiences of this "target" group (Dembo and Miran, 1976).

As with most problems in deviant behavior, the prevention of drug use is not an easy task. The taking and widespread use of drugs is well ingrained in our society, and the line separating legal from illegal drug use is extremely fine in many respects and social situations. The increase in drug awareness and the greater public disapproval of the use of certain drugs, such as crack, may have significant long-term effects in preventing use. However, it must be said that while teaching kids to "say no" to drugs is a laudable goal, there is no agreement on the best mechanisms to insure that refusal. Making potential users aware of the negative physical consequences of using a drug seems to be less influential in preventing use than whether one's friends use a drug. Unfortunately, we do not know how to manipulate friendship patterns of youngsters. The prevention of drug use will probably remain more a matter of informal social control than of formal social control. It should also be remembered that prevention efforts might do better aimed at controlling the two drugs that represent the greatest problem for the most users: alcohol and tobacco.

SUMMARY: MAKING SENSE OF ADDICTION

Addiction has been discussed in terms of the social and social-psychological processes involved in becoming an addict, and the consequences of drug taking. Addiction is essentially a social process that can be viewed independently from the physical properties of drugs and their impact on the human organism. Merton's (1968) theory of anomie, for example, attempts to explain drug addiction in terms of cultural values (or goals) and the availability of illegitimate means (or norms). According to this theory, the addict is one who has pursued high and unrealistic cultural goals, but who has turned to illegitimate means to achieve them when the legitimate means have been blocked. Inability to achieve goals illegitimately or the rejection of illegitimate means leads to a "retreat" from social life and increases the probability of using drugs. A more recent version of this theory, which has been updated to account for the increasing number of addicts in the middle and upper classes, suggests that this new type of addict, rather than being deficient in the intellectual and social skills required for success to reach his or her goal, may be deficient in the personal skills needed for success or may be frustrated by economic realities (Platt, 1986).

Alternatively, Lindesmith's (1968) theory of addiction emphasizes, first, the addict's learning how to use the drug from others, and, second, that continued use of the drug will relieve the withdrawal distress which results from discontinued use. It has been commented upon previously that Lindesmith's theory has difficulty in identifying and operationalizing the personal cognitive elements in this process and accounting for the fact that some opiate use is oriented around the achievement of euphoria, rather than only the elimination of withdrawal distress. But Lindesmith's account of the addiction process as essentially a learning process would seem to make better sense of the information that has been presented.

Addiction begins with others taking the role of "teachers" with those to be initiated into the use of drugs. Persons have to learn about the drug, how to use it, and the reactions it will produce. This information comes most reliably from those who have experienced the drug at some earlier time and whose judgment is trusted by the initiate. The maintenance of addiction is, similarly, best accounted for not with the physical properties of the drug (though the relief of withdrawal distress and the achievement of euphoria are important motivating forces), but with the drug-supporting role of the addict subculture. If learning initial drug use from others begins the process, the subculture perpetuates it.

This interpretation is consistent with the evidence of the changing nature of heroin addiction in the past decade. Trends in the incidence of heroin use show that the rates of addiction may now be leveling off, but that there are substantial increases in large cities through the mid-1970s. This increase spread to smaller cities. These data also suggest a learning interpretation.

The learning of drug use is most strikingly seen in the process of juvenile initiation into drugs, whether marijuana, crack, or some other drug. While some popular opinion holds that youth drug use is the result of "pushers" who coerce or trick youngsters into experimental use, the induction into juvenile drug use is actually a complex social process embodying the availability of tutors and the opportunity to take drugs. Rather than drug use being an escape from reality, much adolescent drug use is a means of embracing the reality many adolescents experience. This reality is a social reality, one where drug use is an integral part of the activities of the group. For youth, drug subcultures do not arise from use, but use comes about through participation in the subculture.

SELECTED REFERENCES

Adler, Patricia A. 1985. *Wheeling and Dealing: An Ethnography of an Upper-Level Drug Dealing and Smuggling Community*. New York: Columbia University Press.

An ethnographic study of a drug-using and -dealing community in southern California. Adler presents an interesting account of the relationships among marijuana and cocaine dealers and the subculture in which this activity takes place.

Bennett, Trevor. 1986. "A Decision-Making Approach to Opioid Addiction." Pp. 83–102 in *The Reasoning Criminal: Rational Choice Perspectives on Offending.* Edited by Derek B. Cornish and Ronald V. Clarke. New York: Springer-Verlag.

The application of a "rational-choice" model to heroin addiction. The author presents data from British addicts to show the processes of becoming an addict and stopping the addiction process. A nice antidote to the medical model of addiction.

Biernacki, Patricia. 1986. *Pathways from Heroin Addiction: Recovery Without Treatment*. Philadelphia: Temple University Press.

A study of how most addicts are able to terminate their heroin addiction largely without outside intervention. Interesting case materials from addicts and readable accounts of conceiving the addiction—and de-addiction—processes in sociological terms.

Goode, Erich. 1984. *Drugs in American Society*. 2nd edition. New York: Knopf.

> A widely used summary of the social use of drugs in the United States. It contains little material on the physiological consequences of drugs, and the psychological and sociological aspects are emphasized.

Inciardi, James A. 1986. *The War on Drugs*. Palo Alto, CA.: Mayfield.

> A discussion of many issues related to drugs, including the trafficking of drugs such as heroin and cocaine. The author advocates more active legal intervention into trafficking efforts, particularly the smuggling of illicit drugs.

Peele, Stanton. 1990. *The Diseasing of America: Addiction Treatment Out of Control*. Lexington, MA.: Lexington Books.

> An interesting account of the meaning of "addiction," a term that has become increasingly broad in meaning.

Stephens, Richard C. 1991. *The Street Addict Role: A Theory of Heroin Addiction*. Albany: State University of New York Press.

> A systematic account of the street addict role combining sociological perspectives and empirical data. In spite of the greater public concern over cocaine, heroin remains an almost intractable problem. Stephens' book tells us why.

Trebach, Arnold S. 1987. *The Great Drug War*. New York: Macmillan.

> Commentary on the recent (since 1986) U.S. national concern with drugs, especially cocaine. The author suggests that media attention and celebrity endorsements have helped create a more sensational atmosphere with respect to drugs.

Drunkenness and Alcoholism

9

Although increased concern has been expressed about the use of "drugs"—especially cocaine—by segments of the public, and through the mass media, alcohol is by far the most widely used mood-altering drug in the United States today. In fact, the drugs of alcohol and tobacco actually cause more physical, medical, social, and psychological problems than the use of other drugs (Trebach, 1987). Alcohol is used by more persons in the United States than any other drug, including tobacco. In a national survey of persons aged 12 and over, nearly 75 percent reported drinking alcohol during the past year, compared to 36 percent who reported smoking cigarettes (National Institute of Drug Abuse, 1988). It is important, therefore, to consider alcohol as a drug separate from others, such as heroin, cocaine, or marijuana.

THE DEVIANT NATURE OF DRINKING

Not all drinking is considered deviant and conceptions of deviant drinking may differ from group to group. There are

232

norms that identify when drinking "steps over the line" between the social or acceptable and becomes deviant or problematic. It is not drinking that is deviant but the conditions under which drinking is done: the physical situation, the age of the drinkers, perhaps the type of beverage (Klein and Pittman, 1990), the conduct of the parties, etc. Since norms are situational, drinking norms are tied to conceptions of "ought" and "ought not" in situations. For this reason, we need more information than the presence of alcohol to make a determination of deviance; we need more information about the conditions under which drinking is occurring.

To make matters slightly more confusing, there may be little agreement about those conditions. Many people in the United States have a fundamental ambivalence about drinking. Many regard drinking as permissible in many social situations and at many different times; other people regard drinking as impermissible virtually all the time and in every situation. Many people will laugh at jokes about drunkenness and condemn public drunkenness. It is as though many of these people have not yet come to grips—morally, socially, and interpersonally—with the use of alcohol.

This chapter attempts to provide an overview of problem drinking and alcoholism as a form of deviance. First, the physical and behavioral consequences of alcohol consumption are discussed. Then the prevalence and social patterning of drinking is discussed, followed by the prevalence and social patterning of alcoholism. Ideas about the etiology of alcoholism are then discussed in the context of the social control of alcoholism.

PHYSIOLOGICAL AND BEHAVIORAL ASPECTS OF ALCOHOL

Problem drinking can lead to a number of well-documented problems, including physical and psychological dependency, various illnesses, impaired social relationships, and poor work performance. But alcohol is not physiologically habit-forming quite in the sense that certain other drugs are. One does not become a chronic drinker as a result of the first, twentieth, or even the one hundredth drink. Furthermore, it is still to be demonstrated that the excessive consumption of alcohol is inherited. There are, however, physical and psychological consequences to the consumption of alcoholic beverages.

Alcohol is a chemical substance derived through a process of fermentation or by distillation. Although the process of distillation of alcohol from grains, such as barley, corn, wheat and others, is of fairly recent origin, nearly all societies have made some form of fermented beverages such as wine and beer for thousands of years (Patrick, 1952: 12–39). Following the intake of alcoholic beverages, a certain amount of alcohol is absorbed into the small intestine. It is carried in the blood to the liver and then disseminated in diluted form to every part of the body. Although there can never be more than one percent of alcohol in the bloodstream, there is some evidence to suggest that even relatively small amounts can affect the brain, as shown in brain X-rays, computer-assisted tomography scans (CT scans), and other recent medical research. In fact, the range and complexity of the effects of alcohol are such that virtually every organ system in the body is affected by alcohol, either directly or indirectly (Secretary of Health and Human Services, 1987: 134–147).

Physiological Dimensions

The immediate effects of alcohol are determined by the rate of its absorption into the body and characteristics of the individual drinker. The rate of absorption, in turn, depends on the kind of beverage consumed, the proportion of alcohol contained, the speed with which it is drunk, and the amount and type of food in the stomach, as well as on certain individual physiological differences, such as the person's weight. In moderate quantities, alcohol has relatively little effect on a person, but large quantities disturb the activities in the organs controlled by the brain and cause "drunkenness." In addition, there are differing levels of susceptibility among individuals to physiological consequences of alcohol, suggesting that the physiological consequences of alcohol consumption can differ between two people even if their consumption is the same (Secretary of Health and Human Services, 1990: Chapter 5).

The consumption of quantities of alcohol over long periods of time may have a number of health-related consequences. Alcohol consumption has been implicated in numerous health hazards, such as accidents and traffic fatalities (see Hingson and Howland, 1987). Chronic consumption of alcoholic beverages has also been linked with various gastrointestinal disorders, pancreatitis, liver disease, nutritional deficiency, impairments of the central nervous system, disorders of the endocrine system, cardiovascular defects, myopathy, certain birth defects, and several types of cancer (Eckhardt, et al., 1981). In fact, persons diagnosed as alcoholics have greater risk of mortality from numerous physiological disorders than persons who are not alcoholics (Taylor, et al, 1983).

Other Health-Related Effects

It has been estimated that one-half of all traffic fatalities in the United States are alcohol related (Secretary of Health and Human Services, 1990: 163). Although the number of traffic fatalities has declined in recent years, the proportion that were alcohol related has increased. Alcohol is significantly associated with emergency room cases in the United States and in other countries. A comparison of emergency room casualties in the U.S. and Mexico reported a high rate of heavy drinking, drunkenness, and alcohol related problems (Cherpitel and Rosovsky, 1990).

Psychological Effects

Alcohol produces a number of psychological effects on emotional and overt behavior (see the review in Pandina, 1982). In moderate quantities, alcohol can lessen tensions and worry and, in general, may ease the fatigue associated with anxiety. The use of alcohol presents an illusion of being a stimulant because it reduces or alters the cortical control over action. Alcohol has a negative effect on task performance, although this is dependent on the nature of the task and the experience of the drinker. The inexperienced drinker tends to overreact to the sensation of alcohol and may be merely fulfilling what he or she perceives to be socially expected behavior in response to drinking. Such a reaction is commonly seen in groups of teenagers who can behave as if intoxicated when they are under the influence of only small quantities of alcohol.

Many people may believe that alcohol "releases inhibitions" that are already present in the individual or that the person acts out of control under the influence of alcohol. While certain mental and motor skills are clearly impaired as a result of alcohol consumption, considerable evidence supports the view that much of so-called *drunken behavior* or drunken comportment, is the product of socialization and not the result of the alcohol itself that automatically releases behavior from controls. An anthropological survey of the way different cultural groups react to drunkenness shows that among some cultures, the normal inhibitions remain in effect. In many societies, alcohol is often consumed in very large quantities without producing appreciable changes in behavior except for a progressive impairment in the drinker's sensorimotor capabilities, such as coordination (MacAndrew and Edgerton, 1969:36). For example, among the Onitsha of Nigeria, the ability to appear sober in spite of heavy drinking is respected, while drunken behavior brings shame (Umanna, 1967). In other societies, people may display considerable physical aggression when drunk, while in yet other societies drinkers may appear "euphoric" or happy but neither sexually "loose" nor aggressive. In still others, the opposite may occur, aggression becoming "both rampant and unbridled, but without any changes whatsoever occurring in one's sexual comportment" (MacAndrew and Edgerton, 1969: 172). In short, there does not appear to be a universal behavioral consequence to drinking alcoholic beverages. Drunken behavior is largely learned behavior.

PREVALENCE OF DRINKING IN THE UNITED STATES

In the United States, the drinking of alcohol, in order of amount consumed and cost, consists chiefly of beer, followed by wine and distilled spirits. Over the past century, there has been a downward trend in the drinking of distilled spirits and an increase in beer and wine consumption, as measured by tax receipts, sales from state-controlled stores, and estimates from the alcoholic beverage industry. In 1850, almost 90 percent of the alcohol consumed was in the form of distilled spirits and nearly 7 percent was beer; by 1960 only 38 percent was in the form of spirits, 51 percent in the form of beer (Keller and Efron, 1961: 3). Over time, there was a sharp increase in alcohol consumption after the repeal of Prohibition through the 1940s, followed by a flattening out of the consumption curve through the 1950s. After a steady rise through the 1960s and 1970s, the total estimated amount of alcohol consumed declined after 1981. The estimated consumption in 1984 was 2.65 gallons per person and this was the lowest figure since 1977, the first time since Prohibition that consumption declined for three consecutive years (Williams, et al., 1986).

Consumption of both beer and distilled spirits declined by 1984, although wine consumption increased. Consumption of spirits continued a long decline, begun in 1970, dropping to a new low of 0.94 gallons per person in 1984 (Secretary of Health and Human Services, 1987: 2) and to .83 gallons in 1987 (Secretary of Health and Human Services, 1990: 14). If this average consumption is compared to other countries, the U.S. per capita consumption of alcohol is less than that of France and Italy, two heavy wine-drinking countries. France also consumes large quantities of stronger spirits such as brandy or cognac. It appears that the Soviet Union also has an extremely high per capita consumption (Traml, 1975).

But while consumption patterns changed, the volume of drinks consumed did not. The proportion of abstainers between 1967 and 1984 was slightly higher (the increase was found mainly among males) but the proportion of those who experienced some kind of difficulty as a result of their drinking was about the same during those 17 years (Hilton and Clark, 1987). There has been little change in the numbers of moderate and heavy drinkers (Secretary of Health and Human Services, 1987: 4).

The proportion of drinkers in the United States has remained fairly stable over the past two decades. National surveys of drinking practices report that approximately one-third of the U.S. population ages 18 and over are abstainers, one-third are light drinkers, and one-third are moderate to heavy drinkers (see the summaries in Secretary of Health and Human Services, 1987: 4 and Secretary of Health and Human Services, 1990: 2). In every age group, more men than women are drinkers and, among those who drink, there are more heavy drinkers among men than women. Among different racial and ethnic groups, whites of both sexes are the least likely to be abstainers.

The frequency and quantity of adolescent drinking has changed over the past decade or so. Annual national surveys show that the proportion of high school seniors who have ever consumed alcohol is high (about 92 percent), but the proportion of seniors who have used alcohol during the month preceding the survey has declined from 72 percent in 1966 and 66 percent in 1985 (Johnston, et al., 1986) to 64 percent in 1988 (Johnston, et al., 1989). Daily use of alcohol declined from a peak of 6.9 percent in 1979 to 4.8 percent in 1985.

Drinking is more common among college students and there have been increases in recent years, not only in terms of the proportion who engage in daily drinking but also in more occasions for heavy drinking. At the same time, however, there has been a reduction in the proportion of students who report daily drinking (Meilman, Stone, Gaylor, and Turco, 1990). In sum, heavy party drinking, especially among college males, is still common, but there may be fewer students who engage in this behavior on a daily basis. Not all of those who drink heavily continue their drinking into adulthood. One study found that those classified as problem drinkers as adolescents or college students tended to be nonproblem drinkers as young adults 5 years later (Donavan, et al., 1983).

TYPES OF DRINKERS

Drinkers can be classified in terms of the deviation from norms of drinking behavior within a culture and dependence on alcohol in the life organization of the individual. Norms relating to the consumption of alcoholic beverages indicate to drinkers which beverages are appropriate at any given time, how much should be consumed, and what kind of behavior will be tolerated after consumption. The classification of types of drinkers involves information about the frequency and quantity of alcohol consumed. Estimates of the average daily consumption can be obtained (Hilton, 1988), as can the "volmax," or the volume of monthly intake with maximum amount consumed per occasion (Hilton and Clark, 1987). An estimate of the frequency of getting drunk can also be obtained.

Various typologies of drinkers might be constructed depending on different combinations of these conditions. A crude typology might make the following distinctions: A *social or controlled drinker* drinks for reasons of sociability, conviviality and conventionality. Social drinkers may or may not like the taste of alcohol and the effects alcohol produce. Above all else, they are able to take alcohol or leave it alone. They refrain frequently and use alcohol only in certain social circumstances. A *heavy drinker* uses alcohol more frequently. In addition, heavy drinkers may consume such quantities that they become occasionally intoxicated. *Alcoholics* are often defined as persons whose frequent and repeated drinking of alcoholic beverages is in excess of dietary and social usages in the community and is of such an extent that it interferes with health or social or economic functioning.

There are no agreed-upon definitions of the terms "social," "heavy," or "alcoholic" drinker that all observers would subscribe to, and these terms are usually defined operationally. That is, kinds of drinkers are defined slightly differently depending on whose definition it is. A heavy drinker may be defined as one who consumes more than one ounce of alcohol a day (Secretary of Health and Human Services, 1981: 19) or those having five or more drinks per occasion at least once a week (Secretary of Health and Human Services, 1987: 2). The distinction between a social and a heavy drinker is also judgmental. Blue-collar workers, for example, seem to base such judgments on an individual's work record and performance. If the individual can perform satisfactorily at work, he is not considered a heavy drinker, no matter how much is consumed (LeMasters, 1975: 161). For females, such judgments are more likely to be based on the neglect of children rather than absence from work.

Another attempt to classify different drinkers involved over 1,500 male drinkers in New Zealand (Martin and Casswell, 1987). "Light drinkers" comprised 43 percent of the sample. These persons drank infrequently and consumed a small quantity of alcohol on these occasions. "Frequent early evening drinkers," which were 28 percent of the sample, drank between the hours of 5:00 pm and 8:00 pm at home and before dinner. "Heavy hotel-tavern drinkers" were 21 percent of the entire sample and they drank in public places two or three times a week. These drinkers were relatively heavy drinkers but they confined their drinking to after 8:00 pm. "Club drinkers" were 4 percent of the sample and these persons belonged to sports or business clubs. They consumed alcohol before dinner. "Solitary drinkers" comprised 2 percent of the sample and they drank virtually every day between 5:00 pm and 8:00 pm or so. Finally, "party drinkers" (2 percent of the sample) drank in the homes of others after 8:00 pm and usually only in the context of a social situation.

There is even substantial disagreement on the term "alcoholism." To some, the term denotes a disease of physiological dependency and uncontrolled drinking (Jellinek, 1960). To others, the term "alcoholism" is very vague and does not describe a uniform phenomenon (Robinson, 1972). Still others have defended the use of the term, indicating that there is some agreement on the condition of alcoholism even if there is little on when a particular person becomes an alcoholic (Keller, 1982).

Partly in response to the definitional confusion, researchers and treatment specialists alike have used another term to denote someone with difficulty with their drinking: *problem drinker*. Although problem drinkers are heavy drinkers, not all heavy drinkers are problem drinkers. Problem drinking is defined in terms of the consequences of drinking rather than characteristics of the persons who drink or the quantity and frequency of alcohol consumed. The notion of problem drinker recognizes that the consumption of alcohol can result in complications in personal living. In addition to ugly hangovers in which they may collapse physically and become filled with remorse and self-disgust, heavy drinkers may experience blackouts, frequent nausea, deteriorating interpersonal relationships with employers, friends, and family, as well as encounters with the police and other social service personnel. Problem drinkers are those people who experience some problem as a result of their drinking, regardless of how much they consume or the circumstances surrounding that consumption. Often heavy drinkers will deny they have problems as a result of their drinking. One study, for example, reported that heavy drinkers believed that drinking heavily resulted in positive experiences, such as euphoria from alcohol and the facilitation of group interaction (McCarthy, et al., 1983). Such beliefs may have been rationalizations for drinking or they may indicate a consistency between one's attitudes toward drinking and one's drinking behavior.

Chronic alcoholics are persons who have consumed large quantities of alcohol over long periods of time. They usually have a "compulsion" to drink continually. They are also characterized by solitary drinking, morning drinking, and general physical deterioration. Chronic alcoholics are entirely preoccupied with alcohol and they may devise ingenious methods to safeguard their supply of alcohol. It is difficult to describe the alcoholic's terror at being without a drink, so the person resorts to hiding his or her liquor under pillows, under porches, and in any place where they may go undetected. Their day may start with 8 ounces of gin or whiskey, and each day the alcoholic may consume quantities of alcohol far in excess of the customary amounts of his drinking group. One study showed that alcoholics had usually developed their patterns of drinking over a period of 20 years or so of drinking, having become intoxicated for the first time at about 18 years of age and, by the age of 30, had already experienced "blackouts" during intoxication (Trice and Wahl, 1958). By 36 years of age, they were drinking in the morning, and within a year later, on the average, they began drinking alone on a regular basis. By 38, they were first protecting their supply of alcohol, and by 39, they were having their first tremors. By the time they are in their 40s, alcoholics have experienced considerable difficulty with their drinking: they have had to change jobs, face the loss of family and friends, and have started to develop health problems as a result of their drinking. Some alcoholics may even die from the chronic consumption of alcohol in the form of physical diseases, accidents, falls, fires, suicide, and poisoning (Secretary of Health and Human Services, 1990: 163–171).

An alcoholic's problems often do not stop after he or she has terminated drinking. During sober periods, alcoholics are aware of the physical problems they have and the social behavior they display to others when they are drunk. This makes them self-conscious and tense as they try to cope with the symptoms of their alcoholism while, at the same time, attempting to improve their social relationships with others, such as family and friends (Wiseman, 1981).

DRINKING AS A SOCIAL AND GROUP ACTIVITY

From the beginning, drinking has been an integral part of American social life. Ale and beer were part of the daily diet in New England at the time of the Mayflower. In fact, it has been reported that the Pilgrim community of Plymouth in 1621 suffered through a beer shortage that was so severe that the ship's captain gave some of the ship's stores to the colonists before sailing back to England. That winter, the Pilgrim's beer supply disappeared, leading to local manufacture of beer and some distilled spirits (Lender and Martin, 1982: Chapter 1). Likewise, wine was part of the usual daily diet of the Spaniards who settled the Southwest and California. Yet, from the earliest days drunkenness was frowned upon and persons who drank excessively were punished.

Many contemporary drinking patterns have been handed down to us from previous generations. The knowledge, ideas, norms, and values involved in the use of alcoholic beverages, since they are passed down from generation to generation, have maintained the continuity of an alcoholic subculture. All drinking patterns are learned, as are all behavior patterns. Just as there does not appear to be any universal way of behaving under the influence of alcohol, it is also the case that there are no universal patterns of drinking. People in the United States will drink under some circumstances but not others. For example, drinking is not usually part of the funeral ritual in the United States, but it is in Ireland where wakes are common and the consumption of alcoholic beverages a usual part of those activities.

Drinking plays a significant role in everyday life. Alcohol is used by many people to celebrate national holidays and to rejoice in victories, whether those of the football field, war, or ballot box; the bride and groom are toasted, and the father of a new child may celebrate with a drink for everyone present; promotions, anniversaries, and other important social events often call for a drink. Business deals may be consummated over a drink, with a toast serving the same symbolic (and legal) function as a handshake on the agreement. Even some religious ceremonies and, on occasion, as mentioned above, the bereavement of death as in an Irish wake, are accompanied by alcoholic beverages. People are also exposed to alcohol in a number of ordinary contexts. A study of alcohol references on prime-time television found that 80 percent of the network programs examined had either references to alcohol or the appearance of alcohol on the program (Wallack, 1987). Alcohol was consumed on 60 percent of the shows.

Despite its widespread use in the United States, the value system implicit in American drinking patterns differs from those found in Europe. Rather than alcohol drinking being a "vice" or a social problem in many European countries, it has remained just one element of a traditional recreational practice. In other words, where drinking in Europe is merely one aspect of a group's coming together, drinking in the United States has all too frequently been the occasion for a group's coming together. This has surely contributed to the ambivalence that many persons have about drinking (Lender and Martin, 1982: 190–195).

Variations in Drinking Behavior

To say that drinking is social behavior is to say that drinking is socially patterned, that there is variation in drinking frequency by age, education, income, size of community, marital status, and religion.

Sex.   National surveys indicate that males and females have different drinking patterns. Men drink more frequently and a greater quantity than women. While 25 percent of the males reported abstaining, 40 percent of the females reported abstaining; moreover, in the heavier drinking categories, males (14 percent) outnumbered females (4 percent) (Secretary of Health and Human Services, 1981: 20). For those who drank, heavier drinking appeared to peak at age 21–34 for males and at age 35–49 for females; thereafter, drinking declined for both males and females (Secretary of Health and Human Services, 1981: 20; also see Cahalan, 1982).

Social Background.   Drinking also varies by education, income, religion, and region of the country. Generally, those adults with a higher level of education drink more than those with less education. Drinking also increases directly with income. With reference to religion, the proportion of those who drink is highest among Jews, Catholics, and Lutherans, and lowest among other large Protestant denominations. Drinking is more common in larger cities than smaller towns, and among unmarried persons than those who are married. Regional differences in drinking reflect interesting patterns. Alcohol consumption has been reported to be lower in the South than other regions of the country, and highest in the West. But, calculated for drinkers only, the pattern changes and shows that the South consumes more alcohol than other regions. These data suggest that although there are more abstainers in the South, persons in the South who drink tend to be heavier drinkers than persons in the other geographic regions. Conversely, the Northeast United States has a relatively high proportion of drinkers but they consume less alcohol than those in the South (Secretary of Health and Human Services, 1987: 6–8).

There are variations in drinking among various racial and ethnic groups (Welte and Barnes, 1987). There is a high proportion of heavy drinkers among American Indian adolescents and adults. Rates of heavy drinking are relatively low among African American adolescents compared to Hispanics and whites. Among Oriental youth, rates of heavy drinking are high among males but low among females.

Public Drinking Houses

A large proportion of drinking is done in a group context in public drinking houses, which are found in most of the world today. Public drinking houses are more than places where alcoholic beverages are sold for consumption on the premises. They have several important characteristics: (1) they involve group drinking; (2) this drinking is commercial in the sense that anyone can buy a drink as contrasted with bars of private clubs; (3) they serve alcohol and can thus be distinguished from coffee houses or teahouses; (4) their bartenders serve as functionaries of the institution and around them, in part, the drinking gravitates; and (5) many customs are associated with them, including the physical surroundings, types of drinks, hours of sale, and kinds of behavior that are considered appropriate (Clark, 1981). There are over 200,000 bars and taverns in the United States alone and they are widely patronized.

Taverns serving alcohol have existed for thousands of years. They date back easily to Greek and Roman times and were an integral part of early American life in colonial times—in part because it was believed that drinking not done in public was likely to be excessive, whereas the sale of liquor in a tavern could be regulated. The Puritan authorities in Massachusetts in 1656 even enacted a law making towns liable of a fine for not maintaining an "ordinary" (tavern) (Field, 1897: 11–12; see also Firebaugh, 1928). As the Industrial Revolution brought thousands of immigrants, particularly single men, to the cities to work in the factories, a new type of public drinking house, the "saloon," replaced the wayside tavern, a facility that served drinks as well as tending for travelers in colonial America. The saloon became common in urban areas and was characterized by strictly male patronage, drinking at an elaborate bar with "free lunches," and a special family entrance. The modern tavern made its appearance after Prohibition. Women were generally permitted, the surroundings were more attractive, and drinking usually took place at tables rather than at a bar. At least five different varieties of public drinking houses emerged, the types largely associated with different areas of the city (and the socioeconomic status of the persons who resided there): the skid row bar, the downtown cocktail lounge and bar, dine and dance-type establishments (also singles' bars), the night club, and the neighborhood tavern. The last type of tavern is the most numerous, comprising perhaps three fourths of all taverns.

The relationship between going to taverns and excessive drinking has been controversial with some writers claiming that tavern attendance does not lead to excessive drinking (Popham, 1962: 22), and others claiming that they do (Clark, 1981). It does appear that tavern patronage is associated with excessive drinking, although the explanation for this relationship is not agreed upon. It should not be surprising that tavern patronage should be related to drinking—that, after all, is a major reason for the existence of the tavern and the more persons go to taverns and bars, the more they are exposed to norms and values that favor and promote heavy drinking. Eventually, excessive drinking could become a habitual mode of drinking in life in general and not just at taverns and bars. Those persons who have been found to be most likely to frequent taverns and to drink heavily are young, unattached, unchurched, gregarious males (Nusbaumer, et al., 1982). A survey of tavern-goers in Canada reported that while most Canadians go to taverns (pubs), few are regulars (Cosper, et al., 1987). The regulars tend to be young, unmarried persons who are heavier-than-average drinkers.

Tavern drinking is not, of course, the only evidence of the social and group aspects of the consumption of alcoholic beverages. Middle- and upper-class drinking, which may not be centered around a public drinking house, is similarly social in nature and centers around various group events, the cocktail party and the bar in the country club or dining establishment. Such occasions for drinking are usually a good deal more private than public drinking houses, although group processes are no less evident.

Ethnic Differences in Excessive Drinking

There are pronounced differences in the extent of excessive drinking in various ethnic groups. These differences are related to some of the factors discussed above (such as religious preference) and these differences point to the importance of drinking norms and the learning of drinking behavior in these groups.

The Irish.   Excessive drinking has long been associated with the Irish, and many of these drinking patterns have carried over to Irish-Americans. It is believed that their rates of alcoholism probably exceed those of any other single ethnic group. Irish men drink because their culture permits and sometimes prescribes general drinking, particularly strong beer and whiskey. Their drinking is not confined to ceremonial purposes. Typically, Irish society has been characterized by a pattern of high socioeconomic aspirations, late marriage, and much emphasis on the virtue of abstaining from sex until after marriage. As a result of these limitations on marriage and the consequent sex segregation, "bachelor groups" are an important feature of Irish social structure. Married men are also a part of the bachelor group, and they are often leaders in socializing young men into their drinking practices.

> A boy became a man upon initiation into the bachelor group, that is, when first offered a drink in the company of older men in the local public house. Farm and marriage might be a source of male identity for a few, but hard drinking was a more democratic means of achieving manhood. (Stivers, 1976: 165)

Drinking is often a source of prestige and esteem among men who live in a male segregated world. But the hard drinker in Irish society was, and still is, seldom a persistent drunkard, because chronic drunkenness has not been culturally sanctioned. Even today the rate of alcoholism in Ireland is comparatively low; but the "heavy drinker" stereotype followed Irish immigrants to the United States where it was translated not into heavy drinking, but drunkenness and alcoholism.

The Italians. A reverse situation exists among Italians who have always had a tradition of using wine with meals. Despite their extensive use of alcohol, the Italians have a very low incidence of alcoholism. Although the rate of alcoholism is also low among Italian-Americans, it appears to be higher than in Italy, even though the total consumption of alcoholic beverages is higher among Italians in Italy. Many Italian-Americans have retained the tradition of drinking wine with meals and regarding the practice as healthful; such an attitude helps to prevent alcoholic excess and addiction. Italians are first introduced to wine very early in life, both men and women drink wine, and there is very little opposition to the drinking of wine by young persons. Alcohol is seldom used for "escape" purposes (Lolli, et al., 1958: 79).

These drinking patterns, in general, have been found to be only partially present in Italian-Americans, and their absence leads to excessive drinking and alcoholism. For example, while 70 percent of the Italian men and 94 percent of the Italian women did all their drinking at mealtime, only 7 percent of first-generation Italian-American men and 16 percent of Italian-American women did so.

The French. Persons in both Italy and France drink about the same large quantities of alcohol each day, but the rates of alcoholism are much greater in France, which is thought to have one of the highest alcoholism rates in the world (Sadoun, et al., 1965). This difference can be explained by a number of factors. (1) Nearly all the alcohol intake in Italy is wine, consumed at mealtime, while in France a substantial amount of the alcohol intake is in the form of distilled spirits and aperitifs between and after meals. In fact, in France, alcoholism rates are lower in the southern part of the country where wine is largely consumed and mostly at mealtime. (2) Exposure to alcohol in childhood is viewed quite differently in the two countries. The French have rigid parental attitudes either favoring or opposing drinking among children, while most Italians accept the drinking of wine in childhood as a "natural" part of a child's development. (3) The Italians have a much lower "safe limit" for amounts of alcohol than the French, and they tend to view drunkenness as a personal and family disgrace. (4) The French view drinking, particularly of copious amounts, as associated with virility, while the Italians do not.

Asian Americans. In the United States, Asian Americans, who comprise about 2 percent of the population, have very low levels of alcohol use. Research has indicated that Asian Americans of both sexes drink less than whites, African Americans, or Hispanics (Klatsky, et al., 1983). A comparison of several Asian groups detected considerable differences in drinking among Asian groups. Rates of abstention were particularly high among Koreans (67 percent), Chinese (55 percent), and Japanese (47 percent), compared to 13 percent in the general population of California (Kitano, et al., 1985; Sue, et al., 1985).

Among those Asians who drink, the Chinese have especially lower rates of heavy drinking. For many Asian groups, alcohol is largely consumed as a part of social functions, public drunkenness is disapproved, and children are educated to observe these patterns. In this way, drinking is socially-controlled. Drinking may also be controlled by the "flushing response," a physiological reaction among Asian people characterized by "facial flushing, which is often accompanied by headaches, dizziness, rapid heart rate, itching, and other symptoms of discomfort" (Secretary of Health and Human Services, 1990: 35).

Because of the low incidence of both drinking and problem drinking among Asian Americans, there has been little research on the drinking patterns that are found in these groups. Sue (1987) reports that although Asian Americans are more likely to be abstainers than any other ethnic group, the frequency and amount of drinking appears to be increasing.

Native Americans. There is a great diversity in the drinking patterns among Native Americans. Some tribes are mainly abstinent, others are characterized by high rates of excessive drinking. One observer has estimated that a smaller percentage of Native Americans drink than is found in the general population, but the percentage of problem drinkers among those who do drink is greater (Lemert, 1982). Drinking was heavy on the American frontier during the nineteenth century among farmers, trappers, and cowboys. It was these groups that the Indians came into contact with and from whom they secured their first alcohol. Also, most Indians were unfamiliar with the use of alcohol and this, with the encouragement of white men, led to excessive drinking. When the excessive use of alcohol became common among certain tribes, federal laws were enacted beginning with a general law in 1802 and a final, more specific one in 1893 (and again in 1938) which made it an offense to serve intoxicants to a Native Ameican. These laws were not repealed until 1953.

Because different tribal groups exhibit great variations in drinking behavior, it is not possible to make general statements about drinking among Native Americans. In a survey of more than 280 tribes, some were found to be characterized by binge drinking followed by periods of sobriety, but other groups were almost entirely abstinent (Lex, 1985). In still other groups, moderate drinking is typical. The Pueblo, for example, of the Southwestern United States are abstainers, and drinking found among some other tribes does not exceed that found in the rest of American society. There are some tribes, however, where there is a large proportion of heavy drinkers, such as the Ojibwa, where researchers classified 42 percent of the adults as heavy drinkers (defined as being drunk 2–5 times a week) (Longclaws, et al., 1980).

Native American drinking patterns are reflected mainly in the behavior of young, unemployed males. Drinking is done in small groups, usually with wine. These groups are similar to the "bottle gangs" found in skid row areas where the object of drinking is to get drunk as quickly as possible. There is little cultural disapproval of this behavior which some observers have interpreted to indicate that Native American excessive drinking is a form of protest against the abuses of living in a white man's society (Lemert, 1982).

Deaths from alcohol-related causes are especially prevalent among Native Americans. The mortality rate for chronic liver disease among Native Americans and Alaskan Natives was reported to be 29.2 per 100,000 compared to 9.7 per 100,000 for the U.S. as a whole (Secretary of Health and Human Services, 1990: 36). And, although Native American women drink considerably less than men, they are more susceptible to health-related problems; women account for nearly half of the liver cirrhosis deaths among Native Americans.

African Americans.   Despite the fact that most studies suggest that alcoholism—as well as problem drinking—rates are generally higher for African Americans than for whites, relatively few studies have been made of the drinking patterns among African Americans (Sterne, 1967; Lex, 1985). On the basis of available evidence, African American women have a higher rate of alcoholism than do white women; furthermore, rates of alcoholism are higher among African American men compared to all blacks than among white men compared to all whites (Secretary of Health and Human Services, 1987: 35–37). Rates of heavy drinking for African American males are inversely related to income; rates of heavy drinking are higher in lower paying jobs.

White males between the ages of 18 and 29 appear to be at the highest risk for alcohol problems (for example, medical, social, occupational), but African American men were at the lowest risk in this age group (Herd, 1989). For men in their thirties, the rates of problems decreased sharply for whites but increased for African Americans. The rates for problems from drinking remained higher for African Americans than whites throughout middle and older ages.

There have been other surveys comparing African American and white drinking patterns. In a sample involving 723 African Americans and 743 whites from a national sample, Herd (1990) found that at aggregate level, both groups had very similar drinking patterns in terms of the proportion of abstainers and frequent and heavier drinkers. But these larger group differences masked some important differences in the circumstances of drinking. For example, Herd found that frequent heavy drinking among whites was associated with youthfulness, high-income status, and residing in "wetter" areas, while frequent heavy drinking patterns among African Americans were reversed. Overall drinking levels were lower among blacks than whites, but African American males reported higher rates of drinking-related problems than white males. These findings suggest that there are important cultural differences in the drinking behavior of African Americans and whites.

A major difference between African Americans and whites concerns reported drunk driving. White men have drunk driving rates 2.5 times higher than those for African American men, and white women have drunk driving rates 5 times higher than African American women (Herd, 1989).

Rates of abstention are higher among African American youth than white youth (Lowman, et al., 1983), suggesting that blacks come to heavy alcohol later than do whites. Despite this later onset of drinking, however, African Americans have been shown to enter treatment at younger ages than whites (Herd, 1985). African Americans also suffer from other negative consequences of heavy alcohol use. For example, African Americans have twice the cirrhosis death rate of white males and nearly four times the rate for white females (Secretary of Health and Human Services, 1987: 37).

The determining factor in the way alcoholic beverages are consumed among African Americans is cultural, not racial or biological (Larkin, 1965). There are pronounced differences between lower-class black and middle- and upper-class African American drinking patterns and rates of alcoholism, much like there are differences among whites by social class. Differences include middle- and upper-class African Americans being more moderate in their use of alcohol than those in the lower-class, and lower-class African Americans drinking more often in public rather than in private places. The fact that many lower-class African Americans drink in public places may also explain their higher arrest rates for drinking.

Hispanics. Hispanics are a heterogeneous group with diverse cultural, national, and racial backgrounds. They comprise about 6 percent of the population in the United States. Hispanic men have relatively high rates of alcohol use and heavy drinking, while Hispanic women have high rates of abstention. The first large-scale national survey of drinking patterns among Hispanics reported that nearly half (47 percent) of Hispanic women abstained from alcohol use and an additional 24 percent drank less than once a month (Secretary of Health and Human Services, 1987: 39). In contrast, only 22 percent of Hispanic men were abstainers, and 36 percent drank heavily (drank at least once a week and consumed at least five or more drinks at a sitting). Hispanic men drank more heavily in their 30s than in their 20s, and their consumption declined only after age 40. Drinking levels increased with increasing education and income, with persons in the higher income brackets and with higher levels of education having lower rates of abstention and higher rates of heavy drinking (Caetano, 1984). Finally, Mexican American men had the highest rates of abstention and heavy drinking when compared with Hispanics of Cuban, Puerto Rican, or other Latin American descent.

A high proportion of Hispanics have reported problems with drinking. About 18 percent of Hispanic men and 6 percent of Hispanic women experienced at least one alcohol-related problem during a one-year period prior to being surveyed (Caetano, 1989), and the prevalence of alcohol-related problems is higher among Hispanic men than among African American or white men. Little is known of the specific risk factors that face Hispanics with respect to problem drinking, although it appears that drinking and excessive drinking are associated with acculturation (the degree to which one has accepted and adapted to the social and cultural norms of a new environment) to life in the United States. Thus, drinking and excessive drinking are high among acculturated American Hispanics as compared to Hispanics in Mexico or Spain (Secretary of Health and Human Services, 1990: 34–35). The proportion of abstainers is higher among Hispanics who have not acculturated.

ALCOHOLISM AND PROBLEM DRINKING

Determining the extent of alcoholism and problem drinking is different from determining the extent of drinking. Because there is no standard definition of "alcoholic" or "problem drinker," it is impossible to obtain completely adequate estimates of such persons in the United States.

The Extent of Alcoholism and Problem Drinking

Although overall drinking patterns have changed in recent years—with the decrease in consumption of distilled spirits and a slight increase in the proportion of abstainers—the proportion of heavy drinkers has remained about the same in the past two or three decades (Hilton and Clark, 1987). Although two thirds of the U.S. population drinks, actual consumption of alcohol is very unevenly distributed throughout the drinking population. The 10 percent of the drinkers (6.5 percent of the total population) who drink the most heavily account for fully half of all alcohol consumed. The other half of the alcohol consumed is accounted for by the cumulative 90 percent of the drinking population who are infrequent, light, or moderate drinkers (Secretary of Health and Human Services, 1987: 4).

If problem drinking is defined only in terms of alcohol consumption, a fairly large proportion of persons report being problem drinkers. When, for example, heavier consumption was defined as 120 or more drinks per month, 15 percent of adult male drinkers and 3 percent of adult female drinkers in a national survey reported drinking at this level (Secretary of Health and Human Services, 1981: 37). Clearly, these persons are in danger of developing either alcoholism or serious problem drinking. Taking into account the fact that definitions of alcoholism vary and that most drinkers are best considered social or infrequent drinkers, the National Institute of Alcohol Abuse and Alcoholism (NIAAA) of the federal government estimates that about 10 percent of adult American drinkers are likely to experience either alcoholism or problem drinking at some point in their lives (Secretary of Health and Human Services, 1981: 38).

Sometimes, the number of alcoholics is estimated by looking at mortality figures, particularly deaths from cirrhosis of the liver. Such figures are complicated, however, by the fact that rates of cirrhosis are only partially related to alcoholism, and less than 10 percent of all alcohol-related deaths are due to cirrhosis (Secretary of Health and Human Services, 1987: 11–12). Still other techniques for estimating the number of alcoholics include examining admissions to voluntary treatment programs and interviewing a sample of the general population about their preoccupation with alcohol or whether they have had any negative consequences to their drinking. All such techniques invariably will disclose different estimates of alcoholism.

Countries with the highest estimates of alcoholism are nearly all European nations. France has the highest known rate, followed by Chile, Portugal, and the United States. Somewhat lower but still high rates are found in Australia, Sweden, Switzerland, the Union of South Africa, and Yugoslavia (Keller and Efron, 1955: 634). Alcoholism and problem drinking in the Soviet Union is thought to be a major problem, although there are no comparable figures to those published officially in other countries (Traml, 1975).

Costs of Alcoholism

Industry loses large sums of money due to excessive alcohol consumption in the form of absenteeism, inefficiency on the job, and accidents. It has been estimated that between 3 and 4 percent of an average work force will be deviant drinkers at one time, and that the costs of such drinking must include not only shoddy work performance, but also the costs of trying to do something about the problem in the form of alcohol treatment programs (Trice and Roman, 1972: 2).

The annual economic cost of alcohol-related problems has been roughly estimated at nearly $90 billion for calendar year 1980 (Secretary of Health and Human Services, 1987: 43). This figure is conservative and includes estimates for lost production, health care expenditures, violent crimes where alcohol has been implicated, fire losses, research on alcohol problems, and the costs for the social responses to alcohol problems in the form of social welfare programs for problem drinkers and their families. Lost employment and reduced productivity accounted for more than one half of this amount. Health care for accidents related to alcohol abuse, including alcoholism, liver cirrhosis, cancer, and diseases of the pancreas, was estimated to cost $10.5 billion.

We have already mentioned the very high accident rate and general mortality rates for problem drinkers. Problem drinkers are more likely to be injured in accidents, to suffer diseases, to be killed in accidents or crimes, and to commit suicide than either nonproblem drinkers or abstainers (Secretary of Health and Human Services, 1981; Secretary of Health and Human

Services, 1987; Eckhardt, et al., 1981; Hingson and Howland, 1987). In addition, there are other, more subtle costs that are difficult to link with precise dollar amounts but whose damage is nonetheless real. These costs include problems of infant and child care that occur when one or both parents drink too much; marital problems and the costs of counseling married couples struggling with alcohol problems; the role of alcohol in the committing of acts of violence against children, spouses and others; the problems of school children who underachieve or get into trouble because their drinking parents do not care or are unable to supervise them; the costs of the children of problem drinkers who may themselves become problem drinkers when they grow older; and the simple but poignant human costs of emotional investments in relationships with persons who drink too much (Straus, 1982: 146).

Alcohol-Related Crime

Arrests for public drunkenness and alcohol-related crime constitute about 30 percent of *all* arrests made by the police in the United States (Federal Bureau of Investigation, 1990: 172). Arrests are much more likely of lower-class persons than middle- or upper-class persons, and members of some minority groups are much more likely to be arrested for public drunkenness. Persons who are arrested for public drunkenness have a very high recidivism rate. The term *revolving door* has been applied to the flow of public drunkenness cases through the criminal justice system. The term refers to the seemingly never-ending cycle of arrest, jail, release without treatment, and rearrest (Pittman and Gordon, 1958). Thus, the high volume of arrests masks the fact that many of those arrested are the same persons over and over again.

In addition to drinking by some persons being a crime in itself, alcohol has been implicated in certain other crimes as well (Room, 1983). Drunkenness is undoubtedly of some significance in some violent personal crimes such as homicide and aggravated assault, but the precise role of alcohol in these crimes is not known. In Wolfgang's study of homicide in Philadelphia, nearly two thirds of the offenders had been drinking at the time of the crime, and many other types of violent offenders had been drinking before their crimes as well (Wolfgang, 1958).

A Department of Justice study found that 54 percent of persons arrested for violent crimes in 1983 had been drinking before the crime (Bureau of Justice Statistics, 1985). The highest proportion of offenders who had been drinking before their crimes was with assault (62 percent) and manslaughter (68 percent). It is unclear, however, what role drinking played in the commission of these crimes. There is no direct "causal" relationship between alcohol use and violent crimes, and alcohol use does not inevitably lead to aggression. It is probable that, in most cases where alcohol is associated with criminal behavior, the drug acts as a depressant and makes the person temporarily less cognizant of the consequences of the act. Even when crimes are committed under the influence of alcohol, no single pattern of criminality is seen as being associated with drinking. The relationship between drinking and crime also reflects the fact that both crime and heavy alcohol consumption are most prevalent among certain groups, especially young males (Room, 1983).

Federal statistics indicate that jail and prison populations have a high proportion of drinking problems, and that persons with drinking problems are more likely to engage in criminal behavior than people without such problems (Secretary of Health and Human Services, 1990: 171–172). There is also a connection between drinking and family violence.

Drunk Driving

The exact role of alcohol in cases of drunk driving is clear. Traffic crashes are the greatest single cause of death in the United States for people under the age of 34 (National Highway Traffic Safety Administration, 1988). Alcohol consumption plays a major role in traffic fatalities. During 1984, more than 44,000 traffic fatalities occurred on U.S. highways and street. The blood alcohol levels (BAL) of about one third of the drivers involved in these fatalities are known (Secretary of Health and Human Services, 1987: 13–14). Comparing the percentage of those who were intoxicated in 1981 with the comparable figure for 1984, the proportion of fatally injured drivers who were legally intoxicated dropped from 46 percent in 1980 to 38 percent in 1987 (Secretary of Health and Human Services, 1990: 164). This trend is also evident in accidents involving teenagers. It has been estimated that the number of fatal teenage crashes involving drunk drivers decreased from 2,187 in 1982 to 1,494 in 1987 (National Highway Traffic Safety Administration, 1988). It is not possible to say that such a reduction is due to any given program of driver education and/or deterrence, but it appears that both of these possibilities may be responsible for the decrease.

Public interest groups, such as Mothers Against Drunk Driving (MADD) and Remove Intoxicated Drivers (RID), both organized by women who lost husbands or children in traffic accidents involving drunk driving, have been very visible in public debates on drunk driving. These organizations have pressed for stronger sanctions against drunk drivers and greater public awareness of the problems associated with drinking drivers. Local chapters of these groups have sprung up throughout the country, and these groups have attempted to influence police law enforcement practices as well as initiate legislation for stronger penalties against drunk drivers. Leaders in these local groups have often experienced victimization, either personally or in their family, but it is also the case that many leaders have not experienced victimization. One analysis of the leadership in local MADD organizations reported victimization was an important factor and that the leaders often had a previous background in activist organizations and that

"MADD tends to be run by activists who have been victimized rather than victims who have become activists" (Weed, 1990: 469).

In spite of these efforts, in many communities drunk driving is dealt with as a misdemeanor; if a person is injured as a result, in many states a sentence of up to five years is not uncommon, and in the case of a death, the person can be charged with negligent manslaughter. Often these sentences are mitigated, although license suspensions are automatic for periods of up to one year in many states. The arrest statistics for drunk driving show that nearly 90 percent of all persons arrested for this offense are male, and most of them are in their 20s (Federal Bureau of Investigation, 1990), a pattern that conforms to research on drunk driving (Bradstock, et al., 1987). Yet, short jail sentences and license revocations have not been proven to be effective long-term deterrents to drunk driving. Furthermore, there is no evidence that private programs of "alcohol education" for drunk drivers who are convicted (or about to be convicted) are effective in reducing recidivism (Jacobs, 1989).

More than 120 million Americans are licensed drivers and close to 100 million of them drink, probably constituting billions of occasions when alcohol is in the body of the person driving. One observer has estimated that the absolute risk of a crash for the drunk driver is about 1 in 1,000, and the risk of causing injury is much lower; the risk of that crash causing death is even lower yet (Ross, 1982: 107). Because only a small number of drunken driving episodes end in such tragedy, it would be inaccurate to say that the average trip of a drunk driver is always life-threatening. The problem of drunk driving is complicated by the fact that the legal definition of "drunk driving" is arbitrary. It is now the case that a blood alcohol level (BAL) of 0.10 percent is the legal limit in most states, but with the actions of groups like MADD and RID, some state legislatures have considered lowering the legal level. This would, of course, increase the number of drunk drivers on the road by definition. By the beginning of 1991, several states—Oregon, Vermont, Utah, Maine, and California—had lowered their legal limits to .08 percent (*USA Today*, December 28, 1990, p. 10A.)

It appears that increased legal penalties can influence this conduct, but only in the short run. In the most careful assessment of this issue, Ross has concluded that increases in the potency of legal threats, particularly enhancement of the perceived certainty of apprehension, does produce a statistically significant decline in drunk driving, but that such effects are temporary since efforts at enforcement can rarely achieve or maintain the needed level of intensity (Ross, 1982 and 1987). Increases in the severity of legal punishments and active programs of law enforcement may influence driver perceptions of legal risk which, in turn, are related to reductions in drunk driving (Secretary of Health and Human Services, 1990: 250–252). Yet, no program has been able to sustain such perceptions over time. The recent use of sobriety checkpoints seem to lead to significant reductions in alcohol-related crashes and traffic deaths (Ross, 1985), but this method too provides diminishing returns over time as drivers are able to anticipate them.

One of the most innovative strategies is one that attempts to block the opportunity of drunk driving. Using an apparatus called an "Interlock System," some jurisdictions are experimenting with a device that prevents drunk drivers from operating their cars (Jacobs, 1989: 170–171). The Interlock System requires factory installation and consists of a tube into which the driver breathes for 4 minutes. If the BAL is acceptable, a green light signals permission to start the car; if the BAL is marginal, a yellow light appears; and, if the BAL exceeds .10 (or some other, lower, preset figure), a red light appears. The car can not be started until the light is green. The system even "codes" the driver's breath so that someone else can not start the car. The device must be checked for accuracy every month or the car will not operate properly.

GROUP AND SUBCULTURAL FACTORS IN EXCESSIVE DRINKING

Group associations and subcultural factors play important roles in determining who becomes an excessive drinker and who does not. Different drinking customs are evident in each modern society and also in the subcultural groups of the society. Subgroups differ in the way in which alcohol is used, in the extent of drinking, and in attitudes toward alcoholism and drunkenness. Some people believe that frequent drinking will lead to alcoholism; yet, those groups with relatively high frequencies of drinking, such as Jewish Americans and Italian-Americans, have low rates of alcoholism (Snyder, 1978; Lolli, et al., 1958). Still, it makes intuitive sense that the frequency of drunkenness does lead to alcoholism; however, the Aleuts, the Andean Indians, and those of the Pacific Northwest, among whom drunkenness is common, appear to have little alcoholism (Washburne, 1961; Berreman, 1956; Mangin, 1957; Lemert, 1954). Clearly, there is more to alcoholism than frequency of drinking and getting drunk.

Rates of alcoholism and problem drinking may be due partially to the degree to which drinking behavior patterns are integrated into the culture or subculture. That is, if conformity to drinking standards is supported by the entire culture or subculture, and the values and sanctions are well established, consistent, and known and agreed to by all, rates of problem drinking will tend to be low. High rates of alcoholism seem to be associated with cultures where there is conflict or ambivalence over its use, where children are not introduced to alcohol early in social and dietary contexts, where alcohol is drunk outside of meals, and where it is drunk for personal reasons and not as part of the ritual, ceremony, or family life.

In some modern societies, such as the United States, Ireland, France, and Sweden, marked ambivalence is seen regarding alcohol use; this results in conflicting values and norms which contributes to high rates of alcoholism (Lender and Martin, 1982; but see Room, 1976). In other societies with low rates of alcoholism, like Spain, Italy, and Japan, and among Jewish groups, attitudes about alcohol are permissive, positive, and consistent. Attitudes that stress positive and negative aspects of drinking may fail to regulate excessive drinking. The overall consumption of alcohol has declined in many countries world-wide. In one survey of 25 nations, nearly two thirds of the countries had experienced a decline in per capita alcohol consumption between 1979 and 1984 (Horgan, Sparrow, and Brazeau, 1986), although consumption has continued to increase in developing countries (Hilton and Johnstone, 1988).

Ambivalence is reflected in differing opinions about the appropriateness of drinking—the circumstances, the amount, and the motivation of the drinker. In the United States, many people are ambivalent about alcohol. For example, 91 percent of a sample of California residents indicated that alcoholism is an illness, but 40 percent also believed that alcoholics drink because they want to (Caetano, 1987). Presumably, few people would be able to reconcile these ideas completely.

The role of group and subcultural factors in producing excessive drinking and alcoholism can be seen in relation to (1) gender differences in heavy drinking, (2) companions and excessive drinking, (3) skid row drinking, (4) occupation and excessive drinking, (5) religious differences in excessive drinking, and (6) ethnic differences in excessive drinking. In each instance, it is clear that excessive drinking, rather than being random or strictly individual behavior, is socially patterned or structured. Such differences point to the importance of social learning and cultural values as dominant forces in problem drinking. What a person does with alcohol and what he or she thinks about it are functions of group membership and feelings of identification with these groups. In those groups where there is agreement about drinking customs and values, and where there are social supports for moderate drinking and negative sanctions for excessive drinking, low rates of problem drinking will be found; where these factors are absent, high rates should be expected.

Gender Differences in Excessive Drinking

Data from two national surveys, one in 1971, the other in 1981, show that there were no increases either in drinking or heavy drinking among women during that decade (Wilsnack, et al., 1987: 97). Heavy drinking was found to a greater extent in certain groups: aged 21–34, never married and divorced/separated, unemployed, and women who were cohabitants. A majority of women drinkers "reported that drinking reduced their sexual inhibitions and helped them feel closer to and more open with others" (Wilsnack, et al., 1987: 99). Of those women who had the heaviest drinking, most suffered from role deprivation. Many had lost family roles (through divorce or abandoning their children) and occupational roles (through unemployment). Female drinking is strongly related to the drinking of their husbands or partners, or other close friends (Wilsnack, et al., 1987: 105). Women were more influenced by their husband's drinking than the husbands were of their wives' drinking.

Female alcoholics appear to have backgrounds that are not dissimilar to those of male alcoholics. Women alcoholics generally appear to have more family disruption histories than do women as a whole, and they have a higher incidence of alcoholism in their families (Beckman, 1975; Bromet and Moos, 1976).

More men than women drink alcohol and male alcoholics outnumber female alcoholics. Women are more likely to abstain from alcohol than men, and men are more likely to be heavier drinkers than women. Typically, men consume more alcohol more frequently than do women. Because men generally weigh more than women, they must consume more alcohol than women to feel the effects of drinking. This tends to reinforce drinking behavior. Men are more likely to frequent public drinking houses and to consume larger quantities of alcohol while there (Nusbaumer, et al., 1982). The drinking habits of males and females are quite different (Wanberg, 1970). Several reasons in addition to those mentioned may account for the differences in the rates of problem drinking between men and women. (1) Proportionately fewer women than men drink. (2) There is greater social stigma attached to women drinking than there is to men drinking. (3) Female homemakers have historically not faced the same occupational drinking hazards (for example, "three martini lunches") as do their working husbands. (4) A woman's self-image is based more on role performance that is not known to large numbers of persons, while men can fail in both their private (family) and public (occupation) roles and, therefore, suffer greater social damage (McCord and McCord, 1961: 10–11). Sex differences in excessive drinking may be related to differing expectations for men and women in our society. Men are more likely to engage in drinking at an earlier age and to engage in more social drinking than women. Among the women, there is a relationship between heavy drinking and marital instability, but it is not clear which is the cause and which the effect.

There has been some research on the possible convergence in the rates of excessive drinking among men and women. With changes in sex roles in society generally, some have thought that female rates would approach those of men. Recent evidence, however, suggests this has not taken place. Hasin, Grant and Harford (1990) compared male and female liver cirrhosis rates over a 25 year period (1961–1985). They found that there was no evidence of convergence of rates during this time in spite of changes in sex roles. However, several surveys have indicated that there has been an increase in heavy drinking among younger women (e.g., Wilsnack, et al., 1987; Hilton, 1988), suggesting that a closer examination of age-specific drinking may be necessary.

Companions and Excessive Drinking

Drinking generally takes place in small groups, and within these groups drinking norms tend to develop. (In fact, the person who is the only one drinking alone in the presence of others may be regarded as a deviant even if the others are drinking.) A close relationship exists between the development of alcoholism and the types of companions one has. Drinking norms appear to conform closely to those of age contemporaries and persons with whom one comes in contact in the context of an occupation. Friends, spouses, and co-workers may be very influential in transmitting norms and attitudes about drinking and teaching drinking customs.

This can be seen in processes of adolescent drinking. According to national surveys, about 25 percent of 10th-12th graders reported abstention from alcohol, 7.6 percent reported infrequent drinking, and 18.8 percent reported light drinking. Heavier drinkers constituted about 15 percent of that population (Secretary of Health and Human Services, 1987: 33; Johnston, et al., 1986). Heavier drinking appears to taper off after age 17. In terms of alcohol misuse (which was defined as self-reports of drunkenness at least six times a year or negative consequences from drinking in two of five areas of social life), 31.2 percent of the adolescents were classified as alcohol misusers. Most adolescent drinking is group drinking, so it should be unsurprising that studies of youth drinking find that peers play a crucial role in the drinking process. The relationship between adolescent drinking and the percentage of one's friends (peers) who drink is strong and reciprocal. That is, a person's alcohol use affects that person's peers' drinking patterns, and peer drinking patterns affect the individual's drinking. This relationship, however, is limited to close friends (Downs, 1987). The number of friends who drink and perceive approval/disapproval from peers are significant predictors of adolescent alcohol use (Jessor and Jessor, 1978; Burkett and Carrithers, 1980). In fact, a writer, after reviewing much of the literature on the importance of peers with respect to illicit drug use as well as adolescent alcohol use, concludes that:

> The most consistent and reproducible finding in drug research is the strong relationship between an individual's drug behavior and the concurrent drug use of his friends, either as perceived by the adolescent or as reported by the friends. . . . Peer related factors are consistently the strongest predictors of subsequent alcohol . . . use, even when other factors are [taken into account]. (Kandel, 1980: 269)

Drinking among the Homeless and Skid Row Drinking

Estimates of the size of the homeless population in the United States are rough, but one government agency put the figure at between 250,000 and 350,000 individuals (U.S. General Accounting Office, 1985; Secretary of

Health and Human Services, 1990: 30). Although the homeless have traditionally been viewed as mostly alcoholics, drug addicts, and transients, there are increasing proportions of elderly, women, unemployed, children, minorities, and mentally ill individuals who are homeless. There are many reasons why a person is homeless, but alcohol abuse is the single most important factor (U.S. General Accounting Office, 1985). Alcohol abuse among the homeless is also often associated with the use of other drugs. It appears that as many as one quarter of alcohol-abusing women and about one fifth of alcohol-abusing men also use other drugs (Secretary of Health and Human Services, 1990:31).

Group drinking plays a major role in the lives of "homeless" men on skid row. Although alcohol is a major preoccupation with them, not all skid row drinkers are problem drinkers. Only about one third of skid row residents may be problem drinkers, another third moderate drinkers, and the remaining third appear to drink little or not at all (Bahr, 1973: 103). More recent estimates suggest that between 20 and 45 percent of the homeless may be heavy drinkers (Mulkern and Spence, 1984; Secretary of Health and Human Services, 1987: 34–35; Secretary of Health and Human Services, 1990:31).

The drinking of alcoholic beverages to become intoxicated is well institutionalized in skid row drinking. Drinking is a symbol of social solidarity and friendship, and group drinking and collective drunkenness are completely acceptable within the culture (Spradley, 1970: 117). The most important primary group among them is the "bottle gang," and no one who wishes to join is turned down. The major problem facing these people is finding a suitable location for drinking and often they wind up drinking, and getting arrested, in a public place. The major function of these drinking groups is to provide social and psychological support for the drinking among its members; these groups also facilitate interaction among persons who may be unknown to each other. So great are the pressures from such groups that an individual who wishes to deal effectively with his or her alcoholism must leave the area.

Occupation and Excessive Drinking

There is a tendency for the percentage of drinkers to increase with occupational status, although this is not the case with problem drinkers. That is, as one moves up from low- to high-status occupations, one finds a greater percentage of persons who drink; but while more persons in high-status occupations drink, they are not all problem drinkers. One study reported that the highest percentage of drinkers was to be found in the top occupational categories, such as lawyers and doctors, but the percentage of problem drinkers in these categories was among the lowest of any occupational group (Mulford, 1964; also see Biegel and Chertner, 1977).

The social patterns of some occupations call for more immoderate drinking than others. Certain business occupations are associated with frequent and heavy drinking. Businessmen and women may initiate and close deals over cocktails and in other settings where alcohol is present and used. In fact, ". . . work histories of sales managers, purchasing agents, and representatives of labor unions who have become alcoholics strongly suggest that their organizations tacitly approve and expect them to use alcohol to accomplish their purposes effectively" (Trice, 1966: 79). Many such persons had little or no previous history of heavy drinking prior to joining the organization; their drinking was learned on the job and reinforced by the company through promotions and bonuses for being a "good" businessman or woman.

There are other occupations that have been thought to be "heavy drinking occupations," but this does not mean that there are a high proportion of problem drinkers in these occupations. Merchant seamen have a high rate of drinking and problem drinking. Life at sea is monotonous, frustrating, and socially isolating; alcohol provides some relief and entertainment. In the tradition of their occupation, some form "bottle gangs" which tend to function much like those found on skid row. Other occupations which are male-dominated, such as the military and construction and building trades, are also said to have a high proportion of those who drink. The distinction between having a high percentage of drinkers and having a high percentage of problem drinkers is an important one, as illustrated in a study of two groups in supposed heavy-drinking occupations: Naval officers and journalists. It was discovered that while Naval officers and journalists drank more frequently than many other occupations, they did not consume greater quantities of alcohol (Cospers and Hughes, 1983).

Religious Differences in Excessive Drinking

Marked differences are found in drinking patterns among persons who are religious and those who are not, and between members of different religious groups. Generally, persons who go to church, who perceive themselves as being "religious," and who believe that drinking is sinful tend not to drink, at least as adolescents (Burkett, 1980). Among adults too, those who declare a religious preference have more abstainers and fewer heavy drinkers than those who declare no such preference (Cahalan, 1982: 112).

The proportion of abstainers is quite high among Fundamentalist Protestant groups, with Catholics, Liberal Protestants, and Jews having fewer abstainers. In spite of the fact that drinking is quite pervasive among Jews, their rates of alcoholism fall far below what one might expect (Snyder, 1978; Biegel and Ghertner, 1977: 206). The appropriate and inappropriate uses of alcohol appear to be far less clear among a larger proportion of American Anglo-Saxon Protestants than among Jews. Even when agreement about drinking behavior is apparent, it is not usually deeply rooted in the culture, and seldom is it free from conflicting attitudes (Plaut, 1967: 126–127). Certain generalizations may be made, even though there are differences between denominations. Among Protestants of northern European descent who drink, drinking is usually not associated with other activities, having the specific purpose of escape or to have a good time.

Orthodox Jews use alcohol differently. Wine drinking is almost universal among Orthodox Jews, as nearly all occasions--births, deaths, bar mitzvahs, religious holidays--require it by both prescription and tradition. Jews thus become used to alcohol in moderation; they start using it in childhood, and drinking is done mostly in ritualistic contexts. Orthodox Jews have another shield against problem drinking: Jews hold powerful moral sentiments and anxieties against intoxication because of their widely held belief that sobriety is a Jewish virtue, while drunkenness is a vice of Gentiles (Snyder, 1978: 182).

SOCIAL CONTROL OF ALCOHOL USE AND ALCOHOLISM

Alcoholism and problem drinking violate norms concerning the moderate and otherwise appropriate use of alcohol. Drinking behavior that goes beyond the accepted practices of the group is likely to be sanctioned. This is the process of social control. Social control efforts with respect to alcohol have been fragmented and generally unsystematic because of the public's ambivalence concerning appropriate alcohol use. This has created disagreement concerning (1) the most effective means by which to regulate the manufacture, sale, and consumption of alcoholic beverages, and (2) the most accurate and useful theory of alcoholism. As a result, there are several theories of alcoholism and no less than five models of social control of alcohol use have been tried—prohibition, legal regulation, education about alcohol use and the use of substitutes, and comprehensive programs to prevent the misuse of alcohol (see Lemert, 1972). First, the major strategies for regulating alcohol are identified, followed by some of the major theories of alcoholism.

Strategies of Social Control

Prohibition constitutes a system of laws and coercive measures that make it illegal to manufacture, distribute, or consume alcoholic beverages. This means of control has had a noteworthy history in the United States. The passage of the Eighteenth Amendment forbidding the manufacture and sale of alcoholic beverages resulted from the actions of various reform groups which identified abstinence as a middle-class symbol worthy of protection under the law (Gusfield, 1963). The Prohibition era began in January 1920 and lasted until the repeal of this legislation in 1933 with the passage of the

Twenty-First Amendment. From the standpoint of limiting access to alcohol, Prohibition seems to have been successful in reducing alcohol consumption in spite of illegal bootlegging of liquor from Canada and the existence of illicit taverns called "speakeasies" (Lender and Martin, 1982: 136–147). A return to a nationwide prohibition on alcohol seems unlikely in view of the increasingly permissive attitudes toward the moderate use of alcohol.

A second strategy of social control is the *legal regulation* of the kinds of liquor consumed, monetary costs, methods of distribution, the time and place of drinking, and the availability of alcoholic beverages to consumers by age, sex, and various socioeconomic characteristics. Unlike a prohibition strategy, the law is used here to establish standards, backed by legal sanctions, for the manufacture, distribution, sale, and consumption of alcohol. Thus, one may not consume alcoholic beverages under a certain age, one can not purchase alcoholic beverages in certain places on certain days or at certain times. If there is any national strategy concerning the use of alcohol in the United States it is one of legal regulation.

A third control strategy is a system of *education* about the consequences of alcohol use, leading to more moderate drinking or even abstinence. This approach depends greatly on a program presenting factual information about the dangers of drinking for potentially heavy or problem drinkers. It is uncertain if these educational programs can reach, and then convince, a public that derives most of its information and values about alcohol use from family and friends. As with formal programs designed for educating people about the use of other drugs, the danger of programs dealing with alcohol is that they are artificial in the sense that they are often divorced from drinking situations.

A fourth control strategy is a program that emphasizes *alcohol substitution*, such as the use of beer with a reduced alcohol content, greater soft drink consumption, and even marijuana usage. Some people might justify the use of marijuana on the grounds that it is less harmful, as far as we know, in the long run than either alcohol or cigarettes, but its proposal as a substitute would probably encounter great resistance even today. Too, with contemporary discoveries about possible cancer-causing chemicals in soft drinks, particularly diet soft drinks that contain saccharine as a sugar substitute, such an argument might be more persuasive on intellectual but probably not emotional grounds.

Finally, there are broad *programs directed at the prevention of alcohol abuse* through changing attitudes toward the nature and amount of drinking alcoholic beverages. Historically, as we have pointed out, Americans have been ambivalent toward drinking. On the one hand, drinking is frowned upon as being inconsistent with certain religious doctrines and principles of abstinence, while, on the other hand, alcohol serves an important function at many social gatherings and abstinence may even be frowned upon on such occasions as a wedding celebration or New Year's Eve. A program that dealt directly with this ambivalence would concentrate on providing public information about the consequences of alcohol consumption, the appropriate contexts (times, places, events) in which drinking should or might take place, and information about the signs displayed by one who is experiencing drinking problems. Such programs have been carried out throughout the United States by civic organizations, volunteer groups, and service clubs. Programs directed at youth have been instituted in schools. Most of these programs have relied on a "public health model" of alcohol prevention, a model that has been successfully applied to various health problems (Secretary of Health and Human Services, 1981: 104). Its application to alcohol problems has not been without controversy with critics claiming that alcohol problems are not really health or medical problems. But it is now conventional wisdom that "alcoholism is a disease" and to understand this "medicalization" of problem drinking and alcoholism, we must now consider the medical model as applied to alcoholism.

Models of Alcoholism

There are many theories of excessive drinking and alcoholism. As with other forms of deviance, some theories concentrate on features or or characteristics of the individual deviant (psychological or biological factors), while other theories focus on influences or causes outside of the individual deviant (sociological factors). A great deal of research in recent years has concentrated on issues of "chemical dependency" not only because alcohol is a drug and patterns of excessive alcohol use parallel those of several other kinds of drugs, but also because many persons who report drinking problems also use other kinds of drugs as well.

Alcoholism can be viewed theoretically in the context of a number of models of explanation (Ward, 1983). These models present a conception or perspective of alcoholism and provide an explanation for its occurrence. While there are many such models, only a few can be briefly mentioned here. (A fuller understanding of some models can be found in Chapters 3 and 4.)

Psychoanalytic Model.   Alcoholism in a psychoanalytic framework is a symptom of some underlying personality disorder. Many psychoanalysts argue that alcoholism is the consequence of some unresolved conflict between id and ego that has its roots in early childhood experiences. According to psychoanalysis, people develop through a series of stages and if something happens to arrest that development, one can become fixated at a particular stage and no further progress can take place. Excessive drinking is sometimes taken in the psychoanalytic model as evidence that the individual was fixated at the oral stage of development. Only when these mental conflicts are resolved will the individual stop drinking. The treatment implication from this model is a lengthy process of individual in-depth psychotherapy.

Family Interaction Model. In the family interaction model, alcoholism is viewed as a family, not individual, problem (Jacob, 1987). This means exploring the web of relationships in the family, rather than looking only at individuals, such as the alcoholic's spouse or the alcoholic's children. One important source of family pressure on drinking behavior is stress, and some research shows that high levels of stress within families, as well as normative systems that promote drinking to relieve that stress, are related to drinking behavior (see Linsky, et al., 1986). The relationship between stress and drinking has been documented at the state as well as family level (Linsky, et al., 1987).

In the family interaction model, it is not so much the original causes of the excessive drinking that matter as those family relationships and interpersonal forces that keep alcohol a problem in the family. Given the importance of family factors in which the alcoholism is found, treatment that involves the whole family must be undertaken. "The goal is to help each family member recognize the degree to which they contribute to the circular and degenerative alcoholic process" (Ward, 1983: 9).

Behavioral Model. The behavioral model originates in the thinking of behaviorist psychologists who conceive of alcoholism as a behavior (or set of behaviors), not a disease. Like the family interaction model, the behavioral model is more interested in those mechanisms that sustain drinking. In general, heavy drinking continues for an individual because it is reinforced in some way. The process of reinforcement may include, among other things, approval from peers for drinking, the desire to achieve euphoria, and the maintenance of certain kinds of relationships with others. Excessive drinking and alcoholism, in other words, are learned. Behaviorists believe that just as one can learn through the manipulation of reinforcements and punishments to become an alcoholic, so too can one learn to become an abstainer. They also believe that an alcoholic can learn to become a social or moderate drinker, and there is some evidence that former problem drinkers can and do become social drinkers under certain circumstances (Nordstrom and Berglund, 1987).

Biological Model. One of the most active areas of research on alcoholism today is that searching for biological antecedents to alcoholism. This research has not found a clear biological mechanism that would explain alcoholism and excessive drinking, but a number of studies have reported the possibility of some kind of biological predisposition to alcoholism. Much of this research has concentrated on the genetic molecular variations in alcohol-metabolizing enzymes because of their effect in removing alcohol from the body. The conclusion from an extensive review of this research emphasizes the possibility of locating some inherited biological mechanism, but it also stresses the importance of other, nonbiological, factors (Secretary of Health and Human Services, 1990). Studies of biological factors in alcoholism have identified two kinds of predisposition: *male-limited* susceptibility which occurs only in males, is highly inheritable, and gives rise to severe early-onset drinking, and *milieu-limited* susceptibility (the more common case), which effects both sexes and requires some kind of environmental provocation. In either instance, biological forces make a person more or less "vulnerable" to alcoholism (Hill, et al., 1987). Looking at alcoholism as a result of biological factors is consistent with what might be called a "medical model" of alcoholism that conceives of this behavior in medical terms. So far, the research on biological antecedents has not been conclusive and no study to date has provided a means by which to explain the tremendous variability in rates of alcoholism and excessive drinking among various groups in society.

Medical Model. The medical profession has adopted a medical model within which alcoholism is likened to any other disease to be treated by medical measures (Jellinek, 1960). As was discussed in Chapter 2, medical models are usually associated with a perspective that finds the cause of deviance within individuals, their biology or psychology.

Is Alcoholism a Disease? Two Positions

Yes, Alcoholism is a Disease	*No*, Alcoholism is Not a Disease
Alcoholism is as much a disease as other diseases that physicians treat. No one would claim that a physician should not be involved in the treatment of ulcers brought about by stress and no one should be surprised that they are involved in the treatment of alcoholism. "I would agree that a behavior, an 'activity,' even a central one is not a disease. But I would also think that a persistent, irrational, self-destructive activity is symptomatic of a disease" (p. 86).	The medical model does not fit alcoholism or problem drinking. Alcoholism is only heavy drinking and heavy drinking is not a disease. No one denies that there are medical (for example, cirrhosis of the liver) and other physical (for example, a high rate of accidents and injuries) consequences to heavy drinking; and no one denies that heavy drinking can become a central activity for some people that can dominate their lives. The question is whether such heavy drinking is a *disease*.
Source: Keller, Mark. 1990. "Review of Fingarette, Herbert. 1988. *Heavy Drinking: The Myth of Alcoholism as a Disease.*" in *Journal of Studies on Alcohol*, 51: 86–87.	Source: Fingarette, Herbert. 1988. *Heavy Drinking: The Myth of Alcoholism as a Disease.* Berkeley: University of California Press.

There is, to be sure, an appropriate context within which discussions of alcohol should include medical terms and problems: the physiological and medical consequences of sustained drinking. There is, however, a second dimension to conceiving alcoholism as a disease that goes beyond the strict boundaries of medicine. The temperance movement of the 19th Century found useful a conception of alcoholism as chronic drinking by someone who had a disease (Conrad and Schnieder, 1980: 73–109; Peele, 1990). This conception was deemed useful because it conveyed an appropriate moral condemnation of inebriety while holding forth the promise of treatment and change for the drinker; it was a more palatable conception of alcoholism than one that viewed it as the result of "bad" people engaging in "sinful" behavior. The disease concept is still a popular way of thinking about alcoholism, although it is a controversial one.

The adequacy of the medical model of alcoholism depends ultimately on the meaning of "disease." If alcoholism is a disease, it is unlike any other presently known to medicine, having not only physical organic characteristics but also psychological and sociological ones as well. At present, furthermore, it is not known whether the consumption of alcohol is a symptom of the disease or the disease itself.

The popularity of the medical model of alcoholism will probably continue in view of the public's inclination to adopt medical perspectives on alcoholism. The medical model approach does have the advantage of providing alcoholics with a more humane setting, for example, treatment in a medical facility rather than confinement in a jail. If this model is adopted, and problem drinkers are regarded as "sick" persons, it is then logical to conceive of them as not responsible for their behavior, however deviant (Orcutt, 1976).

But, as with many aspects of alcohol consumption and alcoholism, people are ambivalent about the medical model. Most persons believe that alcoholism is a disease but they also believe that alcoholics drink because they want to (Caetano, 1987). Such ambivalence may account for the reluctance of many persons to adopt fully the medical model as applied to alcoholism.

Combined Perspectives.

Persistent heavy drinking is a phenomenon that may require a broader perspective than any of these views individually. Some claim that alcoholism can not be understood without a comprehensive understanding of alcohol problems must use a framework that includes biological, psychological, familial, social, and sociocultural risks (Trice and Sonnenstuhl, 1990). This is not to say that alcoholism can not be controlled or that these kinds of risks are immune from manipulation. Trice and Sonnenstuhl, for example, suggest the development of drinking norms in work contexts and the use of constructive confrontations with problem drinkers to help reduce the impact of such risks. But such measures require a comprehensive understanding of public policy and broader control measures.

PUBLIC POLICY WITH PUBLIC DRUNKENNESS AND ALCOHOLISM

Individual drunkenness and alcoholism can be dealt with in a number of ways that might alleviate the problem (see the overviews in Secretary of Health and Human Services, 1981: Chapters 6 and 7; Secretary of Health and Human Services, 1987: Chapter 6 and 7; Secretary of Health and Human Services, 1990; and Sugerman, 1982). Some of these techniques include community-based treatment programs and Alcoholics Anonymous. There is some evidence that certain alcoholics may be able to drink again, but that claim has opened a controversy in the care of alcoholics.

Community-Based Treatment Programs

Concern over problem drinking resulted in 1971 in the creation of the National Institute on Alcohol Abuse and Alcoholism, and the funding of alcohol treatment programs in local communities. Since their inception on a nationwide basis, these treatment programs have provided services for thousands of problem drinkers. In recent years, a substantial public health movement has arisen involving community-based referral and treatment centers for problem drinkers, some providing counseling, some hospitalization. Alcoholism treatment services and persons in these services continue to increase. American Hospital Association surveys indicate that alcoholism and drug treatment units increased from 465 in 1978 to 829 in 1984, a 78 percent increase, while total in-patient hospital beds for drug and alcohol clients increased 62 percent over the same period (Secretary of Health and Human Services, 1987: 243). At the end of September, 1984, over 540,000 patients were in treatment in both alcohol only and combined alcohol and drug units. It has been estimated that there are over 2,500 work organizations that also provide programs for alcoholic employees (Roman, 1981).

In spite of these gains, it has been estimated that these programs reach only a relatively small proportion of the population of alcoholics and problem drinkers. The remaining problem drinkers are either in other programs or not receiving any treatment whatsoever. Typically, community-based treatment programs service more lower-class drinkers than middle- or upper-class persons, and they provide mainly out-patient services. Persons who come for help in such programs are those who perceive drinking as causing problems in their lives in the areas of health, social relationships, or work (Hingson, et al., 1982).

Community-based treatment facilities in some communities were increased as a result of the Uniform Alcoholism and Intoxication Treatment

Act of 1971, a set of recommendations developed by the National Conference of Commissioners on Uniform State Laws. To date, over twenty states have adopted the recommendations, the most important of which is the decriminalization of public drunkenness and the setting up of public detoxification centers. The central idea beyond these ideas was that, as in the medical model as applied to problem drinking, alcoholism is a disease rather than a crime. While handling public drunks, particularly the skid row type of drinker, is more humane under such circumstances, one evaluation of the Act in Seattle has found that recidivism rates increased four-fold over what they had been when public drunks were arrested and jailed (Fagin and Mauss, 1978). This increase was found mainly in the form of self-referrals from problem drinkers who also find detoxification centers more humane and attractive than drunk tanks in jails.

Other alcohol treatment centers and programs that serve communities may be private, such as a residential "alcohol hospital" or the out-patient services available to problem drinkers and their families. Such programs may use such services as counseling, referrals to medical facilities, the use of behavioral modification techniques, aversion drugs, and others. It is difficult to evaluate the success of these programs. Since they are private and expensive, they do not serve many of the total population of problem drinkers, and those persons who enter and complete such programs may be very highly motivated to do something about their drinking problem. These are hardly representative of all problem drinkers (Chafetz, 1983).

Alcoholics Anonymous

Of all treatment methods, Alcoholics Anonymous (AA) is the most widely known. It also appears to be one of the most successful approaches to alcoholism. Alcoholics Anonymous was founded in Cleveland more than a half-century ago by two alcoholics who felt that their mutual fellowship had helped them with their drinking problems. AA adopts the medical model of problem drinking that conceives of alcoholism as a disease, and the only way to insure not catching the disease again is never to drink alcohol. AA has no formal organization as such; it is organized into local chapters. There are no officers or dues, although it does maintain a central office in New York City and publishes a journal called *A.A. Grapevine.* It is a strictly voluntary organization with more than 10,000 groups nationwide, as well as chapters in other countries. Total U.S. membership was reported to be 630,679 in 1983, up from 476,000 in 1980 (Secretary of Health and Human Services, 1987: 259). Of this figure, about 30 percent are women, and 31 percent of the membership attends for reasons relating to drug as well as alcohol problems.

The object of AA is to "delabel" the alcoholic and move that person back into society as a contributing, independent individual. Toward that end, the program breaks down the alcoholic's social isolation from the rest

Alcoholics Anonymous: The 12 Steps Theory

The philosophy of Alcoholics Anonymous is presented in the "12 steps."

We . . .

1. admitted that we were powerless over alcohol—that our lives had become unmanageable.
2. came to believe that a Power greater than ourselves could restore us to sanity.
3. made a decision to turn our will and our lives over to the care of God as we understood Him.
4. made a searching and fearless moral inventory of ourselves.
5. admitted to God, to ourselves, and to another human being the exact nature of our wrongs.
6. were entirely ready to have God remove all these defects of character.
7. humbly asked Him to remove our shortcomings.
8. made a list of all persons we had harmed, and became willing to make amends to them all.
9. made direct amends to such people wherever possible, except when to do so would injure them or others.
10. continued to take personal inventory and when we were wrong promptly admitted it.
11. sought through prayer and meditation to improve our conscious contact with God as we understood Him, praying only for knowledge of His will for us and the power to carry that out.
12. having had a spiritual awakening as the result of these steps, tried to carry this message to alcoholics, and to practice these principles in all our affairs.

of the community. Life stories are told at meetings and each member is assigned a sponsor—someone who has been coping with his or her drinking problem and has made sufficient progress to help someone else. Reciprocal obligations are particularly important in AA. The special relationship between the sponsor and "baby" (the term for the new member) creates solidarity and identification with the group. This aspect of AA is similar to other self-help programs such as those found for drug addicts, mental patients, or the obese, all of which use "ex's" to help cure persons with conditions similar to their own. The process of using these persons as change agents and the general outcome of such efforts has been described as "*retroflexive reformation*" by Cressey (1955). Essentially, person A, the ex-deviant, in attempting to reform person B, the present deviant, is more apt to be rehabilitated than person B because A had to learn and internalize rehabilitation values and attitudes in order to convey them to B. In this respect, perhaps the most therapeutic period in the program is when a new member

Alcoholics Anonymous: The 12 Steps Practice

What is A.A.? A.A. was begun in 1935 in Ohio by two recovering alcoholics and is now found in almost all cities in the U.S. There is a headquarters in New York City that prints a newsletter. There are about 2 million members in 63,000 groups in 113 countries. About one half of these groups are in the U.S. There are no dues and members use only their first names in meetings. Local chapters are self-sustaining and there is no national organization that maintains membership rolls or annual meetings. The most important organizational characteristics of A.A. are local autonomy and member confidentiality. The national organization promotes local decision-making and authority.

How Effective is A.A.? This means, among other things, that researchers have trouble making generalizations about the effectiveness of A.A. from organizational records: there are none. Evaluations of A.A., therefore, must be done on the local level, from groups that do not maintain records themselves and which stress member privacy. And, there are different kinds of groups that help different alcoholics. In any given large city, for example, there are meetings for airline pilots, attorneys, non-smokers, senior citizens, young adults, for drinkers only, and for people with a combination of alcohol and drug problems. Although generalizations are dangerous, A.A. does appear to help some alcoholics, especially those who stay with the program for some time. There is no other drug or alcohol program that provides the degree of interpersonal support that A.A. does, and continued contact with the organization (by attending meetings) can surely help reinforce anti-drinking attitudes.

What Are The Meetings Like? Local chapters, or home groups, of A.A. hold meetings almost anywhere and at any time. Some groups prefer breakfast meetings, others meet at lunchtime, others at night. Most chapters meet weekly. The meetings last up to about 90 minutes or so. Attendance is not taken and the only universal rule is that no one is admitted under the influence of alcohol or drugs, unless it is their first time and they are seeking help. The meetings begin with the A.A. serenity prayer, a moment of silence for fellow alcoholics, a reading from the Big Book of Alcoholics Anonymous, and a request from the chairperson for a recovery-related topic. The topic could be almost anything that has to do with drinking. Everyone can speak and, at the end, members join hands and recite the Lord's Prayer in unison. There is a good deal of personal sharing and it is not unusual that a common bond develops among members who may not have known one another prior to the meeting. Prayers are optional and most meetings appear to discourage explicitly religious conversation; all that is necessary is that members recognize some higher authority or power than their own will.

begins to bring others to AA meetings and takes some responsibility for their drinking behavior. It has even been claimed that the general philosophy of AA is similar to other bodies of philosophy that stress increasing dependency on others, vulnerability, and an accepting of one's self (Kurtz, 1982).

It is difficult to evaluate the success of this program. No records are maintained on individuals who attend AA meetings, first names only are used at those meetings, and the confidentiality of a member's private life is safeguarded. A review of the literature on outcomes in AA suggests an overall abstinence rate of between 26 percent and 50 percent at 1 year, which compares favorably with the results of other approaches (Miller and Hester, 1980). Alcoholics Anonymous does not serve all problem drinkers. There seems to be a significant difference between problem drinkers who attend AA meetings and those who do not. A study reports that AA problem drinkers tend to regard themselves, even before they attended a meeting, as persons who often shared their troubles with others; they tended less frequently to have known persons who they "believed" stopped drinking through will power; and they had lost long-time drinking companions and had exposure to positive communications about AA (Trice, 1959).

Basic to the program in AA is that members recognize the existence of a spiritual higher power and that they admit to themselves that they are powerless to cope with their own alcoholism. These positions have more recently been challenged by a number of alcoholic groups much like AA but who feel that the spiritual basis is not necessary for recovery, nor is the idea that an alcoholic must be submissive and powerless (Marchant, 1990). The major group with this approach is Methods of Moderation and Abstinence (MOMA) which has chapters in 89 major cities, all states, and seven foreign countries. Others are Secular Organization for Sobriety (SOS), International Association for Secular Recovery Organizations (IASRO) and Rational Recovery (RR). These groups try to protect the individual from AA's doctrine of submissiveness which they regard as damaging to the self-esteem of the recovering alcoholic. One of the MOMA's alternatives to AA's 12 steps is for a member to "assume responsibility for one's own life, though at times choosing to seek the help of others." Among Rational Recovery's 11 tenets is "the idea that I need something greater than myself upon which to rely is only another dependency idea, and dependency is my original problem."

THE CONTROVERSY: CAN ALCOHOLICS EVER RETURN TO DRINKING?

Alcoholics Anonymous claims "once an alcoholic, always an alcoholic," even though persons may remain sober for very long periods of time. This particular claim, in fact, has been the object of much interest and debate, and the source of a controversy regarding whether alcoholics can ever return to drinking without again becoming an alcoholic.

If the position of Alcoholics Anonymous is correct, an alcoholic may never return to drinking again without becoming a problem drinker. Indeed, it has become conventional wisdom for treatment specialists that persons must refrain *completely* from drinking in order to insure that the person will not become a drinking alcoholic again. This belief was challenged in a study of 40 alcoholics who were voluntarily hospitalized and subsequently served as subjects in a controlled drinking experiment (Sobell and Sobell, 1973). The subjects participated in a behavior therapy program designed to have them drink responsibly and without getting drunk. The setting was a room in the hospital that was a simulated bar. Then, through the use of aversive stimuli (such as drugs that produce nausea or electrical shocks), 20 alcoholics participated in 17 treatment sessions, while a control group of the remaining 20 alcoholics received group therapy and other services. The investigators reported that those who went through the behavior therapy program did significantly better after release from the hospital than those who did not. The findings seemed to confirm the possibility that some alcoholics could be taught to be controlled drinkers, something that is inconsistent with the medical model or the disease model of alcoholism.

In a follow-up of this study, however, using the same subjects several years later, it was found that there was no evidence that alcoholics could return safely to drinking after participating in the controlled drinking program described above (Pendery, et al., 1982). In the follow-up, eight of the 20 alcoholics who went through the program were found to have been drinking excessively, six were abstaining completely by the end of the follow-up, one seemed to have really controlled his drinking, four died alcohol-related deaths, and one of the alcoholics could not be found for the follow-up study. These results did not support the idea that alcoholics can be taught to drink in a controlled manner.

But the issue may be more complicated than even these studies imply. If alcoholics can be taught to drink moderately, this would suggest that alcoholism is not biologically determined, as some claim, even in its later stages. Rather, such data would suggest that environmental factors are powerful influences over heavy drinking patterns. There is some research dealing with other addictions, such as smoking cigarettes and the use of heroin, that supports the idea that some addicts may be able to refrain from the addicting substance without treatment (Waldorf, 1983; Biernacki, 1986), and others who may be able to continue to use the addicting substance without again becoming addicted (Glascow, Klesges, and Vasey, 1983).

SUMMARY

Alcohol is the most commonly used mood-altering drug in the United States. The consumption of alcohol is integral to many social situations and is actually prescribed or socially required on many occasions. Drinking may be deviant depending on the norms of the groups to which an individual belongs. People are socialized into the drinking norms of their groups and this socialization process explains the differences in drinking behavior among different groups, such as those between males and females, and among various ethnic groups.

There is no agreed upon definition of alcoholism and the term "problem drinker" is now used by many treatment specialists. Alcoholics are heavy drinkers who consume significant quantities of alcohol frequently over a long period of time. Problem drinkers are those who experience some difficulty as a result of their drinking—in their jobs, family relationships, or in some other area of their lives. Whereas the term alcoholic is usually defined in relation to a combination of the amount and frequency of drinking, the notion of problem drinker is defined in terms of the consequences of drinking, regardless of the amount or frequency consumed. It is still not known exactly when someone becomes an alcoholic, and there may be individual differences such that it takes one person a longer or shorter time to reach this state than another person.

There are a number of subcultural and group influences on excessive drinking. Companions are particularly important for adolescent and homeless persons, although for different reasons. The percentage of drinkers increases with occupational prestige and the percentage of abstainers is greater among adults who declare a religious preference, with fundamentalist Protestant groups having many abstainers and Catholics, liberal Protestants, and Jews having fewer abstainers. Rates of drinking are not always strongly related to rates of alcoholism; rates of alcoholism are high among the Irish and French, but low among Italians (where rates of drinking are very high) and Asian-Americans. Men are more likely than women to drink and to drink more heavily.

The control of drinking in the United States reflects a fundamental ambivalence about the use of alcohol. Drinking is socially valued in some situations but not in others, and where drinking is permissible excessive drinking may not be. This ambivalence has not been resolved by various thoughts about the origins of alcoholism. The major perspectives include the view that alcoholism is a disease, a view bolstered by recent biological research, and the behaviorist perspective which views alcoholism as learned behavior. Each of these models has implications for the control of problem drinking. Regardless of the merits of these views, the public has now accepted a disease model of alcoholism and views treatment as a more appropriate response to problem drinking than punishment. At the same time, many forms of drinking are illegal and alcohol has been implicated in the causation of a number of crimes.

There are a number of specific treatment approaches to the control of alcoholism, including community treatment, in-patient treatment programs, and Alcoholics Anonymous. The former two rely on a combination of counseling and detoxification, while the latter is an informal, voluntary program. It is difficult to determine the precise success rates of any given program, but even successful programs suffer from the fact that alcohol treatment programs deal with only a small percentage of all problem drinkers. Subsequent research dealing with the causes of alcoholism, including the interaction of biological and sociological factors, may point to more effective means of dealing with excessive drinking and alcoholism.

SELECTED REFERENCES

Fingarette, Herbert. 1988. *Heavy Drinking: The Myth of Alcoholism as a Disease.* Berkeley: University of California Press.

> There are many places to find a disease-conception of alcoholism, but Fingarette makes a strong case that alcoholism is behavior: drinking heavily. When heavy drinking becomes a central activity for persons, they are regarded as alcoholic. At the same time, they are likely to have medical problems as a result of heavy drinking, but these are not the same thing as a "disease" of alcoholism.

Hilton, Michael E. and Walter B. Clark. 1987. "Changes in American Drinking Patterns and Problems, 1967–1984." *Journal of Studies on Alcohol*, 48: 515–522.

> A recent survey of the epidemiology of drinking patterns in the United States over a 15 year period.

Jacobs, James B. 1989. *Drunk Driving: An American Dilemma.* Chicago: University of Chicago Press.

> This is an excellent account of the behavior and crime of drunk driving. Jacobs is sensitive to a variety of social, political, and legal issues surrounding this behavior and his account brings them all together.

MacAndrew, Craig and Robert B. Edgerton. 1969. *Drunken Comportment: A Social Explanation.* Chicago: Aldine.

> A classic study of drunken behavior by two anthropologists. The study concludes that drunken comportment is more the result of cultural than of physiological forces. Evidence is obtained from a number of cultures to reflect cultural diversity and differences in drinking behavior. Fascinating reading.

Secretary of Health and Human Services. 1990. *Alcohol and Health: Seventh Special Report to the U.S. Congress.* Rockville, Maryland: National Institute of Alcohol Abuse and Alcoholism, Government Printing Office.

> This edition of this valuable National Institute of Alcohol Abuse and Alcoholism series updates much information on alcohol, causes, behavior, and consequences. The report also documents patterns in drinking behavior among minority groups in the U.S. The previous report (the 6th, published in 1987) also contains much on the biological basis of alcoholism. Since

each of the special reports contains so much information about patterns of alcohol use, it is safe to say that it is an absolutely indispensable reference source for any serious student of the subject.

Wilsnack, Richard W., Sharon C. Wilsnack, and Albert D. Klassen. 1987. "Antecedents and Consequences of Drinking and Drinking Problems in Women: Patterns from a U.S. National Survey." Pp. 85–158 in *Alcohol and Addictive Behavior: Nebraska Symposium on Motivation, 1986.* Edited by P. Clayton Rivers. Lincoln: University of Nebraska Press.

A large-scale national survey of drinking among women, still an understudied subject. Female problem drinking is compared to male drinking and is tied mainly to family and interpersonal relationships.

Heterosexual Deviance

10

Sex, like other forms of human activity, is governed by collections of norms that regulate the type of acceptable sexual behavior and general sexual orientations. The determination of whether a sexual act or condition is deviant must be made in reference to these norms. The next chapter will discuss the nature of "deviant" sexuality regarding preference and behavior with respect to homosexuality. Let us note now, however, that one becomes a heterosexual in the same manner as a homosexual: through the acquisition of "sexual scripts" that surround sex roles which are learned in a process of sexual socialization. Whether something or someone is a sexual stimuli depends on the nature of that socialization and the kinds of responses given the stimuli. Sexuality in general, while it has a biological base, is a social process largely explainable in social terms (Plummer, 1982). A sociological understanding of sexual deviance requires an awareness of sexual norms, their content, and the effects of violating those norms. In this chapter, sexual norms pertaining to the regulation of heterosexual activity are discussed. The importance of sexual norms and

274

major types of deviance from those norms involving heterosexuality are presented.

SEXUAL NORMS

Sexual norms differ from other norms only by virtue of their content. In other respects, they are the same. As the term implies, sexual norms govern sexual behavior. Sexual norms are learned from others by symbolic communication, direct interaction, and example. They elicit conformity through a complex system of rewards and punishments, and they specify what *ought* to take place in given situations. Like other forms of human activity, sexual behavior must be learned. Sexual behavior encompasses a variety of acts and reflects a combination of persons, situations, statuses, and physical surroundings. Sexual intercourse by an unmarried couple might violate some peoples' norms. Sexual intercourse by a married couple would conform to most persons' expectations of appropriate sexual behavior unless it were undertaken in public. Some persons might object to certain sexual acts undertaken by married persons in private, such as sodomy or sado-masochism. As these brief examples show, sexual norms must take into account such factors as the relationship between the two parties (some sex, such as masturbation, can be undertaken, of course, by one person), the physical setting, the social situation, and the precise behavior. The problem is compounded when one considers that sexual intercourse is no longer necessary for procreation (since test-tube babies are now feasible), and some sexual norms, such as those concerning sexuality in the marriage relationship, arose to guarantee this once-necessary process for societal regeneration. Sexual gratification is not the only human preference; there is also the desire for intimate physical contact and communication, which places sexual activity outside, as well as inside, the marriage relationship. Like other relatively scarce and temporary conditions, sexual gratification, then, is not confined solely to those areas prescribed by sexual norms.

Deviating From Sexual Norms

Deviations from sexual norms involve many different types of behavior, some of which are prohibited by law and some of which are negatively reacted to in other ways. They have in common the fact that they may violate the norms of certain groups, legal codes, or both. Many of the offenses do little harm to others, and, in fact, the "victim" may have been a willing participant.

One convenient way to determine whether a sexual act is deviant is to examine social reactions to the act. The content of sexual norms may be reflected in laws and group norms, but that content is made public in such things as instances of public stigmatization, for example, through the mass media, informal sanctioning efforts by individuals, and the existence of organizations set up to promote or discourage sexuality. Actually, the term "sexual deviant" is misleading since sex is often a minor part of a person's total life organization. And, in terms of the time involved in such activities, sex is a brief part of a person's life. Although there are variations in different societies, most sexual deviations involve the following considerations: (1) the degree of consent to a sex act, such as forcible rape; (2) the nature of the object, for example, restricting legitimate sex objects to human beings rather than animals; (3) restricting legitimate sex objects in terms of a certain age and of a defined distance in kinship; (4) restricting the nature of the sexual act to certain kinds of conduct; and, finally, (5) regulating the setting in which the sex act occurs (DeLamater, 1981). What constitutes sexual deviance, of course, may be proscribed by legal codes, but this does not mean that various groups in a society necessarily consider the sexual acts deviant. In other words, as with other areas, some individuals and groups may disagree with the law.

Many prohibitions exist for sexuality, but these prohibitions are far from uniform in large, modern, industrial societies, such as the United States. Some groups subscribe to norms that prohibit the following: forced sexual relations (forcible rape), sex relations with members of one's own family (incest), sexual intercourse with a person under a certain age (statutory rape), sexual molestation of a child, adultery, sex relations between unmarried adults, co-marital sex relations between two or more married couples (swinging), abortion because of an unwanted pregnancy, deliberate exposure of one's sex organs (exhibitionism), watching persons who are undressed or in the act of sexual intercourse (voyeurism), sex relations between persons of the same sex (homosexuality or lesbianism), and sexual intercourse with an animal (bestiality). Even in this long list, there are omissions. For example, some people also have normative prohibitions on the display of the naked human body, presumably because this exposure involves display of the genital organs or women's breasts. The sale of pornographic, and what is called indecent and obscene, material is also often prohibited or condemned.

It may be said that a sexual deviation is an act contrary to the sexual norms of the group in which it occurs. This has a number of limitations, however. Even here it is difficult to draw the line, for some acts are only slightly at variance with the norms. Moreover, sexual norms vary widely from group to group. What is an accepted act with sexual connotations in one group or subcultural context may become a serious breach of law in another, as the following three situations illustrate:

1. A truck driver in a roadside cafe seats himself in a booth, gives the waitress his order, and, as she turns to depart, pats her on the buttocks. The other drivers who witness this are not offended, nor is the waitress, who is either inured to such behavior or interprets it as a slightly flattering pleasantry.

2. The same behavior occurs in a middle-class restaurant. The waitress feels that an indignity has been committed upon her person, and many of the waitresses consider it an offensive display of bad manners. The offender is reprimanded and asked to leave.

3. A man bestows the same pat upon an attractive but unknown woman on a city street. She summons a nearby police officer, some indignant witnesses gather to voice their versions of the offense, and the man is ultimately charged with a sexually-motivated offense (Gebhard, et al., 1965: 2).

Social Change and Sexual Behavior

Sex has become a dominant aspect of life in many societies. Gagnon and Simon (1970b) emphasized the significance of this evolutionary process when they stated that few other topics occupy so much of the leisure time of the waking life, or even perhaps of the dreaming life, of large portions of society. "Entire industries spend much of their time trying to organize presentations around sexual themes or try to hook products onto a potential sexual moment or success. That there has been a radical shift in the quantity and quality of sexual presentations in the society can not be denied" (Gagnon and Simon, 1970b: 1).

In a national survey in 1970, the Kinsey Institute found that with regard to a variety of different forms of sexual expression, most persons were extremely conservative (Klassen, Williams, and Levitt, 1989: Chapter 2). Most respondents disapproved of homosexuality, prostitution, extramarital sex, and most forms of premarital sex. Even masturbation was disapproved by almost half of the sample.

There is reason to believe that during the past few decades there has been a shift in American society, as well as in some European societies, in both the importance of, and the interest in, sex. Increasingly, there is greater freedom in the mass media, and, in particular the motion picture industry, in public discussions of sex, and in the presentation of explicit sexual themes. The theme of homosexuality, for example, is being presented in an increasing number of plays, novels, and motion pictures. Previous bans on forms of sex prohibitions, such as those on premarital sex relations and homosexual behavior, are declining, and the naked body, with genitalia and pubic hair shown, is appearing in various forms of mass media. Some groups have greeted such an openness with enthusiasm, while others condemn the trend as morally unacceptable. Still others are ambivalent about these changes, but are likely to feel uneasy about what is going on.

Changes in the attitudes of specific sexual practices have occurred as well. One set of changes, sometimes called the "sexual revolution," is asso-

Kinsey's Determination of Sexual Orientation

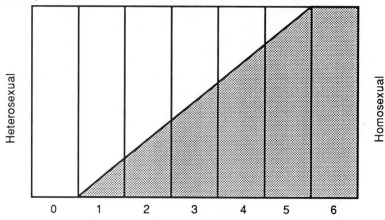

Key:

0 = Exclusively heterosexual with no homosexual
1 = Predominately heterosexual, only incidentally homosexual
2 = Predominately heterosexual, but more than incidentally homosexual
3 = Equally heterosexual and homosexual
4 = Predominately homosexual, but more than incidentally heterosexual
5 = Predominately homosexual, but incidentially heterosexual
6 = Exclusively homosexual

Kinsey defined these points in his study in terms of both behavior and orientation or preference. He also was interested in the self-definition or self-conception of the individuals (how they defined themselves).

Kinsey reported that for all males, 50% were exclusively heterosexual (category 0) throughout their lifetimes, while 4% were exclusively homosexual throughout their lifetimes. The others were somewhere between and the term "bisexual" is appropriate for these. There is considerable change in the proportions over time and within subgroups.

Source: Alfred C. Kinsey, Wardell B. Pomeroy, and Clyde E. Martin, *Sexual Behavior in the Human Male*. (Philadelphia: W. B. Saunders, 1948), p. 638.

ciated with an increased permissiveness concerning a number of sexual acts, including premarital sexual intercourse, cohabitation, spouse-swapping, sexually explicit telephone conversations, open marriages, and sexually oriented nudity. While such a revolution is undoubtedly overdrawn, there have been changes in attitudes about some forms of sexuality, such as sex before marriage. Using data from a national survey conducted over time, Smith reports that the percentage saying that premarital sex was always wrong dropped from 35 percent in 1972 to 25 percent in 1990; and, the percentage saying that it was not wrong at all rose from 26 percent in 1972 to 42 percent in 1985 (Smith, 1990). However, disapproval of extramarital and homosexual relations has shown little if any shift toward more permissiveness.

Many factors account for changes in the contemporary sexual scene. While it appears that there has been a revival of religious feeling in recent years, patterns of church attendance show a long-term decline that has recently reached a plateau (Harris, 1987: 67–71). There has been a steady decline in church attendance since 1958 in the United States, from 49 percent that year to 40 percent in 1980. Church attendance has remained at 40 percent through the late 1980s. Still, it would not be quite accurate to conclude that there has been a general decline in religion in the United States. Most persons believe in God and subscribe to a normative system that regulates clearly many forms of sexuality. These persons, however, may not attempt to promote their norms against those who take a different view.

Changes in sexual norms are reflected in actual behavior as well as other aspects of sexuality. Consider this: women now wear pants, and men wear earrings. Not all women wear pants, and not all men wear earrings, but there has been a broadening of what is acceptable expression of one's gender. One can still find segments of the population that condemn all but the most traditional methods of expressing sexuality. Exceptionally strict or puritanical sex views often encourage censorship of things sexual, and a return to traditional practices and values.

These views are in sharp contrast to the desire to become more individualized and expressive in one's personal life-style. Changing sexual norms are reflected in a number of other developments, such as the increasing demand for equality by women (the women's movement); the shift in sex gender roles and performance, even in such matters as hair style, dress, and the use of cosmetics by heterosexual men; the wider recognition of the importance and necessity for women to experience orgasm; the introduction and widespread use of contraceptive devices which offer greater security from unwanted pregnancies; and greater tolerance by the young of variations and deviance in behavior. The erosion of rigid gender roles has also contributed to a more relaxed atmosphere concerning sexual matters.

The gradual liberalization of sex reflects changes in the United States in how people have defined and understood sex (D'Emilio and Freedman, 1988). The shift in sexual attitudes reflects many broad, social forces, including the effects of urbanization which loosened small town social control and made possible sexual experimentation not previously available. The development of capitalism also contributed to these changes. The economy became increasingly dependent upon the manufacture and sale of consumer goods, and the ethic that encouraged the consumption of such goods also contributed to an acceptance of self-indulgence, pleasure, and personal satisfaction.

Changes in sexual norms are not uniform throughout society; nor are these changes felt in all groups. Some conservative groups today espouse traditional views regarding sexual norms, which they regard as moral edicts, not social guidelines. Sexual norms do not change quickly, and there will always likely be resistance to such changes. Major social changes require time and tolerance, two major preconditions for all normative changes. Thus, while it is relatively easy to document changes in the content of sexual norms, it is more difficult to trace the nature of the change—how it originated, who advocated the change, the relative political power of interested groups, and how the change was eventually made permanent, if it was.

In the sections that follow, several forms of sexual deviance are discussed. The discussion is necessarily selective and largely confined to those areas in which there is sociological research on which to base conclusions regarding theory and social policy.

Heterosexual Deviance as Behavior

Space prohibits a discussion here of many forms of heterosexual deviance. It should be clear, however, from those forms of sexual deviance discussed thus far that sexual deviance is behavior, not a condition. The medical model of deviance has exerted a strong influence over our thinking about sexual matters and has traditionally pointed to the conclusion that sexual deviance is an illness to be treated and cured, much like any other illness, and that it represents a form of mental illness, perhaps coupled with hormonal imbalance. One need not, however, invoke a medical explanation for sexual deviance. Sexual deviance is to be expected with changes in the content of sexual norms and the increasing acceptance of alternative sexual activities. Sexual deviance is ultimately tied to the social context in which it occurs; it is patterned behavior that gives the appearance of being generated from the same general sources as other forms of deviance: the learning of deviant patterns, perhaps after experimenting with minor forms of the behavior previously.

The important indication that sexual norms are changing lies not in whether there are more norm violations (that is, whether more persons are cohabitating today than five years ago, more persons are engaging in extra-marital affairs , and so on), but in the nature of the reactions, or sanctions, of others who are witnesses to or knowledgeable about the behavior. While there has been a general softening of sanctions with respect to premarital sex, stronger sanctions are still usually found with respect to adultery; while previously homosexual practices among consenting adults are less likely to draw strong reactions today from most persons, child molesting is still very strongly disapproved. But, regardless of the severity of sanctions, behavior that violates a group's norms can be considered deviant, whether the content of those norms is sexual or not.

Sex and Nudity

Whispering Pines is a nudist resort in North Carolina. There are a number of conditions that one might think would be conducive to sexuality: nudists camps are located in relatively private settings, members are there voluntarily and presumably these are people with more "liberal" attitudes about being naked, and, perhaps above all, people are unclothed and have plenty of opportunity to engage in sex. Many people have a strong association between nudity and sex.

But like most nudist resorts, there are a number of norms at Whispering Pines that restrict sexual behavior. For example, attendance at the resort is legitimated by a "club" atmosphere. The fees at Whispering Pines are $250 per year, plus $32.50 for annual American Sunbathing Association dues. Members then pay a per-day site rental charge. Whispering Pines allows no more than 10 percent single men and 10 percent single women at one time, thus restricting the proportion of sexually unattached persons. Children, whose socialization to the sexuality is incomplete, must be supervised at all times in the camp. If one acts rude, insists on having binoculars or cameras, or acts lewdly, that person's name and picture could be put on a American Sunbathing Association list which would bar visits to nudist resorts around the country. First-timers, called "Cottontails," must learn these norms.

In addition to these restrictions, there is an explicit rule that sex in public is forbidden. Perhaps because of all these norms the owners of Whispering Pines indicate that open sexuality just doesn't happen. Attendance at Whispering Pines is defined as nonsexual. As one owner put it:

There are no bathing beauties. People are just people. Women are always worrying about fat knees, legs, that sort of thing. You come out here a while, you wouldn't worry about stuff like that. They accept themselves as what they are.

Persons who attend nudist camps go to some length to dissociate the relationship between nudity and sexuality. Nudity is considered natural, not "dirty." But such a conception would be considerably more difficult to maintain if nudist resorts condoned or encouraged more profuse sexuality.

Hill, David. 1990. "Nothing but the Bare Essentials in this Nudists' Camp." *Des Moines Register*, October 18, p. 9E.

SELECTED FORMS OF HETEROSEXUAL DEVIANCE

As we have seen, the meaning of sexual deviance is situational; that is, what is sexually deviant is often a function of the situation in which it is found.

Sexual intercourse between a male and a female may or may not be deviant depending on such factors as the age of the partners, the marital status of the persons, and the time and place of the act. Heterosexual deviance includes such acts as premarital and extramarital sex, and prostitution. As with many instances of sexual deviance, the degree to which one considers a given act as deviant depends on the extent to which the act violates one's norm governing that behavior. Because such evaluations differ, there are disputes about the extent to which some acts, such as premarital sex, are deviant.

1. Extramarital Sex

One of the most important contexts in which sexual norms exist is the marriage relationship. Sexual activities that are permitted persons who are married to each other are sometimes denied to those who are not. Extramarital and co-marital sex each involve having sexual relations outside of the marriage relationship.

Sex Outside of Marriage

Premarital sexual relations represent one area in which rapid changes have been occurring in sexual norms. Not long ago, such behavior would have carried with it relatively severe sanctions, primarily from one's own family and friends. In the Kinsey Institute survey in 1970 (Klassen, Williams, and Levitt, 1989: Chapter 2), the form of heterosexuality that received the greatest degree of disapproval was extramarital sex; this was considered wrong by about 87 percent of the respondents. Respondents did indicate that whether the two parties were in love made a difference in their evaluations. Extramarital sex by persons in love was disapproved less than that without a love component.

Today, however, premarital sex connotes deviance to a lesser extent, not only among more permissive members of society, but among many traditional, conservative segments as well (Sagarin, 1977: 437). Furthermore, nonmarital cohabitation has become more and more common in American society, not only among large segments of the young, but among elderly persons as well. Two forms of extramarital sex—premarital sex and cohabitation—represent areas in which the content of sexual norms has changed, especially in the last two or three decades. These forms of behavior are becoming less deviant, both in terms of the content of the norms and the strength of the sanctions against violators of those norms. For this reason, these two forms of sexual behavior will be discussed first.

Premarital Sex. Over three decades ago, Reiss (1970) observed that premarital sex was deviant. He argued that it was viewed negatively by many persons as going beyond the tolerance limit. In a national sample of adults in 1967, he found that premarital sex was viewed as a norm violation by 77 percent of the sample, many of these adults expressing strong opinions against it. Those opposed to it felt that it resulted in unwanted pregnancies, guilt feelings, and venereal disease, and that it often lead to unwanted or unstable marriages (Reiss, 1967).

A comparison of two sex surveys, with data gathered 30 years apart, shows that premarital sexual intercourse has become more frequent in American society and more tolerated over time. The famous Kinsey reports (Kinsey, et al., 1948 and 1953) were based on data that was obtained between 1938 and 1949, and a more recent survey, by Hunt (1974), used data in the 1970s. Hunt found that premarital sexual intercourse was more common by the 1970s than it had been earlier. Kinsey found that 71 percent of the males reported they had sexual intercourse by age 25, compared to 97 percent of the males in Hunt's survey. The biggest change, however, was in the increase for the proportion of females who reported sexual intercourse; 33 percent of the females reported to Kinsey that they had sexual intercourse by age 25, while 67 percent of the females reported intercourse prior to age 25 to Hunt. In two other surveys of samples of unmarried women, Kantner and Zelnick reported that between 1972 and 1977 the proportion of women who reported having premarital sexual intercourse increased from 27 percent to 35 percent (Kantner and Zelnick, 1972 and Zelnick and Kantner, 1980).

It is sometimes believed that premarital sex is related to a poor or negative self conception, but the relationship is complex. One review summarized the relationship as follows:

> . . . individuals with high self-esteem may decide to become sexually active for reasons different from those of individuals with low self-esteem. For example, an adolescent with high self-esteem may engage in sex because she is involved in a loving relationship, whereas an adolescent with low self-esteem may do so because she is afraid of being rejected. (Crockenberg and Soby, 1989: 155)

These authors also point out that one area in which self esteem has been related to premarital sexuality is through contraception use. Adolescents with high self-esteem are more likely to use effective contraception and are less likely to become pregnant. But it is not the case that girls with high self-esteem are less likely to have premarital sex.

Cohabitation.   There has been an ever-widening extension of intimate sexual relationships to include cohabitation, or "living together." Cohabitation has become a major alternative to the more traditional form of married-and-living-together. Cohabitation can be said to "refer to all those unmarried, heterosexual couples with an intimate, sexual relationship who share a common residence" (Buunk and van Driel, 1989: 46). Until the end of the 1960s, cohabitation was associated largely with lower-class life-styles, but the last few decades have witnessed an increase in this form of behavior and a lessening of sanctions against it.

In 1980, approximately 1.6 million unmarried couples were living together in the United States, more than triple the number since 1970 (Spanier, 1983). By 1981, there were about 1.8 million unmarried couples which represented about 4 percent of all couples. It has also been estimated that about 28 percent of all unmarried couples have one or more children present in the household. The U.S. census in 1985 reported about 4% of the population were cohabitating at the time, but rates of ever-cohabitation are higher since most relationships end with marriage or another cohabitation. This means that perhaps nearly half of all married couples in 1990 had some cohabitation experience (Buunk and van Driel, 1989: 48). The figures in some countries, especially in Western Europe (particularly Sweden and the Netherlands) suggests that as many as 20 percent of all couples living together in a heterosexual relationship are cohabitating.

Almost all cohabitors indicate that they will marry eventually (although not necessarily to the person they are presently cohabitating with); cohabitation is often a temporary living arrangement that can be struck as much out of economic necessity than a personal relationship. Short-term cohabitants can share resources for a time, such as when going to college or in other times of temporary relative poverty, and then split up. As one person put it:

> I think you work harder to maintain your relationship if you're married. If your relationship starts to fall apart, and you're just living together, it's easier to say "Let's split." I'm not sure if this is a good thing or bad. Sometimes a relationship should end. I don't think a relationship should persist at all costs. (Buunk and van Driel, 1989: 61)

Cohabitors tend to be more liberal politically and not as religious when compared to the general married population. They are also more likely to come from disruptive family backgrounds; their families are more likely to be marked by divorce and separation than married couples (Buunk and van Driel, 1989).

In summary, the data suggest that premarital sexual relations and cohabitation are more frequent than earlier periods for both males and females. It is also possible to argue that public attitudes toward these forms of sex have become more lenient and accepting than previously as the practice has become more common. As such, there might be doubt as to the extent to which this behavior is considered deviant by many people. It is probably the case, however, that most persons would not identify cohabitation, especially that which involves sexual relations, as a long-term life-style, as preferable behavior.

Extramarital Sex

No society gives its members complete freedom to engage in extramarital sex, although some are more tolerant of it than are others. Extramarital sex refers to sexual behavior by a married person with someone other than his or her spouse. This is also a form of sexual behavior that is widely disapproved.

Affairs.   One form of this kind of sex is sometimes called an adulterous relationship or "having an affair." Having an affair may or may not mean having sex since "seeing someone" can involve a platonic relationship as well. In one report, married persons who were having affairs were careful to distinguish between a dating relationship and one that involved sex (Frost, 1989). To these participants, the term "cheating" was associated with sex but not dating-type relationships. However, the participants would not inform their spouses they were seeing someone else, even though they were not having sex with them, suggesting that they were able to anticipate the reaction of their spouse.

Kinsey (et al., 1948 and 1953) reported in the 1950s that about half of all married men and one quarter of all married women eventually engaged in extramarital sexual relations. Hunt (1974) found that the figure for males had not changed much twenty-five years later (in fact, it was slightly lower than that reported by Kinsey: 41 percent), but the figures for wives had increased to three times that reported by Kinsey. In 1983, *Playboy* magazine, in a series of issues throughout that year, reported a survey of 100,000 readers who can not be taken as representative of the American public as a whole. The *Playboy* survey reported that 36 percent of the married men and 34 percent of the married women had an extramarital affair. In fact, among the younger age categories (persons 29 years of age and under), more females than males reported extramarital affairs. A different survey that year reported that that 30 percent of the men and 22 percent of the women in their sample of couples who had been married 10 or more years had at least one affair (Blumstein and Schwartz, 1983). The problem with all of these estimates, including Kinsey's, is that they are not based on a random sample of the population; it is therefore difficult to estimate exactly how frequently this behavior actually occurs.

The *Playboy* survey also found that most persons who reported extramarital affairs were over 40 years of age. In fact, by the age of 50, almost 70 percent of the men and 65 percent of the women have had an affair. Most of the respondents indicated that an affair signaled a problem in the marriage. Males felt more frequently that affairs were important because they offered mainly sexual variety, then reassurance of their physical desirability, a change of routine, better sex, and sex without commitment; females, on the other hand, indicated that an affair offered mainly reassurance of their desirability, followed by better sex, change of routine, sexual variety, and sex without commitment.

Persons who engage in extramarital sex are differentiated from married persons in a number of respects (Buunk and van Driel, 1989: 102–105). Males have more permissive attitudes about extramarital sex than do females. Older persons are less permissive than younger persons. Higher educated persons from upper-middle-class backgrounds have more permissible attitudes than do members of other classes, but extramarital sex is found in all social classes. Dissatisfaction in one's marriage is also related to a more permissive view of extramarital sex. Actual involvement in extramarital sex is strongly related to approval of this behavior by friends and acquaintances and the proportion of one's reference group who have actually had such affairs. Physical opportunity is another important consideration, such as temporary separation among married persons.

The degree of acceptance of extramarital affairs is not high. This is a form of sexual behavior that is considered deviant by most persons in the United States, regardless of its frequency. There is little evidence of any change in public attitudes toward extramarital affairs over time.

There are mail order ads from organizations that facilitate extramarital, and other, sex by purporting to offer finders services in arranging meetings between men and women interested in sexual relationships, regardless of their marital status. These organizations gear their publicity to men who want to meet sexual partners. A "club" may send an unsolicited letter containing membership in the club. The announcement and invitation from one such club promised the following services in exchange for an annual membership fee:

1. Twelve monthly "listings of choice women who actually live in your area."
2. Informative tips and proven techniques on how to improve the quality of your relationships and guidelines to meaningful personal development.
3. Free 24-hour Hotline service.
4. One year subscription to three publications on swinging.
5. Our iron-clad solemn promise and total commitment that we will live up to our every promise.

It requires an annual investment of $20 to find out whether the promise was a good one. Another club offered a lifetime membership of $200 to receive a "complete list of the most desirable women who live within your area," as well as special reports on sex clubs, listings of all nude beaches, resorts, and other activities, such as receiving letters from girls "who share your interest."

Phone sex is a more recent variation of sexual stimulation outside marriage. Callers are able to talk with another person about their sexual fantasies and to hear, for a fee that is charged to the telephone bill, sexy talk from a stranger. The auditory incitement, coupled with an active imagination, can produce sexual satisfaction either by itself or in connection with masturbation. The popularity of phone sex during the late 1980s may have been related to the fear of AIDS, the strict impersonality of the encounter, and the degree of control that can be exercised by the caller. There are no commitments between the parties, no need to "perform" in front of another person, no need to engage in foreplay, and the caller controls completely the nature and length of the encounter. Furthermore, this is a relatively guilt-less form of sex since there is no physical contact between the parties. It is sex without touching and that can be called, by individuals who might be inhibited from going to a sex club or having a real extramarital affair, no sex at all, merely stimulation.

Co-marital Sex.   While extramarital sex implies that one of the marriage partners does not approve of the other partner having sex with others, another form of sexual deviance does involve such consent. During the 1960s, a form of nontraditional sexual relationship between individual married couples emerged to public view, involving exchanges of partners among two or more couples and sometimes single persons in a threesome. A number of terms describe this behavior: "co-marital sex," "group sex," "swinging," and "mate" or "wife swapping." Swinging is extramarital sex where the partners are doing it at the same time and generally in the same place. The deviant nature of co-marital sex has as much to do with the emphasis on sexuality as a recreational activity and not linked with a loving relationship or procreation (Gorchos, et al., 1986: 10).

It does not appear that swinging is a widespread form of sexual deviance. Swinging is noteworthy, however, because of two characteristics of its most frequent participants—the general conventionality in other aspects of their lives, and their middle-class status (Little, 1983: 74). Most of the persons who engage in co-marital sex are white, middle-class persons who live in suburbia (Gilmartin, 1978).

The most common way of beginning a co-marital relationship with others is through personal reference with like-minded individuals. Such persons might be identified through a newsletter specifically for that purpose or through mutual friends. These are planned, not spontaneous, events.

Couples may initiate such relationships out of boredom or to attempt to "re-romanticize" their marriage (Fang, 1976). Sexual relations are usually done under a set of implicit norms or understandings concerning the circumstances under which swinging can occur. One common characteristic is that swinging takes place under the surveillance of each partner and, in this way, some control is exercised over the extramarital activity of each partner, thus minimizing the possibility that the activity will take a romantic turn (Bartell, 1971). The avoidance of an emotional relationship with someone other than the swinger's spouse is essential for the participants to continue in this activity. Swingers stress the importance of the marriage relationship and honesty in extramarital affairs; such affairs must include the partner in some way and show respect for all parties (Buunk and van Driel, 1989: 100–102). Couples may also restrict other partners so as to avoid emotional entanglements. Business partners, neighbors, and friends may, for example, be avoided for this reason.

Swinging is far from an accepted form of sexuality. Generally, the swinging activities of a couple will be hidden from nonswingers because of a fear of negative social sanctions. The course that swinging may take has been traced in a study of 136 cases (Palson and Palson, 1972). It begins with considerable superficial curiosity and enthusiasm for a new situation and relationship. While some experiment with swinging, others undoubtedly are influenced by the desires of the mate to try it. Much like the process described for initiating marijuana use, the potential swinger may feel that the spouse would like to try something different. The spouse agrees not out of personal curiosity but out of a desire to fulfill the other's wishes. It is usually the husband who suggests swinging, but the wife also often finds the experience of being desired by other males a pleasurable one. Two women have described their first experiences: "I got turned on, although I hadn't anticipated a thing up to that point. In fact, I still have a hard time accounting for my excitement that first time and the good time which I actually had" (Palson and Palson, 1972: 30). And another: "I *never* experienced anything like that in my whole life. I have never had an experience like that with quite so many. I think in the course of three hours I must have had 11 or 12 men, and one greater than the next. It just kept on getting better every time. It snowballed" (Palson and Palson, 1972: 35).

Another person, a woman, indicated to an interviewer that her sexual experience included being part of a threesome fairly early in her sexual career:

> I had my first threesome when I was twenty-six, with two men. They were
> strangers who were visiting my apartment. After a few drinks, each of
> them had a hand on one of my knees. They had been best friends for years
> but had never even seen each other naked. They just spontaneously
> reached out for me at the same time. It was . . . an organ event; just that,
> no more. Nothing happened between them. I took turns with them. . . .
> That was the first time I had sex with more than one person spontaneously,
> not by arrangement. (Karlen, 1988: 53)

Swinging may be related to other forms of sexual deviance in terms of the social processes that precede the swinging experience. One study, for example, found that married women who had engaged in mate swapping had a higher rate of premarital sex, with more different men and more often with each man, than had women without mate-swapping experience (Bell and Peltz, 1976). The swinging women also evidenced a higher level of sexual interest as reflected in a greater average monthly frequency of marital intercourse.

There has been little effort in the literature on co-marital sex to bring data to bear on hypotheses from a theory of heterosexual deviance. One notable effort by Jenks (1985) reported that swingers were generally less likely than nonswingers to be subject to social controls. The swingers Jencks studied had moved often and were thus living in their communities for fewer years than nonswingers. The swingers also were less likely to have a religious identification. Their involvement in swinging might be predicted by control theory (see Chapter 5) which predicts that deviance is more likely when people are freed from social restraints. The process of becoming a swinger also appears to depend upon the learning of attitudes that would permit sex with a nonspouse, with the spouse present, for purposes of recreation.

2. Prostitution

Prostitution appears to be virtually universal, but at the same time it is generally disapproved in most societies. The extent of prostitution and the reaction to it has fluctuated over many years, but the essential nature of prostitution has remained the same. We can define prostitution as sexual behavior on a promiscuous and mercenary basis, with emotional indifference. Actually, it is difficult to define prostitution precisely because there are different conceptions of this activity (Aday, 1990: 104). To some, prostitution is dehumanizing and a gross violation of the laws of nature, while to others, prostitution is merely an alternative occupation that may perform a useful service to society. Prostitution does involve sex for money, but, depending on one's definition of sex, so too does acting in a sexually explicit movie, going to a sex therapist with one's spouse, and having a sex massage.

Prostitution is generally a crime but in most countries, as well as the United States, it is not the sex act that is a criminal offense but, rather, solicitation for purposes of prostitution. Although some prostitutes may be selective on the basis of age, sex, race, or physical attractiveness of client, most prostitutes will engage in sex with anyone who can afford to pay their price. Most of these relationships are temporary and impersonal, but some are not. Sexual acts will vary in price and may include both oral and anal sex acts, as well as sadistic, masochistic, and exhibitionist acts in addition to traditional forms of heterosexual relations. Prostitutes may also obtain money from clients for engaging in role playing (pretending to be exotic characters) without intercourse or other sexual activity, and for other services in addition to sex, such as simply listening to the troubles of clients.

Prostitution is linked to several important values in American society. The general culture stimulates the importance of sexual values in life, and the satisfaction of these values or desires may be difficult for many men. Advertising is often done with sexual references to attract attention and there is no lack of sexual references and examples in the mass media, let alone in material that is explicitly pornographic. As individuals come under increasing sexual stimulation from a number of everyday sources, it is not surprising that some of that stimulation will be expressed in the need for overt sexual behavior. It is not possible, however, always to act in the sexual manner one wishes, and prostitution may serve as an outlet for expressing these desires.

In a sense, there is no great mystery to what motivates most clients of prostitutes. Prostitutes provide easy and certain sex and no later interpersonal responsibilities (Kinsey, et al., 1948: 607–608). It may afford a solution to an unsatisfactory marital relationship without the necessity for an extra-marital affair. Moreover, prostitutes may be able to provide services that are hard to get elsewhere. Some clients desire variety in their sexual contacts, others wish some special service other than regular intercourse (James, 1977: 402–412). There are also clients who have physical characteristics, such as disfigurements, that make finding other sexual partners difficult. Also, just being the client of a prostitute is sexually exciting to some men since it may represent a sexual fantasy they have had.

Not all clients, however, are motivated entirely by physical sexual attraction. Holzman and Pines (1982) interviewed 30 clients of prostitutes to determine the meaning of the act for them. Many of the clients reported that they were more concerned about the prostitute's personality than her physical appearance. They looked for "personal warmth and friendliness. Although they wanted to pay for sex, it seemed that they did not want to deal with someone whose demeanor constantly reminded them of this fact" (Holzman and Pines, 1982: 112). Among other things, information from clients tells us that prostitution is a desired commodity for which there is widespread demand with sometimes limited supply.

Nature and Extent

The nature of prostitution involves sex for money. The range of sexual acts appears bounded only by the imagination and the financial resources of the client. Clients want as much as they can get for as little a price. Prostitutes are engaging in a business and, like most businesses, "time is money." The more time they spend with each client, the less time they have for additional clients. So, prostitutes are interested in making their contacts with clients as short as possible unless the client is willing to pay for the extra time.

A sample of 72 Southern California prostitutes were asked about their activities and income (Bellis, 1990). The sample as a whole served about 560 customers a day, ranging from 2 to 30 with an average of 8 clients. Fees ranged from $20 to $100, the most common being $30. The most popular act was a combination of intercourse and fellatio (called "half and half"), which was requested by 75% of the customers.

It is not possible to ascertain exactly how many persons obtain at least part of their living from selling sex. Some prostitutes earn all or most of their income from prostitution, while others work only part-time. Those who obtain some of their income from prostitution use other means to generate income as well, such as welfare, part-time jobs, and engaging in other illicit activities. Prostitution is also something that does not require one to participate for long periods of time. Prostitutes may engage in this activity for a time and return to prostitution only out of economic necessity. During the interim, they may work a legitimate job.

It is difficult to provide some numerical estimates of the extent of prostitution. A 1971 study of 2,000 prostitutes estimated that prostitution involves between 100,000 and 500,000 women in the United States and that they earn over $1 billion annually (Winick and Kinsie, 1971). The 500,000 figure has also been reported in a book on deviance (Little, 1983: 35). While such figures represent nothing more than guess, they may represent as reliable an estimate as exists. Arrest statistics provide another source of information. In 1989, 107,400 persons were arrested by the police in the United States for prostitution and commercialized vice (Federal Bureau of Investigation, 1990: 172). Most of these arrests are of female prostitutes, but a number of arrests involved males. Some of these arrests are of male prostitutes but, remembering that the actual crime involved in most jurisdictions is solicitation, most were clients of prostitutes. Most of those arrested were in their early 20s and most (about 60 percent) were white.

While arrest data provide some statistical data and suggest some relationships, they are notoriously poor indicators of the number of prostitutes. Some prostitutes are arrested only once, while others are arrested frequently. Moreover, some prostitutes are arrested under charges other than prostitution, such as vagrancy, disorderly conduct, loitering, or some other sex offense. Prostitution arrests are also very susceptible to manipulation by decisions on the part of the police to "crack-down" on prostitution for short periods of time. This produces high arrest rates which may give the appearance that prostitution is increasing when, really, the only thing that is increasing is police activity.

Although prostitution is extensive today, it appears to have declined steadily throughout the past four or five decades, except for periodic increases in wartime. Kinsey (et al., 1948: 597) estimated that prostitution accounted for less than 10 percent of the total nonmarital sexual outlet for males, and the figure is undoubtedly much lower today. He reported that not more than one percent of extramarital sexual intercourse is with prostitutes. Kinsey (et al., 1953: 300) also estimated that the frequency of American male visits to prostitutes had been reduced in 1950 to about one half of what it was prior to World War I. Much of this decrease in the use of prostitutes seems attributable to increased sexual freedom on the part of women, thereby creating greater sexual access for males.

International Aspects

In addition to the forms of prostitution one sees in the United States, there is also an international trade in prostitution. The trafficking of women from country to country is less visible than the trade in other illicit commodities, such as drugs, but it is nevertheless pervasive. Women may be coerced, persuaded with money, or just kidnapped against their will and sent to another country. There, they are sold to private parties as sort of sexual slaves or to individuals who use them as prostitutes from whom to derive income.

Women who have this experience usually later have opportunities to escape from their captors, but often they do not. Many of them need what little money they might make and others have been socialized to accept the decisions of men, even if those decisions are against the woman's self-interest. Government inattention to the problem has not helped. As one observer said:

> Usually, many factors coalesce to create conditions of female sexual slavery. Often but not always, the conditions of poverty combine with female role socialization to create vulnerability that makes young girls and women susceptible to procurers. Social attitudes that tolerate the abuse and enslavement of women are reinforced by governmental neglect, toleration, or even sanction. At levels of government and international authority where action could be taken against the slave trade, one finds at best suppression of evidence and at worst complicity in it. (Barry, 1984: 67)

Women, particularly those who are widowed early or from poor countries such as India and other Southeast Asian nations, may be tempted by the opportunity to make money through prostitution for their families. Sometimes, those opportunities are encouraged by relatives of the girls and women who sell them into sexual slavery for an initial sum as well as a percentage of profits later. Many of the child prostitutes in Thailand, a country that is said to have more than 30,000 child prostitutes, work there against their will in a sex industry that is big business. Thailand is known for its extensive child prostitution and many foreigners, tourists and businessmen, ask about child prostitution. "It is an accepted investment for even respected businessmen, just as frequenting brothels is an accepted pastime" (Kunstel and Albrights, 1987: 9). Most of these girls are under 15 years of age and some are 11 years old and younger. In Thailand, poor girls from the economically deprived northern part of the country are pressured to go south to engage in prostitution with the expectation that they will send back their earnings to their families. Depending on the house of prostitution where they work, they might not be permitted to leave the house, always being "on call" waiting for a customer. In some houses, when a customer arrives, a bell is rung and the girls must leave whatever they are doing and go to a main room. There, they arrange themselves before the customer in a specially lit area of the room to permit better visibility. The girls all have numbers pinned to their clothing and the customer selects the prostitute he wishes by calling out the number.

Types of Prostitutes

Prostitutes can generally be classified according to their methods of operation, the degree of privacy they have in their work, and their income. There are three major kinds of prostitutes, and several other types that work in particular settings. A *streetwalker* is a prostitute who solicits clients directly in a more or less public place, such as a street corner or bus station. Most prostitutes are streetwalkers. This type of prostitute is found mainly in large urban areas and often has a record of many arrests. Streetwalkers are often the cheapest prostitute clients can find. Streetwalkers operate within a given territory and have contacts among persons in the local community, such as the desk clerk at a cheap hotel or motel.

A *bar girl* is a prostitute who solicits clients in a public place, but one more protected from general public view than soliciting on a street corner. This type of prostitute may solicit her clients alone. Sometimes she will work with another person, sometimes a pimp, but more usually someone who works in the bar, such as a bartender. Usually, bar girls have to be more attractive to get clients because they are likely to encounter a smaller number of potential clients in taverns than on the streets.

A *call girl* is the highest status type of prostitute. She works out of an apartment or hotel room and takes clients strictly on a referral basis. While she may take customers in her place of living, usually the client is met in his hotel room or place of residence. She is also the best paid type of prostitute and enjoys the greatest immunity from arrest and the stigma that goes with being a prostitute. She may have a wealthy clientele and may provide interpersonal services other than sex, such as "counseling" or just listening. Increasingly, the call girl is becoming a more common type of prostitute. This type of prostitution permits women more privacy from the police as well as from conventional persons, and also better permits part-time work.

Other prostitutes may work in massage parlors, photographic studios, or commercial escort agencies. Not all women who work in such places are prostitutes, but the nature of these businesses are conducive to conducting prostitution along with providing legitimate services as well. Prostitutes who work for escort services can work when they wish and for how long they wish, thus providing them with greater flexibility. Some prostitutes may travel with a specific group of clients. These later prostitutes are sometimes called "road whores" and they may cater to working-class migrant labor camps or work urban conventions (James, 1977). There are reports of prostitution associated with truck stops where the clients are long-distance truckers and other travelers who know what is actually going on (Diana, 1985; Aday, 1990: 113–115).

One report identified three kinds of prostitutes who work in business offices (Forsyth and Fournet, 1987). "Party girls" have sex with clients for money. "Mistresses" have sex with their boss, principally with the motivation of obtaining job security. The "career climber" has sex with a series of bosses and her behavior is motivated by concerns about career mobility and advancement.

Organized houses of prostitution, which once flourished in "red-light" districts of many cities, are not common today (Heyl, 1979). Nevada is the only state with local county options for legalized prostitution; many counties have legalized such prostitution although not the county in which the state's most populated city—Las Vegas—is located (Reynolds, 1986: 86–128). Customers travel out of town to trailer complexes, called "ranches," located in rural areas where the women work, paying from 50 to 60 percent of their income to the trailer owner. One of the best known of these brothels, the Mustang Ranch, filed for bankruptcy in 1990. Thousands of people attended an auction that sold parts of the brothel (pictures, furniture, etc.) as souvenirs. It is doubtful that legalized prostitution will ever replace completely the illegal prostitution that takes place in Las Vegas, if only because some men will probably always want prostitutes in their hotel rooms.

Prostitution and Deviant Street Networks

Prostitution does not take place within a social vacuum. There are discernable sets of relationships, called "deviant street networks," within which much prostitution takes place (Cohen, 1980). These networks are composed of persons who are engaged in a number of illicit activities, or "hustles," one of which is prostitution. Other network activities might include petty theft, forgery, credit card fraud, embezzlement, drug trafficking, burglary, and robbery. The prostitutes in these networks are hustlers. They engage in whatever activity is likely to yield them the most income at the time, though most networks are organized to specialize in only one or two of these crimes (Miller, 1986: 36). The street network is not only a source of information about criminal opportunities, but also a source of support, self-esteem, and courage for its participants.

Miller (1986) studied a series of deviant street networks in Milwaukee, and Cohen (1980) studied similar networks in New York City. The controlling members of the networks were mostly African American males, most of whom had extensive police records. These men were not really pimps for the women. Each man had one, two, or three women who worked for him, each group forming something of a pseudo-family. These men functioned more like "deviant managers" than pimps (Cohen, 1980: 55–59) and provided on-the-spot protection and supervision, working closely with the women. These networks are relatively autonomous from one another.

Women are recruited to these networks from a number of sources. Black women are largely recruited from a family structure and background that is, in some respects, similar to the deviant street network. Black recruits were often from families that were more like domestic networks—that is, families composed of kin, near-kin, and unrelated persons who live together mainly for economic reasons (Stack, 1974; and Angel and Tienda, 1982). Most of the groups are headed by females and traditional sources of family authority and supervision are absent. Some black females actually begin various forms of hustles before they enter the deviant street network (Valentine, 1978). Girls of 16 or 17 years become acquainted with such activities as shoplifting, against their parents wishes. Later they might be invited to participate in prostitution.

> Often the initial recruitment as well as recruitment thereafter is described by women as rather low-key and offhand. Just as someone might approach a group chatting in a kitchen to ask for help in moving a newly acquired couch into an apartment in exchange for a beer or simply as a gesture of friendship, so might someone be asked to help lug copper tubing pilfered from an abandoned building in exchange for a share of the profits or to help sell some stolen merchandise or some marijuana in exchange for a portion of what was being sold. (Miller, 1986: 79)

Other girls are recruited from the population of runaways in the city. These girls are offered shelter and money in exchange for their labor. They become part of the extended "family." Prostitution is one activity, but the women may also engage in the other forms of crime as well, such as theft and drug dealing. In this manner, prostitution exists not by itself but as part of a larger web of illegal activities and persons in the deviant street network.

The Process of Becoming a Prostitute

Inasmuch as youth and some degree of physical attractiveness are important prerequisites for being a successful prostitute, most are usually between 17 and 24 or so. The peak earning age is probably around 22. Some prostitutes are older, but most of these have taken up or continued the profession for special reasons, such as drug addiction, alcoholism, or some other expensive habit. Other than the fact that most prostitutes appear to come primarily from the lower socioeconomic classes and often from inner city areas, there is no evidence that they enter this profession because of poverty itself, even though most desire to improve their economic position. At one time, there was a disproportionate number of prostitutes from disadvantaged foreign-born groups; today there is a disproportionate percentage from racial minorities. It is clear that prostitutes are economically motivated and engage in this activity mostly for the money.

Some 50 or 60 years ago, it was widely believed that the prostitute was often the victim of a "white slaver" who had induced a sexually inexperienced woman to enter the profession against her will. Studies of prostitutes, however, indicate that the process of becoming a prostitute is quite different from this old stereotype. A study of 30 prostitutes, for example, found that early sexual activity, a history of delinquencies, and, frequently, prior commitment to a training school for girls were all important precursors to prostitution (Davis, 1981). The mean age for first intercourse for this sample was about 13 years, although this fact in itself does not account for prostitution. It is not uncommon for the potential prostitute to experience geographic mobility and disruption prior to entry into prostitution, and a number of prostitutes report family difficulties such as divorce. While the transition to prostitution may be traumatic for some women, others may find a much more stabilizing life than that they left behind (Gagnon and Simon, 1970a: Chapter 7).

Child Prostitutes.   Child prostitutes, or "baby pros" (Bracey, 1979), represent a little understood and studied area of prostitution. Inciardi's (1984) study of child prostitutes, aged 8 to 12, uncovered that the girls became introduced to prostitution through their parents or other family members. None of the girls engaged in prostitution on a full-time basis and all were attending elementary school at the time; none were runaways. Their backgrounds frequently contained the presence of nudity, sexual promiscuity, pornography, and prostitution among family members. As one of the baby pros put it:

> My sister would take me to work with her [to a massage parlor] sometimes
> when she couldn't get a baby sitter. I can't remember the first time I saw
> a dude get on top of her, but it didn't seem to bother her. She said it
> was fun and felt good too. (Inciardi, 1984: 75)

In general, Inciardi reports that the girls conveyed an almost cavalier attitude about participating in prostitution. They continued their involvement because they were paid well and many seemed to fear the rejection from their family if they stopped. This was especially true for girls whose parents were involved in the production of pornography. "Now I'm used to it," one girl reported, "and the spending money is real nice" (Inciardi, 1984: 76).

Adolescent Prostitutes.   Adolescent prostitutes begin their careers at about the same age: 14 or so (Weisberg, 1985: 94). Most come from dysfunctional homes marked by family separation, divorce, conflict, a lack of parental affection, and substantial sexual abuse in the family. Many adolescent prostitutes report they have had a background of being in trouble in a number of social settings: in school, at home, in the community. Many girls who begin careers in prostitution seem to sort of drift into this activity. Because of their marginality, some girls may associate with others who are also marginal or stigmatized. While a number of girls report unconventional early sexual experiences, there is little evidence that these experiences determined their later deviance. One prostitute whose initial sexual experience was with her father explained that:

> You don't have words to say how it felt. I really didn't feel anything. But
> your own father doing that to you—I felt like dirt for years. It didn't
> hurt but it didn't feel good either. At that age (11 years old) I wasn't sure
> exactly what was going on. . . . It was no big deal. I didn't exactly like it but
> I didn't exactly not like it either. After we did it for a while . . . it still was
> no big deal. (Diana, 1985: 65).

For most males, entrance into prostitution is often unplanned and takes place almost accidentally. Some describe the process in almost fatalistic terms. One youth recalled his first experience:

> Then this man came up and he offered me $25 if I would do his little thing with him. And I said 'sure.' That was the first one I ever did. I thought, "That was great for 20 minutes of my time, to get that and then go party on something." (Weisberg, 1985: 52).

There are two subcultures for male prostitutes. One is a peer-delinquent subculture where prostitution is part of a larger routine or life-style of hustling through all sorts of illegal activities. The other subculture is a gay subculture where male prostitution represents more the sexuality of the prostitute. In either case, almost all of the clients of male prostitutes are other males. Some of these male prostitutes engage in this behavior only occasionally, while others are heavily involved in inner city street life and prostitution (Weisberg, 1985:40).

For females, running away from home and the need to support oneself are strong influences. Like older prostitutes, the prime motivation for this behavior is monetary. Female adolescent prostitutes have had male parental relationships that are more likely to be experienced as negative (Weisberg, 1985: 89). Like older female prostitutes, many adolescent female prostitutes have negative attitudes toward men.

The customers for male and female adolescent prostitutes are similar: 30–50 years old, usually white and from a variety of social and occupational backgrounds (Weisberg, 1985: 161). Male prostitutes generally use a car or a client's residence rather than a hotel since that is suspicious. Females prefer a place near the street, like hotel or a close-by apartment.

Prostitution as a Career. Most prostitutes do not begin their involvement in childhood, and child prostitution is less common than prostitution involving adults. An important factor in becoming a prostitute is usually association with persons who are on the fringe of prostitution. In order to participate fully in "the life," one must learn to become a prostitute from others who know the trade. In the United States, contacts with persons in or on the fringes of prostitution are largely with women who themselves are practitioners. While some prostitutes are influenced by pimps, this is not the usual mode of entering the profession. Most prostitutes who have pimps obtain them after entering the profession. Nor is it the case that novice prostitutes are sexually inexperienced. Most girls who eventually become prostitutes did not begin to associate with other prostitutes or pimps until after they were sexually active and had explored the idea of selling sex (Gray, 1973).

The developmental career of a call girl includes at least three stages: the entrance into the career, the apprenticeship, and the development of contacts. The mere desire to be a call girl is insufficient; there must be training and a systematic arrangement for contacts. One call girl said, "You can not just say get an apartment and get a phone and everything and say, 'Well, I'm gonna start business,' because you gotta get clients from somewhere. There has to be a contact" (Bryan, 1965: 289). One study has concluded that "the selection of prostitution as an occupation from alternatives must be sought in the individual prostitute's interaction with others over a considerable time span" (Jackman, et al., 1963: 160). In a Los Angeles study of 33 call girls, half had had initial contact with a call girl, some over a long period of time, others for shorter times (Bryan, 1965). Some were solicited by a pimp with offers of love and managerial experience. When a call girl has agreed to aid a novice, she assumes responsibility for her training; women who are brought into prostitution by a pimp may either be trained by him or referred to another call girl.

Once contact has been made and the woman decides to become a prostitute, the apprenticeship begins. This is hardly a formal process of instruction, but some women report spending an average of two or three months in an apartment in training. A woman is performing as a prostitute during this time. The content of the training consists of learning the value structure of the profession of prostitution and other "dos" and "don'ts" in problematic situations. The acquisition of a set of norms and values serves to create solidarity within the group and to alienate the apprentice from "square" society. Among the norms that are transmitted is the belief that prostitution is simply a more honest sexual behavior than that of most people; moreover, the learning includes the beliefs that men are corrupt and exploitative, that dealings with other prostitutes should be "fair," and fidelity to the pimp.

The rules governing interpersonal contacts with the customers include what to say on the phone during a solicitation—a "line" such as needing money to pay rent, buy a car or cover doctor bills; social interaction in obtaining fees; the nature of specific customer preferences and what types of customers to avoid; how to converse with a customer; caution in the use of alcohol; and knowledge of physical problems associated with prostitution. Little instruction seems to be given on sexual techniques. In spite of the importance of such rules, one study found that there was considerable variation in the adoption of such rules (Bryan, 1965).

Contacts. Since a call girl must have access to a clientele, an equally important aspect of training is the acquisition of contacts. This is done during the apprenticeship period. Books or "lists" can be purchased from other call girls or pimps, but some are unreliable. Most frequently, names are secured through contacts developed during the apprenticeship period. For an initial fee of 40 to 50 percent, the trainer call girl will refer customers to the apprentice and oversee her. This fee becomes the pay of the "teacher" along with the convenience of having another woman available to meet the demand or take care of her own contacts. Over time, it is important that the new call girl be able to develop her own list of clients.

Self-Concept

Prostitution requires a new conception of self. Societal reaction, arrests, and association with other prostitutes serve to change the self-concept of the prostitute. The self-image of the urban prostitute has been found to be related to the degree of social isolation, the more isolated women tending to define their behavior in a more acceptable light (Jackman, et al., 1963: 150–162). Some prostitutes, however, particularly if they cater to wealthy clientele and have another occupation, such as a secretary or model, may look down upon "prostitutes," and associate that word with only the streetwalker type of prostitute.

As with other forms of deviance, prostitutes are aware of the reactions by others to their work, but the work is often justified in three ways: (1) prostitutes are no worse than other people and often less hypocritical; (2) prostitutes achieve certain of the dominant values in society such as financial success and the support of others who are dependent upon them; and (3) prostitutes perform an important and necessary social function. Research on the philosophies of 52 call girls with an average age of 22 and length of experience of 27 months found that virtually all respondents maintained that prostitution was important because of the varied and extensive sexual needs of men and the necessity to protect social institutions (Bryan, 1965: 287–297). These reasons are similar to those given by sociologists who argue that the existence of prostitution can be explained by the functions they provide (Davis, 1937).

The favorable self-concept developed by the prostitute is supported particularly by the theory that clients should be exploited. Most prostitutes develop anti-masculine attitudes as part of their training and these attitudes are reinforced on the job. In exploiting them, the prostitute regards herself as being no more immoral than her customers and the rest of the world. Another view is that in essence most interpersonal relations between the sexes are acts of prostitution. Wives and others use sex to achieve other often materialistic objectives, whereas prostitutes are at least honest about this deception. This is not to say that prostitutes have a positive image about their work. Most adolescent prostitutes, for example, have negative attitudes about prostitution (Weisberg, 1986:163).

An interview with a prostitute in Spokane illustrates these points (Murphey, 1987). She works alone, without participating in a street network and without a pimp. She was not sexually active until 18 and had a high school diploma. She is 23 and has been working the streets for two years. She has no illusions about her job. She is in it for the money, and nothing else. (She admitted to an income of between $4,000 and $5,000 per month—all of which, of course, is tax free.) Her entry into prostitution was partially justified with the belief that prostitution was not much different from what a lot of women were doing in the singles bar scene. "I'm just not giving it away," she said.

Like most of her coworkers, she learned the ropes from others to permit her to operate. The basics for her own method of operating include no contact without a condom, not allowing men to tie her up, not having oral contact of any kind, and exercising judgment before getting into anyone's car or going to their room. Like virtually all prostitutes, she has a jaundiced view of men. "When I look at these guys, they're like such fools to me, you know? Anybody who would have to go out and pay for it must really be out of it, or really be hurting or something." This prostitute's feeling were echoed by one of the baby pros that Inciardi (1984: 77) interviewed: "You have to be awfully fucked up to want to be pissed on or screwed by a kid. . . ." The female adolescent prostitutes interviewed by Weisberg (1986: 89) also had negative attitudes toward men.

This is not to say that prostitutes do not develop personal relationships with clients; many prostitutes report that some clients are actually a kind of friend.

> Not infrequently, personal friendships with customers are reported: "Some of them are nice clients who become very good friends of mine." On the other hand, while friendships are formed with "squares," personal disputations with colleagues are frequent. Speaking of her colleagues, one call girl says that most "could cut your throat." Respondents frequently mentioned that they had been robbed, conned, or otherwise exploited by their call girl friends. (Bryan, 1966: 445)

Reasons for this difference between the ideology and actual beliefs among prostitutes may include the lack of cohesiveness among prostitutes and possibly the fact that the stigma of the occupation is more than the ideology implies. Nevertheless, the learning of the ideology, as well as the other norms and values associated with prostitution, constitutes an important part of becoming a prostitute.

Some prostitutes engage in this form of deviance without developing a deviant self-image, without progressing in the acquisition of deviant norms and values, and without extensive identification with prostitution as a career. The transition to career prostitution is accomplished as a person comes to acquire the self-conception, ideology, social role, and language of prostitution.

After going into prostitution, the women tend to develop attitudes and behavior patterns that are a part of the social role they play. In this connection, they develop an "argot," or special language, for their work, special acts and services, patterns of bartering with their customers, and a large number of rationalizations for their activities. While many prostitutes are able to leave this occupation for marriage or for other employment, a few others are able to achieve a high standard of living and to maintain it. For others, however, age, venereal disease, alcoholism, or drug addiction result in a derelict life, punctuated more or less regularly with arrests and jail terms.

For many prostitutes, there is a strong economic motivation to continue prostitution. Other inducements in one study included loneliness, entrapment because of pimp control, and drug addiction (Davis, 1981: 312). For those women who participate in a deviant street network, prostitution becomes a part of the "family" activities and simply another hustle. Undoubtedly, many women continue in prostitution because of a lack of other more legitimate and higher-paying alternatives. These women may feel trapped because of economic necessity and an absence of more attractive alternatives. Thus, the extent to which prostitutes are "willing" participants in this occupation can be questioned (Hobson, 1987).

Prostitution and AIDS

The contemporary concern about AIDS has implications for prostitution. Although most AIDS cases in the United States involve homosexuals, AIDS is associated with intravenous drug use. Most prostitutes who contact AIDS do so from their experience with such drug use and many prostitutes appear to believe that they can avoid AIDS. The prostitute in Spokane, mentioned earlier, for example, denied that she was at any greater risk for AIDS because she saw a physician regularly. Clearly, such behavior would not prevent the disease, although it might lead to an early detection. This would, of course, be of little consolation either to her or her clients.

Rates of AIDS among prostitutes vary from 4 percent in Los Angeles to over 80 percent in some eastern U.S. cities (Bellis, 1990: 26). The risk that prostitutes incur is generally not from sex with strangers but from the relationship between prostitution and intravenous drug use. In fact, most women with AIDS have contacted the disease through intravenous drug use. In Bellis's study of 72 prostitutes, he found that although concerned and knowledgeable about AIDS, the prostitutes did nothing to protect themselves or customers from it by, for example, abstaining from heroin, by relying on unused needles, or requiring condom use. "Yeah, I'm concerned," reported one, "until I stick the needle in. When I'm hurting, dope's the only thing on my mind" (p. 30).

The prostitutes were driven by the need for money to supply their drug habits. As another put it: "I'm afraid of AIDS but I've got a drug problem. Drugs drive me. If it weren't for heroin, I wouldn't be out here doing this. Dope pushes everything else out of my mind." (Bellis, 1990: 30).

Prostitution and Social Control

Attitudes toward prostitution have varied historically and today vary in different countries. The attitudes toward prostitution depend in part upon the role of women in society and whether the prostitute provides services in addition to sexual activity. In ancient Greece, for example, prostitutes were generally highly respected. Similarly, the devadasis, or dancing girls, were connected with the temples of India for centuries; besides singing and dancing, they practiced temple prostitution. And, Japanese geishas, trained as they have always been in the arts and music, and in conversation and social entertaining, can be cited as examples of women who could often engage in prostitution yet still have high status in society.

Through the middle ages, prostitution was not regarded as criminal activity—rather, it was seen as a necessary evil. Both civil and religious officials attempted to regulate prostitution, but the degree of condemnation found among some segments of modern American society was entirely absent.

Prostitution, particularly soliciting, has been strongly disapproved under Anglo-American criminal law. Such strident attitudes were largely derived from the Protestant Reformation. Even today, many Catholic countries, such as those in Latin America, have a comparatively tolerant view of prostitution. Where it is illegal, prostitution represents an effort to control certain private moral behavior by punitive social control. Undoubtedly, only a small proportion of acts of prostitution are ever uncovered.

Prostitution is opposed on many grounds because (1) it involves a high degree of promiscuity, particularly with strangers; (2) the prostitute is willing to sell and commercialize her sexual participation with emotional indifference outside of marriage, one participating for pleasure and the other for money; (3) the social effects on the women who engage in it are unwholesome; (4) it is a threat to public health in that it facilitates the spread of venereal diseases; (5) it needs police protection in order to operate and thus reduces the quality of general law enforcement; and (6) sexual acts with a prostitute are generally such that there is no possibility for marriage and procreation and, for this reason, are different from ordinary premarital sex relations.

Law enforcement is often a sordid business in cases of prostitution. Because there is seldom a complainant to such cases, the police must be more aggressive in detecting these crimes. The police may resort to questionable enforcement practices in the zeal to enforce these laws. The demeanor of a prostitute to a police officer has a great deal to do with her vulnerability to arrest (Skolnick, 1975: 112). A prostitute is frequently arrested as the result of solicitation by the police or a "lure" provided by them, and sometimes informers are used to locate the rooms being used. In order to "buy" her way out of an arrest, she may offer to serve as an informant in the apprehension of her pimp or a narcotics peddler. Although the males who frequent prostitutes are also technically breaking the law, at present in many jurisdictions, it is still the prostitute who is of primary concern to police officials. Rarely are customers also arrested.

Laws against prostitution generally discriminate against women. Many persons claim that a woman should be able to engage in intercourse for money if she so desires. A national organization of prostitutes in the United States (called COYOTE—Call Off Your Old Tired Ethics), founded in 1973 in San Francisco, along with one in France, attempted to change public attitudes toward prostitution by calling for the decriminalization of laws dealing with prostitution. It began publishing a newsletter called *COYOTE Howls*. The COYOTE group called for the repeal of laws against prostitution maintaining that prostitution was an activity that could actually benefit a community by providing an outlet for male sexual activity. The group claimed a membership of 20,000 in 1979.

COYOTE attempts to dissociate prostitution from its historical association with sin and crime and place prostitution in a context of work and civil rights (Jenness, 1990). By attempting to repeal existing laws and engage community leaders in debate over questionable enforcement practices, COYOTE members have attempted to redefine prostitution from a social problem facing the community to merely an occupational choice that is exercised by some women. These claims are underscored by the claims that (1) not all prostitution is forced, (2) prostitution represents merely a service occupation in the community, and (3) denial of a woman the choice to engage in prostitution is a violation of her civil rights to work as she pleases. These arguments have not led, however, to wholesale repeal of laws against prostitution.

A group of prostitutes formed a newer organization in 1985 called WHISPER (Women Hurt in Systems of Prostitution Engaged in Revolt). The purpose of this organization was slightly different from COYOTE. WHISPER depicts the down side of prostitution with graphic accounts of women bound in chains and ropes by their customers, burned with cigarettes by their pimps, and generally degraded sexually for the money they need to live. WHISPER de-emphasizes the notion that prostitution is entirely voluntary; it is maintained that one needs money and prostitution is one way to get it (Hobson, 1987: 221–222), and that women who engage in prostitution are victimized.

3. Other Forms of Heterosexual Deviance: The Example of Transvestitism

Because of the wide range of sexual norms (which specify appropriate sexual objects, times, physical situations, and motivations for sex), there are many forms of heterosexuality that can be defined as deviant. One such form is transvestitism, or cross-dressing. A transvestite is a man or woman who not only cross-dresses, but does so for reasons that include sexual satisfaction (not for money, entertainment, etc.). Because the distinguishing dimension of transvestitism is the element of sexuality, transvestites are most often men. When women cross-dress today, they most often do so for fashion, not to "pass" in public as a man. Many transvestites identify with the opposite sex and wish to look like that sex.

Transvestitism has been noted for centuries and is not a new sexual form. Dekker and van de Pol (1989) describe a number of cases of women who cross-dressed for a variety of reasons (patriotic, economic, role-related) including sexual satisfaction. Transvestitism was used by lesbians during the early 18th Century because sex in the 18th century was almost exclusively conceived to be intercourse with a penis; therefore, some lesbians cross-dressed and used artificial penises to have relations with their lovers.

Transvestitism in modern times is no more visible than 300 years ago. Most persons who cross-dress do so under conditions of secrecy and privacy, although many male transvestites are married. The reactions of the wives of transvestites, upon learning of their husbands preferences, are instructive of the deviant nature of this behavior. In one account, June, an English woman, who has been married to her husband, George, for 21 years describes their relations:

> First of all when he told me, I let him have the go-ahead and he could dress. The children were a lot younger then, so when they'd gone to bed, he could come down dressed; but then something inside you rebels and is repulsed and says: "This isn't right." You married a man and you've got this man dressed as a woman and enjoying the role. I just cracked up and kept crying. And it was more and more tablets [anti-depressant pills]. (Woodhouse, 1989: 103)

The deviant nature of transvestitism may be clear to persons who are removed from the behavior but it is often not to persons close to the transvestite, especially when the behavior begins. Some elements of cross-dressing are common, even among small children and condoned by parents. Even in adulthood, women today often wear men's clothing. Because of wide variations in fashion, cross-dressing is even encouraged under certain circumstances, such as when a "unisex" look was popular among both men and women.

Many wives of transvestites seem not to regard the behavior as terribly important at first, but their sense that their husbands are engaged in deviant behavior increases with time. Polly, another transvestite's wife, reported that:

> At first, I couldn't see anything odd in it at all, because I dressed as I wanted to—jeans or dress, I couldn't see what the big hang-up was. Obviously, I realize it now, but at the time it seemed very cut-and-dried. So he puts on a dress occasionally, what's that to me? It's when you learn you get a change of personality at times as well then it starts to worry you. (Woodhouse, 1989: 106)

There is an element of public ambivalence about transvestitism. On the one hand, the behavior is, in and of itself, harmless and it can be said to involve men who are merely extending the frontiers of masculinity and engaging in some gender-bending. But, on the other hand, transvestitism, sooner or later, comes into conflict with a variety of sexual norms both of the transvestite and his or her partner. The degree of cross-dressing and, especially, deriving sexual satisfaction from such behavior is particularly troublesome for some persons. The element of motivation is a main differentiation between the motivations of men and women to cross-dress. Many women will dress like men, partially or completely, for purposes of fashion, while men who cross-dress tend to do so for reasons of sexual satisfaction. Transvestites want to be feminine, but it is a kind of femininity that is divorced from the everyday lives of most women as mothers and wives; it is a fantasy world that even women can not identify with. The wives of transvestites, quoted above, have difficulty accepting their husband's behavior partly for this reason.

Perhaps the main reason for the deviant nature of transvestitism is not the wearing of the clothing of the other sex, but that transvestites derive sexual pleasure primarily from their role and not others as their source of satisfaction. If correct, transvestitism is considered deviant by many because it is essentially antisocial. There are other heterosexual activities that draw attention for the same reason. Masturbation is another form of antisocial sexuality, but masturbation is usually a short term behavior and not a primary means of satisfaction for most persons given a choice. Under these conditions, masturbation can be regarded as natural behavior. When, however, it becomes the only source of sexual expression, it too is likely to be regarded as deviant

PORNOGRAPHY

It has been extremely difficult to decide legally exactly what is obscene and, therefore, pornographic. In the 1973 Miller case, the Supreme Court ruled that states can ban material as being obscene if it (1) appeals to "pruriency," (2) contains descriptions "patently offensive" to community standards, and (3) lacks, as a whole, serious literary, artistic, political, or scientific value. All three conditions must be present before the material is considered obscene. An important aspect of the Supreme Court decision was that it put back into the community the final decision as to what is obscene. Local courts, using local community standards rather than national standards, were to be the final judges. Obviously, the courts of Las Vegas have different standards from those of Topeka, Kansas. What is unclear, however, is the best manner to gauge this community feeling. In a 1978 decision, the Supreme Court ruled that juries sitting in judgment over a publication or film must "determine the collective view of the community as best it can be done." Furthermore, the Supreme Court also declared that children's sensitivity was not to be taken into account when the publication or film was directed at adults only.

Although not itself a form of deviance, pornography is thought to be linked with forms of sexual deviance as well as posing other problems for society. "Pornography" is a word that derives from the Greek work *porne* (prostitute) and referred originally to a description of prostitutes and their trade (Barry, 1984: 205). Clearly, pornography no longer is limited to descriptions of the behavior of prostitutes and their clients but has broadened to include virtually any sexually explicit material. Some pornography has a violent quality and presents certain recurring themes, such as sexual slavery. Pornographic depictions can be found in motion pictures, magazines, written works such as books and short articles, and newer forms of pornography, such as "Dial-A-Porn," a device where persons can call a telephone number and receive sexual messages. The popularity of sexually explicit materials is hard to judge, but one estimate has placed the adult male readership of the two most popular sexually oriented magazines (*Playboy* and *Penthouse*) as exceeding the combined readership of the two most popular news magazines (*Time* and *Newsweek*) (cited in Malamuth and Donnerstein, 1984: xv). Even if these publications do not fit a strict definition of pornography, such figures, if true, reflect a high level of interest in sexually-oriented materials.

Social and political concern over the possibly harmful effects of "obscenity" (a legal term) or "pornography" (a popular term) has increased over time. Interest in the possibly harmful effects of pornographic movies and magazines during the late 1960s and early 1970s led to the formation of a national commission to study the nature and effects of pornography (Commission on Obscenity and Pornography, 1970), as well as Supreme Court decisions, to be discussed below, dealing with the regulation of pornography. More recent interest in the 1980s also resulted in a national commission to examine pornography since the time of the last commission and to make recommendations regarding the uses and regulation of pornography (Attorney General's Commission on Pornography, 1986).

The 1970 national pornography commission concluded that many of the problems of pornography were the result of a less than open atmosphere regarding sexual matters. That body recommended more systematic public discussion of the subject as well as a program of sex education in the schools. This tone is conveyed below:

> The Commission believes that much of the "problem" regarding materials which depict explicit sexual activity stems from the inability or reluctance of people in our society to be open and direct in dealing with sexual matters. This most often manifests itself in the inhibition of talking openly and directly about sex. Professionals use highly technical language when they discuss sex; others of us escape by using euphemisms—or by not talking about sex at all. Direct and open conversation about sex between parent and child is too rare in our society. Failure to talk openly and directly about sex has several consequences. It overemphasizes sex, gives it a magical, nonnatural quality, making it more attractive and fascinating. It diverts the

expression of sexual interest out of more legitimate channels. Such failure makes teaching children and adolescents to become fully and adequately functioning sexual adults a more difficult task. And it clogs legitimate channels for transmitting sexual information and forces people to use clandestine and unreliable sources. (Commission on Obscenity and Pornography, 1970: 53)

The more recent pornography commission did not concur with this conclusion and the tone of the 1986 report took a more serious view of pornography, with numerous statements about the possible harmful effects of exposure to pornography and the need for greater, not less, regulation. The differences of opinion suggested in these two national commissions can be best understood with an understanding of the larger context of pornography and the social reactions to it.

The Nature of Regulating Pornography

Explicit sexual references for purposes of entertainment or arousal can be found throughout time in Greek and Roman mosaics, poetry, and drama, Indian writings such as the *Kamasutra*, medieval ballads and poems by Chaucer, French farces of the fourteenth and fifteenth centuries, Elizabethan poetry and art, and into the present time. What is fairly recent is the regulation of such references and themes by law (Attorney General's Commission on Pornography, 1986: 235). The earliest regulation was that generated by medieval religious institutions—such as the church—and that regulation was limited only to descriptions of sex that were combined with attacks on religion or religious authorities. Even common law courts in England were reluctant to take on the issue of pornography in a direct manner.

Contemporary legal concern with pornography dates from the early 1800s in England. As printing became more economical and therefore more available to the masses, the kinds of sexually explicit materials that circulated in a limited fashion up to that point were circulated more widely. The increase in audience generated an increased demand, which prompted an increase in the supply (Attorney General's Commission on Pornography, 1986: 241). This occurred at a time, right before the Victorian era, when issues of sexual morality were perceived to be increasingly condemned. The development of citizen's groups, such as the Organization for the Reformation of Manners and its successor, the Society for the Suppression of Vice, paralleled this social concern. In the United States, the New York Society for the Suppression of Vice developed from the same concerns, and this group pressed for more stringent legislation against pornography. Such groups were successful in securing legislation against sexually explicit materials and, through the first part of the twentieth century, the manufacture and marketing of pornography were almost entirely clandestine. Subsequent legal skirmishes involved issues dealing with First Amendment protection of "free speech" and the most recent laws governing pornography have reflected these issues. They also reflect deep ambivalence regarding the nature of pornography and how or whether to regulate it.

U.S. Supreme Court decisions on pornography have usually been close decisions among the Justices, suggesting that the disputes among segments of the public about what is and what is not pornography also pervades jurists. In the United States, Massachusetts enacted a statute prohibiting the distribution of pornography in 1711, and Vermont followed suit in 1821. The first federal statute prohibiting the importation of pictorial pornography was enacted in 1842. Obscenity was first prohibited in 1865 in a federal statute forbidding its distribution in the mail. Continued legislation coincided with the decline of the direct influence of religion over community life, the spread of free universal education, and increases in literacy. Modern legal opinion is most recently represented in two Supreme Court decisions, both from 1973. In *Paris Adult Theaters I v. Slaton* (413 U.S. 49, 1973) and *Miller v. California* (413 U.S. 15, 1973), the Court decided that pornography in public places was an appropriate candidate for regulation, as opposed to one's own home, and that local community standards ought to be employed in making a determination of offensiveness. These decisions threw back to local jurisdictions the determination of what is and what is not pornography and what to do about it. As a result, some jurisdictions experienced prosecutions, but others did not, for the same material. Controversies continue over such matters as how those local community standards ought to be discovered and the "true" meaning and definition of pornography.

Pornography and Everyday Life

The United States, like many other western nations, experiences much sexual explicitness in everyday life. Portrayals in the mass media, such as films, novels, television, periodicals, and newspapers contain graphic descriptions of sexuality not found in the media during previous years. The motion picture industry instituted a rating system in 1968 to self-regulate the nature of movies and to alert potential viewers to the sexual and violent nature of movies. In an effort to keep up changing social sentiment, the motion picture industry changed its rating system in 1990 to designate films "NC-17," or not suitable for children under 17. Network television has also had some form of self-regulation in the form of in-house censors who are charged with complying with federal regulations. Cable television, on the other hand, which is not subject to the same Federal Communications Commission regulations as broadcast or network television, permits more latitude in the presentation of sexually explicit material.

The spread of cable television, along with the expansion of video cassette recorders (VCRs), has broadened the market of pornography. The availability of X-rated and NC-17 rated video cassettes means that pornography can be a part of any dwelling that has a VCR. Many video retail outlets have an "adult" section that contains sexually explicit cassettes. What the development of the print medium did to expand the market of pornography a couple of hundred years ago, the development of the home video system and cable television did to spread the availability of pornography during the last decade.

The industry that manufactures pornography is extensive. The manufacture of films, video tapes, and magazines is heavily concentrated in certain areas (80 percent of the industry is located in and around Los Angeles; Attorney General's Commission on Pornography, 1986: 285) and, although not illegal, is substantially "underground." Production of sexually explicit motion pictures is done on a low budget and it is not uncommon for the writer, producer, and director to be the same person. There is little overlap between this industry and the mainstream film industry, and it is very rare when an X-rated movie personality becomes well-known for mainstream movies as well. The production of sexually explicit novels often involves a team of writers, rather than a single individual, who pool different sections of the book, often using the same material altered slightly to fit historical and social circumstances and characters.

There is a connection between the pornography industry and other forms of vice, and also between the pornography industry and organized criminal syndicates. Many of the pornography retail stores visited in one observational study in Philadelphia also provided prostitution, gambling services, and illicit drug sales (Potter, 1989). In nearly 40 percent of the stores, customers could gamble or be referred to gambling and in 70 percent of the stores customers could obtain drugs or information on where to get drugs. Prostitution was frequently obtained on-site and the owners of pornography shops were also likely to be involved with similar businesses, including a sex tabloid which carried explicit personal advertisements, a massage parlor, and an escort service.

The Effects of Pornography

The effects of pornography have been said to be of two kinds: direct and indirect. Direct effects would include whether recipients of pornography are aroused as a result of that exposure or whether the recipient's behavior changed as a result of the exposure. Studies that have examined the relationship between exposure to pornography and sex crimes, for example, have looked for the existence of direct effects. Indirect effects would include subtle long-term effects, such as the redefinition of sexual objects or sexual accessibility. One example of a possible long-term consequence is that women are seen more as objects for sex or violence than as persons, and that the context for sexual relations is divorced from emotions and feelings. That is, one long-term danger is that sex is reduced simply to a physical act rather than a component in a relationship between people.

Harmful Effects. The issue of harm is a difficult one because there is little agreement on what constitutes harm and how it might be measured. Too, it is not always possible to say that some harm is caused by exposure to pornography. As a result of these problems, social science research to date has not been able to provide definitive answers to the many questions about the harmful effects of exposure to pornography. The results of laboratory studies on exposure to pornography and subsequent acts of aggression are best described as inconclusive, but primarily negative. Exposure to nonviolent pornography does not seem to lead to instances of aggression (Donnerstein, et al., 1987: 38–60; Smith and Hand, 1987).

Much of the attention of the 1986 national pornography commission was on sexually violent material and for this reason, the commission says, the conclusions of the 1986 commission differed from that of the 1970 commission. In 1970, the pornography commission concluded that exposure to pornography does not result in increased sexually aggressive behavior, either in the form of interpersonal violence or in increased incidence of sexual crimes (Commission on Obscenity and Pornography, 1970: 32). This commission did not conclude that there were no effects; indeed, most consumers of pornography reported being aroused as a result of that experience.

The 1986 commission spent considerably more time assessing material that contained both highly explicit sexual and violent content. "It is with respect to material of this variety," the 1986 commission concluded, "that the scientific findings and ultimate conclusions of the 1970 Commission are least reliable for today, precisely because material of this variety was largely absent from the Commission's inquiries" (Attorney General's Commission on Pornography, 1986: 324). It is specifically with respect to sexually violent material that the Attorney General's Commission on Pornography (1986: 324) concluded that: "In both clinical and experimental settings, exposure to sexually violent materials has indicated an increase in the likelihood of aggression," especially aggression toward women. Other research has supported this contention (Donnerstein, et al., 1987).

The commission qualified its conclusions, however, in the following manner:

> We are not saying that everyone exposed to [sexually violent] material . . . has his attitude about sexual violence changed. We are saying only that evidence supports the conclusion that substantial exposure to degrading material increases the likelihood for an individual and the incidence over a large population that these attitudinal changes will occur. And we are not saying that everyone with these attitudes will commit an act of sexual violence or sexual coercion. We are saying that such attitudes will increase the likelihood for an individual and the incidence for a population that acts of sexual violence, sexual coercion, or unwanted sexual aggression will occur (Attorney General's Commission on Pornography, 1986: 333).

Material that contains sexual violence includes material that contains sado-masochistic themes, such as the use of whips, chains, and devices of torture. This material also involves the recurrent theme of a man making some sort of sexual advance to a woman, being rebuffed, and then raping the woman or in some other way forcing himself violently on her. In most of this material, whether in magazine or motion picture form, the woman is depicted as eventually becoming sexually aroused and ecstatic about the subsequent sexual activity. Exposure to such material, the commission suggests, not only may lead directly to variations of this sort of behavior, but it also might result in attitudinal changes that are conducive to the perpetuation of the "rape myth" (Attorney General's Commission on Pornography, 1986: 329). In the rape myth, the woman says "no," but really means "yes" and men are therefore justified in acting upon the woman's "no" as really meaning "yes." After all, the myth continues, even if the woman really means "no" at the beginning, once the forced sexual activity begins, she will enjoy it and change her mind.

Other effects mentioned by the Attorney General's Commission on Pornography are perhaps less obvious but no less important. One such effect is the "degrading" manner in which women in pornography are usually depicted. In such material, persons, usually women, are depicted as existing solely for the sexual satisfaction of men. Other material depicts women in decidedly subordinate roles, or in sexual practices that many would consider humiliating.

Positive Effects, or the Possible Functions of Pornography. The existence of pornography in modern society suggests to some sociologists that it serves an important social and personal function. One sociologist likened the function of pornography to that of prostitution, namely that in a society that negatively labels impersonal nonmarital sex there are largely two roads, prostitution with a real sex object and pornography, which can lead to "masturbating, imagined intercourse with a fantasy object" (Polsky, 1967: 195). If such an interpretation is correct, one might suppose that the frequency and availability of pornography would decrease as the stigma of recreational sex lessens and its frequency becomes greater. There is, however, no evidence that this is presently the case.

Furthermore, a number of studies have suggested that although both male and female consumers of pornography are sexually aroused by that experience, no negative consequences follow from that condition (see the reviews in Commission on Obscenity and Pornography, 1970). It is even possible that exposure to pornography is healthy and a crime preventive. Eysenck (1972), for example, reports that sex criminals are several years older than noncriminals before they first view a picture of intercourse. Other studies suggest that sex criminals are more likely to be raised in sexually restrictive families and, as a result, are likely to have less information and exposure to sexual matters (Goldstein, et al., 1974). Rapists, in particular, were found to have come from sexually repressive environments. Other research has found that the availability of pornography, including violent pornography, is not necessarily related to aggressive criminality, such as forcible rape (Abramson and Hayashi, 1984), although other research has concluded that sales of pornographic magazines in the United States is related to rates of reported crimes against women in these states (Baron and Straus, 1984). Data like these, as well as other studies, have led one observer to conclude that:

> Contrary to what common sense might suggest, there is a negative correlation between exposure to erotica and development of a preference for a deviant form of sexuality. The evidence even indicates that exposure to erotica is salutary, probably providing one of the few sources in society for education in sexual matters. (Muekeking, 1977: 483)

It is probably safe to say that portrayals of sex and violence in the media do have an effect on some people beyond sexual arousal. It is not possible at present to predict such effects because of individual and cultural differences, and not all of those effects are negative, such as instances of sexual aggression. There are conflicting research findings about both short- and long-term effects of pornography (see Malamuth and Donnerstein, 1984), and it is likely to be some time before we are able to take into account all of the many factors that go into the relationship between pornography and subsequent behavior. The more subtle effects of pornography—including the imagery of women, sex, and physical relationships without an emotional context—may be the more important consequences of pornography because they are long-lasting. Clearly, we need to know much more than we do at present. In the meantime, there is hardly any dispute about the fact that pornography is not entirely good (beneficial) or bad (harmful). Exposure to it produces a range of effects depending on the person and the situation, and absolutism from either censors or zealous libertarians should be avoided in reaching civil compromises on the manufacture, sale, and distribution of pornographic materials (see also Downs, 1989).

SUMMARY

Sexual norms represent the guidelines by which the determination of sexual deviance can be made. The content of many sexual norms represents complicated combinations of appropriate sexual objects, times, places, and circumstances for sexual behavior. Sexual norms can change over time, as shown in the increased tolerance for premarital sexual relations in the United States. Other forms of heterosexuality, however, such as co-marital sex, are less tolerated. Transvestitism represents another example of sexual behavior that violates subtle norms about gender-appropriate clothing and behavior. Because sexual norms are usually specific to particular groups, and because such norms change over time, there is disagreement from group to group on the deviant nature of some acts.

Prostitution is a form of sexual deviance that has long been regulated, often unsuccessfully, by both sexual norms and formal prohibitions, such as law. The attraction of prostitutes for clients includes the desire for sex without subsequent responsibilities and entanglements, and these desires are evidently international since prostitution is found in most countries. There are different types of prostitutes and these types form a stratification system on the dimensions of income and privacy. Much prostitution takes place in a larger urban context of "hustling" where prostitution is only one illicit means by which to earn a living. Prostitutes who are members of deviant street networks will participate in other forms of illegal, as well as legal, work to supplement their incomes.

Many women, and some men, engage in prostitution on a part-time basis, but the transition to becoming a prostitute involves elements of learning and opportunity. Many prostitutes have sexual experiences prior to becoming a prostitute, but the learning of prostitution involves the acquisition of a particular set of attitudes and values. These attitudes pertain to being able to use one's body for the pleasure of others in exchange for money, as well as certain attitudes toward clients and the law. Other attitudes include the belief that the prostitute is doing what other women are doing in singles' bars and on dates, only the prostitute is being more honest about what is going on. The maintenance of a nondeviant self-concept relies heavily on such attitudes. Prostitutes also must learn how to develop a set of contacts with clients, as well as how to take care of themselves in terms of disease and the law.

Sexually oriented materials and messages are common media experiences. The definition of what actually constitutes obscene material is difficult to determine. The first kind of regulation of pornography was religious, but legal regulation has been much more recent. It appears that pornography may be more prevalent today than even a decade ago, and concern over an increasing market for pornography has prompted national inquiry into the issue of pornography. The widespread use of video cassette recorders has increased the opportunity for more pornography into homes and other private places. Interest in pornography reflects a concern that there are possibly harmful effects from exposure and that tighter regulation of pornography would eliminate or reduce these harmful effects. Actually, some observers have claimed that there might be beneficial as well as harmful effects, and two national pornography commissions, as well as much social and behavioral science research, have disagreed on the nature of these effects.

SELECTED REFERENCES

Attorney General's Commission on Pornography. 1986. *Final Report.* Washington, DC: Government Printing Office.

The most recent federal government commission on pornography. Its conclusions refute those of the earlier (1970) national commission by reporting that prolonged exposure to violent pornography may be related to aggressive behavior, especially against women.

Downs, Donald Alexander. 1989. *The New Politics of Pornography.* Chicago: University of Chicago Press.

A highly readable account of the politics of pornography: the groups involved, their interests, and apparent relative power.

D'Emilio, John and Estelle B. Freedman. 1988. *Intimate Matters: A History of Sexuality in America.* New York: Harper and Row.

The subtitle tells it all. A very readable account of the evolution of sexual norms and practices. The more sexually conservative "good old days," the authors report, may not have been so conservative after all.

Donnerstein, Edward, Daniel Linz, and Steven Penrod. 1987. *The Question of Pornography: Research Findings and Policy Implications.* New York: Free Press.

A compendium of research findings related to the nature and impact of pornography from a behavioral science perspective. Donnerstein's research has been widely discussed and argued.

Klassen, Albert D., Colin J. Williams, and Eugene E. Levitt. 1989. *Sex and Morality in the U.S.* Middletown, CT: Wesleyan University Press.

A report from the Kinsey Institute of Indiana University on sexual behavior using data from a national sample. The account helps document the extent of some behavior as well as attitudes.

Miller, Eleanor M. 1986. *Street Woman.* Philadelphia: Temple University Press.

A description of prostitution that differs from conventional wisdom on the subject. Prostitutes are street hustlers who engage in other forms of crime and deviance as well. They are without pimps but heavily immersed in a deviant subculture.

Two Reactions to Homosexuality

Anti-homosexual feelings are motivated by a number of sources, including the attitude that homosexuality is a sin, and it is inappropriate even if it is not a sin. The two reactions below illustrate each of these rationales:

1. From a person reacting to a series of newspaper articles on homosexuality: "In the beginning, God . . . created Adam and Eve—not Adam and Steve." (*Des Moines Register*, October 26, 1990, p. 3T)

2. From a nationally-known advice giver:

Dear Ann Landers:
My partner and I made a formal commitment five years ago. Although our union is not recognized legally, the wedding ceremony was deeply spiritual. All our friends and family members shared in our happiness. We are gay.
"Denny" and I are invited everywhere as a couple. Everyone thinks of us that way. The problem we have been struggling with came to a head a few weeks ago at the wedding of Denny's brother.
Denny is a great dancer. He never sits out a number. The women love to dance with him, which means I am left alone lot. At his brother's wedding, I insisted that Denny dance with me and he finally did. No one reacted, at least we didn't notice any stares.
We enjoy dancing together but we don't want to make others uncomfortable. Do you think it's OK? We need an outside opinion.
Jerry in D.C.

Ann Says: In Eastern European countries, men traditionally dance together. Nobody thinks it's strange. If you and Denny want to dance together in the company of family and friends who are aware of and accept your relationship, I see nothing wrong with it. I assume, of course, that you mean conventional dancing—no Lambada, no cheek to cheek and no slow dancing with erotic overtones.

(*Des Moines Register*, October 26, 1990, p. 3T)

demnation of homosexuality. The leader of these efforts was Rev. Jerry Falwell, whose group, "The Moral Majority," in the early 1980s represented not only a religious organization but also a group interested in lobbying for anti-homosexual laws. These groups constituted a significant backlash to the prevailing trend of increased toleration of homosexuality. Such groups were responsible for public forums that debated the issue of hiring and retaining homosexual teachers and other public servants. Some of these initiatives appeared on political ballots in some states as referendums, others were the topic of public conversation. Throughout the 1980s and into the 1990s, the fear of AIDS polarized opinion about homosexuality in a number of communities.

Some people and groups are more tolerant of homosexuality than others. Being female, having known homosexuals, and having parents who had an accepting attitude toward homosexuality have been found to be associated with more tolerance (Glassner and Owen, 1976). Within families, mothers are more likely to be able to tolerate their son's homosexuality than are fathers, and among other groups greater rejection of homosexuality is found among working- and lower-class persons, religious fundamentalists, and persons without a college education (Hammersmith, 1987). Evidently, those persons who have had nonsexual contact with homosexuals in the past and those from backgrounds that permit diversity of social roles are more likely to have a chance to find homosexual stereotypes misleading or incorrect. Society's increasing tolerance of homosexuality, the freer circulation of information about homosexuality, and the militancy and openness of gay and lesbian organizations are important trends now altering the stereotype of the homosexual.

Social Dimensions of Homophobia

The term "homophobia" refers to the fear and dislike of lesbians and gay men. Many persons believe that homosexuality is sinful, "sick," or just wrong, and that it should therefore be against the law. In a Gallup poll conducted in 1986, 54 percent of a national sample of adults agreed that homosexual relations between consenting adults should be illegal and 51 percent believed that the Constitution of the United States does not protect private homosexual acts (Gallup Poll, 1986). There seems little doubt that the AIDS epidemic has increased fear about homosexuality since AIDS is always fatal and has been tied closely to the sexual practices of gay men.

Homophobia takes many forms, both attitudinal and behavioral. Many persons who have homophobic attitudes have little difficulty identifying and expressing those attitudes. They are, in other words, attitudes about which persons are often aware and about which they feel strongly. Other persons may be more reluctant to express homophobic attitudes but they may display some kind of homophobic behavior. Homophobic behavior includes avoidance of homosexuals and lesbians and things that are associated with homosexuality. Such behavior might also include overtly discriminatory behavior, such as failing to give a job to a qualified homosexual.

The origins of homophobia have been said to include the link between homosexuality, religion, and psychological maladjustment. The association between strong Christian beliefs and intolerance toward homosexuality is a common observation. There is some dispute about the origins of such an association. Greenberg (1968) argues that an important cornerstone of the development of the modern church was the rejection of homosexuality. This rejection was conducive to the strengthening of the Christian community at a time when the group was struggling. The psychological position tends to view homosexuality as an immature, undeveloped form of adult sexuality. One observer argues that homophobia represents a reaction and defense against repressed affections for the same sex (Herek, 1984). While everyone has homoerotic feeling at some time, these are repressed in the course of "normal" sexual development. The successful resolution of the Oedipus Complex (see Chapter 2) leads a person to such a "normal" psychological state; an unsuccessful outcome may lead to a homosexual position.

Empirical efforts to study homophobia have examined such antecedents as religious background, degree of religiosity, and political conservatism. One researcher who examined these factors adds the factor of "homosociality," which is defined as the social preference for one's own gender but not necessarily an erotic attraction (Britton , 1990). In this study, general religious and social conservatism relate strongly to homophobia and that "conservatism about the proper roles of men and women seem to be the source of this relationship" (p. 436) between conservatism and homophobia. These finds were confirmed in another study by Ficarrotto (1990) who found that sexual conservatism and social prejudice (by race and sex) were independent and equal predictors of antihomosexual sentiment. Furthermore, Britton finds that persons who favor the maintenance of sex-segregated institutions (for example, Boy Scouts for boys only, Girl Scouts for girls only, all male social organizations such as lodges and clubs) tend to be the most homophobic but only against men. In this case, homophobia may serve as an important boundary that helps maintain the distinction between appropriate social and sexual interaction in these settings. Even persons who are religiously and socially conservative—persons who are likely to oppose homosexuality in general—may thus exhibit greater tolerance for female than male homosexuality.

The Attribution of Homosexuality

A common myth is that adult male homosexuals can be readily recognized as physically effeminate persons and that lesbians are masculine in manners and appearance. Actually, most homosexuals are indistinguishable physically from heterosexuals. Where they are socially visible—and many are—it is because they perform a homosexual role. While some homosexuals report that they can identify other homosexuals, this is often because they are looking for them; heterosexuals often wish not to find or notice them. In a study involving questionnaire responses from 1,900 gay men and 1,000 lesbians, it was found that only 6 percent of the lesbians felt that others could tell they were a lesbian; 68 percent felt others could not tell, and 27 percent were unsure (Jay and Young, 1979). As one lesbian put it: "Don't be silly. I can't tell other lesbians are lesbians. How can *most people*, which would indicate *straight people*, tell I'm a lesbian?" (Jay and Young, 1979: 188). This point was confirmed in a more recent study that used videotaped interviews with both homosexual and heterosexual men and women (Berger, et al., 1987). The interviews were shown to 143 male and female raters and the raters were asked to identify the homosexuals. More than 80 percent of the raters could not identify the homosexuals in the interviews.

It is difficult to generalize about a possible homosexual appearance because many exceptions can be found. Some indication of the great variations in the social and physical characteristics of homosexuals can be seen in the long list of important historical figures who were homosexuals—philosophers, military leaders, artists, musicians, and writers.

> Socrates and Plato made no bones about their homosexuality; Catullus wrote a love poem to a young man whose "honeysweet lips he wanted to kiss; Virgil and Horace wrote erotic poems about men; Michelangelo's great love sonnets were addressed to a young man, and so were Shakespeare's. There seems to be evidence that Alexander the Great was homosexual, and Julius Caeser certainly was—the Roman Senator Curio called Caeser "every woman's man and every man's woman." So were Charles XII of Sweden and Frederick the Great. Several English monarchs have been homosexual. . . . About some individuals of widely differing kinds, from William of Orange to Lawrence of Arabia, there is running controversy which may never reach a definite conclusion. About others—Marlowe, Tchaikovsky, Whitman, Kitchener, Rimbaud, Verlaine, Proust, Gide, Wilde, and many more—there is no reasonable doubt. (Magee, 1966: 46)

Persons are often attributed or imputed to be homosexual as a result of how they are defined by others, whether the person has a sexual preference for his or her own sex or not. Regardless of what evidence of homosexuality exists, it may become documented by retrospective interpretation of the person's behavior, a process that consists of a review of past interactions with the individual. Here, past interactions are reinterpreted in view of other evidence of homosexuality; a search is made for those subtle cues and behavior that might provide evidence to justify the attribution of homosexuality. Indirect evidence includes rumor, general information about the person's behavior, associates, or sexual preferences, or the experiences an acquaintance might have had with the person even though it might not have been verified. Direct observation includes behavior that "everyone knows" as evidence of homosexuality, such as effeminate appearance and manners for a male, and masculine appearance and manners for a female which may or may not be true.

SEX ROLE SOCIALIZATION AND BECOMING A HOMOSEXUAL

A complete understanding of homosexuality requires information about the meaning of "gender" and the processes by which individuals come to identify with one gender over another. These processes lead to feelings of sexual preference and identity.

Sexual Development

Although it has a biological basis, sexuality is learned. Sexuality is a social construction "that has been learned in interaction with others" (Plummer, 1975: 30). It is not dictated by body chemistry, but social situation and expectations. "Male" and "female" are socially constructed categories and so too is the conduct that arises from these roles. One learns to be aroused by some persons or objects but not by others. One also learns at what age one is supposed to be capable of arousal and sexual intercourse. In fact, one can learn that virtually anything is a sexual stimulus if paired with a sexual response. The sex drive, in other words, "is neither powerful nor weak; it can be almost anything we make it" (Goode and Troiden, 1974: 15). The social meaning of sexuality, then, is one attained in the manner as other social acts, as part of the overall socialization process. Sexuality is learned not all at once but over a period of time, and according to the principles of learning and social interaction that have been discussed in other chapters. Persons learn to become homosexuals through the same general learning processes by which persons learn to become heterosexual. It is the content of the learning that is different.

The typical conception of sexuality, however, is quite different from this sociological portrait. Many people have become accustomed to thinking about sexuality as being totally innate and exclusively dependent upon certain vague, biological determinants. Actually, while there are unmistakable biological limits within which sexual development proceeds, sexuality is probably better conceived to be largely socially determined on a continuum from very masculine on one end, to very feminine on the other. In fact, Kinsey's (1948) conception of sexuality was operationalized to be on a seven point scale with completely heterosexual at one end, and completely homosexual at the other. In between are different orientations that are distinguished from one another by the different socialization experiences of the persons. Much of that socialization takes place within the context of sex roles.

Sex roles (sometimes called gender roles) are those collections of norms that define male and female behavior. It is useful that sex roles be very distinct from one another because sex roles form an important part of our individual and group identities. One of the reasons that homosexuals and transvestites may be so condemned by some in our culture is that they threaten this distinction and, therefore, threaten an important part of our identity. In cultures where homosexuality is more accepted (e.g., ancient Greece), there was no strong collective boundary that needed maintaining. The Greeks had no need for their collective identity to be maintained by a strict moral code. In our culture, a Judeo-Christian ethic maintains strong boundaries between male and female to reinforce the historically strong emphasis on a separate Jewish identity (Davies, 1982). This concern, articulated in the Old Testament, was meant to help insure the survival of the Jews at a time when their continued existence was in doubt.

The learning of sex roles begins at birth with the behavioral expectations of parents and others. It begins as early as the color of the baby's blanket which is meant to convey certain expectations about the sex role of the infant. The significance of a blue or pink blanket may be lost on the baby, at that moment, but not on others who will react to the baby more on the basis of the blanket color than anything else. If the blanket is blue, people will begin immediately to detect (expect) masculine attributes, while if the blanket is pink, people will detect and expect feminine attributes. Boys and girls have their nurseries decorated differently, boys with "boy things," such as footballs or other sporting equipment, and girls with "girl things," such as dolls. Over time, boys will be expected to act more aggressively and girls more passively to the sexual "scripts" to which they have been presented. Behavior that conforms to the expectations of others about the child's sex will be reinforced, those that do not will be punished. Evidently, if there are any inherited biological tendencies, they are modifiable to a large extent over time, even in cases where parents raise children the opposite of their sex either because the parents wanted one sex and had another or because the child was born with sex characteristics of both sexes. Thus, whether individuals come to identify themselves as males or females is a result of the sex roles to which they have been assigned and have come to perform.

Similarly, the learning of sexual behavior begins early in life (Akers, 1985: 184–185). Persons can learn that virtually any object or person can provide sexual satisfaction, but sexual behavior is always embedded in a web of normative constraints that define only certain objects and persons as acceptable. Rewards and punishments from early childhood help the individual to define acceptable sexuality. Most persons learn to adopt heterosexual roles and to derive sexual satisfaction from objects and persons that are considered to be "conventional," that is, within the norms of their group.

But, the sexual socialization process is sometimes not perfect, and some individuals will come to derive sexual satisfaction from objects and persons outside the group's normative structure. This is to be expected for at least two reasons. First, the area of eroticism is an ambiguous one for socializers. Many parents and others feel uncomfortable providing sex education that includes sex-appropriate satisfaction information. For most socializers, the topic of sex is an embarrassing one. Second, the area of sexuality covers much ground, from appropriate partners, to appropriate time, objects, places, and ages. In fact, sexual norms are among the most complicated because of the different combinations of contingencies that one must learn. It is unsurprising, therefore, that there are instances where the socialization process fails to prepare an individual adequately for sexual growth and maturation. Some individuals will find themselves open to sexual alternatives, such as using a prostitute for sexual gratification or engaging in unusual sexual practices, such as sadism or masochism. For the same reason, it is also unsurprising that some persons come to be attracted to members of the same sex. Even taking into account the complexity of sexual norms and ambiguity of the socialization process, by far most persons are socialized "appropriately" and become heterosexual.

Becoming a Homosexual

The general theoretical perspective presented here views the development of sexual preference in the larger context of sexual socialization (see also Plummer, 1981). Persons come to develop a sexual preference, or orientation, because they have learned this preference, or, conversely, they have not learned other alternatives.

It is necessary again to distinguish *homosexual behavior*, which refers to sexual practices with one's own sex, from *homosexual preference*, which refers to the subjective feelings that a person of the same sex is more sexually attractive than a person of the opposite sex. These terms refer to different things. A person may engage in homosexual activities but still be primarily attracted to persons of the opposite sex. On the other hand, some married males may feel more attracted to persons of their own sex and find most of their sexual stimulation coming from them. The degree to which a person combines both a high level of homosexual attraction and homosexual behavior may be a product of participating in a homosexual subculture and the extent to which the person is a member of that subculture. For this reason, there is no such thing as *the* homosexual, but differing degrees of involvement with homosexuality at the levels of behavior and attraction.

Large numbers of children engage in experimental sex play of a homosexual nature, particularly when experimentation with members of the opposite sex is difficult or impossible. The very first homosexual experience among a group of homosexuals in Great Britain was found to be usually with a school boy of the same age (Westwood, 1960). These experiences, however, did not necessarily lead to homosexuality or a pattern of sex behavior because such sex behavior among boys may have little emotional feeling. The first "significant homosexual experience" can be defined as one carried out with an adult or repeated acts carried out with the same boy over a year or so. Over two thirds of such experiences were with another boy. Only 20 percent were first introduced to homosexuality as boys by adults, and another ll percent had no experience of any sort until they were adults; and in all such cases, the partner was an adult. Contrary to the popular view, "seduction" was not an important factor with these boys. Another British study that involved six groups of 50 homosexuals each found that by the time they were adults, nearly all the individuals who later became homosexuals had had at least one exposure to sex (Schofield, 1965). Three fourths of the men in three groups had had their first exposure before 16, and 16 percent had had it with an adult. With most homosexuals, there is a long period during which they fight against their homosexual activity before recognizing it as a permanent behavior and assuming a conception of themselves as homosexuals.

It has been emphasized that sex roles are learned. Behavior patterns associated with masculinity and femininity are learned as part of one's sex role; they are not biologically inherited. Homosexuality and heterosexuality can thus be understood within three concepts: (1) sex-role adoption; (2) sex-role preference; and (3) sex-role identification. *Sex-role adoption* refers to the active adoption of the behavior characteristic of one sex or the other, not simply the desire to adopt such behavior. This is sometimes called sex-role or gender role nonconformity. *Sex-role preference* is the desire to adopt behavior associated with one sex or the other, or the perception of this behavior as being preferable. Finally, *sex-role identification*, which is crucial in homosexuality, is the actual incorporation of a given sex role and the unthinking reaction that is characteristic of that role. In other words, the person comes to internalize the sex role and develop a self-concept consistent with that role. Some people may identify with the opposite sex but also adopt many of the behavior characteristics of the opposite sex.

It appears that persons who eventually become homosexuals are likely to be those who have acquired a sex-role identification or sex-role assimilation toward members of their own gender in childhood. Some research has suggested that the effectiveness of traditional sex-role learning in children is associated with the sex of the siblings, absence of father, or birth-order position in the family. One study, for example, reported that most homosexuals in the sample came from backgrounds where the father was either physically or emotionally absent (Saghir and Robins, 1973). In a comparison study of a group of homosexuals and a group of heterosexuals, 84 percent of the homosexuals and only 18 percent of the heterosexuals reported that their fathers were indifferent to them. Only 13 percent of the homosexuals, but two thirds of the heterosexuals, identified with their fathers; and while 18 percent of the homosexuals had what they would call a satisfactory relationship with their fathers, 82 percent of the heterosexuals reported a satisfactory relationship (Saghir and Robins, 1973: 144–145). One homosexual, when asked about the causes of his homosexuality, implicated the absence of a father figure:

> Well, for one, because I was never raised around a man, and I never had my father there, you know. My brothers were there off and on, very more off than on. And like, when I was away, I was with my grandmother and my auntie. When I came out here to L.A., I was with my mother and grandmother. I never really had a male image to enforce in me this and that, you know, so I guess that might have had a strong influence on the future. (Green, 1987: 355)

Other research has failed to document the supposition that homosexuality comes about because of an identification with the parent of the opposite sex (Bell, et al., 1981). Most research, however, does point to the importance of childhood sex role development as being important, and especially any behavior that deviates from sex role expectations. This sex role nonconformity, in the form of "sissy" behavior in boys and "tomboy" behavior in girls, has been noted with large numbers of persons who later developed homosexual preferences (Green, 1987; Bell, et al., 1981). In one study of homosexual couples, 75 percent reported that they had been called "sissy" as a boy (McWhirter and Mattison, 1984: 130). But methodological problems are severe in these studies and one such design—that by Green cited above—experienced a sample loss of one third of the total sample before the study was completed (Paul, 1990).

Early childhood experiences, however, do not by themselves determine the individual's eventual sexual orientation. Sexual learning continues throughout adolescence and into early adulthood. Adolescence is particularly important because it is during this period that young persons are changing from "homosocial" contacts (contacts primarily with one's own sex) to "heterosocial" contacts (contacts with the opposite sex). By the end of adolescence, they are fully aware of these contexts—which persons are sexually desirable, when and with whom it is appropriate to have sexual relations, and so on. By this period in their lives, most people have developed a sexual identity. This identity is deep-seated and reflected in feelings of sexual preference and orientation, regardless of the individual's behavior (Harry, 1984). In this sense, adult homosexuality is "just a continuation of the earlier homosexual feelings and behaviors from which it can be so successfully predicted" (Bell, et al., 1981: 186). As some individuals learn to identify with "male," some also learn to identify with "female." The development process is the same; the content of what is learned is different.

The web of specific influences in the determination of sexual orientation is difficult to identify completely, but the definitions of the situation offered by others must be added to that web. People define certain objects and situations as "male" and "female" and these definitions temper their expectations for behavior in those situations. Surely, Berger (1986: 179) is correct when he notes:

> Sexual orientation is a complex phenomenon. Becoming homosexual is the result of both personal and social variables and is determined in part by how one's behavior is labeled by others.

Many homosexuals later report that the formation of their sexual identity was a confusing and lonely process, and that the reactions of others—either real or anticipated—were an important part of the process. Mike, a 19-year-old British male, conveys the isolation he felt through this period.

> I went through such hell. I thought I was going to have a breakdown. Gradually you attach the label gay to yourself because if you don't you

really crack up. I did it gradually after years of torment, but still hated myself for it. Accepting that it could be real was the hardest part of my life. I felt lonely, couldn't turn to anyone through fear of what would happen to me. I didn't know any gays so how could I know that we are just ordinary people? I felt I would only be alone as I wasn't straight but also I wasn't the kind of gay my mates used to laugh and joke about. (quoted in Plummer, 1989: 207)

Developing a Homosexual Identity

People acquire homosexual identities in what is often a long interactive process that depends heavily on the reactions and actions of others. This can be illustrated in the distinction of career or secondary homosexuals who perform a homosexual role from those who engage in homosexual behavior without having that behavior reflect a homosexual self-concept. As pointed out in Chapter 4, one important difference between primary and secondary deviance is the extent to which deviant behavior is committed by someone with a deviant self-concept. Primary deviants commit deviant acts but have no deviant self-concept, while secondary deviants do have deviant self-concepts. Primary homosexual behavior is more likely to be the result of a particular situation and may occur, for example, in one-sex communities like prisons, isolated military posts, naval ships, and boarding schools. There are also those male prostitutes who may commit homosexual acts for money (Luckenbill, 1986). Such persons who commit homosexual acts are not homosexuals in the full sociological sense; they are not career deviants, and the central reason they are not is that they have failed to develop a homosexual self-concept.

Secondary homosexuals tend to seek sexual gratification predominantly and continually with members of the same sex. They have developed a self-concept and play a homosexual role in connection with these acts. In fact, Goffman limits the term "homosexual" to "individuals who participate in a special community of understanding wherein members of one's own sex are defined as the most desirable sexual objects and sociability is energetically organized around the pursuit and entertainment of these objects" (Goffman, 1963: 143–144). An important aspect of career homosexuality is association with other homosexuals, particularly in gay bars which tend to serve as important meeting places for homosexuals in many cities. The development of a homosexual self-concept is almost inevitable after association with other homosexuals, in both sexual and nonsexual contexts, for a period of time.

Gay bars facilitate the maintenance of a homosexual self-concept by managing contacts with nongays and by reinforcing homosexual life in a situation controlled by homosexuals (Reitzes and Diver, 1982). Gay bars are also important in facilitating homosexual liaisons, both of short- and long-term nature. In one survey of 92 homosexual couples, the gay bar was the most common meeting place for the couples (Berger, 1990). Such places serve as locations for social support as well.

Several crucial factors are related to the development of a homosexual identity. Involved in this process are the expectations of others, the degree of identity with the role models available, and the reactions of others—the attribution or imputation of homosexuality to him. Generally, homosexual identity grows initially from the realization of a homosexual preference and subsequently out of continued participation in one-sex activities and environments. Official definitions of the person as a homosexual by medical doctors, psychiatrists, or even the police may also serve to influence the development of a homosexual identity.

The realization of a particular sexual orientation can take place at almost any time, but early adolescence is common. The precipitating conditions, those events that immediately proceed the realization, also vary from person to person. A 40-year-old male reported that he achieved insight about his sexual identity from reading a chapter in a book on sexual development at age 14:

> When I read the chapter, I knew immediately that's who I was. I'll never forget it, as it was one of the most traumatic evenings I've ever had in my life. *I just knew.* I had to go through this entirely alone. There was simply no one to talk to. Oh, I sort of considered briefly discussing it with the family doctor. I just felt very alone. I wondered if there were anyone else like me. After that night, I continued to participate in school activities and to date and all that. But it was all a facade and I knew it. (Lynch, 1987: 40)

When a person says to himself that "I am a homosexual" it is quite different from his actually engaging in a homosexual act. A person can engage in a homosexual act and think of himself as a homosexual, heterosexual, or bisexual, just as one who engages in a heterosexual act thinks of himself as heterosexual, homosexual, or bisexual. The recognition of a homosexual identity is extremely important, and when the person does fully change to this homosexual identity, and make that identity public, he is said to have "come out."

The "Coming Out" Process

Coming out involves the public declaration of a homosexual identity (Dank, 1971) and finally conveying only that identity to nonhomosexuals (Plummer, 1975). Coming out is a process, not an announcement. It involves elements of recognizing one's sexual preferences, experiences with others in sex-role socialization, a process of realization regarding sexual identity, and behavioral commitments to a homosexual life-style (see Dank, 1971 and Coleman, 1981–2). The coming out process takes place over many years, and the process is a tenuous one in that not all homosexuals progress from one fixed point to another. Descriptions of the coming out process emphasize the continuing importance of sex-role socialization and the expectations of others.

Stages in Coming Out   Troiden (1989), for example, conceives of the coming out process in four stages: (1) sensitization, (2) identity confusion, (3) identity assumption, and (4) commitment. In the *sensitization stage*, the individual is becoming aware that he or she is different from others of the same sex and that by high school the individual had a distinct sense that he or she was not the same as others of the same sex. The individual will feel marginal but because of his or her age will be unable to understand these feelings.

The *identity confusion stage* represents a separation of one's sexual feelings and recognition of one's sexual orientation from one's behavior. Here, the individual experiences sexual attraction for others of the same sex, but either fails to act upon those feelings or tries to deny them. By middle or late adolescence, a perception of self as "probably" homosexual begins to emerge. Gay males, for example, begin to suspect they "might" be gay at an average age of 17 (Troiden, 1989: 53). The stigma of homosexuality discourages an open discussion of these changes, and ignorance and a lack of awareness of others help to continue the confusion.

The important events in the *identity assumption stage* include defining the self as homosexual and presenting the self to other homosexuals. This is the first step in coming out. The individual may have some contact with homosexual subcultures (e.g., gay bars) at this point. Initial contacts with other homosexuals are important here for resolving some of the internal conflicts the individual faces. Coming out involves a clear self-definition as homosexual, initial involvement in a homosexual subculture, and redefinition of homosexuality as a positive and viable life-style.

In the final stage, *commitment*, homosexuality is taken on as a way of life. This might mean combining homosexuality with emotionality, such as having a homosexual relationship with a single partner (Warren, 1972). Taking a homosexual lover is one way to confirm a gay identity. Another is disclosure of one's homosexual identity to nonhomosexuals.

The process of coming out involves some difficult decision-making by the individual and one factor is the reaction of others to this decision. In one publication two authors noted that the reaction of one's parents can vary. Coming out to one's family may elicit various reactions:

- Your parents will accept you as you are. This is not common, and with the best intentions in the world they may take a long time to come to terms with your situation.
- Your parents will try to understand but the news will make them feel guilty, as if your gayness is their fault. They probably will think your life is headed for ruin if you persist in your homosexuality, and therefore will pressure you to change.
- Your parents will not react. They will refuse to believe you, and the subject will never be brought up again.

- Your parents will reject you. Melodramatic as this may seem, gay people do get thrown out of their family home, disowned, and told never to come back. (Muchmore and Hanson, 1989: 73–74)

There are, however, many individual differences in the coming out process. One study reported that 18 percent of a sample of 199 homosexuals labeled themselves as "homosexual" without overt sex with another male; 22 percent gained their homosexual identity while participating in a long-term relationship with another male; and 23 percent developed their identity only after involvement in such a relationship (McDonald, 1982). Generally, the social contexts for coming out are in the presence of other homosexuals. This is why it is important to consider homosexual subcultures.

Homosexual Subcultures

Like other subcultures, homosexual subcultures are collections of norms and values. In this case the subculture permits or condones homosexuality. Persons come to learn these norms as part of the coming out process and continue to be exposed to them in social situations that involve other homosexuals. Most persons participate in a gay community to a greater or lesser extent. Even heavy participation in a homosexual or "gay" community does not mean that homosexuals have only contacts with other homosexuals. Homosexuals also have contacts with the "straight" world, with family members and employers being only two of the more frequent nonhomosexual contacts.

There is no one homosexual subculture or gay community, just as there is no single homosexual life-style (Bell and Weinberg, 1978). What exists are variations on a common theme—the protection and facilitation of homosexual relations through the common bonding of like others around the homosexual role. In a local homosexual community, secret ("closet queens") and overt homosexuals may be linked through sex and friendship, and these groups, which often cut across social class and occupational lines, serve to relieve anxiety and to furnish social acceptance.

The actual number of such gay communities depends on the region and outside community's tolerance of homosexuality. There are gay communities in many cities, large and small, and rural areas as well (Miller, 1989). New York City and San Francisco have well-developed gay communities; other cities have gay subcultures that are less visible to outsiders. Some of these communities are well-organized, others are not; some are interracial, such as the Black and White Men Together (BWMT) organization. Regardless of place, gay communities are concerned about such issues as external stigma, legal sanctions, AIDS, providing support to members, and the maintenance of homosexual relationships.

The development of gay communities seems to depend in large part on the lack of tolerance for homosexuality and the necessity for the community to reduce the stigmatization of homosexuals by the outside society. In this sense, gay communities are very functional for the participants—they provide a "training ground" for norms and values, a milieu in which people may live every day, and social support as well as an information medium for members. The strength of gay subcultures may have increased to the point where they can be considered "cultures," or dominant systems of norms and values, in some cities in the United States (Humphreys and Miller, 1980). Homosexuals in the United States rely more on subcultures than do those in, for example, the Netherlands or Denmark where the gay communities were less well developed because of lower levels of repression and negative attitudes toward homosexuality (Weinberg and Williams, 1974: 382–384). Gay communities, as such, seem to develop when they are needed to provide a supportive and learning environment to members.

Another subcultural aspect of homosexuality is the increasing number of formal homosexual organizations and gay clubs in many parts of the world. Gay gathering places are found in many countries and some are even provided for gay tourists (Whitam and Mathy, 1985). The homosexual, or "homophile," movement in the United States began on the West Coast after World War II. The first major organization was the Mattachine Foundation, established in 1950 in Los Angeles as a secret club to promote discussion and education about homosexuality. The club was later headquartered in San Francisco and the name changed to One, Inc. National organization of homophile societies approached reality in 1966 with the establishment of the North American Conference of Homophile Organizations which, among other things, organized meetings for local clubs. Some homophile organizations have been militant, such as the New York Gay Liberation Front, and the Gay Activist Alliance, also in New York (Humphreys, 1972). There are now thousands of grass roots gay liberation groups in the United States.

Contrary to other organizations of deviants, such as Alcoholics Anonymous and Synanon for drug addicts, homosexual organizations espouse no desire to change the behavior of their members. Such organizations wish to ease some of the social and legal stigma surrounding homosexuality, and in this sense they wish to reinforce and legitimize homosexuality. These organizations engage in a number of activities that are both educational and political. They furnish information, distribute literature, and publish periodicals that are of interest to homosexuals. Homosexual organizations reject vigorously any idea that homosexual behavior represents a "sickness" or pathology and most argue that homosexuality is not "deviant" in any sense.

PREVALENCE OF MALE HOMOSEXUALITY AND HOMOPHOBIA

The number of homosexuals depends in large part on how homosexuality is defined and how we then can count persons who fit this definition. Estimates of the frequency of homosexuality have varied for these two reasons.

Prevalence of Homosexuality

To date, estimates as to incidence, prevalence, and increase or decrease of homosexuality have been based on inadequate data. The principle reason for this is that there are different kinds of homosexuals, with differing degrees of commitment to homosexuality, and estimates will vary depending on who is counted as a homosexual. A study conducted in 1989, for example, reported that 1.4 percent of men have had adult homosexual contacts whose frequency was characterized as being "fairly often" (Fumento, 1990: 207–208). If everyone who has ever had sexual contact with the same sex is counted, the estimate will be much higher; if only those persons who have publicly declared themselves to be homosexual are counted, the estimate is likely to be much lower because there are many persons with strong homosexual orientation who have not publicly declared themselves to be homosexual. Other observers have estimated that persons who have homosexual identities number in the several million, and "according to which definition of 'homosexual' one uses, homosexuals represent the first, second, or third most common minority in the United States today" (Paul and Weinrich, 1982:26).

Various attempts have been made to estimate the size of the male homosexual population, particularly in the United States and in Great Britain. Earlier and contemporary estimates agree that in general about 4 to 5 percent of the male population are homosexuals. As reported above, Kinsey reported that at some time between adolescence and old age about 37 percent of the white male population had had homosexual experiences (homosexual behavior) to the point of orgasm. But only about 4 or 5 percent considered themselves to be exclusively homosexual throughout their adult lifetimes. Only half as many females as males had had homosexual experiences in the Kinsey (et al., 1953) study. Another earlier study found that a conservative estimate of the extent of homosexuality in the United States was roughly 4 to 6 percent of the total male population over 16 (Lindner, 1963: 61).

In a more recent national sample of the male population, Harry (1990: 94) asked the following question in a telephone interview:

> I have only one question. You may consider it somewhat personal to answer but most people have been willing to answer it once we remind them that this is a totally confidential survey. We reached you on the phone simply by chance and don't know your identity. Here's the question: Would you say that you are sexually attracted to members of the opposite sex or members of your own sex?

In this survey, 3.7 percent indicated they were homosexual or bisexual.

Estimates of homosexuality in other societies is even more difficult, although some observers have ventured guesses. Whitam and Mathy (1985) assert that homosexuals have existed in all cultures and at all historical moments in relatively small incidences (4 to 5 percent of the total male population). This is contradicted by other estimates that claim near-universal participation in homosexuality among certain cultural groups in New Guinea, while in other cultures, except for the very rare report, no homosexuals seem to appear (Herdt, 1981). There does appear to be substantial variation from culture to culture.

Variations of Male Homosexuality

Just as heterosexuality has many social and behavioral variations, so too does homosexuality. Homosexuality varies in the social structure and among certain individuals.

Some evidence exists that homosexual behavior is more frequent in certain occupations, but the evidence is inconclusive: homosexuals are found in almost every occupation and among persons at all educational levels. Some occupations may attract homosexuals because they need not conceal their behavior, other persons accepting the special nature of their occupational roles. Once in the occupation, others may accept the definition of themselves by others as "effeminate." Schofield concludes that:

> For whatever reason, it is in fact now probable that there is a higher proportion of revealed homosexuality in certain job categories—such as interior decoration, ballet and chorus dancing, hairdressing, and fashion design—than in others. The adjective *revealed* is important, because the true proportions for those occupations in which greater concealment is necessary is not known. (Schofield, 1965: 209)

Short-Term Relationships.

Many male homosexuals, however, are likely to have a widespread sexual life, their relations with other homosexuals confined generally to brief and relatively transitory sexual encounters. Homosexual males go through a variety of relationships before settling down, in contrast to lesbians who appear to form more long-lasting relationships (Troiden, 1989). For many male homosexuals, relationships are less likely to be permanent, but even less lengthy affectional-sexual ties in homosexual life are likely to be overshadowed by the predominant pattern of "cruising" and one-night stands. There are indications,however, that "cruising" is declining due to the more likely possibility of acquiring AIDS through this promiscuous behavior.

Because of the relatively impersonal nature of such sex relations, certain male juveniles who are likely not to be homosexuals themselves, offer their services on a monetary basis (Reiss, 1961). Transitory sexual relations can also be arranged with a homosexual prostitute, a "hustler" for other homosexuals who provides, particularly for those who are less attractive physically and older, those services that might be difficult to obtain without great effort (Luckenbill, 1986). The adult homosexual prostitute is a part of homosexual life. He learns his behavior role—such as gestures, vocabulary, clothing, and even makeup—in the same sense that heterosexual prostitutes become a part of heterosexual life (Rechy, 1963: 36).

Some homosexuals become acquainted initially in such public places as toilets, bars, parks, clubs, cafes, baths, hotels, beaches, movie theaters, and other public places (Weinberg and Williams, 1975). A large part of these sexual relations are highly impersonal and may even be carried out in a "tearoom," or "T-room," a homosexual term for public toilet (Humphreys, 1975). These tearooms are readily accessible to the male population, being located near public gathering places—department stores, bus stations, libraries, hotels, YMCAs, and, in particular, in isolated parts of public parks. Usually a third person may act both as voyeur and lookout, and frequently there is little conversation among the participants.

Persons who have studied homosexuals have found that a substantial proportion of them have been heterosexually married at one time or another, the figure in one study being as high as one fourth (Dank, 1972; Lewin and Lyons, 1982). Admitted homosexuals marry because of social pressures, a flight from homosexuality, or for commitment to a home-centered life (Ross, 1971). It appears that in some cases, the person may view his marriage to a woman as an act which will demonstrate to himself and to others that he is very much like a "normal" person. Most of them did not conceive of themselves as homosexuals at the time they were married, in their twenties, even though they may have engaged in homosexuals acts; only later did they develop this identity.

Many such married persons who engage in impersonal sexual relations with other men can not be called homosexuals sociologically in terms of a homosexual identity. Humphreys' (1975) study of tearoom participants showed that the largest group of participants (38 percent) were married or were previously married men, largely truck drivers, machine operators, or clerical workers. Most of them did not want a homosexual experience, but rather a quick orgasm, which was more satisfying than masturbation, less involved than a love affair, and less expensive than a prostitute. The second group were "ambisexuals," mostly better educated and members of the middle and upper classes, many of them married or otherwise heterosexual, who liked the "kicks" of such unusual sex experiences. The gay group of openly confessed homosexuals constituted only 14 percent, and the last group of "closet queens" made up an even smaller proportion of tearoom trade. (Closet queens are homosexuals, unmarried or married, who keep their homosexuality a secret.)

Long-Term Commitments. This is not to say that homosexuals are incapable or not desirous of forming more or less permanent bonds with other homosexuals. A study of 190 gay men found that men do establish long-lasting love relationships with other men (Silverstein, 1981). A more extensive study assigned 485 male homosexuals to one of five different types, depending on a number of variables (Bell and Weinberg, 1978: 132–134). The 67 homosexuals assigned to the *close-coupled* type were living with another homosexual in a quasi-marriage kind of relationship. They tended not to have other sexual partners, nor to engage in cruising. They also had few sexual problems and few regrets over being homosexual. The 120 individuals assigned to the *open-coupled* type were living with another homosexual, but were engaged in cruising and had an active sex life outside of that relationship. The 102 individuals assigned to the *functional* type were "single" (not "coupled") and were active sexually with a number of sexual partners. They also experienced few sexual problems and few regrets about being homosexual. The 86 persons in the *dysfunctional* type had frequent homosexual relations but experienced many sexual problems and regrets over their homosexuality. The 110 persons in the final type, the *asexual* type, had few homosexual contacts, and experienced many problems and regrets over their sexual orientation.

For male homosexuals who live together as partners in a more or less permanent union, the homosexual relationship may be quite stable, their sexuality having become integrated into long-standing affectional, personal, and social patterns. A study of the relationships of 156 male couples revealed that, one third had lived together for more than 10 years (McWhirter and Mattison, 1984). Those couples who had stayed together the longest had the largest differences in age. All of the couples who stayed together longer than 30 years (8 couples) had age differences between 5 and 16 years. Just as in heterosexual relationships, the reasons why couples stay together initially are different from the reasons couples stay together in the fifth, tenth, or twentieth year. While physical attraction, sexuality, and compatibility were important initially, companionship, economic benefits, and lack of possessiveness were important later on. To preserve these benefits legally, particularly if one partner dies, gay couples are agitating for the right to a civil marriage ceremony and, if not, a religious one.

One study of 92 male homosexual couples reported that relationships in the couples lasted from one to 35 years (Berger, 1990). Few of the couples had any commitment ceremony although many wanted one. Most of the couples' close friends were also gay and about two thirds of their families were supportive of the relationship. The most persistent conflicts among the couples had to do with money and relations with family members, issues that plague heterosexual couples as well.

FEMALE HOMOSEXUALITY (LESBIANISM)

The term "lesbianism" is derived from the Greek island of Lesbos where the Greek poetess Sappho (600 B.C.) made herself the leader of a group of women whose relations were characterized by homosexual feelings and behavior. In contrast with male homosexuality, less research has been done on female homosexuality, or lesbianism. It may be because of less scientific interest or because lesbianism is harder to study than male homosexuality.

In many ways, female homosexual activities resemble those of males, but in certain ways there are differences.

The Nature of Lesbianism

Like male homosexuals, lesbians to a degree are subject to public stigma and social rejection. Unlike male homosexuality, however, lesbians seldom fear the law. Sexual acts between women usually are not specifically designated as illegal, and where they might be brought under other laws, this is rarely done. In fact, according to Kinsey et al. (1953: 484), not a single case of a female convicted of homosexual activity was recorded in the United States from 1696 to 1952. In his large sample of women who had had a homosexual experience, only four had had any difficulties with the police. Still, for many lesbians, some fear of disclosure is always present when they are on the job and even when they are among their nonhomosexual friends.

The two worlds, the "straight" and the lesbian, present a particular problem for lesbians who work. Many female homosexuals appear to be more committed to their jobs than most women because they do not depend on a male for financial support. In such situations, many lesbians will attempt to manage their employment settings by pretending to be heterosexual in manners, behavior, and expressions. In this respect, they may be similar to male homosexuals.

The attribution of lesbianism is different for female homosexuals. Lesbians are not as often identifiable as are male homosexuals and any adverse public opinion against them usually becomes manifest only when, for example, they proclaim a unique style of dress or association. Close relations and a certain amount of limited physical demonstration are common among close women acquaintances, and the line between that demonstrative behavior and deviant lesbian relations is more difficult to draw than in male homosexual relations. Females are likely to be more bisexual and inconsistent in their sexual behavior than male homosexuals (Blumstein and Schwartz, 1974). Lesbians, however display the same range of behavior and identity as do male homosexuals. Some lesbians are married, but primarily oriented sexually toward other women; others are unmarried but bisexual; still others are unmarried, but strongly lesbian.

Extent

While it is difficult to estimate the number of male homosexuals, it is far more difficult to estimate the number of lesbians, largely because the lesbian subculture has been much less open and less cohesive. This may be changing as the women's movement has served as a source of organization and focus to many local lesbian groups. In any case, female homosexuals more frequently maintain a seeming front of heterosexuality, so their identification is extremely difficult. Over thirty years ago, Kinsey et al. (1953: 512) estimated that one fifth of single women and 5 percent of all women, at the time of marriage, had had a homosexual relation leading to orgasm, a figure that was lower than that for men. Most sexual studies do not even attempt to estimate the number of lesbians. Perhaps the only agreement about the extent of lesbianism has been the popular speculation that it is more widespread than has been commonly thought, and that it may even be more prevalent than male homosexuality. In a number of respects, lesbianism represents one of the most understudied forms of sexual behavior.

Becoming A Lesbian

It is important to understand how certain values and norms might be conducive to female sexual experimentation, although they might not lead to female homosexuality. Everywhere, women have, to varying degrees, valued themselves as heterosexually desirable, because it is on this basis that they are portrayed extensively in the mass media, and it is on this basis that the expectations of men are developed. Sexuality, therefore, becomes a part of a woman's self-evaluation, and women recognize sexuality, both their own and others', in strongly emotional terms. Traditionally, women have been regarded and treated as "sexual objects" for males, and this attitude may also shift to other females.

Another factor is the set of norms, which differs from that for men, that permits women to touch one another physically and to become emotionally related to one another. Although physical intimacy is usually defined exclusively in social terms, it is both more accepted and more common among women than among men in our society. In some situations, shared discussions of sex and sexual fantasies may lead to behavioral experimentation.

It appears that male models of the stages of "coming out" do not apply well to women; research on the process of becoming a lesbian suggests that there are no uniform stages for lesbians (Risman and Schwartz, 1988: 131). Women more often "drift" into homosexuality than men, starting with vague romantic attachments with other women. Physical contacts with women generally occur before the age of 20, and a large percentage of such sexual contacts before 15. There is little evidence, however, that females are in any sense "seduced" into lesbianism. As with males, female sexual orientations precede lesbian physical relations, and most lesbians have had heterosexual relations before lesbian ones (Bell, et al., 1981). A study of one group of lesbians indicated that while the nature of the first sexual contact varied, in the majority of cases, only manual stimulation was involved. Almost a third of the women studied said that oral sex was part of their first contact, but it was unrelated to the achievement of orgasm at that time (Hedblom, 1972: 56). Despite the fact that most of the lesbians maintained clear-cut boundaries between homosexual and heterosexual worlds, nearly two thirds of them had had sexual relations with men, a third of them within the previous year.

Most lesbians discover their homosexual feelings in late adolescence, often even in early adulthood, and overt homosexual behavior frequently develops as a late stage of an intense emotional relationship. By middle or late adolescence, a perception of self as "probably" homosexual begins to emerge among women who later will form a lesbian identity; lesbians begin to suspect they "might" be gay at an average age of 18 (Troiden, 1989: 53).

The general processes of sexual development and sexual socialization discussed earlier for males applies to females as well. Females learn early in life in most cultures the nature of the female sex role and expectations for that role (Reiss, 1986). The sexual-role behavior patterns of lesbians tend to resemble closely those of heterosexual females. The sexual learning of both homosexual and heterosexual females is close because the cultural expectations of the female role are generally consistent. This does not mean, however, that the early experiences of the two are always the same. One important study found that a significant number of homosexual females displayed "tomboy" attitudes as young girls and behaved as boys (Saghir and Robins, 1973: 192–194). In fact, boy-like behavior in girls who later became lesbians was more common than girl-like behavior in boys who later became homosexuals. The finding that girls who eventually become lesbians are more likely than boys who eventually become homosexuals to display sex role nonconformity has been documented in other studies as well. In one major study, over three quarters of the lesbians displayed boyish behavior and interests (Bell, et al., 1981: 188).

Like many homosexual males, early lesbian experiences may be the result of experimentation and curiosity, which are the major reasons given by females who identify with heterosexuality but who have experienced lesbian relations. One study reported that it was common for some women to have had their first experiences as a result of "male orchestration," that is, situations in which they were part of a "spontaneous threesome" with the male encouraging the two females to engage in sexual intimacies (Blumstein and Schwartz, 1972: 282). Other than situations that involve sexual experimentation, women do not become lesbians because they are seduced into that orientation; rather, the development of a lesbian sexual orientation or preference more often precedes lesbian behavior. There is no evidence that females become lesbians because they have had more negative experiences with men than have heterosexual women (Brannock and Chapman, 1990).

A major difference between male and female homosexuals is that lesbians tend to view themselves as being less promiscuous than the male homosexual (Hedblom, 1972: 55). This self-perception is confirmed in the behavior of the lesbian. She is less likely to "cruise" for sexual partners, even in bars, and is more generally likely to "go with someone" or to be "married" in the homosexual sense, with long-term, more emotional bonds.

> Being a female homosexual is like being a female generally, both sexually and socially. There is a tendency to greater conformity, stability of relationships, and an absence of indiscriminate sexual involvements. There is also a general emphasis on relationships, romantic involvements, and faithfulness in relationships. (Saghir and Robins, 1980: 290)

This difference is woven into the female role in which sexual gratification is placed within the context of an emotional or romantic involvement. The average male homosexual tends to experience more instability in relationships, whereas the lesbian tends to be more reserved and selective in her involvements to the point where she is uninterested in multiple sex partners or varieties of sexual practice. Both male and female homosexuals can and do develop long-term relationships with other homosexuals and lesbians, but the frequency of such relationships seems to be greater among lesbians. Moreover, the nature and frequency of relationship problems among lesbians parallel those for heterosexual married couples; in each instance, the participants in the relationship may need counseling over similar issues, including unequal power, duties, or other complaints in the relationship (Boston Lesbian Psychologies Collective, 1987).

Only limited studies have been made of the role of occupation in the development of lesbianism, although some occupations are suspected of having high rates of lesbian activity. A study of stripteasers found that one fourth of them engaged in lesbian relationships (McCaghy and Skipper, 1969). Moreover, the study estimated that the bisexuality common in this group ranged from 50 to 75 percent. Several factors were thought to contribute to this high percentage, including the limited opportunity stripteasers have for stable sexual relations with males; the negative attitudes they have developed toward men because of the way men act at sex shows lead many of them to prefer to associate primarily with women.

Lesbian Self-Concept

Achieving an identity for women often involves a relational context (Gilligan, 1982). The initial attraction for women with other women is not a sexual but an emotional one of friendship or closeness on the basis of mutual interests. A woman who recognizes her attraction to other women may "try on" the label of lesbian to see how it fits (Browning, 1987). During the course of that process, lesbians are likely to make links on emotional grounds with "special" women (Troiden, 1988 and 1989). These linkages tend to be of longer duration than the relationships among male homosexuals.

Because women tend to emphasize the emotional over the physical aspects of their attraction for one another, this self-labeling process occurs in the context of friendship with another adult woman. The close personal relationship that forms the basis for a lesbian encounter is crucial in the development of a lesbian identity. "The majority pattern appears to be one in which self-identification as a lesbian develops before or during genital contact itself, and as a late stage of a close, affectionate relationship" (Cronin, 1974: 273).

It has long been thought that lesbianism contains elements of both masculinity and femininity (Greenberg, 1988: 373–383). Such stereotypes form the basis for the conception of lesbians for many persons. In this conception, a lesbian must be masculine looking, acting, and/or thinking. When there are two lesbians, one is often socially expected to be feminine but the other to be masculine. In reality, some lesbians may fit such stereotypes, but others do not. Only a minority of lesbians are actually committed to such a "butch" role, even though some may experiment with it, particularly during the "identity crisis" period that takes place when they enter into a homosexual subculture after they "come out."

Nearly all the women interviewed by Simon and Gagnon (1967: 265) wanted to become emotionally and sexually attached to another woman who would respond to them as women. Still, when they abandon the world of men and heterosexual life, lesbians must take on many social responsibilities carried out by men. It is difficult for the homosexual female to develop an acceptable self-concept and identity, although the emergence of the women's movement and the women's gay liberation movement have been helpful in this respect. This is not to say that lesbians are generally unhappy persons who are overwhelmed with personal problems. One study of 127 lesbians, in fact, reported that most of the women studied were happy and satisfied with their lesbian role (Peplau, et al., 1982). This satisfaction was related to characteristics of the relationship they experienced with another woman, such as the equality of involvement and equality of power in the relationship, factors stressed by the women's movement in general.

Lesbian Subcultures

Lesbians do not utilize the gay world or become involved in its subculture as much as do male homosexuals. Homosexual subcultures are functional entities organized to give support and provide a context for social relationships. One reason there are fewer examples of lesbian subcultures is that the role of the lesbian is less alienating than that of the male homosexual. Another reason is that lesbians can mask their sexual deviance behind an assumed asexual response to women to a much greater degree than can men who are assumed to be more sexually active and aggressive. Because lesbians are more likely than male homosexuals to form more permanent relationships, the social and sexual lives of these women is more private than the more public social life found in gay bars and other homosexual meeting places. The longer-lasting relationships suggest less of a turnover of partners for lesbians and thus less need to "make the gay scene" to search for partners. Still, lesbians in long-lasting relationships have many problems (Blumstein and Schwartz, 1983).

The fact that lesbian subcultures are less well developed than those for male homosexuals should not be interpreted as meaning that lesbians are essentially without problems. Lesbians, for example, must deal with issues of homophobia and heterosexism as well as general sexism in society because they are women (Dooley, 1986). The stigma of anti-lesbianism combined with instances of discrimination of the kind that all women experience in occupations, housing, and other areas can result in greater social rejection than faced by males. Male homosexuals, also, tend to be more comfortable economically than lesbians, and economic hardship can be pronounced among some lesbian couples.

The gay liberation process, where gay males have been proclaiming their civil and human rights, has largely overshadowed the lesbian movement. As Adams observes:

> From the beginning of gay liberation, lesbians often found themselves vastly outnumbered by men who were, not surprisingly, preoccupied with their own issues and ignorant of the concerns of women. Many women became increasingly frustrated as gay liberation men set up task groups to counter police entrapment, work for sodomy law reform, or organize dances that turned out to be 90 percent male. (Adams, 1987: 92)

In an effort to meet some of the daily problems of living and lack of political power, certain lesbians do benefit from the functions of a homosexual community (Simon and Gagnon, 1967). Particularly in large cities, some lesbians on occasion tend to congregate in certain bars, usually those patronized by male homosexuals, and these places facilitate lesbian sexual relationships. Research suggests that lesbians begin to participate in a gay community around the ages of 21–23 (Troiden, 1989: 59). The gay community can also provide contacts for females who have no ongoing relationships at the moment. In addition to providing opportunities for sexual contact, such meeting places can also provide social support. In this milieu, the lesbian can express herself fully and openly with persons who have had feelings and experiences much like her own. As with the male homosexual subculture, a special language and ideology are also provided, and these help to provide members with attitudes and rationalizations with which to resist negative social attitudes and stigma.

As noted earlier, homosexual subcultures arise to meet the personal and social needs of homosexuals. Because lesbians are less stigmatized and have less need for an organized subculture, the lesbian subcultures that do exist are fewer in number and less organized than those for males. Few female groups have been organized on the national level, and only one—the Daughters of Bilitis—has any claim to a national representation. A number of lesbian organizations, however, have come into existence at the local level whose main functions are educational and counseling (Simpson, 1976). Often such groups are associated with those organized for male homosexuals.

The affiliation of gay women with the women's movement of the 1970s has further reduced the need for a distinct, well-developed female homosexual culture. As the women's movement helps to carry forward the political interests of lesbians along with those of heterosexual women, the need for separate organizations to perform this function is greatly reduced. The women's movement has provided important sources of ideas about female sexuality. Among other things, it has suggested that female-female associations may be a welcome alternative to unsatisfactory heterosexual relationships. Sexual experimentation, while not overtly suggested by the ideology of the women's movement, may result from membership and commitment to the ideology. In fact, some women may feel that it is important to have at least one sexual experience with another woman in order to widen sexual and political liberation. One woman reported: "I wanted to bed with a radical lesbian; I just had to know what it was like" (Blumstein and Schwartz, 1974: 287). Some of the more militant members have even suggested that heterosexual relations are politically incompatible with the ideology of the movement. It has been said that:

> The purpose of feminist analysis is to provide women with an awareness of their servitude as a class so that they can unite and rise up against it. The problem now for strictly heterosexually conditioned women is how to obtain the sexual gratification they think they need from the sex who remains their institutional oppressor. It is the lesbian who unites the personal and political in the struggle to become freed of the oppressive institution. (Johnston, 1973: 275–276).

AIDS AND THE HOMOSEXUAL COMMUNITY

The gay communities have been undergoing great change in recent years. Unlike previous threats, in the form of stigma from nonhomosexuals or laws that discriminate against homosexuals, this threat is more internal and comes in the form of a disease: Acquired immune deficiency syndrome (AIDS). Obviously, having acquired AIDS is not deviant, and our discussion here is meant to identify an important force of change in gay communities rather than to document the full nature and extent of this illness.

The Disease and Its Transmission

People with a well-developed form of AIDS suffer from unusual, life-threatening infections and rare forms of cancer (American College Health Association, 1987). The virus that causes AIDS also produces milder but often debilitating illnesses called AIDS-Related Complex (ARC) that involves enlargement of lymph nodes, chronic fatigue, fever, weight loss, night sweats, and abnormal blood counts. The disease progresses at different rates in different people. Symptoms may take years to show up after initial exposure to the virus, and some persons may not develop symptoms for several years after such an exposure.

Who Gets AIDS?
Cases of AIDS among Adults, United States, 1988

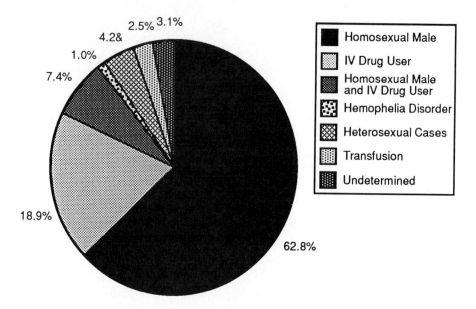

Source: Bureau of the Census. 1990. *Statistical Abstract of the United States, 1990.* Washington, D.C.: Government Printing Office

Many people with ARC improve without treatment and others progress to AIDS itself. The virus that causes AIDS and AIDS-related conditions is called Human Immunodeficiency Virus (HIV). It is a virus that must live and reproduce itself inside human cells. It is present in certain body fluids (notably blood, semen, and vaginal secretions). The only test available for AIDS at this writing is a blood test that looks for the presence of an antibody to HIV. As such, it is not a test for AIDS directly.

About two thirds of the people with AIDS are men who have had sex with other men (American College Health Association, 1987). An additional 17 or so percent are intravenous drug users who have shared needles with other people, and 8 percent are people who fit both categories of homosexual relations and intravenous drug use. A small percent of those with AIDS acquired it through heterosexual sexual contact and most of these are males. In the United States, fewer than one percent of all adults are believed to be infected with the AIDS virus.

There is a relationship between AIDS, race, and class, largely because of the connection between race and certain kinds of drug use. As one observer put it:

In New York City, more than half of the adults with AIDS are African-Americans and Hispanics, largely as a consequence of the racial composition of intravenous drug users. Nine out of ten children who died in 1987 of AIDS in New York were minority children. The Centers for Disease Control has reported that a black child is fifteen times more likely to be born with AIDS than a white child. (Price, 1989: 65)

AIDS is a problem throughout the world. Several countries in Africa have experienced severe increases in AIDS cases. In some cities—including Lusaka, Zambia, and Kampala, Uganda—more than 20 percent of adults are now infected (Eckholm and Tierney, 1990). Unlike the United States, AIDS in Africa is spreading mainly through heterosexual intercourse, facilitated by long-neglected epidemics of venereal diseases that promote viral transmission.

There is no cure for AIDS, but it can be prevented in many cases. Actually, AIDS is not an easy disease to get. HIV is a very fragile virus that requires just the right conditions. Therefore, HIV cannot be transmitted by casual contact. Rather, AIDS is transmitted by intimate sexual contact and by exposure to contaminated blood. Some sexual activities are more dangerous than others and increase the risk of acquiring AIDS. Anal intercourse is especially risky, as is vaginal intercourse.

The first cases in the United States were reported in 1981 (Surgeon General, 1986). The number of persons estimated to have been infected with the AIDS virus in the U.S. now numbers about 1.5 million, all of whom are capable of spreading the disease sexually or by sharing needles. Of this number, perhaps 20 or 30 percent will come down in 5 years with the full-blown AIDS disease. Out of 161,073 AIDS cases reported in this country since June, 1981, 100,777 people have died as of the end of December, 1990 (*Des Moines Register*, January, 25, 1991, p. 7A). Estimates from the Center for Disease Control estimate that as many as 215,000 persons will die from the disease in the next three years.

Patients with AIDS may require little or much medical intervention, depending on the stage of the disease. But all AIDS patients, regardless of stage, have emotional and psychological needs that may extend to their friends and families. Health care professionals are becoming more sensitive to these psychological needs but we have much to learn about the psychological and social impacts of AIDS (Baum and Temoshok, 1990).

IMPACT OF AIDS

It was not until October 2, 1985 that the word "AIDS" became familiar to almost every household in the Western world. On that day, Rock Hudson died. This is not to say that AIDS was unknown before this time—indeed, by the time that Hudson had died, 12,000 people had died or were dying from the disease. Rather, until Hudson's death, AIDS seemed little more than a localized disease that was largely confined to a few marginal groups in society. With the death of a famous movie star, the disease loomed large to all persons. Shortly thereafter, other celebrities would be diagnosed as having AIDS, although many would be too embarrassed to admit it (Shilts, 1987: 585–586). Broadway choreographer Michael Bennett was said at first to be suffering from heart problems. A spokesman for designer Perry Ellis said he was dying from sleeping sickness. Lawyer Roy Cohn insisted he had liver cancer, while conservative fund-raiser Terry Dolan said he was dying of diabetes. When Liberace was on his deathbed, a spokesperson said he was suffering the ill effects of a watermelon poisoning. Only later, when the disease became more prevalent and when others were willing to discuss the disease, would victims admit to having AIDS.

The Effects of AIDS

Since about two thirds of those with AIDS are homosexuals, one might predict that the impact on gay life has been enormous. Observations of

AIDS, Women and Minorities

AIDS cases do not strike randomly but they appear more frequently in minority populations than one would expect based on the percentage of minority group members in the total U.S. population.
In 1989, the ethnic breakdown was as follows:

Ethnic Group	Percent of All AIDS Cases
White	56.4%
African American	30 %
Hispanic	12.4%
Other	1.2%

Female cases of AIDS increased over 50 percent through the decade of the 1980s.

Percent of AIDS Cases by Year and Sex

	1981	1989
Male	93.2%	89.4%
Female	6.8%	10.6%

Source: U.S. Centers for Disease Control, as reported in Bureau of the Census 1990. *Statistical Abstract of the United States, 1990.* Washington, DC: Government Printing Office, page 117.

homosexual activities in many cities, reports from homosexuals, and other sources indicate that this is the case.

The Impact on Sexual Behavior.

One study has reported that the fear of AIDS has caused some change in the sexual behavior of gays, specifically in terms of the reduction of casual sexual contact (Quadland and Shattis, 1987). As such, the biggest changes had to do with a reduction of sexual relations among homosexuals who did not know one another, primarily as these persons locate one another in gay bars or other places.

There was no evidence in this study, however, of a reduction in monogamous sex or sexual relationships in general. Persons involved in these couple relationships reported that their sexual behavior had not changed as a result of AIDS. These results suggest that the extensive programs of sex education that have been undertaken in some communities have had some impact on those in high risk categories for AIDS. Similar results have been reported in studies of homosexual behavior in other communities (Coates, Stall, and Hoff, 1990).

At the same time, there is some research to suggest that gay men tend to underestimate the riskiness of their sexual practices. One study reported that 83% of the gay males who engaged in at least one high-risk behavior in a typical month evaluated their behavior as relatively safe (Bauman and Siegel, 1990). This high percentage was attributed to a misperception that "reducing" the number of partners and "knowing" one's sexual partners greatly reduced the risk of AIDS. While this general perception is correct, no amount of "reducing" or "knowing" can reduce the absolute risk of unprotected anal intercourse. In another study of 92 gay couples, Berger (1990) found that virtually all of the homosexuals described their relationship as being "monogamous," but only about half of the couples practiced safe sex.

A Political Impact.

The other major impact of the AIDS epidemic was a political one. From the early days of 1980 when the first gay men began to fall ill, five years passed before the major institutions dealing with a serious health crisis—medicine, public health, federal and private scientific research establishments, mass media, and the gay community's leadership—recognized the problem (Shilts, 1987). Funding for AIDS research and for medical facilities for the victims of AIDS did not arrive until late 1985, and it was not until the next year that the Surgeon General of the United States sought to take action. In the early days, AIDS was seen as a local problem that was largely confined to deviant groups without an effect on the larger society. When AIDS became a problem for heterosexuals, official action took place. In 1986, Dr. C. Everett Koop, the U.S. Surgeon General at the time, after interviewing many persons, decided to address the problem directly. Koop's recommendations were to avoid mandatory testing (which he thought would scare away those most at risk) and to call for a program of federally funded research and preventive tactics that included the use of condoms and monogamous relationships. In any case, Koop's program placed the disease in the forefront of a major public health movement.

Public opinion polls in 1987 indicated the extent to which the public has become aware of the AIDS problem. A Gallup poll in November 1987 found that 68 percent of Americans believed AIDS to be the country's most serious health problem (*New York Times*, November 29, 1987, p. 6). This can be compared with the next largest segment, 14 percent, who believed cancer was the most serious health problem and 7 percent who thought it to be heart problems. Just as homosexuals have taken precautions with respect to AIDS, so too are most people. More than half of American adults, according to the poll, (about 55%) were taking specific precautions, such as using condoms, choosing sexual partners more carefully, or avoiding blood transfusions if possible. Evidence of such concern, however, may not be found among sexually active young persons.

It is not possible to identify all of the consequences of the AIDS epidemic. One such impact, however, is clear: the AIDS epidemic has produced a backlash of public opinion about homosexuality. Public opinion polls indicate that there are highly stigmatizing attitudes both toward AIDS patients and gay men, and that the public views AIDS patients and gay persons with similar attitudinal prejudice (St. Lawrence, Husfeldt, Kelly, Hood, and Smith, 1990). Some members of the public "blame" homosexuals for the disease. This may be the case also for health care workers, including doctors and nurses. In one study, images of homosexuality in 59 articles in medical periodicals were categorized (Schwanberg, 1990). The largest proportion (61%) of these images were negative. This represents a change from a previously neutral position.

It is also clear that AIDS represents a personal tragedy for thousands of persons, homosexual and heterosexual, both direct victims and the friends and family of victims. Kelly (1990) describes a "stigma fallout" where intimates and close friends and associates—themselves not infected with the disease—suffer from the stigmatization of the AIDS victim. These persons, some of whom may be gay, may abandon the AIDS sufferer when the signs of the illness are unmistakable. It is difficult for many AIDS sufferers to maintain a supportive social network during this time. Regardless of the other policy consequences, it is clear that AIDS has taken a prominent place in the national policy agenda.

SUMMARY

Homosexuality refers both to sexual behavior and sexual orientation. Homosexual behavior involves sex relations between members of the same sex. Homosexual orientation is an attitude or feeling of preferring one's own sex to the other for purposes of sexual gratification. There is much more homosexual behavior than there are persons with homosexual orientations. Normative codes and laws prohibiting homosexuality are ancient, but public opinion in recent years has reflected a more accepting attitude toward homosexuality. Homosexual organizations have been successful in many states in pressing for antidiscrimination legislation, although there has been a conservative backlash in some areas. While some believe that homosexuality is not deviant, the degree of public stigmatization is significant and probably most people in the United States consider homosexuality to be deviant.

The development of a homosexual orientation takes place initially within a biological context, but the full meaning of such an orientation must be sought in the sexual socialization process and the acquisition of and identification with sex roles. Sexual socialization is a complex process that begins with the learning of very complicated norms. Sexual norms identify appropriate sexual objects, times, places, and situations. There are many possible combinations of these contingencies and room for error in socialization, given that this tends to be an embarrassing topic for many socializers. Sexual preference appears to be formed by adolescence, although few homosexuals report homosexual behavior until later. The acquisition of a homosexual identity occurs even later after a process that involves increasing homosexual behavior and possibly participation in a homosexual subculture, or gay community. The "coming out" process entails a series of stages that represent increasing commitment to homosexuality. Sociologically, a homosexual is one who has a homosexual identity.

Lesbianism is acquired in the same general manner as homosexuality among males—through the acquisition of a lesbian sexual orientation. Lesbians are more likely than male homosexuals to develop their sexual orientation in the context of a friendship relationship with another woman. Lesbian sex is more likely, in other words, to take place in the context of an ongoing relationship with another woman. Lesbian relationships of some duration are common. This is not to say that male homosexuals do not develop such relationships; they do. However, the research suggests that lesbians participate in casual sex with strangers less often than males, and they participate less frequently in a lesbian subculture, unlike male homosexuals. Lesbianism, by its nature, is more private. Lesbians, therefore, have less to fear from social stigma or the law because their behavior and orientation is less visible to others. And, for these reasons, lesbians have less need for the supporting atmosphere of lesbian subcultures.

The AIDS epidemic will continue to produce changes in gay communities and homosexuals. There is some evidence that the threat of AIDS has changed the behavior of many homosexuals in reducing the frequency of casual sex, but there is little evidence of changes in the behavior of homosexuals who are participating in monogamous relationships. It has only been since 1986 that the federal government, including public health officials and researchers, have addressed the disease. It was only when the disease began to affect celebrities and heterosexuals other than intravenous drug users that AIDS was seen as a national problem.

SELECTED REFERENCES

Bell, Alan P. and Martin S. Weinberg. 1978. *Homosexualities: A Study of Diversity among Men and Women*. New York: Simon and Schuster.

> A study of homosexuals that reports tremendous diversity among homosexuals, their backgrounds, extent of commitment to a homosexual subculture, and involvement in a homosexual life-style.

Greenberg, David F. 1988. *The Construction of Homosexuality*. Chicago: University of Chicago Press.

> The central theme concerns how homosexuality came to be defined as deviant. No serious student of homosexuality can ignore this work. Greenberg's scope is wide, and the level of detail impressive. This is a major reference on the subject.

Kelly, Robert J. 1990. "AIDS and the Societal Reaction." Pp. 47–61 in *Perspectives on Deviance: Dominance, Degradation, and Denigration*. Edited by Robert J. Kelly and Donal E. J. MacNamara. Cincinnati: Anderson.

> A sociological account of how the stigma associated with having AIDS is associated with the stigma associated with homosexuality.

Langevin, Ron, ed. 1985. *Erotic Preference, Gender Identity, and Aggression in Men: New Research Studies*. Edited by Ron Langevin. Hillsdale, New Jersey: Lawrence Erlbaum Associates.

> A collection of papers concerning the processes of developing erotic preference and gender identity.

McWhirter, David P. and Andrew M. Mattison. 1984. *The Male Couple: How Relationships Develop*. Englewood Cliffs, NJ: Prentice-Hall.

> A report of many male homosexual couples, many of whom had been together for long periods of time. The report counteracts the stereotype that much male homosexuality is for the purposes of short-term sexual liaisons only.

Shilts, Randy. 1987. *And the Band Played On: Politics, People and the AIDS Epidemic*. New York: St. Martin's Press.

> A journalistic account of the AIDS problem in the United States. Well-written and compelling, Shilts documents the lack of action on the part of the federal government and general public in the early stages of the AIDS epidemic. This book also represents a contribution to the social movement's literature in sociology.

Troiden, Richard R. 1989. "The Formation of Homosexual Identities." *Journal of Homosexuality,* 17: 43–73.

A description of "coming out" from a sociological perspective. The stages and the sequence the author identifies may not be invariant for every individual, but they capture well the process involved.

Mental
Disorders

Individuals who have or are suspected of having mental disorders are likely, depending on the nature of the behavior exhibited, to be strongly stigmatized in our society. Certainly, others are likely to react differently to someone with a mental disorder than to someone with a physical illness. Most people are generally sympathetic toward someone with a physical illness because it is often better understood or more visible, or because their own past experiences make it easier for them to identify with it. Mental disorders, on the other hand, involve many intangible feelings and ideas that others often can not comprehend or that may even make them fearful. Attitudes toward the mentally ill may range from avoidance to ridicule and revulsion. The different reactions to physical illness as opposed to mental illness have to do with the behavioral aspects of mental disorders.

Persons tend to think that persons with mental disorders are somehow more responsible for their conditions than the physically ill. This may explain in part the stigma associated with psychiatric disturbances. Persons with mental disorders are also stigmatized because their behavior is often socially disruptive and unpredictable, even threatening or frightening to other persons. Usually, persons who are physically ill do not threaten the community in the same ways as do some persons with mental disorders.

The behavior exhibited by a person with a mental disorder may be inappropriate to the situation and even bizarre. A mentally disturbed person may not be able to meet normal expectations in particular situations or in everyday life. These social expectations, or norms, form the base of the assessment of deviance and the subsequent administration of social sanctions. Persons with mental disorders may be sanctioned, either by being labeled mentally ill, insane, "crazy" or, in more extreme cases, by commitment to a mental hospital. While most observers consider mental disorders to be a serious form of deviance because it affects other persons as well as the person with the disorder, efforts to understand further the nature of mental disorders, their causes and cures, have been hampered by the extremely difficult problem of socially defining mental disorder. As sociologists and social anthropologists increasingly give their attention to this topic, social and cultural factors are being implicated in the explanation of the origins of mental disorders and the distribution of disorders from group to group and place to place. In order to understand more fully the broad sociological approaches to this form of deviance, it is necessary to have some understanding of the various types of mental disorder.

PSYCHIATRIC TYPES OF MENTAL DISORDERS

There are many problems with the psychiatric conception of mental disorder, as well as specific psychiatric definitions of particular mental disorders that will be discussed later in detail. There is, for example, a heated dispute among psychiatrists about whether mental disorders are medical problems. Some psychiatrists believe that what we call mental disorders are really "problems in living" or instances of aberrant behavior (Szasz, 1974) This view does not deny that people may act oddly, but that such behavior, often called "insanity," is not a medical problem (Szasz, 1987). Other psychiatrists claim that mental disorders are as real as any other medical problem and therefore a medical model or psychiatric approach to them is appropriate (Roth and Kroll, 1986).

Because such disputes are far from solved, one might well question the relevance of a brief description of various psychiatric types of mental disorder. First, they are terms widely used by psychiatrists, who are after all the persons chiefly responsible for the treatment of mental disorders. Second, they are terms used by others, including, frequently, sociologists.

It is therefore necessary to become familiar with the terms and their use. It should be understood, however, that there is considerable ambiguity about many of these terms. Psychiatrists do not always agree among themselves about their meaning and the meaning of many terms change over time. Many "normal" persons also may exhibit the behavior described in each type, although in all probability to a lesser extent and in a manner that does not provoke much societal reaction. Probably all persons have to some degree exhibited such behavior as hallucinations, phobias, persecution complexes, and emotional extremes of elation and depression. And, almost everyone at one time or another has had irrational fears, daydreams, flights of ideas, and lapses of memory. By themselves, such experiences do not reflect a mental disorder, but when coupled with other experiences, or when they interfere with basic social processes, they may be reflected in psychiatric diagnoses of mental disorders. Unfortunately, the reliability of psychiatric diagnoses has been poor, so that different psychiatrists faced with the same or similar symptoms might reach different conclusions about diagnoses.

The Diagnostic and Statistical Manual

In an effort to deal with the problem of shifting definitions and criteria for accurate diagnoses, the American Psychiatric Association (1986) has developed a manual, called the *Diagnostic and Statistical Manual* or DSM-III-R, that attempts to standardize psychiatric diagnoses of specific mental disorders. Many changes have been made in psychiatric language over the years of the DSM. The term "neurosis" which refers to a minor psychiatric ailment that does not interfere with daily functioning, has been dropped, as has "homosexuality" as a diagnostic category. Of course, the words neurosis and homosexuality are still used but they do not constitute precise diagnostic categories. The manual has over the years replaced the expression "mental illness" with the phrase "mental disorder" Rather than providing specific criteria by which each diagnosis can be made, DSM-III-R identifies broad correlates of the various disorders. These correlates are features of the disorder that are frequently associated with it. In this manner, practitioners are led not to conclude that a person is, say, a schizophrenic, but that a person has some of the characteristics of schizophrenia.

The DSM-III-R, like its predecessors, is hardly a dispassionate, scientific document. In late 1985, for example, the group of psychiatrists charged with the revision of DSM-III (which become DSM-III-R) proposed a new diagnosis called "paraphilic rapism" or rape. The idea that rape is a disease so upset feminist leaders, because it might provide an insanity defense for rapists, that they promptly protested, causing the head of the psychiatric committee to say: "We probably will withdraw the diagnosis of rapism . . ." (Goleman, 1985). It is difficult to conceive of a sound medical diagnosis being withdrawn when threatened with a law suit. This is not the first time that political considerations affected the content of DSM. In 1973, homosexuality was withdrawn as a disorder after lobbying by various groups.

Many psychiatrists are not unhappy with other features of the DSM. DSM-III-R, for example, recognizes two kinds of depression: major depression and dysthymic depression, but each covers a variety of syndromes and symptoms. When examined closely, the nature of the symptoms and treatment for depression is fraught with difficulty. As one psychiatrist put it:

> Today's depression classification is as confusing as it used to be 30 years ago. All things considered, the present situation is worse. Then, psychiatrists were at least aware that diagnostic chaos reigned and many of them had no high opinion of diagnosis anyway. Now, the chaos is codified [in the DSM] and thus much more hidden. (Van Praag, 1990: 149; see also Zimmerman, 1988)

According to the traditional classification of psychiatry, there are two basic types of mental disorders—those with an organic basis and the functional, or nonorganic, disorders. The organic types are usually caused by some organism, by a brain injury, or by some other physiological factor, including possibly an inherited condition. These types clearly have a biological or physiological basis and are properly medical conditions. The *functional disorders*, on the other hand, have no organic or physical basis and are less specifically broken down into various categories.

Organic Mental Disorders

Organic disorders are all brought about because of organic, or physiological, problems. They include the senile (or old age) psychoses, paresis, and the alcoholic psychoses. Senility accounts for about one fourth of all state hospital admissions. Some of the cases are marked by arteriosclerosis (or, hardening of the arteries) and thus result from poor circulation that affects the brain. These disorders are characterized by a loss of memory, particularly for recent events, inability to concentrate, and certain delusional thoughts. Alzheimer's Disease is a kind of senility. Some symptoms may be brought on by social accompaniments to old age—such as loss of status and social isolation—rather than from organic sources.

Paresis is a condition caused by syphilis. It develops at least 10 years after the initial syphilitic infection and there is often progressive brain degeneration in those who go untreated. The mental symptoms are often a complete alteration in personality characteristics—for example, a neat, well-dressed person becomes slovenly. Eventually, memory about time and place becomes defective and, in some cases, there is depression. The prolonged existence of chronic alcoholism, associated with vitamin and nutritional deficiencies, may in some cases produce such deterioration in physical and psychological behavior that alcoholic psychoses result. The *psychoses result-*

ing from alcoholism are not definitely organic, but they are so classified. Relatively few alcoholics develop psychoses.

Functional Mental Disorders

The disorders of primary interest to social scientists are the functional disorders, which, for purposes of our discussion, can be subdivided into neuroses, psychoses, manic depressive psychoses, and paranoia. These functional disorders will be referred to as mental disorder in the sections that follow.

According to many psychiatrists, the functional mental disorders "function" to adjust the individual to his or her particular difficulties; hence, the term "functional" For most of the functional mental disorders, there is no conclusive evidence that they arise from organic causes—either from heredity, physiological disorders, or from some other organic deficiency. Although there is a great deal of confusion in the psychiatric profession as to whether mental disorder is actually a disease, most psychiatrists, because of their medical training, think about mental disorders in disease terms and in terms of medical diagnoses and treatment (Goffman, 1959: 320–386). But unlike many other areas in medicine, research on functional mental disorders has been severely hampered by the lack of a standard test by which to make a diagnosis. In the absence of an objective test, the diagnosis must depend upon the clinician's professional judgment; this, in turn, is affected by his or her own conception of mental disorder and the explanation for its origins that he or she holds, as well as the values and attitudes of the patient.

The lack of reliable diagnoses with respect to functional disorders can best be seen in various minor psychiatric conditions. *Compulsive behavior*, for example, involves behavior over which people think they have little or no control, such as stepping on, or avoiding, cracks in the sidewalk, excessive hand-washing or bathing, counting telephone poles or other objects, dressing in a set manner, and insisting on certain meticulous order, such as all drawers carefully closed or shoes and other objects lined up in order. Neuroses may also be reflected in *obsession*, or persistent ideas, often representing emotional fears of objects, acts, or situations. *Phobias* are often of a general nature, such as fear of confinement (claustrophobia) or its opposite, fear of open places (agoraphobia).

Manic-Depressive Behavior. Manic-depressive behavior ranges from extreme elation, in the manic stage, to extreme depression, although manic-depressives do not necessarily pass through cyclical stages of mania and depression. The shift from one mood to the next is very quick. In the manic stage, the person talks quickly, moving from one topic to another in an understandable manner. In the depressed stage, there is much brooding but little serious mental deterioration. Activity is minimized and talk turns to feelings of dejection, sadness, and self-depreciation. Contact with reality is maintained, as are memory and place-time orientation.

Paranoid Behavior.   Paranoia is not widely used as a diagnostic category. Most persons suffering from paranoid disorders are now considered to exhibit a form of schizophrenic behavior. Paranoid persons are extremely suspicious and have ideas of persecution, with an intellectual defense that often appears to have some basis in reality. Their delusions are usually limited to a few areas and may even center around a single person. Most paranoids live normal lives and their personalities do not deteriorate over time. They do not hallucinate.

Schizophrenic Behavior

The diagnosis of persons with schizophrenic behavior is made more frequently than any other serious functional disorder and, for this reason, it warrants more than brief mention.

Schizophrenia accounts for about one fourth of all first admissions to mental hospitals, and one half to three fifths of persons occupying hospital beds at a given time. This disorder, sometimes called dementia praecox, develops primarily between the ages of 15 and 30, with few persons developing it after 50. Partly because of its nature and extent, social science research and writing on mental disorders are probably more concerned with schizophrenia than with any other disorder.

The condition of schizophrenia is said to involve a detachment of the emotional self from the intellectual self, and it is for this reason the term "schizophrenia"—or "split personality" as it is sometimes called—is used. There are different classifications of schizophrenia, but the classifications are unreliable. The most characteristic symptom of a schizophrenic is withdrawal from contact with the world and an inability to play expected roles. An imaginary world may be built up, including false perceptions and hallucinations of all kinds, such as ideas, voices, and forces that enter daily living and that can not be controlled. Schizophrenics may become exceedingly careless in their personal appearance, manners and speech; sometimes their behavior involves pronounced silliness with a good deal of situationally inappropriate behavior.

It is widely held in the psychiatric community that schizophrenia is a disorder that is caused basically by biological mechanisms but also affected by social factors. Some claim that schizophrenia has its origin in a certain chemical in the brain (dopamine) that is released under conditions of stress or when activated by another drug, such as amphetamines (see Szasz, 1987: 73–74).

Studies of the biological basis of schizophrenia have failed so far to disclose precise chemical or organic sources of this disorder. Research, however, continues and subsequent work may eventually identify the grounds on which physical and social factors combine. Other researchers have claimed that because of certain hereditary genes, schizophrenics are persons who are vulnerable to certain social situations (Heston, 1970; Gottesman, 1972). In this connection, geneticists have attempted to demonstrate that heredity plays a leading role in schizophrenic disorders. For example, Kallman (1938 and 1946) has been the leading proponent of the theory that schizophrenia is inherited, and the conclusions from his studies, even though done almost 50 years ago, are often still cited as proof that genetic factors predispose certain persons to schizophrenic behavior. More recent studies have also explored the relationship between hereditary factors and schizophrenia. A British researcher compared the adjustment of 47 adults born to mothers diagnosed as schizophrenics with a matched control group of adults born to nonmentally ill mothers (Heston, 1966). Both groups had been reared in foster homes. The investigator found a higher incidence of schizophrenia and other pathologies among the group born to the schizophrenic mothers than among the matched control group. This type of genetic study, however, does not support a hereditary etiology for schizophrenia. As Mechanic (1989: 63) notes, "[b]ecause the subjects of the study were removed from their schizophrenic mothers shortly after birth, we can not conclude that the higher rate of pathology was a result of interaction with a schizophrenic mother." However, this general theory requires studies on twins and a close examination of the results from those studies. Results of that research so far have been ambiguous in establishing a causal connection between inheritance factors and schizophrenia, for a number of reasons.

Criticisms of Inheritance of Schizophrenia.   Efforts to establish a genetic link to schizophrenia have been severely hampered for several reasons. Several factors can be mentioned here:

1. The problem of diagnosis of schizophrenia itself is the greatest obstacle in the genetic study of this disorder. Diagnoses are so unreliable, in fact, that one psychiatrist has even suggested that the only meaning of the term resides in the vague conceptions within psychiatrists themselves (Szasz, 1976). Two other researchers, after an extensive review of the literature, conclude that schizophrenia represents more of a moral pronouncement than reliable medical diagnosis (Sarbin and Mancuso, 1980). A survey of psychiatrists in the United States discovered little agreement on the signs or symptoms of schizophrenia (Lipkowitz and Idupuganti, 1983). Having a standard reference source with lists of symptoms, such as that found in the DSM-III-R has thus far failed to resolve such disagreements.

2. Cultural factors appear to play a role in the symptoms of schizophrenia. Differences in the nature of schizophrenic symptoms have been found, for example, between persons from Irish and Italian subcultures, the former favoring fantasy and withdrawal to the extent of paranoid reactions,

while the latter suffered from poor emotional and impulse control (Opler, 1959; Waxler, 1974). Among some cultural groups, as, for example, the Hutterites, a tightly knit religious group living in the Western United States and Canada, schizophrenia is a rare disease (Eaton and Weil, 1955). Furthermore, the kind and amounts of disorders found in such diverse groups as Hindus, Chinese, and Spanish Americans varies depending on their cultural experiences, not their biology (Murphey, 1959; Jaco, 1959).

3. Diseases or physical illnesses of various types may run in families without necessarily having a genetic basis. Studies that have attempted to use improved diagnostic criteria and care in controlling for extraneous variables have been unable to find a family link in the transmission of schizophrenia (Abrams and Taylor, 1983).

Of course, even if genetic factors were implicated in the etiology of schizophrenia, this would not rule out the importance of environmental, psychological, and sociological influences. In any case, a sociological perspective would be necessary to explain the patterns of the disorder that are not amenable to a biological or physical explanation.

A Tentative Conclusion. At the present time, there are different theories about schizophrenia and no single theory has proven most useful. Some psychiatrists claim that all schizophrenics suffer from a single pathophysiological process but that there are individual variations, while others believe that schizophrenia is a syndrome consisting of a number of different diseases, each with its own pathophysiological process (see Kirkpatrick and Buchanan, 1990). Until such time as there is agreement on a theory of schizophrenia, we will be unable to interpret accurately a number of dimensions of this disorder. For example, it has been reported that there is a definite association between being related to a schizophrenic and manifesting thought disorder (Romney, 1990). Those who are biologically oriented might interpret this finding as providing support for the notion that schizophrenia is found in both persons because of a shared biology. However, another interpretation might find the meaning of such a finding in the learning opportunities afforded such persons; we learn from relatives certain kinds of acting and thinking, some of which might be considered "disordered." Only an agreed-upon theory of schizophrenia can distinguish such interpretations.

PROBLEMS OF DEFINITION

It is difficult to assess the deviant nature of mental disorder because what is meant both by mental health and, consequently, mental disorder, has not been adequately defined. While usually defined in terms of some deviation from "normality," it is not a simple matter to define normality. Certainly some individuals behave in strange or inappropriate ways, or they verbalize bizarre thoughts or rationales for their actions; but actions that may be normal in some situations may not be normal in others. For example, an individual may wash his or her hands 50 times or more during the day, which might be appropriate and necessary for, say, a dentist, but would be considered extremely odd and "compulsive" for an officer worker or business executive.

Consequently, there is no single way to define mental illness. A number of broad approaches have been utilized including (1) the statistical, (2) the clinical, and (3) in terms of residual norms, a sociological definition.

1. Statistical Conception

Although mental health is not the same as the statistically normal in terms of averages, if it were to be viewed in this way, it might be said that the "mental health" of those persons in the "middle" would represent what might be termed normality. Moreover, just as statistical definitions of deviance suffer from their inability to take into account norms, statistical definitions of mental illness neglect an important component of disordered behavior—it is behavior that is thought strange, inappropriate, or wrong, given the circumstances. Just like statistical conceptions of deviance, statistical conceptions of mental illness can only tell us what most persons do, not what they ought to do.

2. Clinical Conception

A clinical conception of mental health would be whatever a clinician says it is; that is, behavior would represent "mental illness" if a mental health professional, such as a psychiatrist, said it did. Even with the aid of such diagnostic devices as DSM-III-R, there are problems in identifying mental disorders. Rosenhan (1973), for example, reported a study in which normal persons were admitted to psychiatric wards and were diagnosed as schizophrenics after they falsely told the emergency room doctor that they were hearing voices or other sounds. Once on the ward, they behaved normally but the psychiatric staff continued to perceive their behavior as evidence of the disorder. Clinical judgments can be influenced by previous conceptions of a disorder, even when subsequent behavior contradicts those conceptions.

In clinical or physical medicine, the problem of defining the "normality" of health is often difficult and complicated, but it can not compare with the complexities in clinically defining "mental health" where there are problems connected with normative definitions and value judgments.

Normative Definitions.   Redlich (1957) has presented three criteria that must be met before behavior can be clinically labeled as normal or abnormal. (1) The motivation of the behavior must be taken into account, such as "normal" washing of the hands and a neurotic washing compulsion. (2) The situation in which the behavior occurs must also be considered. Wearing swimming trunks on an Alaskan street in winter is one thing; on a summer bathing beach, another. (3) By whom is the judgment made that the behavior is clinically abnormal—the experts, such as the psychiatrist, or the general public? Since there are no universal clinical criteria, many propositions regarding the normality of behavior are perceptibly lacking in both reliability and validity and are subject to challenge by the public.

The very behavior that may be contrary to the ideal values of "mental health" may often be considered normal in another society. In some cultures, it is believed that personal transgressions are related to disease. In other cultures, people may fear "irrational" things, such as humans being transformed into cannibals. To an outsider, these fears appear to be "neurotic" mental disorder in the sense that they are sheer fantasy. One must, however, distinguish between individual fears and culturally induced fears. This problem is prevalent in the clinical diagnosis of mental disorder in more complex societies with numerous and varied subcultures and social classes, where people play roles that are normal in their own group but are considered clinically abnormal to a psychiatrist. In commenting on the artificial line between mental normality and mental disorder, a French sociologist has remarked:

> The dividing line between the two realms varies ... from group to group within the same society. Thus it is never entirely possible to escape from relativity. The function of the psychiatrist is to search for the "causes," to report on the "whys" of the illness, but society decides who the patients will be. There is a subtle play of influences between doctor and the public. The doctor, through the mass media or other agencies, tends to enlarge the field of mental illness, to make the public more aware of disturbances that are minor and have been until then attributed to "oddness" or "eccentricity." On the other hand, he accepts the lay definition of mental illness, and his work is limited to refining or making more explicit this definition by introducing categories of "insanity".... But these categories never extend beyond the boundaries of insanity as defined by public opinion. (Bastide, 1972: 60)

Value Judgments.   The clinical definition of "mental health" leads into the area of value judgments. Among the definitions of mental health used by leading psychiatric writers are the striving for happiness, effectiveness, and sensitive social relations, freedom from symptoms and unhampered by conflict, the capacity to love another person, to mention a few. Menninger's still widely quoted definition states:

> Let us define mental health as the adjustment of human beings to the world and to each other with a maximum of effectiveness and happiness. Not just efficiency, or just contentment, or the grace of obeying the rules of the game cheerfully. It is all of these together. It is the ability to maintain an

even temper, an alert intelligence, socially considerate behavior, and a happy disposition. This, I think, is a healthy mind. (Menninger, 1946: 1)

With such criteria, it is difficult to see how anyone can be regarded as normal. A state of emotional health is thus regarded as "par," if one can use the golf term, for the upper levels of health attainment. They are ideals, and they are often contradictory. These contradictions are evident in the widely varying estimates of the prevalence of "mental disorder" in the general population using clinical definitions. A review of 25 studies has reported percentages ranging from 1 to over 60 (Dohrenwend and Dohrenwend, 1969). Moreover, the median rate of disorder for studies after 1950 was seven times higher than for studies before 1950, a difference not likely to have been due to a real difference in the trends but in the nature of the criteria used to make the estimate. In the Midtown Manhattan survey, for example, one of the best known national assessments of the need for psychiatric services, questions were asked in interviews about psychiatric disorders, or feelings of "nervousness" and "restlessness," and difficulties in interpersonal relationships (Srole, et al., 1962). This information was then abstracted and given to a team of psychiatrists who rated the person and the amount of "impairment" in psychiatric terms. The Midtown Manhattan survey estimated that about four fifths of the population was experiencing mental disorder. Clearly, only the most broad conception of mental disorder would permit such a large number of persons to be counted as being mentally ill. Other, more reasonable, assessments have placed the figure at between 16 and 25 percent of adults under 60 years of age (Dohrenwend, et al., 1980).

3. Residual Norms and Societal Reaction

A sociological definition of mental disorder can be stated in terms of norms and sanctions. Groups have norms that designate as deviance behavioral acts that are called crime, sexual deviations, drunkenness, bad manners, and other more specific behavior. Each of these norms, in other words, applies to specific behavior. Mental disorder likewise can be viewed in terms of norm. It constitutes residual norm breaking or residual deviance in that it designates deviant normative behavior not covered by other behavioral terms (Scheff, 1984 and 1975). Various forms of "mental disorder" include such behavior as withdrawal from association with others, hallucinations, muttering, posturing, depression, excited behavior, compulsions, obsessions, and auditory states. As with other normative violations, mental disorders involve a violation of normative expectations. Unlike other violations, however, it is difficult to specify ahead of time the nature of that violation. It is "residual," or left over, for this reason.

Such residual deviance must be viewed as normative behavior not only in and of itself but also in terms of the social context in which it occurs. Talking to spirits within the religious context of spiritualism, for example, would not be considered residual deviance; Nor would seeing visions of events not present when under the influence of hallucinogenic drugs. The normative violations are called mental disorders because of the context in which they occur. More specifically, they are violations because they are "out of context."

An operational definition of "mental abnormality" would ask "normal for what" and "normal for whom." These considerations seem to be helpful in any adequate definition of "mental disorder." This is why it is hard to draw a sharp, operational line between mental health and mental ill-health except in terms of norms. It is really a problem of the social limits of eccentricity. An English writer stated that there appears to be no clear-cut criterion of what actually constitutes a psychiatric case, for whether the person is thought to need treatment is always "a function of his behavior *and* the attitudes of his fellows in society" (Carstairs, 1959: 337). Persons may be impaired or not depending on the way their behavior is evaluated by others. Some psychiatrists have even gone so far as to deny that mental disorder is an "illness" but rather that it merely represents defective strategies for handling life situations that the individual finds difficult (Szasz, 1974; Torrey, 1974).

The more closely persons' behavior conforms to the expectations of others, the more favorably they are evaluated by those persons. On the other hand, when behavior is not within the expected range, it is likely to be evaluated negatively. Collective action, then, on the part of a family, neighborhood, or community to hospitalize an individual as being mentally ill will always be a product of interaction between the nature of the behavior and the tolerance of the group for such behavior.

These alternative definitions have led to different ways of thinking about and explaining mental disorders. There are disputes about every aspect of mental disorders, including their conceptualization and explanation. Some favor a biologically-based explanation to reflect recent advances in biochemical research on behavior, while others prefer to examine disorders from the perspective of mental conflicts. These medical and psychological perspectives are in contrast to a sociological approach that views mental disorders from the reference of social roles and normative expectations.

SOCIAL STRATIFICATION AND MENTAL DISORDER

Social factors are important in understanding mental disorders. This is indicated by the fact that mental disorder, either as a whole or by type of disorder, has not been found to be distributed randomly in the population. Like other forms of deviance, mental disorders are socially patterned. The evidence indicates that diagnosed mental disorders are related to differences in social class, sex, and occupation.

Social Class

The highest rates of diagnosed severe psychiatric disorders have been found to be disproportionately concentrated in the lowest social classes (see Little, 1983: 96–98). The midtown Manhattan survey, mentioned earlier, found that one third of those in the higher socioeconomic status groups were rated "well," while less than 5 percent of the lowest strata were so rated. In the highest group, only 12.5 percent were considered "impaired," while nearly one half of the lowest strata were so rated (Srole, et al., 1962: 138).

A study of nearly all persons in New Haven, Connecticut who were patients either of a psychiatrist or a psychiatric clinic, or who were in psychiatric institutes, revealed rather decided class differences (Hollingshead and Redlich, 1958). The group of 1,891 patients was compared with a 5 percent random sample of the normal population, or 11,522 persons. When both groups were divided into give classes and compared, Class I at the top and Class V at the bottom, it was found that the lower the socioeconomic class, the more prevalent the diagnosis of disorder. Class I contained 3.1 percent of the population and only 1.0 percent of the mental patients, whereas the lowest group, with 17.8 percent of the population, had almost twice as many mental patients. When sex, age, race, religion, and marital status were considered, social class was still found to be the most important factor. Diagnoses of minor psychiatric ailments, however, were found to be more prevalent at the upper socioeconomic levels.

Part of the observed difference in disorders by social class was undoubtedly due in part to differential diagnosis on the part of psychiatrists. Psychiatrists are recruited largely from the upper socioeconomic class, and it should be unsurprising that they interpret behavior from that perspective. Moreover, a number of Class I patients were not included in the study because some private psychiatrists refused to cooperate. Some studies have challenged the relation of schizophrenia to social class on the ground that there is little difference between classes if the cases are examined not by the patients' class but by that of their father's (see Dunham, 1965).

Most evidence, however, points to the conclusion that the more serious disorders, especially schizophrenia and serious forms of depression, are more prevalent in the lower classes (Dohrenwend, et al., 1980). In the midtown Manhattan study, for example, Srole and his associates (1962; and Srole, et al., 1977) demonstrated that the mental health of respondents was positively related to their socioeconomic status, and that downwardly mobile individuals were more apt than upwardly mobile individuals to be psychiatrically impaired.

Although the relationship between social class and mental disorder is fairly clear, the interpretation of this finding is not. A strong negative relationship between social class and mental disorders might mean a greater incidence of mental disorders among lower-class persons, or that disordered lower-class persons are more apt than their middle- or upper-class (but equally disordered counterparts) to be diagnosed and hospitalized. Some emphasize the importance of genetic factors combined with childhood socialization patterns and conflicts of lower-class life itself. Some writers suggest that the greater incidence of schizophrenia in the lower class is the result of genetic selection and either downward social mobility or a person's failure to move upward with his or her peers as a result of the debilitating consequences of the disorder (Mechanic, 1972). A more likely explanation is that the relationship is due to the nature and stresses of lower-class life, differential socialization, social mobility, differential psychiatric diagnoses, and the differential treatment of mental disorders.

The relationship between social class and mental disorders suggests that economic factors play a role in the distribution of this form of deviance. Brenner (1973) found that employment rates were related to mental hospital admissions. The relation between the employment rate and rate of hospitalization was particularly strong among the poorly educated. Catalano and Dooley found various measures of the economy to be good predictors of admissions to mental hospitals in Kansas City (Catalano and Dooley, 1977), as well as in a more rural area in Maryland (Catalano, et al., 1981).

Sex Differences

Not only do mental disorders vary by social class, but the nature and frequency of mental disorders also differ by sex. There appear to be no consistent sex differences in rates of functional psychoses in general, but females do have higher rates of manic-depressive disorders than do men. A thorough review by Zigler and Glick (1986: 240–250) reports that men, on the other hand, are likely to have higher rates of some types of disorders than women. Depressive symptoms are more common among female than male schizophrenics, and females are also more likely to be diagnosed as paranoid than males. Explanations for the higher rates of depression among women have been phrased in terms of the greater status pressures on women, particularly their tendency to find marriage less satisfying than males (Gove, 1972).

Other studies have indicated that marital status, per se, may not be as important as previously thought, but the fact that a person is a female does seem to be significant in the likelihood of the person seeking treatment for a condition (Warheit, et al., 1976). Females may be more likely than males to seek help for their psychological maladies. It also appears that a higher proportion of females in Western societies beyond the United States have more psychiatric disorders than do men, but one study has estimated that only between 10 and 30 percent of this excess may be due to greater willingness to seek help among females (Kessler, et al., 1981). Studies in the United States, on the other hand, have consistently indicated that females are more likely than males to contact medical personnel, even ordinary physicians, about mental health problems (Leaf and Bruce, 1987). The likely explanation for this difference is that men feel more strongly than women about being labeled with mental disorder if they seek professional help.

Age, Race, and Regional Differences

There are no consistent differences in mental disorders by age; no particular age group shows a consistently higher rate for various types of mental disorders than another age group (Dohrenwend and Dohrenwend, 1969). Although race differences have not been studied extensively, the available evidence suggests that no specific race is more likely than another to experience more mental disorders (Dohrenwend and Dohrenwend, 1969). Surveys of mental disorders in rural versus urban areas show that rates of mental disorders are higher in urban areas, but only slightly so. Rural populations are more likely to be resistant to the use of treatment facilities and are more likely to tolerate longer a person who may be labeled as mentally disturbed than in an urban setting (Eaton, 1974).

SOCIAL STRESS AND MENTAL DISORDERS

At several points, we have indicated that the distribution of mental disorders in society may be related to, among other things, the amount of stress that persons experience. Occasional emotional stress is useful and quite normal in many ways to the individual. In contrast to healthy stress, however, intense and persistent stress of a social nature, associated with anger, fear, frustration, worry, and so forth, can threaten physical and emotional health. In the field of medicine, for example, much interest has been shown in the relation between excessive social stress and such physical conditions as hypertension and digestive disorders.

Our concern here is with the relation between social stress to mental disorder. It appears that social stress is directly related to behavior that is frequently defined as mental disorder. Such stressful situations in life as marriage, divorce, the illness or death of a close relative or friend, as well as more minor but still stressful events such as marital disputes, coping with troublesome children, or even minor but annoying conflicts with other persons, have been used to predict mental disorder. These various situations and others like them have been incorporated into a scale and weighted according to the degree of stress that each is likely to generate. A review of studies in which these scales have been used has led to the conclusion that they clearly indicate that certain stressful "life events tend to occur to an extent greater than chance expectation before a variety of psychiatric disorders" (Paykel, 1974: 147). Certain types of disorders are the result of a higher proportion, while others are the result of a lower number, of stressful life events. Thus, persons who attempt suicide report the greatest number

Relative Weights of Stressful Life Events

Below is an example of some life events and their relative degree of stress. There are other scales with different weights but the idea is the same: to gauge the extent to which the accumulation of eventful things in one's life adds up to a certain level of stress. Note that even "positive" events produce stress.

Event	Life Change Score
Married	500
Widowed	771
Divorced	593
Separated	516
Pregnancy	284
Birth of child	337
Illness or injury	416
Death of loved one or other important person	469
Start of school or job	191
Graduation from school or training	191
Retirement from work	361
Change of residence	140
Taking a vacation	74

Source: Adapted from Robert E. Markush and Rachel V. Favero. 1974. "Epidemiologic Assessment of Stressful Life Events, Depressed Mood, and Psychophysiological Symptoms—A Preliminary Report." In Barbara Snell Dohrenwend and Bruce P. Dohrenwend, eds., *Stressful Life Events: Their Nature and Effects.* New York: Wiley.

of these events, depressives the next highest, and then schizophrenics (Paykel, 1974: 148).

Stress and Anxiety

In social interaction, persons frequently encounter conflict situations that may produce stress, particularly if the situation threatens the person's self-image, roles, or values. Stress factors tend to produce a certain amount of anxiety (Blazer, et al., 1987). Anxiety resembles fear in many ways; like fear, it is an emotional reaction produced by stimulation with which one is unable to deal. Unlike fear reactions that call forth avoidance and even flight from a real danger, however, the emotional reaction in anxiety does not go on to completion. Fear is overt whereas anxiety is covert. It leaves persons in an undefined emotional state with which they would like to cope but can not.

Stress is seen in neurotic compulsive behavior, such as orderliness and obsessional ideas that help to relieve anxiety. The acts, words, and thoughts involved in the compulsive relief of stress and anxiety may include preoccupation with certain obsessions, tapping, counting, or saying set words. In hypochondria, for example, constant preoccupation with health simply constitutes solutions in which this preoccupation diverts and releases anxiety. It has been said that the "fruit" of resisting the compulsion is mounting anxiety, while the "reward" of indulging the compulsion provides only a temporary respite (Cameron, 1947: 277). The societal reaction of others to certain noticeable forms of neurotic behavior may tend to increase stress and anxiety and thus compound the problem. The following illustrates such a case.

> A 33-year-old married woman discovered some beetles while cleaning out an old cupboard in her house. [This upset her.] She immediately had to wash her hands and to repeat the washing three times. Each time she cleaned and dusted the house she began to wash her hands three times, and thereafter in increasing multiples of three. She was soon washing her hands hundreds of times a day and thereafter felt compelled to bathe herself between six and nine times daily. All the time, she recognized that these compulsions were morbid but felt helpless against them. In the next stage of the disorder, she developed the belief that every object that might have come into contact with hair had become contaminated. She began to dispose of her own and her husband's personal possessions and thereafter to sell articles of furniture ridiculously cheaply. At the time of her admission [to a psychiatric hospital], her entire suite of furniture . . . had been sold and the patient came in covered by an unused bedsheet, the only uncontaminated object in the home that could be used to cover her naked body. (Roth and Kroll, 1986: 9)

Stress and Social Situations

Many functional mental disorders appear to arise out of a continuous series of events, often over a long period of time, but stress situations often act as precipitants, bringing the process to a climax. The effect of an immediate stress situation is particularly important in the manic-depressive disorders. The hypothesis has been advanced, particularly in connection with disorders that do not interfere with the individual's ability to function in society that the intense striving for material goods and the competitive emphasis in present-day industrial urban society lead many persons to irreconcilable conflicts. In a well-known and still widely respected study, Horney (1937) characterized life in modern Western societies as being highly individualistic, with great competitive striving for achievement and social status. According to her, this leads to conflicts between competitive, materialistic desires and their possible fulfillment and between competitive striving and the desire for the affection of others, all of which produce stress and neuroses.

Psychological stress affecting mental disorder appears to be linked to social class, and hence helps explain the relationship between class and certain mental disorders. Lower-class persons have been said to experience more unpleasant events and also experience the most difficulty in dealing with them (Myers, et al., 1974). In the mid-town Manhattan study of mental disorder *the number* of stressful factors, but not their nature, was found to be associated with mental disorder, and low-status groups were found to encounter the most stress (Srole, et al., 1962 and 1977). A review of results from eight epidemiological surveys underscore the importance of the relationship between stress and social class. Kessler reports that various indicators of social class—income, education, and occupational status—are related to stress but differentially depending on the population (Kessler, 1982). Income is the strongest predictor of stress among men, while education is the best predictor among women in the labor force and among homemakers. Clearly, the relationship between stress and social class is more than merely an economic one, as reflected in the importance of noneconomic variables with respect to stress among women.

Stress can be generated by role-conflict as well. In a study of a group of schizophrenic married women, it was concluded that these women had repeatedly experienced severe difficulties over the years in their marital situations (Rogler and Hollingshead, 1965). In fact, husbands or wives who were schizophrenic presented "no evidence that they were exposed to greater hardships, more economic deprivation, more physical illness, or personal dilemmas from birth until they entered their present marriage than do the mentally healthy men and women" (Rogler and Hollingshead, 1965: 404). The intensive study of schizophrenics and nonschizophrenics in this representative sample of lower-class husbands and wives, between the ages of 20 and 39, in the slum and housing projects (*caserios*) of San Juan, Puerto Rico, showed that such disorders could not be explained by childhood experiences, social isolation, or occupational history. Rather, they were due to the stress created by conflicts and problems associated with lower-class life and neighborhood situations. The schizophrenics had many more, as well as more severe, problems than the nonschizophrenics. The culture and their lower socioeconomic status in the society present some persons with tension points. Such problems in this Spanish culture of Puerto Rico include courtship, women's adjustment to sexual and other roles in marriage, the disparity between achieved and desired levels of living, conflict with neighbors, and the absence of privacy in the housing projects, as well as various problems of role fulfillment and performance. These stress problems continue to mount, imposing contradictory claims and leading to conflict, mutual withdrawal, and alienation of neighbors, until individuals reach a breaking point in which they are "trapped."

Evaluating Stress as a Precondition for Mental Disorders

There is considerable evidence that stress is linked with much mental disorder, particularly the neuroses. It is not known, however, how different people respond to stress and in what manner stress-fed situations are perceived or reacted to by the individual. Stress, in itself, often even of a severe nature, does not inevitably produce mental disorder, as is seen in studies of the effects of stressful modern living, of the stress of wartime civilian bombings, and of the stress situations of soldiers under combat, prisoners in Nazi concentration camps, and persons with several physical illnesses or injuries.

As Mechanic has observed:

> In its simplest form, stress conceptions suggest that all people have a breaking point and that mental illness and psychiatric disability are the products of the cumulation of misfortune that overwhelm their constitutional makeup, their personal resources, and their coping abilities. Stated in this way, the perspective is not very useful for it can not successfully predict who will break down but in retrospect can explain everything. (Mechanic, 1989: 73)

Clearly, there are many persons who can withstand considerable stress without significant mental difficulties. Another question is whether all life changes, favorable or positive ones as well as unfavorable or negative ones, can reasonably be called "stressful." Some theorists believe that life changes play a role in producing mental disorders by triggering a process to which individuals are vulnerable. One study, for example, found that life events play a significant causal role in the occurrence of depression, but that only certain types of events, those involving long-term threats, are important (Brown and Harris, 1978). A related view is that events in general, not only adverse ones, contribute to mental disorders, particularly schizophrenia. Some persons may be especially prone or vulnerable to the effects of stressful life events (Dohrenwend and Dohrenwend, 1980).

It is still unclear how life events are perceived by persons and how they may be handled in given social situations. Some persons, for example, have more social support than do others and they are better able to cope with life events, particularly the adverse ones, than others. For this reason, coping strategies are important factors that may intervene between life events and mental disorders.

Coping and Social Adaptation

Some persons ward off stress because they are better able to cope with it (see Williams, et al., 1981). Successful adaptation involves a number of aspects. First, the person must have the capabilities and skills to deal with social and environmental demands; these are called coping capabilities. Coping involves being able to anticipate as well as react to environmental demands; it also involves the capacity to influence demands when possible (Mechanic, 1989: Chapter 7). Second, people must be motivated to act and react to demands. Persons can escape anxiety by withdrawing or lowering aspirations, but such behavior may be costly to the individual and may be constrained by social expectations and roles.

Successful coping also depends on resources, both physical and social. Economic resources can provide relief from many sources of stress. Social resources, in the form of social support from family, friends, neighbors, and others, (in the form, for example, of specialized support groups such as Alcoholics Anonymous) can alleviate other sources of stress by providing an atmosphere that is conducive to solving interpersonal problems.

An adequate explanation of mental disorders must consider the interaction of stress and individual coping skills (Wheaton, 1983). Chronic stressors may be more important than acute ones, suggesting the period of time one experiences stressors is important. If so, this makes long-term individual coping skills and more or less constant social support all the more important to ward off the disabling effects of stress. There is clearly more to learn about the nature of stress and how best to deal with this problem. Evidence is now accumulating that stress is more related to social roles than to other conditions such as gender or marital status (Thoits, 1987). Thus, it is not just being a woman or being married that carries with it certain stressors that may lead to mental distress, but the combination of roles that certain people—such as married women—perform.

SOCIOLOGICAL PERSPECTIVES OF MENTAL DISORDERS

Mental disorders have been explained in terms of biological, personality-type, and behavioristic factors related to the sociocultural setting. Currently, there are three general theoretical perspectives from which it is possible to examine the nature of mental disorder. First, one can regard it as resulting from some biological defect or genetically-based deficit, but because of the limitations already noted, the studies done in this area will not concern us here. Attention has been called to their existence here because of the tendency among sociologists to ignore biological considerations when they interpret their findings (Gove, 1987).

Second, mental disorder can be regarded as the outcome of certain types of personality that have been formed from conditioning or learning experiences. Such types display behavior that is inappropriate to situations, and they are triggered into social recognition by the impact of various kinds of interpersonal relations and cultural patterns on them. From this perspective, disorders may result from mental conflicts, superego defects, or traumatic events in early childhood.

A third more sociological possibility is to view mental disorder largely in behavioristic terms, seeing it as a kind of behavior that has become defined as deviant and unacceptable by the significant others who surround the person. Here, mental disorder as such becomes tied to the values and social preferences operating in a given cultural system. From this perspective, what is considered mental disorder in a society is extremely changeable, for it will tend to vary as cultural values, normative expectations, and social preferences change.

Fourth, mental disorder may also be conceived sociologically in terms of social roles, primarily in terms of an inability to shift roles, as playing the role of a mentally ill person, and as self-reactions. As was pointed out in Chapter 2, the adequate performance of social roles is basic to the assessment of deviance. Inadequate role performance violates normative expectations, thus increasing the probability that a negative sanction will be imposed. Like other forms of deviance, mental disorders elicit negative sanctions from a number of sources, including family, friends, employers, and relatives, as well as from such outside sources as the police and mental health professionals. Moreover, in addition to reactions from others, self-reactions are also likely.

Inability to Shift Roles

Many persons who develop mental disorders appear to be unable to shift easily or at all from one social role to another. Everyone normally plays many roles, even in a single day, depending on the situation and the expectations of others. Schizophrenics, on the other hand, can be viewed as individuals who find it difficult to play the roles expected of them in normal social relations and therefore tend to be socially isolated. Under stress, they are unable to change their role in social situations. According to Cameron's (1947: 466–467) well-known theory, for example, paranoid behavior appears to be a product of inappropriate role-playing and role-taking. Such persons look at things in an inflexible way; they can not shift roles or see alternative explanations for the behavior of others. Gradually, a private world is built up in which the self as a social object becomes central. These persons develop a "pseudocommunity" which is a product of their unique interpretations of "persecution" in the ordinary behavior of others toward them. They are unable to interpret accurately the roles of others and are therefore not socially competent to interpret their motives and intentions. This interpretation of a "pseudocommunity" has been challenged by Lemert (1972: 242–264), however, who maintains, after studying a number of cases of paranoia, that the community to which the paranoid reacts is real and not a pseudo or symbolic fabrication; he actually is, for example, treated unfairly by others. Lemert argues further that in addition to the inability of the paranoid to shift social roles, the delusion and associated behavior that develop must be understood in the context of a process of exclusion that disrupts the paranoid's social communication with others.

Residual Norms and the Mentally Disordered Role

A sociological theory of mental disorder in terms of playing the role of the mentally ill has been set forth by Scheff (1984). As pointed out previously, Scheff argues that there are many expectations (norms) that govern behavior in social situations. These norms are relatively clear and the violations of these norms are also relatively clear. Violation of a law is a crime, using illicit drugs is considered abuse, and committing suicide violates the norm of wanting to live. But there are other expectations that are vague and merely understood in social groups, such things as "the appropriate length of time for staring into space and the proper way to imagine or fantasize" (Aday, 1990: 134). These expectations constitute residual rules of living. Normative violations that characterize mental disorder such as withdrawal, depression, compulsions, obsessions, and hallucinations are common and, as pointed out earlier, Scheff terms such norm violations "residual rule breaking" to distinguish them from the violations of criminal law or social conventions such as etiquette. They may arise from diverse sources, such as organic difficulties, psychological problems, external stress, or willful acts of defiance against some person or situation.

Residual rule breaking may be very common and much of it goes unnoticed; it is primary deviance. The average person, for example, may have an illusion or hear odd sounds or voices and simply forget it. For most persons, most of the time, Scheff (1984) argues, their deviance is largely unrecognized, ignored, or rationalized depending on the circumstances.

The explicit identification and labeling of such residual behavior by others, however, helps to organize the behavior into a "role of being mentally ill" that has been defined by, and therefore learned from, the culture. Many societies, such as the United States, have shared conceptions of what is meant by insanity, or what Scheff (1974) has termed the "social institution of insanity." Popular conceptions of mental illness (or "crazy" behavior) are perpetuated in everyday conversations and in the mass media, including advertising. All adults probably know how to "act crazy."

Like the typecasting of actors, the playing of the role of a mentally disturbed person can become stabilized because of the expectations of others and role-taking received from others. Where professional "treatment" by psychiatrists, psychologists, counselors, and others tends to attach the label of mental disorder to a person, it may enhance the stability of the mentally ill role to the person. Once labeled as mentally disturbed, persons may have difficulty in turning to other, more socially acceptable, roles. They may adopt the deviant role as the only one available for them.

There is much to recommend this approach to mental illness: (1) it takes into account the normative aspects of mental disorder as deviance in terms of residual norms; (2) it focuses attention on how persons can become aware of these norms and the imagery associated with "crazy" behavior; and (3) it suggests that mental disorders can be viewed in terms of role-playing that is expected of persons who have been labeled mentally disturbed.

The theory set forth be Scheff (1984) conceives of labeling by others as a necessary condition for the explanation. This view has some support. For example, one study showed that on the whole people tend to reject persons as having a "mental disorder" according to the source of the "help" to which they turn. The rejection scores given for identical descriptive cases of mental disorder by a sample of persons interviewed were found to be lowest for those who sought no help, followed by those who received help from a clergyman, a physician, and a psychiatrist, and then finally by going to a mental hospital (Phillips, 1968). Another study indicated that the views held toward patients by others were positively related to the length of hospitalization (Greenley, 1972). Clearly, the process of institutionalization is itself stigmatizing, regardless of the stigma that may arise from the disorders themselves (Goffman, 1961).

In spite of such supporting evidence, the residual norms perspective on mental disorder has been severely criticized. The labeling aspect has been exaggerated by Scheff and by others who take the position that the public automatically applies a negative stereotype of mental disorder to individuals and that pronounced mental disorder, or secondary deviation, results from this labeling. Gove (1970) asserts that the vast majority of people who become psychiatric patients have real, serious disturbances, and they are not arbitrarily labeled as mentally ill, since most families tend to deny mental illness until its recognition is unavoidable. In fact, the nature of the psychiatric symptoms were found by Gove to be more important than the attitude of the patient's family in predicting the response of others. Moreover, Gove indicates that the evidence on stigma, while it is far from being conclusive, suggests that for most ex-patients stigma does not present a serious problem, and that when it is a problem, it is related more directly to the person's current psychiatric status or to his or her general ineffectiveness than to having been a patient in a mental hospital.

Scheff can also be criticized for other problems in his explanation. Scheff (1984) indicates that mental disorders represent deviance from norms that can not be identified; this is what he means by "residual" norms. However, Scheff also inconsistently maintains that persons learn "residual" norms via interpersonal communication and the mass media, and that there are clear stereotypes of "crazy" behavior. In a comparative study, no support was found for the labeling hypothesis that popular stereotypes of mental disorders are primary determinants of symptomatology (Townsend, 1975). Nor does this conception explain the fact that the types of disorders that some persons have developed in the absence of labeling of any kind, and that these some of these disorders are sufficiently serious as to require hospitalization (Roth and Kroll, 1986: 15–16).

Labeling theorists argue that once individuals have been labeled as deviants by others, particularly psychiatrists, they have experienced an important and frequently irreversible socialization process that leads these persons to acquire a deviant identity as a mentally disordered person. The empirical literature on whether this does indeed occur is sketchy, however, and has been summarized accurately by Gove (1982: 295): "... a careful review of the evidence demonstrates that the labeling theory of mental illness is substantially invalid, especially as a general theory of mental illness."

Self-Reactions and Social Roles

All persons have a self-reaction to their appearance, status, and conduct. They come to conceive of themselves not only as physical objects, but as social objects as well (Shibutani, 1986). This capacity of self-conception, which all persons have, plays an important role in mental disorder. Mentally disordered persons may develop distorted self-conceptions or self-images that are reflections of difficulties in interpersonal relations and continuing anxiety. Some may become less confident and more preoccupied with themselves, while others, without logical reason, may adopt egocentric ideas of being either a great success or a great failure. Where interpersonal relations have been difficult, the mentally disordered person may learn to use self-reactions in fantasy. Such persons may dream of themselves as people they are not in order to overcome conflicts. Self-centered reactions obstruct the capacity to communicate and to relate to others, and this consequently magnifies the person's own concern about symptoms and conflicts so that there is less ability to act with emotional feeling.

What a person does can result in self-approval or self-reproach. People can praise themselves for what they have done or said, or they may be disturbed by what they have done and rebuke themselves, producing frustration and conflict. For adults with a depressive psychosis, this self-punishment, representing an internalization of difficulties with their outside social situations, can become a "tragic melodrama, where the depressed self-accused lashes himself so mercilessly in talk and fantasy that death seems the one promise of penance and relief" (Cameron, 1947: 101). In such mental disorders, the self may become so detached from the individual that it becomes not a social object but a physical object to be mutilated and punished. In certain forms of neurotic behavior involving dissociation, the person may even be able to forget his or her own identity. In some cases of hysteria and amnesia, the person may come to identify with a prior role or with another self; here one attempts to get away from one's conflicts by changing oneself, sometimes with the addition of selves where one self does not know about the others.

Disturbances in language and in meaningful communication, which are often symptoms in schizophrenia, indicate their connection with interpersonal relations. The schizophrenic is able to invent a world of fantasy which lifts the individual in his own estimation. The disorders in thought processes are eventually expressed in language. Language becomes private and not social; whether the other person understands the language is immaterial. The patient may invent words and link them together in such a fashion as to make speech incoherent to others. In response to the question "Why are you in the hospital?" one patient replied:

> I'm a cut donator, donated by double sacrifice. I get two days for every one. That's known as double sacrifice; in other words, standard cut donator. You know, we considered it. He couldn't have anything for the cut, or for these patients. All of them are double sacrifice because it's unlawful for it to be donated any more. (Well, what do you do here?) I do what is known as the double criminal treatment. Something that he badly wanted, he gets that, and seven days criminal protection. That's all he gets, and the rest I do for my friend. (Who is the other person who gets all this?) He's a criminal. He gets so much. He gets twenty years' criminal treatment, would make forty years; and he gets seven days' criminal protection and that makes fourteen days. That's all he gets. (Cameron, 1947: 466–467)

Cultural factors influence the nature of self-reactions. These factors include religion, the degree to which material success is emphasized in the society, and the amount of control one perceives oneself to exercise over events in the world. One study reported that mental patients of Asiatic origin were more likely to perceive of the label of mentally ill in more magical terms, reacting more to the "power" of the label and how that "power" is transferred to the individual (Rotenberg, 1975). Patients with Western cultural backgrounds, on the other hand, were more likely to react to the label as indicating some sign of "differentness" from their group or their self-concept; the Western patients had a self-concept of being "sick." Cultural differences, however, do not explain all differences in self-reactions. Serious mental disorder, such as psychotic schizophrenia, appears to display similar symptoms in a number of widely differing cultures, suggesting that, at least for this malady, cultural molding may have limits (Murphey, 1982).

Mental disorders may represent a continuum that involves personal resources, symptoms, and social expectations (Gove and Hughes, 1989). Mental disorders may be viewed in this sense as a career path along which there are various contingencies that involve a combination of factors. At one level, acute distress may develop, followed by some sort of mental disorganization. Beyond this level, some persons may accept psychotic episodes as real and part of their world. This latter group is quite removed from reality and suffering severe isolation. Gove and Hughes maintain that the boundaries of such a theory begin with certain biological conditions, many of which we seem to know little about. Regardless of the precise role of biological factors, Gove and Hughes say, mental disorders are reflected in real symptoms and they are interpretable within a broader sociological framework.

THE SOCIAL CONTROL OF MENTAL DISORDER

Persons who have mental disorders may voluntarily seek help, or they may have this help forced upon them involuntarily. In either case, the assistance may be in the form of out-patient treatment with an individual psychiatrist, psychologist, or social worker, treatment at a community psychiatric clinic, or residential care in a local hospital or a mental hospital. If the treatment is involuntary, this is most frequently given in a mental hospital. The widespread use of large mental hospitals has shifted in recent decades to the general development in recent years of local voluntary out-patient community facilities for persons with mental disorders.

Civil Rights and Involuntary Commitment

Mentally disordered persons may voluntarily commit themselves to a mental hospital, or they may be committed by the courts. Though the majority of admissions to mental hospitals are voluntary, involuntary commitment depends on the responses of others to the behavior of mentally disordered persons and a willingness to take legal action. Generally, depending on the jurisdiction, there are three conditions that are examined: dangerousness to self, dangerousness to others, and inability to care for one's basic needs for living (Roth and Kroll, 1986: 84). The complainant, who may raise any of these issues, may be the next of kin, such as husband, wife, a neighbor, employer, or some other person such as a psychiatrist, family physician, clergyman, or the police (Szasz, 1981).

The police in most jurisdictions have considerable discretion, on an emergency basis, in arresting and taking persons to mental hospitals, particularly if they have annoyed others or acted strangely or suspiciously in public places. Emergency apprehensions have traditionally occurred under one of five conditions: (1) when the signs of serious psychological disorder are evidenced by particularly agitated behavior or indications of physical violence; (2) when signs of serious disorder are shown by incongruous behavior in physical appearance like odd posturing, nudity, and extreme uncleanliness; (3) when there is evidence that a person has attempted or is attempting suicide, whether or not the individual is mentally disoriented; (4) where persons whose behavior is disoriented create a nuisance in public; and (5) where the police have been summoned by a complainant (Bittner, 1967).

After apprehension, the courts or hospital staff decide whether to keep or to release the person. It is at this point that a person's civil rights may be seriously endangered. Some of the more common problems include the following:

1. Many judges do not take sufficient time to determine adequately whether the person should actually be admitted, simply accepting the statement of medical examiners. In a study of involuntary commitments in a Florida county, it was found that 79 percent of the cases were declared to be mentally incompetent, and that in 39 percent of the cases the examining committees had not met legal requirements (Fein and Miller, 1972). In this sense, the commitment process has the appearance of due process, but it is often a predetermined verdict.

2. Since the person is seldom represented by counsel, he or she often can not interrogate either the "accusers" or the medical examiner. Traditionally, the reason for this has been that the commitment process has been justified only partly on the need to protect the community. More frequently, this decision, like those made in juvenile court, is justified on the ground that a person requires treatment.

3. Many states have such loosely defined commitment laws that there is no precise determination of under what conditions the individual should be hospitalized. This is an important consideration, particularly since psychiatrists have been unable to determine with any objective certainty when a mentally disordered person is potentially dangerous (Monahan, 1981). As one observer put it: "dangerousness, like beauty, is to some extent in the eye of the beholder" (Stone, 1975: 26).

4. Further issues are raised if persons who are committed to state hospitals do not actually receive anything that can reasonably be called "treatment." Psychiatric treatment in mental hospitals is sometimes little more than custodial care. Under these circumstances, there is little benefit that might come from the hospitalization, except to protect the community. Some court cases have substantiated this view as in the 1975 U.S. Supreme Court case of *Donaldson v. O'Connor.* Donaldson had been committed to the Florida state mental hospital in 1957 by his father. During the following 18 years, several organizations had petitioned the hospital for his release. Each request was denied, even though there was no evidence that Donaldson was dangerous either to himself or others. The confinement consisted simply of custodial care and the Court found that the patient's constitutional rights were being denied by this confinement without treatment. A further legal position was the claim that commitment of an involuntary nature is a deprivation of liberty to which the state can not be a party without due process of law, in which case the commitment must be justified in the name of the treatment the individual requires and must receive. Mere custodial care in an institution is not sufficient basis for continued confinement.

This is not to deny that confinement for some persons is unwarranted. There are some people who require confinement, even for short periods of time, by virtue of their behavior. But the basis on which involuntary commitment decisions are made has changed. Throughout the 1960s and 1970s, the criteria of dangerousness to others became the main benchmark used in involuntary proceedings (Stone, 1982). The importance of this standard resides in the inability of psychiatrists (or anyone else) to be able to provide accurate predictions of dangerousness. As a result of the use of this criteria, most large cities now contain a number of largely homeless people who may have difficulty providing for themselves but for whom mental hospitalization is now inappropriate (Herman, 1987). The process of this "deinstitution-alization" is discussed below.

Mental Hospitals

Mental hospitals are institutions of fairly recent origin. For centuries, the traditional method of dealing with pronounced mental disorder was punishment or isolation of the person, or both. Often the mentally disordered person was cruelly treated. Beginning in the eighteenth century, persons were confined in institutions referred to as lunatic or insane asylums, which were later more politely called mental hospitals. Rothman (1971) shows that psychiatrists at the time thought that insanity was the result of chaotic social forces, and that order, discipline, and social stability would produce mental

Where are Persons with Mental Disorders Treated?

Because of the variety of behavioral symptoms and the degree of impairment, a variety of facilities are used to deal with persons with mental disorders. The declining use of mental hospitals has been followed by an increasing use of outpatient facilities.

Types of Facilities for Persons with Mental Disorders, United States, 1987

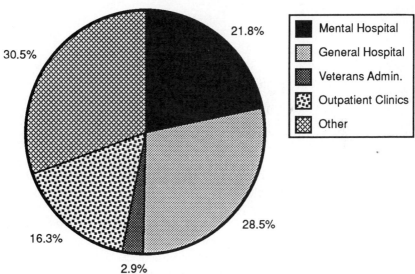

Source: Bureau of the Census. 1990. *Statistical Abstract of the United States. 1990.* Washington, D.C.: Government Printing Office. Page 114.

health. The charge of the asylum, thus, was to bring discipline to those with mental disorders. Psychiatrists lobbied for funds to build asylums and by 1860 almost 85 percent of the states (there were 33 at the time) had built public insane asylums. By the end of the nineteenth century, it was apparent through the intent of the founders and the operation of the facilities that the asylum served as a custodial, and not a treatment, center.

Mental hospital admissions escalated during the first half of the twentieth century. Most mental hospitals were large, overcrowded and custodial in orientation and routine. Pilgrim State Hospital in New York at the time reached a population of about 12,000 by 1955 and that same year total county and state mental hospital population in the United States reached a peak of over 558,000 patients (Conrad and Schneider, 1980: 62). Given both the large number of patients and the small number of trained professionals to administer treatment, it is not surprising that much of what went on in these hospitals had nothing whatsoever to do with psychotherapy or other treatment. In fact, the routines of daily living looked much more like a prison than a medical facility.

By 1956, mental hospital populations had begun to go down. The large reductions in in-patients, however, did not occur until 1965. By 1981, there were 125,000 in-patients in state mental hospitals (Kiesler and Sibulkin, 1987: 46–47). This represented a 78 percent decrease from 1955. By the beginning of the 1990s, there were about 115,000 patients in such hospitals (Mechanic, 1989: 161). Several factors have been said to account for this decline in mental hospital populations. including economic and fiscal reasons (Scull, 1977). Mental hospitals had become too expensive for most counties and states. The large physical plant, combined with expensive daily upkeep, constituted an enormous drain on limited resources.

The rise of psychotropic drugs that exert their principal effect on a person's mind, thought, or behavior in the 1950s rendered mental hospitals unnecessary for a large proportion of patients (Conrad and Schneider, 1980). In 1952, a new drug, chlorpromazine, was developed in France and eventually marketed in the United States under the trade name of Thorazine. This drug, and nearly 1,000 others that were developed since, was greeted enthusiastically by psychiatrists who discovered that wards were quieter, delusions decreased, and institutions ran more smoothly. Moreover, many patients could be maintained at home on drugs rather than being committed to an institution. Psychiatrists believed that these drugs would greatly facilitate psychotherapy, although critics charged that the drugs would merely mask psychiatric symptoms without making any fundamental change in a person's psychiatric condition. As important as the widespread use of chlorpromazine was, significant reductions in state mental hospital in-patients did not occur until the next decade.

For those who remained, the mental hospital appeared to have two main functions: the treatment of patients to enable them to return to normal society, and the provision of "protection" for both patients and society.

Characteristically, a state mental hospital is forbidding in appearance, and contains large numbers of patients whose daily lives are scheduled to fit certain institutional routines. A custodial atmosphere still pervades most public mental hospitals, and the hospitals have often been little more than dumping grounds for the aged with senile disorders, the chronically ill, and the mentally disordered of the lower class. It has been said that persons "are not in the hospital because they are mad, but because they have been rejected by society and have no suitable place in it" (Perucci, 1974: 30). The use of psychotropic drugs, however, did change some aspects of mental hospitalization. Fewer wards were locked and more parts of the hospital could be open to the outside.

It has been almost impossible for some hospitals to provide much effective psychotherapeutic treatment because of inadequate staffs and a general lack of funding. The majority of the patients receive either no treatment or are given somatic treatment, usually in the form of drugs or shock therapies like insulin or electric shock. Where psychotherapy is administered to patients, its effects are hard to determine. Psychiatrists do not typically agree among themselves about the best treatment for specific ailments and the nature of psychotherapy to be employed in specific cases. As a result, it is not unusual to find research that indicates that variations in the patient success rate usually have more to do with the therapist than the type of treatment (Luborsky, et al., 1986).

Social Structure of the Mental Hospital

Mental hospitals often have elaborate physical plants, but the social structure of these institutions is even more varied and complex. Goffman (1961) referred to mental hospitals as "total institutions," which referred to organizations—such as monasteries, prisons, schools, and hospitals—that change people's lives and identities. As with most "total institutions"—prisons, for example—mental hospitals represent unique communities with special social structures in terms of status and decision-making powers. Generally, patients in public mental hospitals make up the "lowest" status group, below attendants and clerical staff. Patients often must undergo a number of degrading experiences such as the "confessional" aspects of group therapy where private, intimate details are revealed to others, and the stark humiliation of failure as represented by hospitalization itself. Patients in most hospitals are exposed to a system that places them under total and continual surveillance by staff. The patient's "status is so low that moral norms, appropriate outside, may be relaxed in the hospital. For example, extramarital encounters may be permitted without censure. Patients are so low on the totem pole that they have little in the way of reputation to lose" (Grusky and Pollner, 1981: 282). While such conditions surely do not describe all mental hospitals, there is a grim reality to patient status in all such institutions.

At the top of the prestige hierarchy are superintendents and professional staff, including psychiatrists, psychologists, occupational therapists and social workers. A major difficulty of this social system of status and power has been that in the treatment of the mentally disordered persons there has often been a breakdown in formal and informal communication among staff persons and between staff and patients (Stanton and Schwartz, 1954: 193–243). Social distance between nursing staff and patients is a common problem that inhibits treatment efforts, and a breakdown in communication among staff members interferes with patient recovery.

Patients in large mental hospitals generally do not have the same kind of relationship with doctors that patients have with their physicians in a large medical hospital. In mental hospitals, patient-doctor relationships appear to be more superficial and impersonal. As one patient has said: "The doctor just comes through one door and goes out the other. He spends no time with the patients" (Weinberg and Dunham, 1960: 41). Other patients have felt that because doctors can shorten or prolong an inmate's stay, they can in a sense exercise great control over their future. Patients therefore try to cultivate the friendship of doctors and even learn to feign symptoms of recovery. This is a process not too dissimilar to that involving prison inmates and parole board members.

Although the situation is improving due to hospital closings and reduced numbers of patients, many public mental hospitals are still inadequately staffed, and there is high attendant turnover. Frequently, the attendant staff lack proper professional motivation and they receive poor training. This is a significant problem because it is with the attendants that patients have their greatest contacts, although their role is largely custodial.

The continual 24-hour experience of being viewed and labeled as a mental patient may encourage the acceptance of the "sick role." This process of role assignment and role-playing increases the likelihood that a mental patient may develop a more or less stabilized identity as a mental patient, what Goffman (1961: 125–169) and others refer to as the "career" of a mental patient. The extent to which the role of a mental patient is the result of staff reactions to the patient or the result of other factors, such as psychiatric symptomology, is disputable (see Gove, 1982). In any case, it is clear that some patients do come to occupy this role and that they have difficulty performing other roles upon release from the mental hospital. Still other patients are better able to make the transition to the outside community. This process of shifting roles, as we have pointed out earlier, is an important process in understanding mental disorders generally, and those persons who experience more difficulty in shifting from the role of mental patient to a nondeviant role may also have the greatest extent of psychiatric impairment.

The Rosenhan (1973) study, mentioned earlier, is another source of insight into the mental hospitalization process. Rosenhan placed eight "normal" persons in twelve different hospitals in five states. None of the patients, after admission to the hospitals, were recognized as a sane person, although they behaved perfectly normally. It appears that the mental hospital itself imposes a special kind of environment that changes the meaning of behavior—the routine activities and behavior of persons in a mental hospital are symptomatic of mental disorders. "Obviously," only mentally disturbed persons are in such places. Even normal behavior is crazy in an environment that is able to interpret that behavior only as "disordered."

The Deinstitutionalization Movement

We noted above the significant reduction in mental hospitalization from the mid-1950s to the present time. As the large mental hospitals have increasingly been found to be of limited value in treating many patients, new methods have been developed. This has often been called the "deinstitutionalization movement," a term that refers to the trend away from the use of large public mental hospitals in the care and treatment of mentally disordered persons. The use of community facilities rather than mental hospitals received their greatest impetus from the proposals of the Joint Commission for Mental Health in 1961 and from President Kennedy's proposals to Congress in 1963 (Joint Commission on Mental Illness and Health, 1961; Kennedy, 1967). The Commission recommended that treatment should be primarily directed at helping persons with mental disorders to sustain themselves in the community, and that most persons in mental hospitals should be returned to the community as quickly as possible to avoid the isolating effects of hospitalization. The main responsibility for persons with mental disorders would remain in the local community. Congress subsequently initiated a program that involved the establishment, with federal aid, of community mental health centers, along with other treatment improvements. These centers, located in the patient's own community, were expected to emphasize prevention as well as treatment.

Out-patients, those who have and those who have not previously been hospitalized, were to be offered a variety of services through these local clinics. These would be staffed by psychiatrists, psychologists, social workers, and nonprofessional workers; the programs include individual counseling and group therapy, as well as some specialized programs, such as occupational therapy, in some of the larger centers. Where necessary, antidepressant drugs may be used. Should a patient require hospitalization, a local general hospital is usually used where the patient can receive various therapies for a short period of time. If more extensive hospitalization is required, transfer to a county or state mental hospital is arranged.

The Reagan administration did not continue the same commitment to community mental health services. Under the Omnibus Budget Reconciliation Act of 1981 (PL 97–35), federal funds for mental health services were cut 25 percent. The law provided for the continuation of mental health services in states but not with the same impact (Kiesler and Sibulkin, 1987:

18). States and local communities had to pick up much of the costs of mental health services, and some jurisdictions were unable to afford these costs.

Along with the development of community resources, there has been an increase in the number of organizations associated with the deinstitutionalization movement, and most of those organizations represent patients. Some organizations—such as Network Against Psychiatric Assault, Mental Patients' Liberation Front, and Mental Health Consumer Concerns—stress the protection of mental patients' legal and social rights. Others, such as Diabasis House, Recovery, Inc., and the American Schizophrenia Association, are interested in or attempt to promote specific therapies or approaches to treatment (Grusky and Pollner, 1981: 356).

The development of community mental health programs has not been without problems. The substitution of alternative care facilities may be as custodial and, in some ways, as repressive as the large mental hospitals they were designed to replace. There is no evidence that community mental health centers are more effective rehabilitators than mental hospitals. While the *intent* of such places was to increase rehabilitative services to persons in need, the *consequences* of such organizations have frequently been different, as seen, for example, with the reintroduction of mental patients into community-based institutions, such as nursing homes, without any obvious therapeutic benefit to the patient. Furthermore, the increase in the numbers of the urban homeless population in recent years suggests that many of these persons are experiencing problems in living that community health centers are not meeting. As a result, many of these persons turn to panhandling or crime to subsist (Herman, 1987).

The deinstitutionalization movement was sped along in the name of civil rights, but trying to protect the rights of mental patients meant, at the same time, that some of the most dangerous persons were also left in the community (Isaac and Armat, 1990). In effect, protecting the rights of mental patients, in some instances, damaged the rights of the community—families, physicians, and others—to decide which persons should be confined for their own good and for the good of others. Taking mental patients out of hospitals, or not sending them there in the first place, did not mean that these persons were placed in contact with treatment specialists in the community. Sometimes in fact, not being hospitalized meant living in a new, sometimes more terrible, place and falling prey to a variety of street people—hustlers, con artists, runaways—who were no less vicious than the callous or violent caretakers they left in the hospital (Johnson, 1990).

In addition to a concern over civil rights, much of the deinstitutionalization movement was based on certain assumptions that have proven dubious: faith in Thorazine and other psychotropic drugs as therapeutic devices, the belief that reduced hospital admissions would also reduce taxpayer costs, and the theory that many patients would be better off in the community apart from their family support systems. Not only are such assumptions untenable, but some have claimed that today's homelessness is largely a result of the deinstitutionalization movement (Isaac and Armat, 1990).

For such reasons, mental health professionals are finding increasingly that mental hospitals are not only appropriate but necessary to care adequately for some kinds of patients. Such institutions may have negative effects on people, but they also may have positive effects. With drugs comprising a large part of the treatment of chronic mental disorders, mental hospitals represent the only alternative for providing adequate care for some patients who need them. Mechanic claims that, "Many patients find their hospitalization experience a relief" (1989: 164). Many enter the hospital from environments that are disruptive and stressful. Many patients are not capable of taking care of themselves, and some are harmful to others. Furthermore, many patients report that their hospitalization is helpful to them and that they appreciate the care they receive there.

Regardless of future solutions, it remains the case that some persons with mental disorders require some kind of confinement while others do not. Failure to provide adequate care of those who are not hospitalized has resulted in the same kind of difficulties as the failure to provide humane treatment of those who require hospitalization: a system in need of reform and, perhaps, re-thought.

SUMMARY

Mental disorder is not easily defined and, for this reason, mental illnesses are related to society's tolerance of eccentricity. A limited number of disorders may have an organic basis, suggesting that medical or chemical intervention is appropriate. The functional disorders, however, have no known physical or biological basis. Such mental disorders are sociologically considered residual normative behavior not covered by other behavioral terms. Behavior that is considered mentally disordered is behavior that is inappropriate to social situations or to the expectations of a group. Mental disorder can also be seen as role behavior; people with mental disorders have difficulty fulfilling role expectations and changing roles. Persons can come to play the role of a mentally disturbed person. In fact, some sociologists believe that once a person has been labeled as having a "mental disorder" or being a "mental patient," he or she may experience difficulty resuming normal roles.

The reactions of others to behavior reflects the degree to which norms are being violated and the amount of tolerance afforded these violations. People who act "strangely" are regarded as such only because our expectations of "normal" behavior are violated. In this sense, what is "mad" depends on our conceptions of appropriate behavior. This is not to deny that people do sometimes act dangerously to themselves or others, or that some people seem unable to take basic care of themselves.

Cultural factors are important in mental disorders. Certain acts may be considered the result of mental disorder in one society but not in another. Mental disorders are structured in society. The more severe disorders are more frequently found in the lower socioeconomic classes, especially various forms of schizophrenia. It is not known, however, whether this reflects differences in actual disorders or differences in diagnoses according to social class. Females have higher rates of certain disorders, such as depression, than males, although part of this is explainable by the fact that females are more likely than males to seek psychiatric help for problems they define as "emotional" or psychiatric.

Many psychiatric disorders are related to aspects of social stress. People who experience more significant life events or changes in their lives are more likely to experience some kinds of disorders than persons who experience fewer of those changes. Stress may be greater in the lower classes, helping to explain observed class differences in disorders. Current research is concentrating on the sources of stress, its relation to particular social roles or combinations of such roles, and coping mechanisms by which stress can be reduced.

The control of mental disorders has shifted from an institutional to a noninstitutional context. With the introduction of mind-altering drugs and an awareness of the economics of mental hospitalization in the 1950s, mental hospital in-patients declined. Many people are still found in mental hospitals, but out-patient care and service is more common. Many have questioned the role of the mental hospital as an institution of healing. Critics have charged it functions more like a prison than a hospital. But, there are still some people who require some form of institutionalization.

The community mental health movement begun in the 1960s has continued to the present day, although funding cutbacks have jeopardized the range and intensity of services. Community mental health centers attempt to use local resources and facilities to deal with those with mental disorders. The increased numbers of homeless in our urban areas, many of whom might have been appropriate candidates for mental hospitalization in earlier decades, indicate that many of the assumptions on which deinstitutionalization and the community mental health movements are based were incorrect.

SELECTED REFERENCES

Abrams, Richard and Michael Alan Taylor. 1983. "The Genetics of Schizophrenia: A Reassessment Using Modern Criteria." *American Journal of Psychiatry*, 140: 171–175.

A study that explores the claim that schizophrenia is inherited from a particular kind of genetic structure. The authors conclude that the evidence in support of such a view is weak.

Gove, Walter R., ed. 1982. *Deviance and Mental Illness*. Beverly Hills, CA.: Sage.

> A collection of papers assessing various aspects of the labeling perspective of mental disorders, including its theoretical and empirical status. The volume adopts the view that most mental disorders reflect real behavioral or psychiatric problems rather than simply the reaction of others to supposed problems.

Isaac, Rael Jean and Virginia C. Armat. 1990. *Madness in the Streets: How Psychiatry and the Law Abandoned the Mentally Ill*. New York: Free Press, and Johnson, Ann Braden. 1990. *Out of Bedlam: The Truth About Deinstitutionalization*. New York: Basic Books.

> Two different accounts of the causes and consequences of the deinstitutionalization movement in mental health. Both are somewhat polemic. Each of these books is different but they come to similar conclusions about the consequences of the movement.

Mechanic, David. 1989. *Mental Health and Social Policy*, 3rd ed. Englewood Cliffs, NJ: Prentice Hall.

> This book has become, in some respects, conventional wisdom concerning sociological approaches to mental disorder—theoretical, empirical, and policy. Mechanic's views may sometimes be disputable, but they are at least reasonable and always worth considering.

Roth, Martin and Jerome Kroll. 1986. *The Reality of Mental Illness*. Cambridge, England: Cambridge University Press.

> A psychiatric rebuttal to the view advanced by Szasz. The authors claim that mental illness is as "real" as other kinds of medical problems and that psychiatrists should be involved in their diagnosis and treatment.

Scheff, Thomas J. 1984. *Being Mentally Ill: A Sociological Theory*, 2nd ed. New York: Aldine.

> One of the most influential sociological views on the causes of mental disorders that adopts a labeling perspective. Scheff argues that mental disorders reflect learned role behavior and that the reactions of others are critical to the assumption of that role.

Suicide

Suicide is the eighth ranking cause of death in the United States (Sanborn, 1990), but the extent to which suicide is a national "problem" depends on one's interpretation of that event. Most people would have different evaluations of suicide depending on the circumstances. The suicide of a teenager is almost universally considered a tragedy, while the suicide of a terminally ill person may be considered entirely understandable or even condoned.

SUICIDAL BEHAVIOR

Suicide is the deliberate destruction of one's self. Suicide is always intentional and it can arise either from deliberate acts on the part of the individual or from a person's failure to prevent death when it is threatened. In his classic study of suicide, Durkheim (1951: 44) included such acts of altruism as religious martyrdom, defining suicide as "all cases of death resulting directly or indirectly from a positive or negative act of the victim himself, which he knows will produce [suicide]." The terms "suicide" and "suicidal" are somewhat ambiguous, however, in view of the wide range of situations to which they can be applied (see Farberow, 1977: 503–505). In some situations, life-taking may actually be obligatory rather than voluntary, as in the old traditional practice of hara-kiri among Japanese nobility and samurai warriors, and in instances where a person directs another to kill himself—as when Nero ordered an attendant to kill him so that his death would not be by his own hand. More recently, there have been cases of terminally ill patients asking their doctors or families to terminate their lives. Euthanasia is a term used to refer to these "medical suicides," or suicides that are motivated by disease or injury.

Suicide may also result from the indirect actions of persons, and some observers have argued that suicide, far from being a relatively quick act, may take place over a long period of time. For example, such behavior and conditions as alcoholism, hyperobesity, the use of certain kinds and quantities of drugs, and cigarette smoking have been discussed as forms of indirect, or slow, self-destructive behavior (see Farberow, 1980). Although suicide in these instances is not the direct objective, such activities are associated with shorter life spans and thereby might be thought by some to constitute indirect suicidal behavior.

Such "indirect suicides" are hardly implausible. It has been estimated that 20 to 36 percent of suicide victims have a history of heavy alcohol use or were drinking right before their suicide (Secretary of Health and Human Services, 1990: 168–170). Some of these suicides reflect a long history of drinking, physical deterioration, and increasing medical problems, while others are more impulsive and not premeditated.

SUICIDE AS DEVIANT BEHAVIOR

So strongly is suicide condemned by Western European peoples that one might assume this attitude to be universal. It is not. Both today and in the past, attitudes toward self-destruction have varied widely.

Historical Background

Prevailing attitudes in Islamic countries strongly condemn suicide. The Koran expressly condemns suicide and in actuality rates of suicide are generally low in Islamic countries, although suicides do still occur there (Headley, 1983). The people of the Orient, however, have not disapproved of suicide under all circumstances. In fact, *suttee*, or the suicide of a widow on her husband's funeral pyre, was common in India until well into the last century, even after it was outlawed in 1829 (Rao, 1983: 212). Priests taught that such a voluntary death would be a passport into heaven, atone for the sins of the husband, and give social distinction to the relatives and children.

Suicide was regarded as acceptable in China; when committed for revenge, it was considered a particularly useful device against an enemy because it not only embarrassed him but enabled the dead man to haunt him from the spirit world. Voluntary death has been given an honorable place in Buddhist countries, but for devout Buddhists, there is neither birth nor death, the individual being expected to be prepared to meet any fate with stoical indifference. For many centuries, suicide was favorably regarded in Japan, and the suicide rate in Japan remains very high today. Among all classes, but particularly among the nobility and military, it was traditionally taught that one must surrender to the demands of duty and honor. Hara-kiri, developed over 1,000 years ago, was originally a ceremonial form of suicide to avoid capture after military defeat. Later it evolved into an act allowing members of the nobility to take their own lives whenever condemned by superiors, in contrast with the fate of ordinary persons, who were hanged in public squares (Tatai, 1983: 18). The suicide pact of lovers who wish to terminate their existence in this world and go to another is still not unknown in Japan, nor is suicide for revenge to protest the actions of an enemy.

The attitudes of contemporary Western European peoples toward suicide originated mainly in the philosophies of the Jewish and, later, the Christian religions. The Talmudic law of the Jewish religion takes a strong position against suicide (Hankoff, 1979). Basic to the Christian condemnation of suicide were the concepts that human life is sacred, that the individual is subordinate to God, and that death should be considered an entrance to a new life in which one's behavior in the old is important. The concept of life after death strengthened the position of the Church.

Although at first Christians sanctioned suicide connected with martyrdom or the protection of virginity, eventually they disapproved of it for any reason, and it became not only a sin in Christian countries, but a crime against the state. The property of a suicide might be confiscated and the corpse subjected to various mutilations. In the Middle Ages, Church leaders denounced suicide, particularly Augustine, who stated in *The City of God* that suicide is never justifiable. He maintained that suicide precludes the possibility of repentance, that it is a form of murder prohibited by the sixth commandment, and that a person who kills himself has done nothing worthy of death. Similarly, Thomas Aquinas opposed it on the grounds that it was unnatural and an offense against the community. Above all, he considered it a usurpation of God's power to grant life and death. Throughout the Middle Ages and well into modern times, the strong religious opposition, the force of condemnatory public opinion, and the severe legal penalties were so effective that few had the temerity to take their lives, despite infrequent sporadic outbreaks of mass suicide on certain occasions, such as epidemics, religious fanaticism to gain martyrdom, or crises (Dublin, 1963).

These views did not go unchallenged, particularly among philosophers in the Age of Enlightenment who stressed the importance of individual choice in all matters concerning life and death. David Hume, in his essay *Suicide*, argued that persons have the right to dispose of their lives without the act being sinful. Other writers, such as Montesque, Voltaire, and Rousseau in France, challenged the laws on suicide and the denial of individual choice about life and death. Other philosophers disagreed. The German philosopher Kant, for example, opposed such views and said that suicide was contrary to reason and therefore wrong. Suicide was punished as a felony in England for centuries, and the suicide's property was forfeited to the Crown. These provisions were not abolished until 1870. In his famous *Commentaries* on the law, Blackstone (1765–1769: 188) had given these reasons for forfeiture: "The suicide is guilty of a double offense; one spiritual, in evading the prerogative of the Almighty and rushing into his immediate presence uncalled for; the other temporal, against the King, who hath an interest in the preservation of all his subjects." In America, a Massachusetts law forbade burial of a suicide in the common burying place of Christians. Instead, burial was in some common highway, with a cartload of stones laid upon the grave, as a brand of infamy, and as a warning to others. This law was repealed in 1823, but it, and others like it, helped shape attitudes toward suicide in the United States.

Public Attitudes Toward Suicide

In addition to suicide, studies also show strong negative attitudes toward suicide attempters. These negative attitudes are most likely to be elicited by attempters who appear to be less serious in their attempts to die, for example, those who did it "only for a gesture" (Ansel and McGee, 1971). Legally, persons who attempted suicide could be prosecuted in New Jersey and in North and South Dakota until 1950. Many attempters, whether serious or not, may endanger the lives of other persons or rescuers, as do those who resort to carbon monoxide gas in rooms and garages, who try to drown themselves, or who use firearms. Attempted suicide is not illegal in any European country, including the Soviet Union. England had such a law from 1854 until its repeal in 1961. Even before its repeal, few persons were prosecuted under the law. Because it was thought that repeal might encourage suicide pacts, the act made it a crime to aid, abet, counsel, or procure the suicide of another.

The acceptance or condemnation of suicide depends on many factors, including religiosity and education. One study reported that religiosity and acceptance of suicide were inversely related, such that the more religious, the less accepting of suicide, regardless of the specific religious preference of the individual (Johnson, et al., 1980). Generally, it was found that younger, better educated males were more accepting of suicide under special circumstances and of euthanasia than were other persons. Some groups may condemn suicide more strongly than others, but the relationship between those attitudes and suicide behavior is sometimes unclear. Markides (1981) found that Mexican Americans were more fatalistic about life and death than were Anglos, but it was not clear that these attitudes were related to suicide rates in each group. Clearly, however, such attitudes are useful in explaining differences in suicide rates generally.

Attempted Suicide

Suicidal attempts in the United States and in the United Kingdom, particularly in urban areas, may be as high as eight to twenty times the total of actual suicide (Davidson and Rokay, 1986). Most suicide attempts are carried out in a setting that often makes intervention by others probable or at least possible, the person remaining near others and thus allowing for the possibility of prevention. This suggests that most attempts are not serious and that intervention is sought.

One study of 5,906 attempted suicides in Los Angeles, as compared with 768 persons who actually did commit suicide, found that the typical (model) suicide attempter was a white female, in her twenties or thirties, either married or single (not divorced or separated), a housewife, native-born, who attempted suicide by barbiturates and gave as a reason marital difficulties or depression (Schneidman and Farberow, 1961). In contrast, the typical person who did commit suicide was a white male, in his forties or older, married, a skilled or unskilled worker, native-born, who committed suicide by gunshot, hanging, or carbon monoxide poisoning and who gave as a reason ill health, depression, or marital difficulties.

Women attempt suicide more frequently than do men. A study of suicide attempts in England found that females outnumbered males in attempted suicide by as much as two-and-a-half times (Hawton and Catalan, 1982: 8). Studies in the U.S. have reported similar results, as well as reporting that females comprise 90 percent of all adolescent suicide attempters (Stephans, 1987: 108). This fact invites at least two interpretations: women are less successful in committing suicide, or, more likely, women more frequently use threats of suicide to accomplish a certain goal. In the English study of suicide attempts, significant differences in the ratio of female to male attempters were found by age category, with younger females much more likely than older females to attempt suicide compared to males (Hawton and Catalan, 1982).

There appear to be differences in the reasons given for the suicide attempt by race. Young black females seem more effected by the loss or threatened loss of a love relationship than are white adolescent females, although, depending on the nature of the relationship, this can be a major crisis for any adolescent (Bush, 1978).

The kinds of persons who just think about suicide and those who actually attempt suicide may be similar. One study of Australian adolescents found that both groups had similar high levels of depression, general anxiety, sleep disorders, and irritability (Kosky, Silburn, and Zubrick, 1990). Attempters were more likely to be associated with chronic family discord and substance abuse, and for boys, the odds of suicide attempts were substantially increased if they experienced loss.

Extent of Suicide

Many persons commit suicide each year, but the number is small compared with the numbers involved in other forms of deviant behavior, such as property crimes, mental disorder, illicit drug use, or problem drinking. Unfortunately, there are very serious problems with statistics on suicide. Official statistics on suicide—those collected and maintained by local, state, and federal government agencies—are the result of a decision-making process by which deaths are classified as suicide, as opposed to some other cause. In many cases, these judgments are not uniform. One study compared the process by which cases were certified as suicides by a sample of 191 coroners in eleven states (Nelson, et al., 1978). It was found that there was extensive variation among the resources, philosophies, procedures, statutes, and backgrounds of the coroners. Thus, a suicide in one jurisdiction might not be recognized and recorded as such in another. It is still not clear what the precise relationship is between officially recorded suicides and actual suicides. As with criminality, not all suicides are known to officials, and some that are known are not recorded. In a survey of 200 medical examiners, more than half felt that the reported number of suicides was possibly less than half of the actual number of suicides (Jobes, et al., 1986). In other countries, such as those in Asia, problems with collecting and maintaining statistics are even greater (Headley, 1983).

Because of these statistical problems, the precise number of suicides can be estimated only with imprecision. In 1987, one official estimated 25,000 suicides occurred annually in the United States (McGinnis, 1987: 21). This figure is close to that of 1981, when 26,010 suicides were recorded in the United States (Hacker, 1983: 70). Undoubtedly, these figures are underestimates of all suicides, perhaps by as much as one fourth to one third, because of the stigma attached to such deaths. Relatives and others may deliberately conceal the true circumstances of a death, and death certificates may be altered to protect the feelings of survivors. A suggestion to standardize the methods and reporting of deaths by medical examiners to improve our knowledge of vital statistics (Jobes, et al., 1987) is obviously only a partial solution to this problem.

Since absolute numbers are misleading, suicide figures are usually reported in terms of rates, that is, the number of suicides for given populations. The suicide rate in the United States has been fairly stable over the past few decades, between 12 and 13 suicides per 100,000 population. The suicide rate is susceptible to fluctuation, and seems particularly responsive to changes in the economy, being generally higher during periods of depression and lower during periods of prosperity. The highest suicide rate in the

United States was 17.4 in 1932; the lowest, 9.8 in 1957. The rates also decline during wartime, a trend noted by Durkheim more than 80 years ago. From 1938 to 1944, during the period of World War II, rates declined from 20 to 50 percent in all the nations that were at war (Sainsbury, 1963: 166; but see Marshall, 1981). In the United States, the rate declined by about one third, from 15.3 in 1938 to 11.2 in 1945. Several factors may account for this decline. The feeling of unity that prevails during most wars is the opposite of social isolation of the typical suicide. Wars that do not generate national support, however, such as the war in Vietnam, are not associated with declines in the suicide rate. Wars may also bring increased economic opportunities, and it has already been noted that suicide is also related to the business cycle. As mentioned earlier, the highest official suicide rate was recorded during 1932, which happened to be the worst year of the Great Depression.

While the overall suicide rate has remained stable in recent years, some investigators argue that this stability masks some interesting and important changes in the nature of suicide. Seiden and Freitas (1980) argue that the steady national rates mask decreases in suicide at the older age categories, and increases in suicide among younger persons (see also McGinnis, 1987). Suicide is strongly related to age, with older persons having higher rates than younger ones, but youthful suicide rates have been increasing significantly during the past two decades.

There are a number of variations in the suicide rate as reflected in differences among countries, as well as within and between social categories, such as age, sex, and race. There are other variations as well, but some of them have not been tied to a theoretical perspective with which to explain them. For example, a study of suicide rates from 1973 to 1979 (over 18,000 suicides) reported that, contrary to popular belief, suicides decline around major national holidays (Phillips and Wills, 1987). Memorial Day, Thanksgiving, and Christmas experienced a decline before, during, and after the holiday, while New Year's, the Fourth of July, and Labor Day had declines before and after the holiday but a normal rate on that day. The variations reported here are patterns of suicide that a sociological theory should be able to explain.

Variations by Country

For many years, Japan had the highest suicide rate in the world, but in recent years this dubious distinction has gone to Hungary with a suicide rate of more than three times that of the United States. The suicide rates in Czechoslovakia, Finland, and Denmark are about twice that of the United States. Because of the unreliability of suicide statistics, international comparisons are difficult. Methods of certifying deaths vary in each country, and some nations have better reputations as careful record-keepers than others. Suicide rates vary from country to country, but there appear to be fewer variations in the reasons for suicide. A Danish writer has attributed the high suicide rate in Denmark to factors outside the individual, such as political and economic factors (Paerregaard, 1980), and similar factors have been implicated in the relatively high suicide rate found in France (Farber,1979), Where the factors revolve around low social integration, as reflected in high rates of alcoholism, a large elderly population, high immigration and low emigration, high urbanization, and comparatively few fears of drying within the population.

In general, predominantly Catholic countries have lower suicide rates, although Austria, a Catholic country, has a high rate. The relationship between religious preference and suicide, as we shall see, is not strong. Among the Scandinavian countries, Finland, Denmark, and Sweden have high rates, while Norway's rate is quite low (Retterstol, 1975). One explanation for differences among Scandinavian countries relates to differences in child-rearing patterns in the countries. It has been claimed that the Norwegian upbringing stresses more openness of emotion and aggressive feelings on the part of children, thereby resulting in less "pent-up" hostility later in life (Hendin,1964). More likely, the greater strength of the primary group in Norway may contribute to relations not conducive to suicide (Farber, 1968).

Social Differentials in Suicide

Two authorities have said that where customs and traditions have accepted or condoned suicide, many individuals will take their own lives, but where it is severely condemned by the state, church, or community it will not occur (Dublin and Bunzel, 1933: 15). Such a generalization about the reaction to suicide, however, can not easily be related to the variations seen within a country, by sex, race, marital status, and so forth. For example, there is no reason to believe that the lower suicide rates among African Americans and the young indicate that suicide is more severely disapproved by them. Moreover, there is no evidence that all the increases or decreases in the rates of the various countries reflect corresponding changes in the differing norms that pertain to suicide (Gibbs, 1971: 302). Differentials in suicide rates are extremely variable. It has been said that there is no social status or condition that generates a constant rate in all populations, and the example has been given of how "an occupation with a high suicide rate in one community may have a low rate in another; and rates for countries or religious groups change substantially over time" (Labovitz, 1968: 72).

Sex. Suicide is much more common among men than among women in Western European countries, generally three to four times higher. And, in the United States, nearly three times as many men as women commit suicide. Of the approximately 26,000 suicides in 1981, 73 percent were committed by males, 27 percent by females (Hacker, 1983: 73). In Finland,

Suicide and Gender

Although females attempt suicide more frequently than males, males are more likely than females to commit suicide in every age group, including younger persons. Consider the following data:

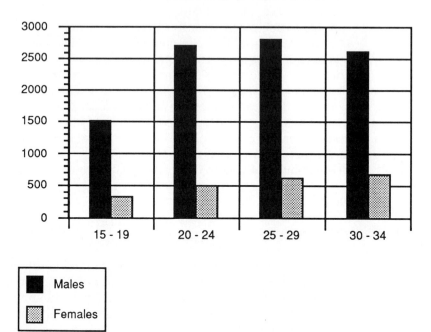

Number of Suicides by Males and
Females in Selected Age Categories, 1986

Source: Sanborn, Charlotte J. 1990. "Gender Socialization and Suicide: American Association of Suicidology Presidential Address, 1989." *Suicide and Life-Threatening Behavior*, 20, p. 148.

almost four times as many men as women commit suicide; in Norway, South Africa, and France, the ratio is about three to one. In Hungary and Austria, countries with very high rates, it is only slightly over two to one. Among older persons, the ratio of men to women suicides is even greater, while among adolescents, the difference is less. Women in Asia commit suicide much more frequently than do men in Western Europe and America; thus, the difference in the ratios is much less. In Japan, for example, the ratio of male to female suicide is only 1.5 to 1.

A number of hypotheses have been advanced to explain these sex differences in the suicide rate. Wilson (1981), for example, argues that failure for males is obvious and clearly defined, but the female sex role is more diffuse and lacks standards for success and failure; failure is less likely to lead to suicide for females because there may be doubt about what constitutes failure. As female roles become more well defined, one should expect a corresponding change in the suicide rate. Looking at changes in suicide by sex over time, Davis (1981) concludes that increased female labor force participation has led to an increase in the female suicide rate, thereby supporting the idea that suicide may have something to do with "traditional" sex role expectations.

Race. Previous research on the distribution of suicide by race concluded that whites have a substantially higher rate than blacks. In a 1969 study, rates for white suicides were found to be twice as high as those for nonwhites (Maris, 1969). One study of suicide among African Americans in the 1970s assessed the oft-stated claim that suicide rates among young African American women had risen sharply during that decade. It was found that black men in their twenties are the most suicide prone group and that there was no substantial increase in suicide among young black women (Davis, 1979). Both African American males and females in the younger age categories through middle age have higher suicide rates than whites, although overall the black rate remains below that of whites (Kirk and Zucher, 1979).

Explanations of why the suicide rate among blacks is highest in the younger age categories as compared to that for whites are varied. Hendin (1969) has suggested that black suicide can be attributed to self-hatred as a result of the black experience in this country. Kirk and Zucker (1979) found that African American consciousness and group cohesion was less among the black suicide attempters they studied, a finding consistent with Hendin's interpretation.

Age. Generally, suicide rates increase with age. The rate for those aged 25–34, for example, is nearly twice as great as for those aged 15–24 (Sanborn, 1990). A Chicago study concluded that the older a person is, the more the individual is socially and physically isolated, with a greater wish to die (Maris, 1969: 15). The rate for persons over 65 in the United States is almost twice as high as for those persons between 25 and 34. A study of patients aged 45–60 concluded that isolation, knowledge of other suicides, pride, belief in an afterlife, and a history of depression were related to suicidal behavior on their part. Moreover, they:

> were more socially isolated; more knew others who had committed suicide; more felt no pride in aging and predicted poor treatment from relatives when they became even older; more approved suicide in some circumstances and did not believe in an afterlife; more had been depressed severely and/ or frequently and had a family history of depression. (Robins, et al., 1977: 20)

Different combinations of age and race show differences in suicide rates. In the U.S. and several other countries, the male suicide rate increases

Suicide and Age

There is a strong relationship between suicide and age: The higher the age, the higher the suicide rate. The data below show this pattern.

Suicide by Age Category, 1988

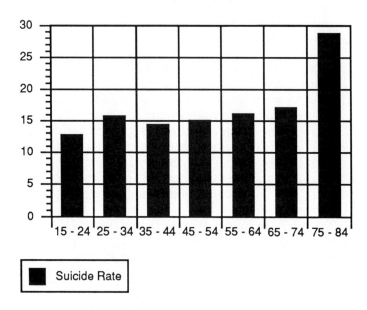

Source: Bureau of the Census. 1990. *Statistical Abstract of the United States, 1990.* Washington, D.C.: Government Printing Office. Page 81

to very old age (e.g., see Stafford and Gibbs, 1988; Headley, 1983), but the rate for white American females increases to a peak at about age 50. The peak rate for suicide for nonwhite females is between 25 and 34, and the peak suicide rate for nonwhite males is between 25 and 29. When looking at the entire population, however, because of the high numbers of whites, the overall suicide rate generally increases with age.

We should find that suicidal behavior is common among the elderly, particular those whose life circumstances may not be to their liking. Administrators of long-term care facilities (either nursing homes or "old age homes") report that there is substantial suicidal behavior among the residents, especially white males (Osgood and Brant, 1990). Refusing to eat or drink and refusing medications were most common suicidal behaviors. The reasons for these behaviors were varied but included depression, loneliness, feelings of family rejection, and loss. Among the most elderly (age 85 and older), the suicide rate for males is 12 times that of females; widowerhood may be a key factor in these suicides (Bould, Sanborn, and Reif, 1989: 67–68).

Marital Status.   Married persons have a lower suicide rate than the single, divorced, or the widowed. The suicide rate is lower for married persons than for single persons regardless of age. About three times as many widowers and five times as many divorced men take their own lives as do married men. Being married is also associated with lower suicide rates among women as well (Gibbs, 1982). It also seems that the nature of marital dissolution and the length of time separated are important variables in this relationship. Specifically, divorce and long-term separation are more strongly related to suicide than short-term separations for whatever reason (Jacobson and Portuges,1978).

As with the relationship between suicide and age, there are racial differences in the relationship between marital status and suicide. Davis and Short (1978) report that although there is an inverse relation between marital status and suicide among African Americans (as with whites), the relationship is not strong and accounts for little variation in suicide.

The relationship between marital dissolution and suicide has been documented extensively in the United States (e.g., Brealt, 1986) but there has been only limited research from other nations. One exception is by Stack (1990), who examined the effects of divorce on suicide in Denmark. The results indicate that although Denmark has a much lower rate of divorce than the United States, the results obtained in the U.S. held in Denmark as well: the greater the divorce rate, the greater that of suicide over time. Specifically, Stack found that a 1 percent increase in the divorce rate was associated with a 0.32 percent increase in the suicide rate, and that divorce trends predicted both adult and youth suicide.

Religion.   Suicide rates among the main religious groups in Western European cultures vary greatly. In general, both in Europe and in the United States, Catholic rates are lower than Protestant, although the rate differentials are not as great as they once were. Formerly, the Jewish rate appears to have been lower than the Catholic, except that, on occasion, when persecutions made their situation particularly difficult or hopeless, waves of suicide occurred. Within recent years, the Jewish suicide rate has risen considerably, perhaps reflecting changes in religious influence and greater participation in the general society. Both Protestant and Catholic rates have increased during the past century (Maris, 1981).

Religious differences in suicide have been interpreted as meaning in part the degree of integration of the various religious groups (Stack, 1982). Protestant religious groups tend to be more individualistic than Catholics. Too, the Catholic position on suicide is more specific with respect to the effect of the suicide on an individual's afterlife. It is difficult, however, to place too much emphasis on religious affiliation alone. The rate of suicide in northern Italy is almost twice that of southern Italy where economic conditions are poorer, there is less education, and adherence to Catholicism seems greater (Ferracuti, 1957: 74). What may be more important than religious affiliation is the degree of one's religious participation and belief system. In general, the more religious a person is, the more that person is apt to find suicide unacceptable (Hoelter, 1979). While strong religious feeling appears to discourage suicide, one study of suicide notes and diaries found evidence that in actuality religion may be used in constructing a justification of suicide (Jacobs, 1970). Mention was often made in these notes of meeting someone in the "hereafter," a "happy reunion," and the freedom of being released from worldly problems and going off "to the final rest." Suicidal persons may use religion to convince themselves that they are blameless and without sin, and that this represents their only choice in dealing with their lives. The following two notes, reported by Jacobs (1970), are typical of this type:

> I am sad and lonely. Oh God, how lonely. I am starving, Oh God, I am ready for the last, last chance. I have taken two already, they were not right. Life was the first chance, marriage the second, and now I am ready for death, the last chance. It can not be worse than it is here.

> My dearest darling Rose: By the time you read this, I will have crossed the divide to wait for you. Don't hurry. Wait until sickness overtakes you, but don't wait until you become senile. I and your other loved ones will have prepared a happy welcome for you.

Occupation and Social Status.   In his classic study of suicide, Durkheim (1951: 257) found that occupational status is linked to suicide, occurring more frequently in the upper ranks of various occupations as well as in positions of higher status. Since that time, however, most research has strongly suggested that suicide is more likely to occur in the lower social classes (Maris, 1969.) One of the most comprehensive studies of differential mortality in the United States found that at least for white males (data for other groups were incomplete), suicide and education, often used as an indicator of social class, were inversely related (Kitagawa and Hauser, 1973). In fact, the least educated group had twice the suicide rate as did the highest group. This relationship has been reported by studies in other countries as well (e.g., Li, 1972).

Suicide rates may be higher in occupations in which there is a great deal of uncertainty and economic insecurity, as well as a less cohesive atmosphere. Those workers who enjoy security of employment and support from other workers generally have a lower suicide rate. In this sense, suicide may have an important subcultural component that operates much like other forms of deviance.

A study of a sample of 166 cases from Chicago reported that occupational status, by itself, does not predict suicide very well (Maris, 1981:

Chapter 6). Citing data from an unpublished study, Maris (1981: 146) indicated that clergy and dentists have roughly the same occupational prestige, but have quite different suicide rates (10.6 and 45.6 per 100,000 respectively). At the lower end of the occupational scale, machinery operators have a suicide rate of 15.7, but mining workers and laborers have a suicide rate of 41.7. It is, however, difficult to generalize from this study since the sample was not random.

TYPES OF SUICIDE

Suicide is related to the type of society, being more common in urban societies. Self-destructive behavior is reported as not occurring among some folk societies. One observer, who asked Australian aborigines about suicide, stated that whenever he interrogated them on this point they invariably laughed at him, treating it as a joke (Westermark, 1908: 220). A similar response was reported from natives of the Caroline Islands. A survey of some 20 sources dealing with the Bushmen and Hottentots of South Africa revealed no references to suicide among these people (Faris, 1948: 148). Suicide does occur among some folk societies, however, some having a much higher rate than others. Suicides have been reported among the natives of Borneo, the Eskimos, and many African tribes. It is also said to have been fairly common among the Dakota, Creek, Cherokee, Mohave, Ojibwa, and Kwakiutl Indians and the Fiji Islanders, the Chuckchee, and the Dobu Islanders.

Durkheim (1951) was particularly interested in suicide among folk societies. As a result of his studies, he classified suicides by type into *altruistic*, *egoistic*, and *anomic*. He then examined the different motives underlying each type of suicide. On the whole, according to Durkheim, suicide in folk societies was considerably different from that found in modern society. To him, suicide was a measure of the degree of social interaction and regulation in a society, the amount of group unity, and the strength of ties binding people together. Suicide was not an individual phenomenon, but was related to features of social organization and structure. Another type of suicide, *fatalistic*, was mentioned by Durkheim, but he made little of it, and those who have commented on his theory have generally ignored it. Suicides of this nature result from excessive regulation in which futures are perceived to be blocked. A good example of fatalistic suicides would be the suicide of slaves.

Altruistic Suicide

Among folk societies, suicides tend to be altruistic in that people take their lives with the idea that by doing so they will benefit others. Individuals in such societies think primarily of the group welfare. When their actions or continued living hurt the group, they may turn to suicide so that the group will have one less mouth to feed or so the group may be protected from the gods. Suicides that may be classified as altruistic are those arising from physical infirmities, those connected with religious rites or warfare, or those related to the expiation for the violation of certain mores, such as taboos. Under such conditions, suicide does not constitute a deviation; in fact, it would be considered a transgression to refrain from the act.

Suicides occur in certain primitive societies where limited food supplies make an old or infirmed person a burden to the tribe. Among the Eskimos and the Chuckchee, for example, old people who could no longer hunt or work killed themselves so that they would not use food needed by other adults in the community who produced it. Some suicide occurs in warfare when persons kill themselves to avoid capture and slavery, or because of their disgrace as warriors. Probably the most common form of altruistic suicide is that done for expiation for a violation of the mores; individuals who fail to commit suicide in atonement for wrongs risk the imposition of other sanctions, such as perpetual public disgrace. The Tobriand Islanders, for example, having violated taboos, generally committed suicide by climbing a palm tree, from which they gave a speech before jumping to their deaths (Malinowski, 1929: 97).

In modern societies, it is possible to know of elderly persons or incurably sick persons who sometimes end their own lives in order not to become a burden on others, though this type of altruistic suicide is generally not approved. A group in the United States, the Hemlock Society, is against suicide except under certain circumstances when they approve of euthanasia. These conditions include when a person is a mature adult who has made a considered decision where medical advice has been rendered (Humphrey, 1987). These types of suicide still invoke considerable ambivalence, with some people strongly condemning them and others approving of them. Individuals in time of war may give their lives in order to accomplish some goal involving group values; sometimes this behavior is approved as being heroic. The Japanese on many occasions during World War II engaged in what was termed suicidal behavior; faced with certain death, large numbers of Japanese troops died to a man in suicidal banzai charges. In the latter days of the war, Japanese kamikaze pilots became legendary for their disregard of their own lives. Loading their planes with explosives, they dived into Allied warships in order to make sure of destroying them (for a general discussion of suicide in Japan, see Tatai, 1983).

Egoistic Suicide

Egoistic suicides are those that occur because of individualistic reasons. These suicides, the most common in modern societies, are a measure of identity (or the lack thereof) with others or a lack of group orientation. In such societies, individualistic motives for suicide are not unusual and are often associated with personal, financial, health, marital, or relationship problems, and occupational difficulties. Egoistic suicides are not the product of a tightly integrated society, as altruistic suicides frequently are, but of one in which interpersonal relations are neither close nor group oriented.

Anomic Suicide

The anomie type of suicide occurs when the individual feels "lost" or norm-less in the face of situations in which the values of a society or group become confused or break down. They also occur where the person is downwardly mobile, or where the individual has achieved everything, so that life has little meaning left. An example of a suicide of this latter type was a wealthy, middle-aged businessman with no apparent financial, health, or marital problems. He had devoted his life to building up his company to achieve something he had always wanted, namely, a merger with a larger company. In this merger, he retained the presidency of his own concern and became the vice-president of the larger company, but after the agreement was con-cluded he immediately went into a depression. After his suicide, it became clear that this was the reaction of a man who had built his business, made the deal he had wanted, and then realized that he no longer was the single, direct owner of the business he had built; he had lost his objectives in life.

Another example of anomie suicide was found in a study of white male suicides in New Orleans (Breed, 1968). Such suicides were associated with substantial work-related problems, as seen in downward mobility, reduced income, unemployment, and other job and business difficulties. In the study of Chicago suicides discussed above, Maris (1969) found that many "suicidal careers" included elements, such as employment difficulties and a sense of hopelessness, usually associated with anomic suicides.

In some instances, anomic suicide arises when the equilibrium of soci-ety has been severely disturbed. A social void exists in which the social order can not adequately satisfy the desires of the person, and the individual does not know which way to turn. Commonly such anomic suicides occur in modern society as an aftermath of severe and sudden economic crises, such as the stock market crash in the United States in 1929, which was followed by a large number of suicides. Similarly, sudden, abrupt changes in the standard or style of living, particularly if the change involves downward mobility, or a breakup of a long-term relationship, such as a marriage, may produce a sense of normlessness and account for the higher rates of suicide among these two groups.

The Special Case of Adolescent Suicide

While the suicide of any person is usually occasion for sorrow, adolescent suicide is generally thought to be an especially sad event. Adolescents are people who are "just starting out" in life and who may be unable to place their immediate difficulties into proper life-cycle perspective. So, what may appear to be a major problem to an adolescent, such as a broken relationship with a boyfriend or girlfriend, may be, from the perspective of older persons, just a normal experience of growing up.

Youthful suicide has been of particularly great concern in recent years as evidence mounts that youth suicide rates have increased dramatically during the past two decades. The five leading causes of death for adolescents and young adults (15–24) are, in descending order, accidents, homicide, suicide, cancer, and heart diseases (U.S. Department of Health and Human Services, 1985). While death rates due to accidents, cancer, and heart disease have declined since 1950, the death rate for homicide has doubled and that for suicide has almost tripled from 1950 to 1982. The youthful suicide rate has declined in recent years, but the trend for the past three decades is disturbing (McGinnis, 1987). The suicide rate in the age group 15–24 is about 9 per 100,000, while the rate for those 15–19 is substantially below that: .4 per 100,000 (Iga, 1981). These figures are in comparison with the national suicide rate of about 12 per 100,000.

The number of adolescent suicides, however, is probably underestimated to an even greater extent than is the number of adult suicides. Children under the age of 10 almost never commit suicide, and only occasionally are there suicides up to age 15. These figures do not mean that children as they are growing up do not on occasion "wish they were dead." That these situations do not end up as suicides seems partly the result of an incomplete formation of self-identity, status, and social roles that are endangered by certain situations. Moreover, childhood crises are usually temporary and there is seldom the long-range "brooding" that accompanies comparable adult crises.

Adolescent suicide rates appear to be high in those countries in which the overall suicide rate is high. In Japan, for example, Iga (1981) reported that 40 percent of male and 60 percent of female junior high school students indicated a wish to die, and an additional 24 percent and 23 percent respectively had entertained the idea of death occasionally. These figures, along with a recorded adolescent suicide rate that is twice that of the United States, was attributed to the system of examinations and higher education for Japanese youth (particularly the one-time entrance examination to college that does much to determine life-chances for Japanese youth) that creates stress, and the fact that suicide is a highly institutionalized adjustment mechanism in Japan.

It is difficult to interpret precisely the significance of increasing adolescent suicide rates. While the pressures of growing up are great, they appear to have always been so. Generational worries about such issues as nuclear holocaust, war (such as Vietnam), the use of illicit drugs, and violence both increase and abate and their relationship to youthful suicide rates is unknown. Undoubtedly, some will blame parents for a lack of moral upbringing and others will blame society for the cumulative effect of its ills, but there is a culture to suicide and that culture is predictive of suicide. We know, for example, that some factors increase the risk of suicide and those risk factors represent a combination of pressures that teenagers experience. Previous suicide attempts, depression, the use of illicit drugs, various kinds of deviant behavior, and suicidal behavior in the family all appear to increase the risk of suicide among teenagers (Gould, Shaffer, and Davies, 1990). What is needed is a theory with which to make sense of these factors.

SOCIOLOGICAL THEORIES OF SUICIDE

Several attempts have been made to formulate general theoretical explanations of suicide: social integration, degrees of social constraint, status integration, status frustration, community migration, and socialization. Each of these perspectives attempt to explain both the patterning of suicide (its distribution by sex, race, age, marital status, and so forth) as well as individual occurrences of suicide.

Social and Religious Integration

In his classic study, Durkheim (1951) stated that the suicide rate in any population could be explained not by the attributes of individuals in the population, but by the varying degrees of social cohesion or social integration. This view is still widely accepted today. Durkheim believed that suicide was related inversely to the stability and integration of social relations among people, whether religious, familial, or other. Many examples were cited by Durkheim as evidence for his thesis, including the lower suicide rates among Catholics compared to Protestants and of married persons as compared with the single, divorced or widowed, all of which he attributed to greater social integration. While he demonstrated his thesis in many ways, Durkheim has been criticized for not establishing a set of rigorous criteria for measuring "social integration" (Pope, 1976).

Maris (1969) has proposed a modification of Durkheim's thesis in terms of suicide varying inversely with the degree of social constraint on the individual. External constraint is low when an individual is not regulated either by other people or by shared ideas; an example is the higher probability of suicide when there is social isolation or role failure in the work situation of men, particularly older men.

Pescosolido (1990) argues that Durkheim's views on the impact of religion on suicide needs to be updated. The integrative effects of religion on suicide has been disputable in the United States (see Stack, 1982), although there is research that suggests that religion is, in fact, related to suicide in ways largely predictable by Durkheim's views (Brealt, 1988). A number of major changes have occurred, however, that may have altered the effect that religion has on suicide. There is greater diversity among Protestant groups than existed in Durkheim's day, and the process of modernization has produced a number of changes in the content of today's religious denominations. One such change is that the integrative strength of religious denominations varies from region to region in the United States. Religious networks, or patterns of interactions, may work differently in each area, so differences in the relationship between religion and suicide can be expected to vary by region.

Status Integration

Somewhat related to Durkheim's theory of social integration is another sociological theory which has attempted, through the use of suicide differentials in the United States, to link suicide to a particular pattern of status occupancy or the degree of status integration in a society (Gibbs and Martin, 1958 and 1964). Suicide varies inversely with the degree of status integration in the population. A person, for example, who occupies conflicting statuses, such as a young person occupying a high status occupation, will experience difficulty in forming and maintaining social relationships; their degree of social integration will thus be less, and their suicide rates will be high.

Fewer suicides occur in populations in which one status position is closely associated with others: as a result, members are less likely to experience role conflict and are more capable of conforming to the demands and expectations of others. In these situations, they are also more capable of maintaining stable and durable social relations with others.

This theory is extremely difficult to test adequately because of all of the possible combinations of status characteristics. Moreover, the theory is usually stated in such a way that all potential statuses are regarded as equally important. For example, one would have to cross-classify combinations of age, sex, race, occupation, marital status, education, and so forth, requiring a very large sample and extensive data on those many persons. The theory has received limited empirical support in a study of white women (Gibbs, 1982), but in a more extensive test, utilizing a more sophisticated measure of status integration with employment, household, marital, and residential statuses and using data from the 1970 census, Stafford and Gibbs (1985) were unable to detect much support for the theory. It appears that not all statuses are equally important and that the importance of some statuses changes over time (Stafford and Gibbs, 1988). Occupational status in particular has been an important status in recent decades, and marital status is more important for males than females.

Status Frustration

Several persons have tried to link suicide and homicide within a framework of different adjustments to status frustrations which produce aggression. According to Henry and Short's (1954) theory of suicide, which is the best known of these theories, the aggression is directed against the self, whereas in homicides, it is directed at others. Specifically, individuals become so frustrated through "the loss of status position relative to others in the same status reference system" that they feel like killing someone (Henry and Short, 1954: 26). In this view, suicide is a form of aggression against the self which is generated by some frustration. Henry and Short speculate that the failure to maintain a constant or rising position in a status hierarchy relative to others is an important frustration arousing aggression.

As evidence of this relation, they explain the lower suicide rate of married persons as compared with that of the nonmarried (divorced, widowed, or single) as being due to the fact that married persons are involved in a stronger relational system in which they must conform more to the demands and expectations of others. The degree of involvement with other people also explains the lower rates of suicide in rural areas, the high rates in the central part of the city, and the general tendency for rates to increase as persons age and have fewer close relations with others.

There is also evidence contrary to Henry and Short's theory. A major portion of their theory rests on the assumption that high-status persons commit suicide more often than low-status persons, but, as has been pointed out earlier, the reverse seems to be the case (also see Levi, 1982). Henry and Short also indicate that women should commit murder more frequently than men, according to their theory, but they admit that the evidence is certainly to the contrary. Still, the relationship between homicide and suicide is, in some theoretical ways, close, and few theories are able to account for these behaviors together.

Community Migration Theory

All of the sociological theories considered to this point emphasize, to different degrees, the integration of individuals with society. Stack (1982) has suggested that integration might be better conceived as a more local, rather than global, concept. Rather than looking at integration with "society," he suggests examining community integration and using migration rates as an index of ties to local communities. He reports that the higher the interstate immigration, the higher the suicide rate, even when controlling for other factors (Stack, 1980). High rates of migration indicate that persons are not developing long-range ties with people or places; they are thus not well integrated into their immediate surroundings. Stack also cites evidence to suggest that the suicide rate for females is particularly susceptible to this process, as might be expected. The male more often is moving to a job that will, in some sense, help off-set the loss of personal relationships; females, on the other hand, are frequently moving *from* something, not *to* something.

Socialization to Suicide

It may initially seem odd to try to think about suicide as a learned behavior. A successful suicide is, after all, not something that can be practiced more than once, and we often associate learning with repeated performances of an act. Yet, suicidal behavior can be learned just as any other behavior is learned.

Akers (1985: Chapter 24) has argued that while suicide can not have been reinforced in the past, acting suicidal can be reinforced. A person can come to commit suicide from a process of social learning.

> There are at least two learning paths to suicide: (1) learning to behave suicidally, but not fatally, and ultimately reaching the point of suicide, and (2) learning about and developing a readiness of suicide and completing it without prior practice in specifically suicidal behavior. (Akers, 1985: 299)

Part of the learning of suicide involves acquiring an acceptance of suicide under certain circumstances and a tolerance of others who commit suicide in those situations.

Suicide is frequently interpreted by persons who work with suicidal persons to be a form of communication, a "cry for help," especially when communicated by children and adolescents (Hawton, 1986). Responding to persons who attempt suicide in a way that does little to provide that help may reinforce suicide as a medium of communication. Certainly, persons can learn that suicide can form an alternative to situations and conditions they perceive to be hopeless. The learning of suicide involves the learning of norms and values that accept suicide as an alternative to some present condition, either for the good of the group (such as in altruistic suicides) or for personal reasons (as in egoistic and anomic suicides).

Part of the learning process involved in suicide can be noticed by observing that well-publicized cases of suicide by public figures, such as Marilyn Monroe, seem to trigger a chain reaction of successful imitations. With data from the United States and Great Britain between 1947 and 1968, Phillips (1974) found a direct relationship between the amount of publicity given a suicide in the newspapers and the suicide rate in those communities where the news stories were carried. The largest increases followed the death of Marilyn Monroe. In the month after her suicide, suicides in the United States increased by 12 percent, and in England and Wales by 10 percent. The suggestion of suicide appears to call into consciousness the option of suicide, giving it a certain legitimacy. In subsequent investigations, Phillips (1979 and 1980) reported a relationship between publicized murder-suicides and increases in airplane accidents, and between publicized suicides and motor vehicle fatalities, particularly of accidents involving only the driver of the vehicle.

Everyone has learned that one way to gain attention or to solicit help for some problem is to be sick or hurt. Techniques for gaining sympathy are learned early in life in experiences with parents and friends, and those techniques that prove to be the most successful—that is, those that receive the most social attention and reward—are retained by the individual. The progressive use of certain techniques, such as deliberately injuring oneself or feigning illness, or even actually producing illness through a controlled nonaccidental taking of drugs, may produce suicide. Obviously, the process is complicated. It involves learning techniques and rationales for using them, and it takes place over a long period of time. This is not to deny the real problems and crises that suicidal persons face; it merely recognizes that not all persons who face the same difficulties and crises turn to suicide.

In the final analysis, the validity of theories of suicide rest on the validity of statistics about suicide. The problems of those statistics are well known. In addition to the assumption that coroners function in the same way and that the same death will be recorded similarly regardless of jurisdiction, perhaps the most significant error made in the use of official statistics has been the same error in the theories themselves. This is the assumption that suicidal actions carry the same meaning throughout Western societies, with officials and theorists using the same definitions and with the officials among themselves using similar definitions. In other words, some uniformity is assumed when actually "an *official* categorization of the cause of death is as much the end result of an *argument* as such a categorization by any other member of society" (Douglas, 1967: 229).

Consequently, the view has been taken by two researchers that suicide notes have more significance for explaining why people commit suicide than do the official statistical analyses (Douglas, 1967 and Jacobs, 1967). One in every 16 suicides leaves a note (Schneidman, 1976). These notes can be analyzed in terms of the perspective of the suicide, what had been experienced, how these experiences were viewed, the social constraints that had restrained the individual from suicide, and how successfully or unsuccessfully they had been overcome.

Schneidman (1976) has challenged the use of suicide notes. He observes that:

> . . . that special state of mind necessary to perform a suicidal act is one which is essentially incompatible with an insightful recitation of what was going on in one's mind that led to the act itself. . . . In order to commit suicide, one can not write a meaningful note; conversely, if one could write a meaningful note, he would not have to commit suicide. (Schneidman, 1976: 266)

To date, because of the controversy regarding the validity and meaning of suicide notes, no systematic use has been made of them in developing or testing theories of suicide. Still, such notes undoubtedly contain important clues concerning an individual suicide.

The Suicide Process

Certainly most people in modern society are aware of the alternative to meeting (or solving) life's problems in the form of personal death. While only a relatively few persons actually commit suicide, many persons have undoubtedly contemplated it. But prolonged frustrations and crises by no means always result in suicide, and it is not clear as yet just why some persons do kill themselves. People face innumerable unpleasant crises in different ways; some become drunk, some seek religion, some make light of the situation, and others evade the issue or even consciously try to avoid it. In the most general sense, the person who commits suicide is generally unable to find a satisfactory alternative solution.

The suicide process involves the search for possible alternatives to deal with problems and then the final decision on death as being the only possible solution. Ringel (1977) has identified three principal components of the suicide syndrome: (1) constricting, or a narrowing of alternatives where problems are all-consuming and the person feels there is "no way out" except suicide; (2) a certain aggression that is directed toward the self, perhaps in the form of blaming oneself for an unfortunate accident or some other trauma in one's life; and (3) the presence of suicidal fantasies where the person constructs and mentally plays out suicidal acts.

Increased social isolation is a consistent feature of suicidal situations (Trout, 1980). Being able to identify this, as well as some of the other conditions that Ringel identifies, and then acting constructively on them, is a crucial dimension for the worker in the suicide prevention center. The process has been summarized in more detail by Jacobs (1970: 23). First, there is usually a long history of problems. Second, there is frequently a more recent escalation of problems such that new difficulties are added to those unsolved old ones. Third, the failure to solve new, as well as old, problems leads to increased social isolation from meaningful social relationships. Fourth, and finally, the individual experiences a profound sense of hopelessness given the nature of his or her problems, the increased sense of failure, and the termination of meaningful social relationships. This process is illustrated in the experience of an adolescent girl who recounted her suicide attempt:

> As a teenager, I basically had no friends, no interests at all. I stayed home. I felt very insecure around people, like I wasn't worthy to be around them. I'd skip classes; I'd be in the john crying. It finally got to the point where I begged my parents to let me quit. My grades were suffering terribly. So my father signed the papers and after that, it's all I heard, "You flukey, jukey bird," from my father because I quit school. Well, I loved my father, but he drank and beat my mother and would bust up the house. She left with us kids several times. Basically, I stayed in my room and I reached the point where I didn't want to be alive. (Stephens, 1987: 113)

Several factors may play an important role in this last stage of a suicide. Some persons become fixed on a goal such that it becomes an obsession. A person whose engagement has been broken, for example, may think that nothing else—parents, career, friends, hobbies, or other interests—is of any consequence. Many suicidal persons also display a lack of objectivity concerning their surroundings; they can see their problems only from their own point of view. Another factor is the interpretation of difficulty by the person. Circumstances such as economic losses or other difficulties that may seriously disturb one person may have little effect on another. All persons come to crises with different personal histories and experiences. The need for the object desired (for example, a relationship or promotion) or the loss of status may be interpreted by a suicidal person as destroying all future hope. The person learns attitudes toward self and others that contribute to depression and withdrawn behavior.

While many suicides represent a pattern of behavior that is increasingly withdrawn, other suicides are the opposite kind of behavior pattern. A study of female suicide attempters, for example, documents a number of suicide attempts by girls who exhibited a pattern of early rebelliousness and defiance at home and in the community (Stephens, 1987). Many of these girls were involved in drugs and sexual promiscuity, and minor delinquencies of other kinds. Like those girls who exhibited the withdrawn pattern, however, their adolescence was not a happy one, and suicide was the result of changes in the person's self-concept and attitudes.

Suicides may, on occasion, be definitely planned without being carried out to completion. Some persons may even have planned to kill themselves on a number of occasions, the final act being prevented by the removal of the original, precipitating cause, an alternative solution, or the reinforcement of some attitude opposed to self-destruction. Other persons may play a role in the potential suicide's definition of the situation.

> An individual comes to feel that his future is devoid of hope; he, or someone else, brings the alternative of suicide into his field. He attempts to communicate his conviction of hopelessness to others, in an effort to gain their assurance that some hope still exists for him. The character of the response at this point is crucial in determining whether or not suicide will take place. For actual suicide to occur, a necessary (although not sufficient) aspect of the field is a response characterized by helplessness and hopelessness (Kobler and Stotland, 1964: 252).

SOCIAL MEANINGS OF SUICIDE

Suicide is a social process that reflects substantial social meanings and antecedents. As is true in other types of behavior, suicidal behavior can be learned, and certain social rewards or advantages can be associated with the behavior. In their notes, suicides often have a substantive advantage in being able to point out to others that "you were wrong about me," or "see, I really do love you," thus portraying a presentation of self in the most dramatic manner possible. Studies of attempted suicide have often noted that the attempt may bring forth helpful reactions from others, reactions more readily obtained in a suicidal situation and most helpful once obtained (Stengel, 1964: 37). One reason for this, of course, is that most persons acknowledge that the voluntary taking of one's own life is deviant conduct. Since this norm is widely held and strongly maintained, the normal reaction when someone attempts to violate the norm is to help them with the problems that "obviously" drove the individual to commit suicide.

Suicidal actions are seldom "irrational." Rather, they are meaningful in the sense that something is fundamentally wrong with the situations of the actor and suicide appears to that person to be a rational, or the only, solution to those problems. The real *meaning* of a suicide is not necessarily that reached by friends, family, or the coroner. Outsiders may regard the suicide as "senseless," and "irrational," because the person was "distraught," "lost," or "depressed." In actuality, studies of suicide notes and diaries, as well as interviews with those who have attempted suicide, indicate suicidal actions through death and the dying process are means of transforming or affirming the essential, "substantial," self of the actor in many ways (Douglas, 1967: 284–319). Suicides generally have a number of patterns of social meanings that they have constructed for themselves and in relation to others, according to Douglas.

1. *Suicide as a means of transforming the soul from this world to another.* Such persons are motivated to end their lives by promises of life after death. Life is spoken of as a property of the individual, not as a part of the person. Suicide notes suggest these persons believe they are returning to God or to a new world.

2. *Other suicides seek to transform the substantial self in this world or another world.* With an act of suicide, the person tries to show others that he or she is quite different from what was thought of them in the past. Through suicide, people can show how committed, loving, trustworthy, and sincere they were; surely, only persons with those qualities would go to the extreme of suicide to prove this, or so the reasoning is sometimes found. These suicides believe they are giving up their lives to ask forgiveness of some wrong, or to prove what kind of people they were after all.

3. *Suicidal actions may be a means of obtaining revenge by blaming others for one's death.* In these suicides, it is important to identify clearly the person to be blamed and to make the connection between the suicide and that person's actions.

A young clerk, 22 years old, killed himself because his bride of four months was not in love with him but with his brother, and wanted a divorce so she could marry the brother. The letters he left showed plainly the suicide's desire to bring unpleasant publicity upon his brother and his wife, and to attract attention to himself and his miserable condition. In them, he described his shattered romance and told reporters to see a friend to whom he had forwarded correspondence. The first sentence in a special message to his wife read: "I used to love you; but I die hating you and my brother, too." This was written in a firm hand but as his suicide diary progressed, the handwriting became more erratic. Some time after turning on the gas he wrote: "Took my panacea for all human ills. It won't be long now. I'll bet Florence and Ed are having uneasy dreams now." An hour later he continues: "Still the same, hope I pass out by 2 a.m. Gee, I love you so much, Florence. I feel very tired and a bit dizzy. My brain is very clear. I can see that my hand is shaking—it is hard to die when one is young. Now I wish oblivion would hurry." The note ended there (Dublin and Bunzel, 1933: 294).

4. *Suicides may also represent escape patterns from the responsibilities of continued life.* In such cases, there is a high degree of restlessness, although the nature of this dissatisfaction is often not specified. The person feels "disgusted with life," or "useless." A married woman of 24, for example, left this suicide note: "I've proved to be a miserable wife, mother and home-maker—not even a decent companion. Johnny and Jane deserve much more than I can ever offer. I can't take it any longer. . . . This is a terrible thing for me to do, but perhaps in the end it will be all for the best" (Schneidman and Farberow, 1957: 43–44).

A divorced man of 50 left this suicide note:

To the Police—
    This is a very simple case of suicide. I owe nothing to anyone, including the World; and I ask nothing from anyone. I'm fifty years old, have lived violently but never committed a crime.
    I've just had enough. Since no one depends upon me, I don't see why I shouldn't do as I please. I've done my duty to my Country in both World Wars, and also I've served well in industry. My papers are in the brown leather wallet in my gray bag.
    If you would be so good as to send these papers to my brother, his address is: John Smith, 100 Main Street.
    I enclose five dollars to cover cost of mailing. Perhaps some of you who belong to the American Legion will honor my request.
    I haven't a thing against anybody. But I've been in three major wars and another little insurrection, and I'm pretty tired.
    This note is in the same large envelope with several other letters—all stamped. Will you please mail them for me? There are no secrets in them. However, if you open them, please seal them up again and send them on. They are to people I love and who love me. Thanks,
George Smith
(Schneidman and Farberow, 1957: 44)

5. *Another suicide pattern is exemplified by those persons who, after killing another person, for example, commit suicide.* In a Philadelphia study, about 4 percent of those who committed homicide then took their own lives (Wolfgang, 1958: 274). Other studies in the United States have shown an incidence of 2 to 9 percent in such suicides. In England and Wales, the proportion is much greater; each year about one third of all murders are followed by suicide, one in every 100 suicides being of this nature (West,

1965). Other crimes where there is a major element of personal disgrace, for example, embezzlement, are also occasionally followed by suicide.

Frequently, such homicides are perceived to be an act of love. Here is a note from a divorced woman who shot her two young sons and took a drug overdose.

> Dear Mommy and Daddy,
>   I'm sorry to do this to you'all but I can't take this life any more. I'm taking my boys with me. Please put one on my right and one on my left side. I love my boys and hope God forgives me and lets me be with them. I know in my heart that my boys will be with God. God please forgive me for I have sinned.
>   I love you, Mother and Daddy,
>   Eileen
>   [P. S.] Please dress the boys in blue. They look good in it. Please put me between them. I love them and want them to be in heaven, God's heaven. Please put with Monty Jay his night, night blanket, one that Mom made. Please put with Jeff his little tiger that he got on his first Christmas on my bed. . . .
> (Daly and Wilson, 1988: 79)

The despondent mother who wrote this note was found in a drug-induced coma and was revived, only to be tried and sentenced to life in prison.

As such notes suggest, a major problem in a suicide is its effect, in terms of guilt feelings on significant others—family, friends, employers, and so forth. Survivors of suicides frequently engage in a number of mechanisms to reduce or manage any guilt feelings that they may experience as a result of the suicide. Believing that the suicide was caused by outside, impersonal forces, believing that the suicide was inevitable, and that the suicide may actually have been a good thing are just some of the ways that survivors can deal with a suicide. These rationalizations also recast the meaning of the suicide into terms that survivors can not only understand, but live with.

It is difficult to substantiate elaborate interpretations of suicide processes from notes alone. Only about 12 to 15 percent of all suicides leave a note (Leenaars, 1988: 35) and generalizing from those that do is risky. We do not know whether those who leave a note are representative of all suicides.

Suicide, Mental Disorders, and Hopelessness

Persons who commit suicide are not generally "mentally deranged" or suffering from "temporary insanity." The idea that suicidal persons are mentally ill developed from the notion that "no one in his right mind" would take his own life. Yet, suicide, as pointed out earlier, may be entirely understandable given a certain set of circumstances. In fact, because suicide is, by definition, intentional, many mental disorders serve to control or inhibit suicide.

Mental disorders or developmental deficiencies that reduce the capacity for
planning and deliberation, and that prevent the psychological organization of
sequential actions, greatly reduce the potential for suicide. (Litman, 1987:
90)

It is difficult sometimes to determine the circumstances surrounding
a suicide, and reports from family or friends about a suicidal person's state
of mind are not necessarily clinically correct. Severe depression seems to be
the most common form of mental disorder associated with suicide. The
percentage of psychotic disturbances, although not large, is great enough to
account for concern, in most cases of attempted suicide, since there may be
present some severe mental disorder which will lead to a repetition of the
attempt unless the disturbance is discovered and treated. Actually, most
suicides are rationally planned and carried out with no more evidence of
mental disorder than would be found in so-called normal persons. The goals
sought by suicides, no matter how exaggerated, generally are real goals; the
personal losses suffered are real losses and are usually not the product of
psychotic hallucinations or delusions having little or no basis in reality.

In Maris' (1981: Chapter 8) study of all officially recorded suicides in
Chicago from 1966 to 1968, it was found that persons who attempted and
completed suicide had a higher percentage of psychological problems than
did a "control" group of persons who died from natural causes. In fact,
about 40 percent of the persons who completed suicide had been hospitalized
sometime in their lives for mental disorders, compared with about 50 percent of those who attempted suicide (but failed) and 3 percent of the persons
who died of natural causes. Furthermore, while depression was related to
suicide in this study, a sense of hopelessness was more important than
general depression. An English study also concluded that certain mental
states were important in adolescent suicide attempters.

The main feelings that appear to precede attempts by adolescents are anger,
feeling lonely or unwanted, and worries about the future. A sense of
hopelessness is a major factor distinguishing depressed adolescents who
make attempts from similar adolescents who do not. (Hawton, 1986: 99)

For these reasons, it would not be accurate to say that all suicidal
persons are suffering from a mental disorder. The relationship seems to
depend on such factors as social stress and one's ability to cope with that
stress.

The precise roles of depression and hopelessness in creating suicidal
thoughts (ideation) or behavior is of current concern. Some reports indicate
that it is really hopelessness that predicts suicide ideation, while others
report that depression is the more important predictor (Rudd, 1990). In
either case, however, it appears that negative life stress, as indicated by such
events as significant loss (the death of someone close) or some sort of failure,
may preceded both depression and hopelessness (Rudd, 1990).

Such experiences are not confined to older persons only. One study of suicide ideation among rural adolescents reported that family characteristics were most predictive of the thought of suicide (Meneese and Yutrzenka, 1990). Families that were "disorganized" were associated with such ideation. These families were without or had little structure concerning general rules in the family, responsibilities of various family members, financial planning, and the like. Stressful life events can impact persons of any age, although the nature of those stresses can vary (home, school, employer, spouse) depending on the age and circumstances of the individual.

Preventing Suicide

The identification of social forces that ultimately produce suicide may eventually point to preventive remedies, largely through changes to communities and entire societies. If suicide is "caused" by the conditions of modern life and stresses placed on individuals in complex, industrialized societies, broader-based remedies should be adopted. In the meantime, most preventive efforts continue on an individualistic basis without necessarily a coherent and valid sociological theory to guide them. Without such an agreed upon theory, policy implications for reducing suicide will be different. For example, the implications for suicide prevention are quite different depending on the theory one subscribes to. A perspective based on control theory (see Chapter 4) that emphasizes social and/or status integration will have different implications than an orientation based on learning theory that emphasizes socialization to suicidal situations and social isolation of suicidal persons.

In spite of the lack of agreement concerning a theory of suicide, a large number of community agencies have been set up in the United States and various other countries to prevent suicides by counseling and offering other assistance. Much of this work was pioneered after World War II. In Vienna, Austria, most of the suicide prevention work is carried out through Caritas, a Catholic organization, which also works with a preventive clinic for attempted suicides at Vienna hospitals. A special suicide prevention telephone service is maintained which makes available persons with whom one can converse about problems. If desired, a social worker can come to the caller's home. Great Britain, too, has a network of volunteer organizations concerned with suicide prevention. The Good Samaritans, established in 1953, combine religion and psychotherapy in counseling suicidal persons.

In the United States, most large communities have some sort of suicide prevention center, usually associated with community mental health clinics. The first such center in the United States was founded in 1958 in Los Angeles (Farberow, 1977: 543). Numerous organizations are now found in most states, dealing with suicide prevention under a variety of names, such as Suicide Prevention Service, Call-For-Help Clinics, Crisis Clinics, Crisis Call Centers, Rescue, Inc., Dial-a-Friend, and Suicides Anonymous. A Center for Studies in Suicide Prevention was established in the National Institute of Mental Health in 1967.

Much of the initial work in suicide prevention centers is done by telephone. Most large cities have telephone numbers of these centers prominently displayed in telephone directories and on public service announcements in the mass media. Efforts are made to encourage persons to call the numbers, to talk about their problems, and to come to the center, if possible. Suicide prevention centers serve as crisis intervention points; that is, the services they offer are geared toward the individual's immediate, rather than long-term, needs, which may require more intensive counseling and advice. They offer short-term services 24 hours-a-day, usually in the form of telephone counseling directed at the situation. As part of this effort, workers answering the phones establish rapport and maintain contact with the caller. The worker then attempts to determine the potential danger in the situation—that is, how close the caller is to committing suicide. Someone who indicates they may kill themselves in the future is further away from that act than one who has a plan and the means available at the moment of calling. The worker must also evaluate the resources available to the person, such as the availability of friends, to help the situation. The final consideration is setting up a treatment plan of some kind, attempting to get the suicidal person to commit him or herself to future activities that will diminish the probability of their suicide.

Persons who call suicide prevention centers fit no convenient stereotype: they present a variety of problems, all of which may require different solutions. As is expected, the suicide rate of callers is substantially higher than that for the population as a whole, being estimated as perhaps as high as 1,000 per 100,000 compared to about 12 or 13 per 100,000 for the general population (Litman, 1972).

One study of callers to the Los Angeles Suicide Prevention Center showed that most of them indicated they were depressed, and about two thirds of them were contemplating suicide, while the others were in the act of suicide when they called (Wold, 1970). Two thirds of the callers were women, and about half had a history of suicide attempts. More than half of them called the center themselves, while the others were represented over the phone by family members, friends, or professional persons. The study reported that the ratings of suicidal potential were high for about 20 percent of the callers, moderate for 40 percent, and low for the remaining 40 percent. In comparing this sample with a group who actually committed suicide even though they had contact with the center (many of those who call are prevented from their acts), it was found that the profiles differed in a number of respects. The proportion of males to females was reversed: two thirds of the completed suicides were male, one third female, and depression was more marked in those who actually committed suicide.

A number of persons who contact such agencies are chronic callers. One study of 67 such chronic callers discovered that 51 percent were drug or alcohol dependent. "The prototype of the chronic caller to the Suicide Prevention Center is a divorced female in her late 30s who is alcohol or drug dependent and intermittently suicidal" (Sawyer and Jameton, 1979: 102).

It is difficult to evaluate precisely the effectiveness of suicide prevention centers, since many callers who do not commit suicide are not found after the call. Several studies have been conducted evaluating the impact of a suicide prevention organization in England, but the results have been mixed (Hawton, 1986: 133–134). Some studies have reported such agencies have been associated with declines in the suicide rate in the area served by the agency, while other studies have reported no change. A study from the United States compared suicide rates in communities that did and those that did not have a suicide prevention center in North Carolina. It was concluded on those criteria that the centers had a minimal effect on suicide rates based on a before-center and after-center comparison of suicide rates (Bridge, et al., 1977). There are those, however, who feel that even if one person is prevented from committing suicide, the services are important and valuable.

One other evaluation is worth mention. Miller (et al., 1984) compared suicide rates between counties in the United States that had suicide prevention centers with those that did not. There was little overall difference in those rates. However, the authors reported that there were significant differences in county suicide rates of young girls (up to 24 years of age). Examining their suicide rates over time, the rates had declined by 55 percent in those counties that had suicide prevention centers and had increased (by an estimated 85 percent) in those counties that had no such agency. The importance of this finding must be assessed by realizing that young girls are by far the most frequent callers to suicide prevention centers.

The formalization of suicide prevention through these centers is an example of how formal social controls serve as a back-up to informal controls. Most persons refrain from committing suicide, if they consider it seriously, because of their strong moral or religious convictions opposed to self-killing. Some persons attempt suicide for reasons other than self-death, for example, to get attention for some problem that might otherwise pass unnoticed or to get revenge on another. Still others are serious in their suicidal behavior but do not have the resources that are available to them in these centers. Suicide prevention centers offer an alternative to the person who does not know where else to turn for help with a serious problem.

SUMMARY

Suicide is strongly condemned in most Western societies but its condemnation has not been universal and for all time. In some societies, and at some times, suicide has been permissible, and even honorable, under certain circumstances. Strong, negative, public attitudes in the United States are associated to religiosity, with younger, better educated males being more accepting of suicide under certain circumstances.

Estimates of suicide often differ from one another because of differences in the way in which deaths are recorded and how the cause of death is determined. As a result, it is not possible to determine accurately how many suicides occur. The suicide rate in the United States has, as far as can be determined, remained about the same over the past several decades at about 12 per 100,000. There are many more attempted suicides than completed ones, perhaps by as much as 20 times. Attempters are more likely to be young women.

Suicide is more common among men than women, and more common among whites than African Americans. Generally, suicide rates increase with age, although black suicide rates peak in the twenties and thirties and decrease in the older age categories. Adolescent suicides are relatively rare (70 percent of all suicides are of older, white males) but the adolescent suicide rate has been increasing during the past three decades. Married people have a lower rate than nonmarried people, and people who consider themselves to be strongly religious (regardless of denomination) have lower suicide rates than those who do not. Generally, suicide rates are higher in the lower classes, but some middle- and upper-income occupational categories have high rates.

There are different types of suicide. Altruistic suicides are committed by people who do so for the perceived benefit of others. Egoistic suicides occur for a variety of personal reasons, including failing health, financial losses, or difficulties in a social or interpersonal relationship. Anomic suicides occur when people lose their purpose for living or feel cut off from their social groups' norms. They may also feel considerable frustration in the face of situations where the values of the group become confused or break down. In the United States, egoistic suicides are the most common.

Sociological theories of suicide have traditionally emphasized the individual's integration with the larger society or social group. Theories of social integration and status integration are examples of such theories. Status frustration and community migration theories, which develop from a social disorganization perspective and a socialization perspective, are more social psychological viewpoints. The usual suicide process involves increasing social isolation and identification of suicide as a possible solution to problems. The meanings attached to a suicide are socially determined and suggest that suicide is an important means by which to affirm or transform a personal identity. People who commit suicide are not necessarily mentally disordered.

The prevention of suicide is seen as an important social responsibility in most communities, and the growth of suicide prevention centers reflects that concern. These agencies are now found in most large cities and in many smaller ones as well. Most of the work of these centers is done by telephone, although personal contact for follow-up of these calls is done wherever possible. There is some dispute about the effectiveness of suicide prevention centers because the identity of the caller is often not known and follow-up can not be performed; plus if no suicide results after a call, some other factor might have accounted for that fact. Some studies, however, have detected some reduction among adolescent suicides that seem to be attributed to the existence of a suicide prevention center.

SELECTED REFERENCES

Akers, Ronald L. 1985. *Deviant Behavior: A Social Learning Approach*, 3rd ed. Belmont, CA: Wadsworth.

> A social learning perspective applied to a number of forms of deviance, including suicide. This selection shows how suicidal behavior can be socially learned, just like other kinds of behavior

Hawton, Keith. 1986. *Suicide and Attempted Suicide among Children and Adolescents*. Beverly Hills, CA: Sage.

> A study of English suicide attempters by one of the best known clinicians in the field. The book also contains information about suicide and attempted suicide in other nations.

Headley, Lee A., ed. 1983. *Suicide in Asia and the Near East*. Berkeley: University of California Press.

> A collection of papers about suicide in a number of countries in Asia and the Near East. Valuable discussions about the statistical picture in each country are presented, as well as the meaning of suicide in these countries.

Maris, Ronald W. 1981. *Pathways to Suicide: A Survey of Self-Destructive Behaviors*. Baltimore: Johns Hopkins University Press.

> A study of suicide in Chicago and a review of previous literature on the subject that finds some evidence that personal pathology may be important in the suicide process.

Stack, Steven. 1982. "Suicide: A Decade Review of the Sociological Literature." *Deviant Behavior*, 4: 41–66.

> A systematic review of the empirical and theoretical literature on suicide. Different explanatory perspectives are identified and evaluated.

Physical Disabilities

A wide range of physical disabilities and impairments result in such severe stigma and discrimination that they may be regarded, and are so regarded here, as being deviant in many, if not most, societies. Goffman (1963) has referred to the blind, the deaf, the mute, the epileptic, the crippled, and the deformed as classic examples of conditions that elicit social stigma. There are others—the mentally retarded, the obese, those afflicted with cerebral palsy, and those who stutter severely.

A large number of persons in the United States have these severe physical disabilities. It has been estimated, for example, that three out of every 100 children born in the United States and Great Britain will be diagnosed as mentally retarded sometime in their lives, and that between 20 and 25 percent of these children will be so severely retarded that they will be diagnosed at birth or in early infancy (Edgerton, 1979: 1). It has also been estimated that there are about two million people in the United States who can neither speak nor hear, and another eleven million persons who have a severe hearing impairment (Higgins, 1980: 33). Moreover, there appear to be about 1.5 million epileptics and more than three million people suffering from orthopedic impairments (Scott, 1969: 42). There are also more than one million blind or partially sightless persons, of which probably 50,000 are totally blind (Scott, 1980: 8).

It is important to distinguish among the terms "impairment," "disability," and "handicap." *Impairment* refers to the loss of some ability usually because of a physical reason. The loss of the use of the optic nerve, a portion of the brain that controls talking, a limb, or hearing from a physical condition present at birth are all examples of impairments. *Disability* refers to the loss of function that accompanies impairment. While impairment refers to a physical condition, disability describes the nature of the loss to the person—eyesight, speech, use of limbs, social relations, and so forth. The term *handicap* refers to the limitation on normal activities of self care and mobility as a result of some impairment (Thomas, 1982: 3–8). Thus, while impairments are physical conditions, the terms disability and handicap have social and behavioral connotations. They refer to one's inability to meet, among other things, social responsibilities and obligations.

Disability is not illness. The sick are exempt from many social role responsibilities, such as normal occupational and familial ones, and they are not considered responsible for their illness. Sick people are expected to want to get well and to seek competent help toward that end. In these respects, sickness has quite different attributes from the more permanent aspects of disability. "The label of sickness, although it may imply severity, also implies a temporary condition which can, through some kind of intervention (usually medical), be made to disappear. It is only the label of disability that carries the connotation of permanency and irreversibility regardless of the degree of severity of the condition" (Safilios-Rothschild, 1970: 71).

There are, of course, degrees of physical disability and impairment. In all probability, the more extreme cases are more likely to be stigmatized to some degree. For example, there is great variability in the extent of blindness, the seriousness of crippling or facial disfigurement, and the severity of mental retardation. While the legal definition of blindness has been put at a 20/200 vision level, Scott (1969: 42) has severely criticized what he terms this arbitrary level as being insensitive to most of the important determinants of the person's functional vision: such a standard "lumps together people who are totally blind and people who have a substantial amount of vision." For this reason, people who are labeled as blind are very diversified, and the "blindness population" is a heterogeneous and shifting one. Similarly, deafness is a matter of degree ranging from those with some slight impairment to those totally deaf. Obesity, too, is a matter of degree and definition, with one study defining the condition as 30 to 40 percent beyond "normal" weight (Cahnman, 1968). Persons can be overweight in varying degrees, but it is the highly visible fat person who is usually termed "obese" and who is likely to suffer discrimination as a result (Millman, 1980). Likewise, a stutterer may have severe and constant speech impairment, or may stutter only under certain conditions. With respect to mental retardation, the range is extremely varied: the American Association on Mental Deficiency lists classifications based on a "normal" IQ (intelligence quotient) of 90–100, of borderline as 70–84, and of various degrees of retardation below 70. About 85 percent of the retardates fall in the category of "mildly retarded" with IQs between 55 and 70 (Edgerton, 1979: 4–5).

DISABILITIES AND THE IDEA OF DEVIANCE

Disability can be regarded as deviant from a number of perspectives. A person with a disability, because of the limitations imposed on the disabled person's range of activities, but chiefly because of the self-reaction to the person's disability, may be thought to occupy a deviant status, much like that of a minority group member. Minority group members may experience prejudice and discriminatory treatment from others. Increasingly, individuals who are physically disabled have defined themselves as members of minority groups and are now organizing for more public awareness of their problems and their right to define their own role in society (Deegan, 1985). Disabilities, whether visible as in the case of the physically disfigured, or not as in the case of the mentally retarded, then, can be considered deviant in that they depart from some normative conception of "normal" and they, the disabled, experience sanctioning processes that lead to social stigmatization.

A handicap is "an imputation of difference from others, more particularly, imputation of an *undesirable* difference. By definition, then, a person said to be handicapped is so defined because he deviates from what he himself or others believe to be normal or appropriate" (Friedson, 1965: 72). Many of the disabled violate the norm of physical well-being and "wholeness." Goffman (1963: 126–130) talks about the importance of *identity norms* which depict "ideal" persons, or shared beliefs about what persons ought to be or look like. In contemporary American society, these ideals are found among persons who are "young, married, white, urban, northern, heterosexual, Protestant, father of college education, fully employed, of good complexion, weight, and height, and a recent record in sports" (Goffman, 1963: 128). Persons with physical disabilities are not deviant because of anything they have necessarily done, but because others impute to them an undesirable difference of not fully matching up to these kinds of expectations. It is in this sense that conditions, as well as behavior, can be considered deviant.

The central difference between physical disabilities and other forms of deviance is that a disability is not a behavior; it is a condition, one over which the individual has no control. While criminals and drug addicts are said to choose these forms of deviance, the visibly physically handicapped, the physiologically obese, and the mentally retarded are accorded a deviant status by factors outside their immediate control. Their statuses are ascribed, not achieved. Physical disabilities are not simply biological facts; they are social constructs as well.

Persons with physical disabilities are often sanctioned and stigmatized. Many have observed that persons with physical disabilities are discriminated against by others. For this reason, a federal act, the "Americans with Disabilities" Act was signed into law in July, 1990. It applies to persons with disabilities and defines as disabled anyone who has a mental or physical impairment limiting "some major life function." It protects the disabled from discrimination in employment, public accommodations, transportation, and telecommunications. The bill, like other civil rights legislation, includes a variety of sanctions for violators. The bill follows civil rights legislation by permitting victims of employment discrimination to seek back pay, reinstatement, and attorney's fees. The concern over providing legal remedies to persons with disabilities reflects the wider concern over the stigmatization of persons with physical handicaps.

Not everyone considers persons with physical disabilities to be deviant. The reaction of others seems to depend on certain characteristics of the person, disability, and social situation. But insofar as disabilities entail departure from social expectations, social stigma is always a possibility for persons with disabilities (Stafford and Scott, 1986). The reactions of others are often very obvious to the handicapped person. A hemophiliac boy, who must not participate in some physical activities and who sometimes must use crutches because of the soreness in his legs, reported:

> I'll be on my crutches sometimes and I'll be walking down the street, and people will get off away from me and walk around me instead of walking beside me because some people think that they might catch something. Nobody ever sat down and told them that they shouldn't be afraid of people with handicaps, that we are people just like them, except that we got a problem. (quoted in Roth, 1981: 93)

Disability as Deviant Status

Disability is a socially defined category of behavior. It is a master status, and, as such, it overrides all other statuses. Those with physical disabilities often experience "a personally discreditable departure from a group's expectations" (Becker, 1973: 33). Groups in general expect a person to have two legs and two arms, ears and eyes that function, and to be able to carry on daily activities in a normal fashion. Persons with physical disabilities do not fit the normative expectations of society. In Sagarin's (1975) words, such persons are "disvalued."

Being disabled is also personally discreditable. Disabled persons find their identities are defined in terms of their handicap. They are no longer individuals, but "cripples," or "deaf-mutes." They become discredited and stigmatized. For this reason, some disabled persons have difficulty securing jobs (even when they could perform the work physically) and engaging in normal social interaction with others. Some may even have to pay for companionship in the form of prostitutes and hostesses in dance halls.

Some persons have argued that deviance and disability, like sickness, differ primarily in the extent of responsibility imputed to individuals for their condition. Deviants can choose to enter a deviant role; the disabled do not choose their disability. These persons argue that deviance is willful and subject, therefore, to the individual's control, and that disability does not fit this model (Haber and Smith, 1971). Thus, while researchers claim that the drug user or criminal chooses to commit the deviant act, this option is not open to the blind, the deaf, or the crippled. Some disabled persons may be seen as being more responsible for their conditions than others, such as car accident victims who were at fault or incurably ill persons whose procrastination in seeking medical care has aggravated their conditions. The obese are rather universally judged, and often unfairly so, as responsible for their overweight, and little sympathy is given them as in the case of the blind or paraplegics. The obese, moreover, tend to internalize that viewpoint, often feeling guilty about their condition.

As we have seen with respect to other forms of deviance, it is not unusual that persons who come to engage in deviant acts take on deviant identities and are sometimes regarded by others as deviant. For example, both the amateur and professional criminal are deviants, but we frequently regard the professional offender as being "more deviant" than the amateur. Participation in deviant acts can lead to deviant identities and statuses. Persons who consume large quantities of alcohol come to be regarded as alcoholics; persons who take narcotic drugs over time come to be addicts; and persons who commit unusual or situationally inappropriate acts are sometimes regarded as mentally ill as a result. A major difference between the physically disabled and these others is that the disabled person is given a deviant status without having committed a deviant act. In this instance, deviant acts do not lead to deviant statuses; the deviant status is conferred on the individual for other reasons. Many of these reasons are found in the concept of the sick role.

The Sick Role

Attitudes toward the physically disabled relate to our conceptions of the "sick role." According to Parsons (1951: 428–479), the sick role consists of two interrelated sets of exemptions. Individuals who are defined as ill are freed from certain obligations and responsibilities due to illness. The illness is not considered their "fault," nor are they expected to improve due to their motivation alone. Sick persons are viewed as persons whose capacity to function normally is temporarily impaired. For these reasons, sick persons are relieved of normal family, occupational, and other duties. In exchange for these exemptions, however, there are certain expectations of persons occupying the sick role. For example, sick persons are expected to define for themselves the sick role as "undesirable" and to do everything within their power to facilitate recovery. In the case of physical illness, persons should obtain and cooperate with medical help, usually a physician. Following doctor's "orders" means precisely that: the advice of a physician on medical matters reflects not only the recommendation of one supposed to know these conditions but also society's expectation that ill persons will try to move to more conventional roles.

So stated, there are four elements of the sick role: (1) an attribution of nonresponsibility toward the individual for his or her condition; (2) an exemption from normal role obligations; (3) a recognition that being sick is undesirable in spite of the benefits of these role exemptions; and (4) an obligation to seek help for the sickness. Thus, persons in the sick role are not normally blamed for their condition and are excused from many obligations on the condition that they attempt to move from the sick role to other roles as quickly as possible. What it comes down to is that persons are given some rewards when sick, but are not supposed to enjoy them. In the final analysis, individuals and groups expect sickness to be a temporary condition.

"Dos and *Don'ts"* for the Sick Role

Dos	*Don'ts*
1. Follow doctors' orders	1. Ignore the advice of health professionals
2. Try to move away from the sick role (try not to perform this role any longer than necessary)	2. Enjoy too much the privileges of the sick role (e.g., release from occupational, educational, and family obligations)
3. Tell others that you do not wish to occupy the role any longer	3. Tell others you wish you could be ill for a longer time
4. Have a legitimate claim on the role before you state a desire to perform it	4. Fake a claim just to enjoy the privileges of the role
5. Conceive of the role as temporary	5. Think that you will be able to enjoy the role on a long-term basis
6. Assume the role because you have no choice	6. Make yourself sick just to be exempt from obligations

The sick role is often legitimated by physicians. There are some persons who seek the sick role and do not obtain it since doctors may refuse to "certify" a person as sick. Those who are successful in obtaining this legitimation are "permitted" to enter the role; those who are denied entry must assume normal roles or face the social stigma of someone who is enjoying the benefits of the sick role illegitimately (Wolinsky and Wolinsky, 1981). The ambivalence of many persons about physical disabilities thus reflects their views that the disabling condition is both unavoidable and, at the same time, undesirable. While illness is defined as a departure from the normal and the desirable, it is not regarded as reprehensible in the same way as is either sin or crime, the sick person neither being blamed nor punished as are those considered to be "sinful or criminal." "So long as he does not abandon himself to illness or eagerly embrace it, but works actively on his own and with medical professionals to improve his condition, he is considered to be responding appropriately, even admirably, to an unfortunate occurrence. Under these conditions, illness is accepted as *legitimate* deviance" (Fox, 1977: 15).

SOCIETAL REACTION AND THE AMBIVALENCE OF DISABILITY

Appearance norms govern conceptions of "ideal" appearance—the size and shape of human bodies and features of those bodies (i.e., functioning body parts). What constitutes a deviant condition is highly variable from culture to culture. In some societies, slimness may be preferred, while in others there is more latitude on size of bodies; in some societies, infirmity is well tolerated, while in others all visible body parts should function or the person is thought to be deviant.

The term "deviant" is not used to refer to a process of conscious, behavioral norm violation but rather, in a more general sense, to refer to conditions that persons feel "ought" to exist. Few regard persons with physical or mental disabilities as voluntary deviants, but these persons are nevertheless sanctioned by others who hold implicit conceptions of "normal" physical characteristics or functioning. Persons who, even involuntarily, violate such conceptions can be considered deviant for purposes of this discussion.

It is this voluntary nature of the deviance, then, that differentiates the physically disabled from other deviants. Deviance is generally considered undesirable, so much so that negative sanctions are applied to those who violate the social norm at issue; but the degree to which it is voluntary is one variable that affects the degree of tolerance of the social audience to a deviant act. The various forms of disability considered here fit this general framework to different extents. The totally blind who have no reasonable hope of recovery, the crippled, and the mentally retarded fit best. The obese and stutterers fit to a lesser extent because there is at least a reasonable chance for them to move from this role to other, more conventional roles, even though the cause of the disability or condition may be outside the individual's control. It is the physically disabled's inability to move from this role, then, that constitutes their "deviance." But unlike other deviants, it is a deviance of status, not acts. It is their condition, not their behavior, that leads to their being stigmatized and considered deviant.

Some kinds of physical disabilities appear to have been socially rejected throughout human history and in almost all cultures (Myerson, 1971), but the concept of disability is essentially culture-bound. Not all impaired persons are regarded as disabled; images of disability vary from culture to culture and from age to age. A facial disfigurement in one country results in stigmatization of the individual. In another, however, it may be regarded as a sign of beauty or even supernatural powers. American adolescents with some physical blemishes may be rejected by their peers, but in some societies scarring marks or tattoos are done for "beauty." The physical accompaniments of old age and obesity may be regarded negatively in contemporary American society, but may be positively evaluated in other societies, such as those where advancing age is regarded as the ultimate stage of human wisdom and power, and where, for example, fat women are encouraged to be so as married women.

The study of physical disabilities illustrates an important conceptual matter in the study of deviance generally. As was seen in Chapter 1, the concept of deviance is best understood by reference to norms, whose violation is called deviance. The discussion of physical disabilities, however, indicates why it is often difficult to know exactly what the norms are. It is for this reason, in fact, that there is considerable ambivalence with respect to physical disabilities. Persons with disabilities are regarded both positively and negatively at different times.

Disability is a social rather than a mere behavioral or biological fact, and it is defined in terms of societal reaction. The extent to which a physical disability is deviant depends on the cultural conception of disability and is reflected in the kind of reaction the disability elicits. The role of cultural conceptions of disability can be illustrated with the blind, the mentally retarded, the visibly physically handicapped, and the obese and persons with other eating disorders. In each of these categories, the importance of cultural stereotypes and societal reaction is quite clear.

The Blind

As we have seen, it is impossible to estimate accurately the number of blind persons in the United States because of the divergent definitions in terms of the degree of blindness. There are also problems of locating and identifying blind individuals. For these reasons, estimates of the number of blind persons range from somewhat less than half a million to more than a million and a half, including persons whose sight is sufficiently impaired to be classified as legally blind, although many included in the higher figure are not so classified (Koestler, 1976: 46). In fact, three categories of blindness are often used: (1) totally blind—the total absence of any light or image perception; (2) legal blindness—central visual acuity of 20/200 in the better eye with corrective lenses and the central visual field so restricted that the individual can see only objects within a 20 degree arc; and (3) functional blindness—inability to read ordinary newspaper print even with perfectly fitted glasses (Koestler, 1976: 45–46). Because these distinctions are somewhat technical, most persons refer to "blind" persons as those who are totally or nearly totally blind (that is, the absence of any light) and the "visually impaired" as those with some lesser but serious visual deficiency.

The blind have long been one of the most conspicuous groups of disabled persons. Since much of human expressive behavior centers on the eyes, it is extremely disturbing to some persons when they confront blind people. Blind people do not give out the same cues as to their psychological or emotional state. They provide less information to others. Various behavior mannerisms increase the social visibility of the blind, such as the use of odd postures, rocking of the head or tilting it at odd angles, touching objects in a groping manner, and the use of distinctive paraphernalia, such as thick glasses, white canes, and seeing-eye dogs.

Historically, the blind have traditionally been relegated to inferior roles as outcasts and beggars; often giving alms to blind beggars was treated as a means of gaining religious merit. In England under Queen Elizabeth I, the blind were grouped with paupers, orphan children, and mentally disordered persons for purposes of poor laws and charities. It was not until the eighteenth century that specialized welfare services were developed for different types of dependents that distinctions were made.

With the urbanization and industrialization of the nineteenth century, the blind experienced a different type of stigmatized reaction.

> The humanitarianism and organized philanthropy of the second half of the century in England and the United States introduced the conception of character defect as a middle-class explanation of pauperism, which had the concomitant result of associating physical defect with personality weakness and lack of self-resolution. . . . At the same time, the humanitarian movement was responsible for the special regard for the blind which gave them a more secure relief status than other dependent groups (Lemert, 1951: 114).

The humanitarian movement also was responsible for attempting to restore the self-confidence and self-reliance of blind persons by creating special schools for the blind. These schools promoted ideas that blind persons could be educated and employed as productive members of society. One blind man reported that being able to attend a blind school provides an important source of mobility.

> [Being blind] broadened my life-style because all of a sudden, I went to the blind schools, I could be free from my family. I didn't have to have my parents say yes or no every time I wanted to go into the city, I could just go. I had a little money in my pocket that I would hustle up by washing dishes in the school, so I would go and do my trip on weekends. I learned my own mobility and my own social skills. (quoted in Roth, 1981: 184)

Today, there are various public stereotypes of the blind, which include beliefs of "helplessness, docility, dependency, melancholia, and serious-mindedness" (Scott, 1969: 21). One writer has talked about a number of recurring themes in literature, historical records, mythology, and folklore concerning blindness and blind people, including the ideas that blind people are: deserving of pity and sympathy, miserable, helpless, useless, compensated for their lack of sight, being punished for some past sin, maladjusted, and mysterious (Nonbeck, 1973: 25). Such ideas form the core of stereotypes about blind people, stereotypes that represent in many instances negative images that reinforce the stigma of blindness. When blind persons encounter sighted persons, the nature of the blind persons' stigma will influence the interaction. Each party in such an interaction may engage in an avoidance reaction.

> The effects of these reactions on a blind man are profound. Even though he thinks of himself as a normal person, he recognizes that most others do not really accept him, nor are they willing or ready to deal with him on an equal footing. . . . The stigma of blindness makes problematic the integrity of the blind man as an acceptable human being. Because those who see impute inferiority, the blind man can not ignore this and is forced to defend himself. (Scott, 1969: 25)

The Mentally Retarded

Mental retardation has to do with limited intelligence. Mental retardation identifies three components: (1) mentally retarded people do not learn as quickly or as much as nonretarded people; (2) mentally retarded persons do not retain or store as much information as persons who are not retarded; and (3) the mentally retarded do not abstract very well, thereby limiting the use of the information they do have (Evans, 1983: 7). Beyond these simple generalizations, there is disagreement even among experts.

The exact number of mentally retarded persons in the United States is not known, nor is it possible to determine exactly which persons could be so classified in view of the wide variations in the meaning of intelligence and how it is currently measured. There is much literature on the nature and meaning of "IQ," and that literature suggests that measures of IQ as found in IQ tests are both "relative" and "fallible." So, in the absence of reliable measurement devices and an agreed upon definition of which IQs are considered "retarded," there exists only educated guesses as to the number of persons who might reasonably be called mentally retarded.

Of all human assets, the ability to think (plan and arrange one's life, manage one's affairs) is probably the greatest, as well as the most cherished. Those who are found wanting in these attributes, because they lack the mental capacity, experience one of the most devastating of all stigmata, at least in modern Western cultures. No other stigma seems as general as that associated with mental retardation in the sense that the mentally retarded person is thought to be lacking in *basic* competence. Other disabled persons may have to refrain from some activities but they retain some competencies, limited though they may be. The retarded person, on the other hand, is, by definition, incompetent to handle any of his or her affairs. To make matters worse for the mentally retarded, this is a permanent condition. "As everyone 'knows,' including the ex-patients, mental retardation is irremediable. There is no cure, no hope, no future. If you are once a mental retardate, you remain one always" (Edgerton, 1967: 207).

Most mentally retarded persons are not physically identifiable as retarded. It is only in certain situations that their retardation may become evident—when faced with a difficult social situation, in trying to carry on a conversation with another of normal intelligence, or in other circumstances that bring them face-to-face with others. Once their incompetence is discovered, several consequences follow. The normal person may "talk down" to the retardate, or may talk more slowly and deliberately. Interaction is reduced to the lowest plane, using the simplest vocabulary and avoiding complexities, such as humor. The normal person may not assume that the retarded person has any knowledge of things that are commonplace, and try to explain even the simplest things during the course of the conversation. Moreover, since the normal person frequently does not wish to embarrass the retarded person, a great deal of tact is exercised. The result of all of this is that interaction is slowed down to the point of virtual cessation; it is not interaction of a kind enjoyed by persons of normal intelligence (Edgerton, 1967: 215–217).

Because of the pervasiveness of the stigma of mental retardation, almost every action of the retarded person is considered to be a result of the retardation. As a result, when a person of low intellectual capacity gets into trouble, it is rather automatically attributed to the retardation. On the other hand, if a mentally retarded person does something for which a person of normal intelligence would be rewarded, this might be attributed to "chance" or random behavior. It is not common that others attribute much plain humanness to the mentally retarded. For example, one observer, a sociologist studying mental retardation, reported a simple truth that opened the door to explaining other social aspects of retardation.

> One of the older female [mentally retarded] students returned from a visit to a beauty shop in a state that could be conservatively described as euphoric. I was surprised. I did not expect them to want to be attractive—an all too common social misconception that militates against many retarded people being given a chance to be attractive. This attitude explains why

some dentists counsel parents against orthodontia for their retarded
children, why retarded people in many institutions are shorn. (Evans, 1983:
120)

The Visibly Physically Handicapped

Persons who have visible physical handicaps from birth or by accident or
illness include individuals born with a physical deformity, persons who have
lost a limb through illness or accident, burn victims, or individuals who
have become partially or totally paralyzed in some manner. Depending upon
the severity of the impairment, the person is likely to be regarded as a
deviant, particularly in a society that tends to place great emphasis on physical health, attractiveness, and basic physical competence. To be unable to
do for oneself in a physical sense is frequently to be reduced to the status
of a child. Given this context, "someone who perceives himself/herself and
is perceived by others as unable to meet the demands or expectations of a
particular situation because of some physical impairment—i.e., an anatom-
ical and/or a physiological abnormality" is considered a physically
handicapped person (Levitin, 1975: 549).

Societies have commonly separated the pronounced cripple from the
"normal" person, and this has resulted in varying degrees of isolation, persecution, and ridicule. The blind, crippled, and lepers traditionally have been
beggars. Among preliterate societies, deformed newly born infants were commonly exposed to the elements to perish, and in ancient Sparta, deformed
children were eliminated from a society that stressed physical perfection.
The attitude of many groups has been to regard physical deformities as a
blight from God, a punishment for having sinned. According to Hebraic
law, for example, physical abnormalities were signs of physical degradation,
and cripples were specifically prohibited from going near the temple. The
Bible contains many specific passages:

> For whatsoever man he be that hath a blemish, he shall not approach: A
> blind man, or a lame, or he that hath a flat nose, or anything superfluous,
> or a man that is broken-footed, or broken-handed or crookback, or a dwarf:
> or that hath a blemish in his eye, or be scurvy or scabbed, or hath his
> bones broken: No man that hath a blemish of the seed of Aaron the priest
> shall come nigh to offer the offerings of the Lord made by fire; he hath a
> blemish; he shall not come nigh to offer the bread of his God. He shall eat
> the bread of his God, both of the most holy, and of the holy: Only he
> shall not go in unto the vail, nor come nigh unto the altar, because he hath
> a blemish: that he profane not my sanctuaries: for I the Lord do sanctify
> them. (Leviticus, 21: 16–23)

Cripples were often treated with ridicule during the Middle Ages and
they frequently served as court jesters. During the sixteenth and seventeenth
centuries, cripples were thought to possess an evil spirit, and many crippled
persons were burned as witches. Cripples were often regarded as evil-doers

or at least as poor wretches, as seen in several of Shakespeare's plays and in subsequent literature such as Victor Hugo's *The Hunchback of Notre Dame.*

Contemporary attitudes toward the visibly physically handicapped still tend to regard crippled persons as people apart from other human beings and persons to be regarded with pity or avoided altogether. Crippled persons face problems with their occupations, their social relationships, and general social participation. Seldom, for example, are crippled persons found as sales personnel in stores or in similar positions where the public might be affected by their external appearance. Even where jobs are possible for someone with a physical handicap, the pay must be enough to offset the costs of having a disability. As one handicapped person put it:

> . . . as a paraplegic, you have to pay for attendants and you have to pay for medical supplies—all the extra needs that the nondisabled person doesn't have. I have to hire people to help me up in the morning and help me into bed at night, and to do the cooking, cleaning, whatever. . . . You also have to buy certain medical supplies, which are outrageous in price. (quoted in Roth, 1981: 45)

Social reactions to persons with physical handicaps often reflect a combination of pity and fear. The source of such attitudes is not difficult to discern. Socialization for children includes an emphasis on independence and the desirability of being healthy and having an attractive appearance. The consequence of such an emphasis is that persons who do not fit our conceptions of "physical success" are regarded with suspicion.

> The more effectively people in general are socialized to respect individual achievement . . . the more likely it is that physical or mental disabilities which limit personal independence and restrict achievements will be seen as a badge of imperfection and inadequacy affecting all spheres of life. (Topliss, 1982: 109)

The Obese

Current appearance norms value thinness for women and muscularity for men (Schur, 1984). Generally speaking, in the Western world whenever obese "people have existed and whenever a literature has reflected aspects of the lives and values of the period, a record has been left of the low regard usually held for the obese by the thinner and clearly more virtuous observer" (Mayer, 1968: 84). Highly visible obese people are frequently perceived as deviants, encountering strong societal reaction; they often suffer great social stigma because they make other members of a group feel "contaminated" by association with them. This may produce a formidable barrier to full, social acceptance (Maddox, et al., 1968). It is only recently, however, that excessive body weight as a source of deviant stigmatization has become a major concern. For example, Goffman (1963), in his classic work on stigma, did not include obesity among the physical stigmata he discussed. But contemporary concern over various eating disorders, such as anorexia nervosa and bulimia, has highlighted the importance of cultural stereotypes of beauty and ideal body shapes in the causation of these disorders.

Attitudes of rejection of obese persons are built into the culture and they are formed at an early age. When various drawings of children with deformities of various types were shown, in one study, to 10- and 11-year-old children, who were then asked to indicate their preferences in selecting these children as friends, the majority selected the obese child as their last choice (Richardson, et al., 1961). The drawings also depicted a normal child, a child with crutches and a brace on one leg, one confined to a wheelchair, one with an amputated forearm, and one with a facial disfigurement. There were, however, significant sex differences in these ratings: boys were more likely to be wary of the amputated child than the obese child, while girls always ranked the obese child last. Feelings of rejection are intensified when even everyday activities become difficult, such as buying clothing. Unable to shop where most people do, obese persons are very limited in terms of selection and styles; thus, sometimes, obese persons look different because of the clothes they wear.

The degree of stigma that is attached to obesity may be a function of the degree to which persons can be blamed or held responsible for their appearance (DeJong, 1980). Contemporary theories of obesity include ascribing the condition to psychosomatic factors, particularly anxiety where overeating is an attempt to reduce anxiety (Kaplan and Kaplan, 1957), although the evidence does not strongly support this theory (Ruderman, 1983).

Attitudes toward obesity have a moral connotation, the individual being presented as "gluttonous" and unwilling to control his behavior regardless of the consequences. Many people feel that obese people are getting what is coming to them, and that they could have prevented the problem. One observer has remarked that "The obese teenager is thus doubly and trebly disadvantaged: (1) because he is discriminated against; (2) because he is made to understand that he deserves it; and (3) because he comes to accept his treatment [by others] as just" (Cahnman, 1969: 294). As a result, obese persons may withdraw to escape negative sanctions, and in so doing, they may add to their difficulties rather than avoiding them.

While substantial, deliberate weight loss is possible with many obese persons, the causes of obesity appear to be both physiological and social. Certainly, the factors of diet, exercise, and self-monitoring (being aware of foods and their relationship to one's weight) are critical to successful weight loss (Colvin and Olson, 1983), but some obese persons seem unable to develop the necessary habits to reduce their weight and to maintain that lower weight.

Eating Disorders

Eating disorders do not appear to encompass an obvious disability, yet these conditions frequently involve an impairment of social functioning. There

In fact, part of the process by which people become socialized to their disability is to deny it, to refuse to adopt the sick role, as though the disability would be of such a short duration that permanent plans taking into account the disability are unnecessary.

One learns the sick role, like any other role. But the major difference for persons suffering from disabilities is that their assumption of the sick role represents a more or less permanent role acquisition. No matter how much they cooperate with medical authorities, no matter how strongly motivated they are to move to more conventional roles, they will be unable to do so. They must resign themselves to a permanent deviant status and to the fact that this status is not of their own choosing.

The development of eating disorders, such as bulimia, takes place over a period of time. The socialization process is a subtle one that involves learning the nature of some "ideal" body size and shape from others. Childhood experiences may include a concern over appropriate gender role behavior regarding dieting (a female, not male activity) and the development of a sense of guilt over the consumption of food (Morgan, Affleck, and Solloway, 1990). Cultural stereotypes found in the mass media are reinforced on playgrounds and in everyday conversations. While some degree of physical victimization is not uncommon among females, bulimia does not appear to be related directly to such experiences (Bailey and Gibbons, 1989). Some women learned as girls to value thinness and to overestimate their weight, regardless of their actual weight. The constant pressure to be thin, coupled with a low rate of success for dieting, results in persistent weight-loss efforts among some women. A desire for slimness often goes beyond a desire for attractiveness and seems to include attempting to gain some control over one's life (Taub and McLorg, 1990).

Disabled persons accommodate their disabilities by going through developmental stages similar to the one suggested by Kubler-Ross (1969) in adjusting to death and dying: denial and isolation, anger, bargaining, depression, and acceptance. These stages last for different periods of time for each patient, and movement from one stage to the next is a complicated process. During this time, medical professionals, as well as family and friends, are important socializers.

It is likely that depression and acceptance continue together for some time after the onset of disability. Physically disabled persons are, as a group, more depressed than nondisabled persons at all age groups and for each sex (Turner and Beiser, 1990). This depression, including major depression, appears to be related to chronic stress produced by the disability. The stress comes not only from the nature of the disability itself, but from worries concerning issues of long-term physical care, such as finding and retaining good medical and physical assistance, insurance coverage for costs, and the prospects for a "normal" life.

Rehabilitation can be conceived as a socialization process as well (Albrecht, 1976). During this time, the disabled person is adjusting to the degree of handicap he or she will experience. There is considerable variability in the rehabilitation process. Various types of disability have different degrees of visibility and therefore different rehabilitative needs. Serious facial burns, for example, are highly visible and are likely to elicit dramatic responses from others; and yet are not terribly disabling in terms of physical functioning. Rehabilitation in this instance would likely focus less on adjusting disabled persons' expectations for physical achievement toward the physical reality, but in matching their expectations for social achievement with the social reality. Moreover, the disabled person must deal not only with the reactions of others, but self-reactions as well.

Self-Reactions of the Disabled

Basically, there are three ways individuals may react to social stigma from their disability (Safilios-Rothschild, 1970). They can deny its existence, they can accept it, or they can seek indirect benefits from the situation. Persons who have always put a singularly high value on their appearances may attempt to deny the existence of an impairment. Deaf persons may pretend to hear. Others may attempt to "mask" the disability, for example, by wearing an artificial leg or arm.

Some may view their incapacitated state as acceptable although not ideal. They are able to accept their condition without being plunged into hopelessness or despair. Often this is not an immediate response to stigma, but one that is reached after experiencing denial for a certain period of time. Persons with different disabilities reach this stage at different times. Some persons may remain at the denial stage for a long period of time before moving to acceptance. The aid of family and friends is important in this transition.

Still a third category of disabled persons adapt to the changes of disability all too eagerly and seek benefits that accrue from it. In these cases, physical limitations and restrictions that could be overcome are maximized, while remaining capacities are minimized. Obese people and those with physical disabilities, for example, may win affection by fawning for attention. At times, there may be financial benefits for incurring a disability. A study of changes in federal legislation concerning disabled persons found that self-reports of disability were strongly related to economic conditions: self-reports of disability were highest when unemployment was high and the general economy depressed (Howards, et al., 1980).

Stereotyping simplifies interaction for the nondisabled when they meet disabled persons, and assessment of the disabled tends to stop at the other's perception of his disability. From then on communication tends to be shaped on that basis. For example, interaction between nondisabled and disabled persons is often characterized by a good deal of spatial and social distance; there is more physical space between the parties, and communications tend to be less personal (Safilios-Rothschild, 1976). This can have a great effect on disabled persons, since it seems to define the situation in which the disabled person should be defensive. Many disabled persons must defend themselves from imputations of moral, psychological, and social inferiority in such interactions.

Disability as a "Career"

Disabled persons are constantly aware of their status of being "different." In essence, they become identified, certified, and derogated just as any other recognized deviant. As such, it is not surprising that being disabled can be conceived in terms of a sociological "career." Career disability, or secondary deviance, is role adaptation rather than a new role formation. Once the disability is legitimated or validated (usually medically), an individual's role expectations may change to correspond with the judged extremity or seriousness of impairment. Societal reaction is crucial in the process of forming a stable or career deviance pattern of disabled behavior, and this social process is instrumental in the creation of a new self-concept.

The stigmatization process is sometimes subtle. It is often not what people say that disturbs, but the way it is said; sometimes, stigma is conveyed in gestures, facial expressions, and behavior. The labeling reaction of others may take the form of being deviant through the kindnesses and concerns of those around the disabled person (Hyman, 1971). On the other hand, sometimes the labeling may occur when disabled persons realize that they can not do something that they must ask others for assistance (Myerson, 1971: 205–210).

Professionals and Agencies

Interactions with professional persons such as doctors, counselors, physical therapists, and social workers are extremely significant in shaping disabled peoples' self-concepts and their movement to career disability. Doctors, for example, have responsibility of revealing the extent of a disability to patients, social workers may work with retarded persons in the community, and counselors may aid paraplegics with employment and personal problems. One study of the mentally retarded points out the dangers of psychologists diagnosing children as mentally retarded based on IQ tests that are constructed in such a way—or, in any case, have the effect—that children who are white and usually from the middle class tend to score higher (Mercer, 1973). If the average scores of the white children are socially defined as "normal," many children from minority groups automatically fall into a lower-than-normal category, thereby increasing their chances of being identified as retarded. Mercer (1973) claims that the test process itself is the most important structural factor contributing to the disproportionate labeling of children as mentally retarded. While the intelligence test may be a diagnostic tool with which to identify retardation, since most operational definitions of retardation select some relatively arbitrary IQ as a cutoff, the detection of the condition should not be confused for the condition itself.

Interactions with professional groups are critical to the disabled person's future role status and self-conception, and the sheer chance selection of the agents may be important in the career of the person. In a study of blindness, Scott (1969: 119) observed that blind persons were often rewarded for conforming to the expectations that the workers at blindness agencies had for them. Blind persons are told they are "insightful" when they describe their problems as their rehabilitators do, and that they are "blocking" or "resisting" when they do not. Gradually, the behavior of blind persons comes to correspond with the beliefs of the workers, particularly in the isolated, sheltered environments that many blind agencies create for their clients. Blind persons who come to live and work in such environments do quite well there, but become maladjusted for life in the community.

Many rehabilitation centers and other agencies involved in the prevention, treatment, and control of disabilities contribute much to the individuals themselves and to the community. Agency support and contribution, however, often depends on the extent of the disabilities they serve; thus, increased numbers of accredited disabled persons strengthen agency requests for additional staff and other resources. It is claimed that many handicaps exist because they are defined as such by treatment personnel (Friedson, 1965: 74). Frequently, the handicapped are aware of these limitations. First-hand accounts from persons with physical handicaps, for example, often contain a number of disparaging references to medical personnel (e.g., Roth, 1981). Handicapped persons report that their initial encounters with doctors and therapists after having incurred the disability did not prepare them for the handicap and that medical personnel were insensitive to their emotional conditions. Agency involvement may not be in the best interest of the disabled. Edgerton (1979: 29) points out that children with Down's Syndrome raised in institutions are less competent than children raised by their parents.

Subcultures and Groups

The formation and growth of subcultures is important for the maintenance of patterns of career deviance. Subcultures institutionalize customs, recruit new members, and provide support to existing members; as a result, they aid in the management of a deviant identity. Some groups are formal, such as those of the blind, the mute, and those with specific physical disabilities. They may hold regular meetings and conventions, and the groups generally serve as quasi-interest groups for their constituent members. Such groups may be formed out of a need for self-defense because disabled persons have been excluded from participating in conventional society.

Some specific functions of disability subcultures are to provide social and recreational outlets, to educate the public about the nature of particular disabilities, to press for favorable legislation for disabled persons, and to offer help in finding marriage partners and jobs. In general, the emphasis is on changing the disabled themselves rather than on changing the society that has labeled and discriminated against them. Other informal groups may arise spontaneously among patients who share a common waiting room in a doctor's office where they share information and offer mutual support.

Participation in a subculture may enable individuals to manage their disability better. For some disabilities, however, this is not possible. For example, obese persons may on occasion belong to an organization with others like themselves, but they often do not closely associate with them and thus are much less likely to form specific subcultures. Moreover, such organizations may be for the purpose of ridding obese persons of their obesity; successful organizations are going to have changes in membership over time that will reduce the cohesiveness of the group (Warren, 1974).

There are many "communities" of disabled persons; however, membership in such communities depends on more than simply having a disability. There are three necessary conditions for membership in deaf communities: (1) identification with the deaf world; (2) shared experiences that come from being hearing impaired; and (3) participation in the community's activities (Higgins, 1982: 38–77). Deaf persons must have accepted their impairment and developed a self-concept that includes deafness. They also must desire to associate with other deaf persons and share aspects of their lives as deaf persons. There are many deaf persons who do not belong to such a community. Higgins points out that there are many such communities, some of whose members are selected by race, age, and preferred method of communication (signers or speakers). Even within the communities of the disabled, there are subcommunities that share problems, deflect stigma from the outside, and experience mutual social support from others.

The nature of a community of disabled people depends, in part, on the participants and their problems. Aged deaf persons, for example, have specialized problems that are better faced with others. In this instance, the community serves as the context in which problems concerning aging, increased social isolation, and facing death are explored and solved (Becker, 1980). Subcultures for disabled persons function much like subcultures for other deviants in providing the locus of social activity.

The Role of Stigma in Disability Careers

There is no invariant career path of deviance for the physically disabled any more than for those who have no disabilities but who are considered deviant by their behavior. Yet, the labels used to denote the physically impaired are pejorative: the mockery of the disabled young by other children, the less active social life of obese adolescents, the misunderstanding of anorexia, the myths and stereotypes about the blind, the ridicule or shunning by others of the mentally retarded and bulimics, and the occupational discrimination against virtually all disabled persons are but a few of the consequences of their conditions for these persons.

Once persons develop an identity around their disability, their interactions with others will be altered. Family and friends will scrutinize the eating habits of anorexics and bulimics, and the reactions of such persons may under some circumstances lead to further instances of the behavior. Some anorexics and bulimics believed that others had certain expectations of them to act anorexic or bulimic. For example, friends of some anorexics never offered them food or drink, assuming continued disinterest on the part of the person.

> While being hospitalized, Denise felt she had to prove to others she was not still vomiting by keeping her bathroom door open. Other bulimics, who lived in dormitories, were hesitant to use the restroom for normal purposes lest several friends be huddling at the door, listening for vomiting. In general, individuals interacted with [anorexics and bulimics] largely on the basis of their eating disorder; in doing so, they reinforced anorexic and bulimic behaviors. (McLorg and Taub, 1987: 185)

Groups and organizations of disabled persons can move some disabled persons away from the sick role to other, more conventional roles. Some groups of disabled persons, for example, use stigma in a positive way, as a means to facilitate normal role acquisition. A study of several groups that aid obese persons lose weight found that the use of stigma or labels encouraged them in their "normalization" processes. Specifically, "groups who used ex's [ex-obese persons, in this case] as change agents, all used strategies of identity stigmatization in order to facilitate normalization of members' behavior" (Laslet and Warren, 1975: 79). The use of stigma in this instance was therapeutic and did not have the effect of pushing the persons further along a deviant career path. Another study dealing with the visibly physically handicapped found that they were active participants in the labeling process and that they were able, through their behavior and verbalization, to negotiate the deviant label (Levitin, 1975). In many instances, the handicapped persons actively resisted the negative labels of others.

Perhaps more clearly than other forms of deviance, the physically disabled are not considered to be deviant only from the actions of formal or informal labels. Stigma alone does not create the physically handicapped person, an objective reality does, whether it is blindness, obesity, being very thin, or a crippling condition or disease. Social support is an important feature of successful adjustment after disability. Abundant evidence is found among cardiac patients, for example, that informal support and higher levels of integration of the individual and his or her family leads to a greater chance of recovery from, or more successful adaptation to, the physical condition (Garrity, 1973). Conversely, if support is lacking, adaptation becomes increasingly difficult.

The ambivalence with which most persons regard the physically disabled thus tempers the conception persons have of these individuals. They are, alternatively, loved and hated, pitied and scorned, feared and accepted, and found attractive and repulsive. While persons make distinctions between "voluntary" and "involuntary" deviants, the fact remains that disabled persons are stigmatized and they must deal with that stigma.

THE MANAGEMENT OF DISABILITY

There are certain inevitable problems of living that confront the disabled as a result of their impairments. These problems depend on the nature and degree of the impairment and pertain to such matters as mobility, securing and retaining suitable employment, dealing with the medical aspects of the impairment, and the physical components of everyday life. Another set of problems, however, comes in the form of the stigma that the disabled experience. The reactions of significant others like family and friends may engender so much stress in disabled persons that they are required to develop management techniques. Thus, the problem of blind individuals in dealing with the reactions of sighted people may be greater than the problem of dealing with the blindness itself.

While the deviant nature of disabilities, because it is a condition, is different from that of deviant behavior, the disabled employ the same general processes of deviance management as other deviants, such as criminals, homosexuals, and survivors of unsuccessful suicides. These management techniques are designed to minimize the stigma that might otherwise result as a result of the deviant condition. Disabled persons may select from a variety of coping techniques (Safilios-Rothschild, 1970). They may be hypersensitive to their condition, deny their status, "normalize" their condition, withdraw from the world as much as possible, identify with the dominant group and hate themselves, become prejudiced against others with the same disability, become militant, attempt to make up for deficiencies by striving in other areas, or may retreat into a mental disorder or alcoholism. This range of responses is similar to that which is available to anyone who is stigmatized for whatever reason.

Two problems emerge when one examines the question of disability management: first, when does disability become a problem for the individual, and, more important, how does the disabled manage stigma from others? Goffman (1963) dealt with the first issue when he differentiated the "discredited" from the "discreditable." While the former refers to individuals whose disability is apparent or readily known, the latter designates those disabled whose condition is neither known or immediately visible. The more visible one's disability, the greater the management problem it presents for the individual. Techniques that can be used to hide the disability from social view are advantageous for the individual. Management techniques are used to suit the particular form of disability and rejection the person encounters. These methods include passing, normalizing, coping, and dissociation.

Passing

Disabled persons can attempt to "pass" and thus avoid playing the deviant role entirely. Passing involves disguising the disability so that others will not notice, and hence not stigmatize the disabled person. There are many ways to pass, but the success of this technique depends upon the visibility of the stigma. A study of the mentally retarded, for example, concluded that the stigma with respect to this disability is so strong that some management of that stigma was essential (Evans, 1983). Many mentally retarded persons have difficulty accepting their diagnosis, and, in fact, some never do accept the official fact of retardation. As a result, many attempt to pass as mentally normal persons, although in many situations this is not successful for them. Some retarded persons will carry pens and pencils, although they can not write, and some will wear wristwatches even though they can not tell time. The most common technique of passing, however, is the communication of unrealistic aspirations (Evans, 1983: 126–127). Some will tell of the cars they will drive or houses they will live in, while others will brag about the professional occupational positions they will have.

Persons with hearing difficulties may deny their impairment. Persons with impaired sight may pretend to see things in the presence of others that they can not really see. Obese persons may wear inconspicuous clothing that will partially hide their condition and prefer to be in dark surroundings. Anorexics, because they do not have an accurate perception of their body weight and size often do not feel conspicuous until their adverse behaviors are pointed out to them. Bulimics have a normal physical appearance and their disorder is evident only in the execution of the behavior (e.g., purging). Bulimics try hard to hide their bulimia by bingeing and purging in private.

Normalizing

Disabled persons can also normalize their deviance. This process requires that disabilities be explained in some manner that is socially acceptable. For instance, physically handicapped persons tell normal persons that they live a "normal" life, or they may avoid taboo words like "cripple" in conversation. They minimize the debilitating effects of their disability and generally disavow their deviant status. An anorexic may make an excuse for missing a meal by deliberately scheduling other commitments at that time, or simply explain that she is just dieting.

This tactic is not without difficulties. Having disavowed deviance and persuaded others to react to the disabled person as normal, the problem becomes one of sustaining the normal role in the face of all of the small amendments, qualifications, and concessions that a disabled person must invariably make (Davis, 1961). In many instances, society is willing to participate in this management technique, to the extent that it can. Special physical arrangements are often made to help physically disabled students in universities, for example, but once admitted, they are expected to perform on the same intellectual level as nondisabled students. Insofar as normalization involves moving away from the sick role, social groups seem to be willing to aid the process.

Coping

Persons with physical disabilities who repeatedly encounter stigma from others develop ways of coping with this situation (Eisenberg, 1982). Wright (1960: 212–217) has outlined three general categories of situations in which the "normal" person is thought by the disabled one to be an intruder. In each instance, coping behavior is called for. First, the recipients feel that the intruder, who is usually a stranger, is interested only in the disability and they wish to retaliate in some manner, usually with biting sarcasm. Second, the disabled person excludes the disability from the situation either through the "ostrich reaction" of completely running away (as children do) or pretending it does not exist, or by redirecting the interaction to another subject. In the third type of situation, the disabled person manages to exclude the condition and preserve the relationship either through good-natured levity or by embarking upon a superficial conversation. In none of these situations does the person wish to talk about the deeper and more personal meanings of his or her condition.

Some persons with physical disorders are able to cope with their condition by admitting the condition and evaluating it positively. Some persons with eating disorders may believe that the stigma of bulimia or anorexia is better than the stigma of being fat.

Dissociation

Another technique of deviance management has been termed "dissociation" (Davis, 1972: 107). Dissociation is a retreat and a passive acceptance of the deviant role, a rejection of conventional roles and activities, which increases the likelihood of interaction with nondeviants. Handicapped children, for example, learn quickly that interaction with other children may be painful and is thus to be avoided if possible. Obese persons may reduce their social activities because they think they are conspicuous to others. Dissociation involves the avoidance of situations in which stigma might be directed at the disabled person. All too often, this involves a degree of social isolation that only adds to the difficulties of the disabled person, because in avoiding social situations of all kinds, the disabled person avoids positive social experiences as well as negative ones.

There are also more general techniques that can be used for managing deviance. They include secrecy, manipulating the physical setting, the use of rationalizations, changing to nondeviance, and participation in deviant subcultures (Elliott, et al., 1982). Bulimics, for example, might believe that their purging was only a temporary weight loss technique. Some of these devices are not applicable for use by persons with physical disabilities. Persons with severe physical impairments, such as blind persons, may have no reasonable hope to change their condition, and hence to become nondeviant. Some disabilities are hard to keep secret because of their physical nature (being confined to a wheelchair), although some disabled persons attempt to conceal the nature of their disability in attempting to "pass" as a non-disabled person. Often dissociation is coupled with participation in a subculture such that disabled persons withdraw from many contacts with nondisabled people and increase contacts with people with similar difficulties to themselves. Whatever management techniques are used, most disabled persons must deflect some of the stigma directed at them by others.

SUMMARY

Persons with disabilities are often stigmatized and reacted to as though they were deviant, but the basis for these reactions is their condition rather than their behavior. As such, disabilities illustrate that deviance can be either a condition or a behavior. Millions in the United States have some kind of physical impairment, although not all of these are regarded as equally deviant. Persons with physical disabilities violate identity or appearance norms of "wholeness" or "health," as well as certain expectations of the sick role. In spite of their involuntary status, people with physical disabilities are often disvalued and must bear substantial social stigma for their conditions.

The sick role is a particular role that exempts persons from certain responsibilities and obligations on the condition that they will move to conventional roles as soon as possible. The physically disabled occupy sick roles more or less permanently, depending on the nature of their impairment. Others are also aware of the physical differences between those who have disabilities and those who do not. This appears to create substantial social ambivalence concerning their physical disabilities.

Because of the stigma attached to disability, disabled people must manage this stigma in some manner. Historically, some disabled, such as the blind and the visibly physically disabled, have been discriminated against and given undesirable social roles, such as that of beggars. Other disabled people are stigmatized more subtly, as in the social isolation and ridicule experienced by the obese and those with anorexia. There are subcultures for the disabled, organizations and other informal groups that provide social support and opportunities for interactions with other similarly stigmatized persons. There are, for example, a number of communities for the deaf, although not all (or even most) deaf people belong to them. The nature of the stigma facing physically disabled people differs with the impairment.

Mental retardation is reacted to with strong stigma since the mentally retarded person is perceived as having little social or mental competence.

While the impairments of the disabled are physical, people must be socialized into the disabled role. Learning to be disabled means learning to accept the sick role on a long-term basis and adjusting one's other social roles around the disability. There may be developmental stages in becoming disabled, much like those described for people who are dying. Such professionals as physicians and social workers are important in this process because they validate the nature of the disability and confer legitimacy on someone accepting the sick role. They do so, however, by sometimes demanding that the disabled person conform to certain social stereotypes they have of disabled people.

Stigma leads to the adoption of certain management techniques by which the stigma can be lessened or deflected. Some stigma is self-imposed since many disabled people occupied normal roles prior to their impairment. Passing as nondisabled, coping with stigma by redefining social situations, and dissociation or retreating from social situations where stigma might be present are three such techniques. Other techniques may work better for people with different disabilities, and many of these stigma management techniques are found among other deviants as well.

SELECTED REFERENCES

Albrecht, Gary L. 1976. "Socialization and the Disability Process." Pp. 3–38 in *The Sociology of Physical Disability and Rehabilitation.* Edited by Gary L. Albrecht. Pittsburgh: University of Pittsburg Press.

> Although somewhat dated, this is still one of the best sociological discussions of the disability process. The book also contains chapters on different sociological aspects of physical disabilities. In all chapters, physical disabilities are seen as a sociological, not individualistic, phenomenon.

Deegan, Mary Jo. 1985. "Multiple Minority Groups: A Case Study of Physically Disabled Women." Pp. 37–55 in *Women and Disability: The Double Handicap.* New Brunswick, NJ: Transaction Books.

> A discussion of the significance of combining two relatively disvalued statuses and the consequences of this combination. A good point at which to begin discussions on the meaning of deviance.

Edgerton, Robert B. 1979. *Mental Retardation.* Cambridge, MA: Harvard University Press.

> An excellent examination of the social, psychological, and some legal issues surrounding mental retardation. The author also examines the social policies that have developed regarding mental retardation.

McLorg, Penelope A. and Diane E. Taub. 1987. "Anorexia Nervosa and Bulimia: The Development of Deviant Identities." *Deviant Behavior,* 8: 177–189.

A sociological study of a new form of deviance that identifies the processes involved in acquiring a deviant identity. Because much of the literature on anorexia and bulimia is still dominated by medical and psychological orientations, this paper is a welcome addition to the literature.

Roth, William. 1981. *The Handicapped Speak*. Jefferson, NC: McFarland.

A collection of personal accounts from handicapped persons regarding the nature and consequences of their disabilities. The experiences of these persons speak more forcefully about their problems and prospects than other third-person accounts.

References

CHAPTER 1—THE NATURE AND MEANING OF DEVIANCE

Alwin, Duane F. 1986. "Religion and Parental Child-Rearing Orientations: Evidence of a Catholic-Protestant Convergence." *American Journal of Sociology*, 92: 412–440.

Becker, George. 1977. *The Genius as Deviant*. Beverly Hills, CA: Sage Publicatons.

Becker, Howard S. 1973. *Outsiders: Studies in the Sociology of Deviance*, enlarged ed. New York: Free Press.

Ben-Yahuda, Nachman. 1990. *The Politics and Morality of Deviance: Moral Panics, Drug Abuse, Deviant Science, and Reverse Discrimination*. Albany: State University of New York Press.

Best, Joel. 1987. "Rhetoric in Claims-Making: Constructing the Missing Children Problem." *Social Problems*, 34: 101–121.

Biddle, B. J. 1986. "Recent Developments in Role Theory." *Annual Review of Sociology*, 12: 67–92.

Birenbaum, Arnold and Edward Sagarin. 1976. *Norms and Human Behavior*. New York: Holt, Rinehart and Winston.

Blake, Judith, and Kingsley Davis. 1964. "Norms, Values, and Sanctions." Pp. 456–484 in *Handbook of Modern Sociology*, edited by Robert E. L. Faris. Chicago: Rand McNally.

Blumstein, Alfred, Jacqueline Cohen, Jeffrey A. Roth, and Christy A. Visher, eds. 1986. *Criminal Careers and "Career Criminals."* Washington, DC: National Academy Press.

Braithwaite, John. 1989. *Crime, Shame, and Reintegration.* Cambridge: Cambridge University Press.

Bryant, Clifton D., ed. 1990. *Deviant Behavior: Readings in the Sociology of Deviant Behavior.* New York: Hemisphere.

Chambliss, William J. 1976. "Functional and Conflict Theories of Crime: The Heritage of Emile Durkheim and Karl Marx." Pp. 1–28 in *Whose Law? Whose Order? A Conflict Approach to Criminology.* Edited by William J. Chambliss and Milton Mankoff. New York: Wiley.

Cohen, Albert K. 1955. *Delinquent Boys: The Culture of the Gang.* New York: Free Press.

———. 1966. *Deviance and Control.* Englewood Cliffs, NJ: Prentice-Hall.

———. 1974. *The Elasticity of Evil: Changes in the Social Definition of Deviance.* Oxford, England: Oxford University Penal Research Unit; Basil Blackwell.

Cohen, Stanley, ed. 1971. *Images of Deviance.* Baltimore: Penguin Books.

Coleman, James William. 1989. *The Criminal Elite: The Sociology of White Collar Crime,* 2nd ed. New York: St. Martin's.

Cox, Jenny. 1989. "Naturist Nudism." Pp. 122–124 in *Degrees of Deviance: Student Accounts of Their Deviant Behavior,* edited by Stuart Henry. Brookfield, VT: Avebury.

Davis, Fred. 1961. "Deviance Disavowal: The Management of Strained Interaction by the Visibly Handicapped," *Social Problems,* 9:120–132.

Dinitz, Simon, Russell Dynes, and Alfred Clarke, eds. 1975. *Deviance: Studies in Definition, Management and Treatment,* 2nd ed. New York: Oxford University Press.

Dodge, David L. 1985. "The Over-Negativized Conceptualization of Deviance: A Programmatic Exploration." *Deviant Behavior,* 6: 17–37.

Douglas, Jack D. ed. 1970. *Understanding Everyday Life.* Chicago: Aldine.

Durkheim, Emile. 1982. *The Rules of Sociological Method.* Edited with an Introduction by Steven Lukes. New York: Free Press; originally published 1895.

Edgerton, Robert. 1976. *Deviance: A Cross-Cultural Perspective.* Menlo Park, CA: Cummings.

Erlanger, Howard S. 1974. "Social Class and Corporal Punishment in Childrearing," *American Sociological Review* 39:68–85.

Faris, Ellsworth. 1937. *The Nature of Human Nature.* New York: McGraw-Hill.

Forsyth, Craig J. and Marion D. Oliver. 1990. "The Theoretical Framing of a Social Problem: Some Conceptual Notes on Satanic Cults." *Deviant Behavior,* 11: 281–292.

Geis, Gilbert and Ivan Bunn. 1990. "And a Child Shall Mislead Them: Notes on Witchcraft and Child Abuse Accusations." Pp. 31–45 in *Perspectives on Deviance: Dominance, Degradation, and Denigration.* Edited by Robert J. Kelly and Donal E. J. MacNamara. Cincinnati: Anderson.

Gelles, Richard. 1985. "Family Violence." *Annual Review of Sociology,* 11: 347–367.

Gibbs, Jack P. 1965. "Norms: The Problem of Definition and Classification," *American Journal of Sociology* 70:586–594.

———. 1981. *Norms, Deviance and Social Control.* New York: Elsevier.

———. 1989. *Control: Sociology's Central Notion.* Urbana: University of Illinois Press.

Goodin, Robert. 1989. *No Smoking: The Ethical Issues.* Chicago: University of Chicago Press.

Gouldner, Alvin W. 1968. "The Sociologist as Partisan: Sociology and the Welfare State," *The American Sociologist* 3:103–116.

Green, Edward C. 1977. "Social Control in Tribal Afro-America," *Anthropological Quarterly* 50: 34–77.

Greenberg, David F. 1988. *The Construction of Homosexuality.* Chicago: University of Chicago Press.

Gusfield, Joseph R. 1963. *Symbolic Crusade.* Urbana, IL: University of Illinois Press.

Hawkins, Richard and Gary Tiedman. 1975. *The Creation of Deviance: Interpersonal and Organizational Determinants.* Columbus, OH: Charles E. Merrill.

Henslin, James M. 1972. "Studying Deviance in Four Settings: Research Experiences with Cabbies, Suicides, Drug Users, and Abortionees." Pp. 35–70 *Research on Deviance,* edited by Jack D. Douglas. New York: Random House.

Herman, Nancy J. 1987. " 'Mixed Nutters' and 'Looney Tuners': The Emergence, Development, Nature, and Functions of Two Informal, Deviant Subcultures of Chronic, Ex-Psychiatric Patients." *Deviant Behavior,* 8: 235–258.

Hicks, Robert D. 1990. "Police Pursuit of Satanic Crime." *Skeptical Inquirer,* 14: 276–286.

Higgins, Paul C. and Richard R. Butler. 1982. *Understanding Deviance.* New York: McGraw-Hill.

Hill, David. 1990. "Nothing but the Bare Essentials in this Nudists' Camp." *Des Moines Register,* October 18, p. 9E.

Horwtiz, Allan V. 1990. *The Logic of Social Control.* New York: Plenum.

Johnson, Robert. 1987. *Hard Time: Understanding and Reforming the Prison.* Monterey, CA: Brooks/Cole.

Karmen, Andrew A. 1981. "Auto Theft and Corporate Irresponsibility," *Contemporary Crises,* 5:63–81.

Kitsuse, John I. 1972. "Deviance, Deviant Behavior, and Deviants: Some Conceptual Problems," Pp. 233–243 in *An Introduction to Deviance,* edited by William J. Filstead. Chicago: Markham.

Lemert, Edwin M. 1951. *Social Pathology.* New York: McGraw-Hill.

———. 1972. *Human Deviance, Social Problems, and Social Control,* 2nd ed. Englewood Cliffs, NJ: Prentice-Hall.

———. 1982. "Issues in the Study of Deviance." Pp. 233–257 *The Sociology of Deviance.* edited by M. Michael Rosenberg, Robert A. Stebbins, and Allan Turowetz. New York: St. Martin's Press.

Lesieur, Henry R. 1977. *The Chase: Career of the Compulsive Gambler.* New York: Doubleday Anchor.

Lewis, Oscar. 1961. *The Children of Sanchez.* New York: Vintage.

Liazos, Alexander. 1972. "The Poverty of the Sociology of Deviance: Nuts, Sluts, and Preverts." *Social Problems,* 20: 102–120.

Lofland, John. 1969. *Deviance and Identity.* Englewood Cliffs, NJ: Prentice-Hall.

Lowman, John, Robert J. Menzies, and T. S. Plays. 1987. "Introduction: Transcarceration and the Modern State of Penality." Pp. 1–15 in *Transcarceration: Essays in the Sociology of Social Control.* Edited by John Lowman, Robert J. Menzies, and T. S. Plays. Aldershot, England: Gower.

Manis, Jerome G. 1976. *Analyzing Social Problems.* New York: Holt, Rinehart and Winston.

Mansnerus, Laura. 1988. "Smoking Becomes 'Deviant Behavior'." *New York Times,* April 24, Section 4: 1, 6.

Meier, Robert F. 1981. "Norms and the Study of Deviance: A Proposed Research Strategy," *Deviant Behavior* 3:1–25.

———. 1982. "Perspectives on the Concept of Social Control," *Annual Review of Sociology* 8:35–65.

———. 1989. "Deviance and Differentiation." Pp. 199–212 in *Theoretical Integration in the Study of Deviance and Crime: Problems and Prospects.* Edited by Steven F. Messner, Marvin D. Krohn, and Allen E. Liska. Albany, NY: State University of New York Press.

Miller, Gale. 1978. *Odd Jobs: The World of Deviant Work.* Englewood Cliffs, NJ: Prentice-Hall.

Quinn, James F. 1987. "Sex Roles and Hedonism among Members of 'Outlaw' Motorcycle Clubs," *Deviant Behavior*, 8: 47–63.

Quinney, Richard. 1981. *Class, State and Crime*, 2nd ed. New York: Longman.

Ranulf, Sven. 1964. *Moral Indignation and Middle Class Psychology: A Sociological Study.* New York: Schoken Books, first published in Denmark, 1938.

Rosecrance, John. 1990. "You Can't Tell the Players Without a Scorecard: A Typology of Horse Players." Pp. 348–369 in Clifton D. Bryant, ed., *Deviant Behavior: Readings in the Sociology of Deviant Behavior.* New York: Hemisphere.

Rossi, Peter H., Emily Waite, Christine E. Bose, and Richard E. Berk. 1974. "The Seriousness of Crime: Normative Structure and Individual Differences." *American Sociological Review*, 39: 224–237.

Rule, James B. 1988. *Theories of Civil Violence.* Berkeley: University of California Press.

Rushing, William A. ed. 1975. *Deviant Behavior and Social Process*, 2nd ed. Chicago: Rand McNally.

Sagarin, Edward. 1975. *Deviants and Deviance: An Introduction to the Study of Disvalued People and Behavior.* New York: Holt, Rinehart and Winston.

———. 1985. "Positive Deviance: An Oxymoron." *Deviant Behavior*, 6: 169–181.

Santee, Richard T. and Jay Jackson. 1977. "Cultural Values as a Source of Normative Sanctions," *Pacific Sociological Review* 20:439–454.

Schwendinger, Herman and Julia. 1977. "Social Class and the Definition of Crime," *Crime and Social Justice* 7:4–13.

Schur, Edwin M. 1984. *Labeling Women Deviant: Gender, Stigma, and Social Control.* New York: Random House.

Scott, John Finley. 1971. *The Internalization of Norms.* Englewood Cliffs, NJ: Prentice-Hall.

Scott, Robert A. and Jack D. Douglas, eds. 1972. *Theoretical Perspectives on Deviance.* New York: Basic Books.

Shapiro, Susan P. 1984. *Wayward Capitalists: Target of the Securities and Exchange Commission.* New Haven, CT: Yale University Press.

Shibutani, Tamotsu. 1986. *Social Processes.* Berkeley: University of California Press.

Spector, Malcolm and John I. Kitsuse. 1979. *Constructing Social Problems.* Menlo Park, CA: Cummings.

Stafford, Mark C. and Richard R. Scott. 1986. "Stigma, Deviance, and Social Control: Some Conceptual Issues." Pp. 77–91 in *The Dilemma of Difference: A Multidisciplinary View of Stigma*, edited by Stephen C. Ainlay, Gaylene Becker, and Lerita M. Coleman. New York: Plenum.

Terry, Robert M. and Darrell J. Steffensmeier. 1988. "Conceptual and Theoretical Issues in the Study of Deviance." *Deviant Behavior*, 9: 55–76.

Thompson, Hunter. 1966. *Hell's Angels*. New York: Ballantine.

Trebach, Arnold S. 1987. *The Great Drug Wa*r. New York: Macmillan.

Troiden, Richard R. 1989. "The Formation of Homosexual Identities." *Journal of Homosexuality,* 17: 43–73.

Troyer, Ronald J. and Gerald E. Markle. 1983. *Cigarettes: The Battle Over Smoking.* New Brunswick, NJ: Rutgers University Press.

Victor, Jeffery S. 1990. "The Spread of Satanic-Cult Rumors." *Skeptical Inquirer,* 14: 287–291.

Watson, J. Mark 1982. "Outlaw Motorcyclists: An Outgrowth of Lower Class Cultural Concerns," *Deviant Behavior* 4:31–48.

Willcock, H. D. and J. Stokes. 1968. *Deterrents and Incentives to Crime among Youths Aged 15–21 Years.* London: Home Office, Government Social Survey.

Wilson, William Julius. 1987. *The Truly Disadvantaged: The Inner-City, the Underclass, and Public Policy.* Chicago: University of Chicago Press.

Winnick, Charles. 1990. "A Paradigm to Clarify the Life Cycle of Changing Attitudes Toward Deviant Behavior." Pp. 1–14 in *Perspectives on Deviance: Dominance, Degradation, and Denigration.* Edited by Robert J. Kelly and Donal E. J. MacNamara. Cincinnati: Anderson.

Wright, James D. 1989. *Address Unknown: The Homeless in America.* New York: Aldine de Gruyter.

Yinger, J. Milton. 1982. *Countercultures: The Promise and Peril of a World Turned Upside Down.* New York: Free Press.

Zeitlin, Marian, Hossein Ghassemi, and Mohammed Mansour. 1990. *Positive Deviance in Child Nuitrition.* New York: United Nations Publications.

CHAPTER 2—BECOMING DEVIANT

Agar, Michael. 1973. *Ripping and Running.* New York: Academic Press.

Becker, Howard S. 1973. *Outsiders: Studies in the Sociology of Deviance*, enlarged ed. New York: Free Press.

Bennett, Trevor. "A Decision-Making Approach to Opioid Addiction." Pp. 83–102 in *The Reasoning Criminal: Rational Choice Perspectives on Offending.* Edited by Derek B. Cornish and Ronald V. Clarke. New York: Springer-Verlag.

Berk, Bernard. 1977. "Face-Saving at the Singles Dance," *Social Problems,* 24: 530–544.

Biddle, B. J. 1986. "Recent Developments in Role Theory." *Annual Review of Sociology,* 12: 67–92.

Byrd, Richard E. 1966. *Alone.* New York: Putnam.

Cohen, Albert K. 1965. "The Sociology of the Deviant Act: Anomie and Beyond," *American Sociological Review* 30:

———. 1966. *Deviance and Control.* Englewood Cliffs, NJ: Prentice-Hall.

Douglas, Jack D. ed. 1970. *Observations on Deviance.* New York: Random House.

———. 1972. *Research on Deviance.* New York: Random House.

Elliott, Gregory C., Herbert L. Ziegler, Barbara M. Altman, and Deborah R. Scott. 1982. "Understanding Stigma: Dimensions of Deviance and Coping," *Deviant Behavior,* 3: 275–300.

Eysenck, Hans. 1977. *Crime and Personality.* London: Routledge and Kegan Paul.

Fishbein, Diane H. 1990. "Biological Perspectives in Criminology." *Criminology*, 28: 27–72.

Gay, Peter. 1988. *Freud: A Life for Our Time*. New York: Norton.

Gecas, Viktor. 1982. "The Self-Concept." *Annual Review of Sociology* 8:1–33.

Gerstel, Naomi. 1987. "Divorce and Stigma." *Social Problems*, 34: 172–186.

Hakeem, Michael. 1984. "The Assumption that Crime is a Product of Individual Characteristics: A Prime Example from Psychiatry." Pp. 197–221 in *Theoretical Methods in Criminology*, edited by Robert F. Meier. Beverly Hills, CA: Sage.

Hanson, Bill, George Beschner, James M. Walters, and Elliott Bovelle. 1985. *Life With Heroin*. Lexington, MA: Lexington Books.

Harding, Christopher and Richard W. Ireland. 1989. *Punishment: Rhetoric, Rule, and Practice*. London: Routledge.

Heiss, Jerold. 1981. "Social Roles." Pp. 94–129 in *Social Psychology: Sociological Perspectives* edited by Morris Rosenberg and Ralph H. Turner. New York: Basic Books.

Herman, Nancy J. 1987. " 'Mixed Nutters' and 'Looney Tuners': The Emergence, Development, Nature, and Functions of Two Informal, Deviant Subcultures of Chronic, Ex-Psychiatric Patients." *Deviant Behavior*, 8: 235–258.

Hogan, Dennis P. and Nan Marie Astone. 1986. "The Transition to Adulthood," *Annual Review of Sociology* 12:109–130.

Inciardi, James A. 1984. "Professional Theft." Pp. 221–243 in *Major Forms of Crime*, edited by Robert F. Meier. Beverly Hills, CA: Sage.

Jeffery, C. Ray. 1967. *Criminal Responsibility and Mental Disease*. Springfield, IL: Charles C Thomas.

Kitsuse, John I. 1980. "Coming Out All Over: Deviants and the Politics of Social Problems." *Social Problems*, 28: 1–13.

Levison, Peter K., Dean R. Gerstein, and Deborah R. Maloff. 1983. *Commonalities in Substance Abuse and Habitual Behavior*. Lexington, MA: Lexington Books.

Lilly, J. Robert, Francis T. Cullen, and Richard A. Ball. 1989. *Criminological Theory: Context and Consequences*. Newbury Park, CA: Sage.

Matza, David. 1969. *Becoming Deviant*. Englewood Cliffs, NJ: Prentice-Hall.

Meier, Robert F. 1989. *Crime and Society*. Boston: Allyn and Bacon.

Merton, Robert K. 1972. "Insiders and Outsiders: A Chapter in the Sociology of Knowledge," *American Journal of Sociology*, 78: 9–47.

Parsons, Talcott. 1951. *The Social System*. New York: Free Press.

Prus, Robert and Styllianoss Irini. 1980. *Hookers, Rounders, and Desk Clerks*. Toronto: Gage.

Rado, Sandor. 1963. "Fighting Narcotic Bondage and Other Forms of Narcotic Disorders." *Comprehensive Psychiatry*, 4: 160–167.

Reitzes, Donald C. and Juliette K. Diver. 1982. "Gay Bars as Deviant Community Organizations: The Management of Interactions with Outsiders," *Deviant Behavior*, 4: 1–18.

Sagarin, Edward. 1975. *Deviants and Deviance: An Introduction to the Study of Disvalued People and Behavior*. New York: Holt, Rinehart and Winston.

Scheff, Thomas A. 1983. "Toward Integration in the Social Psychology of Emotions," *Annual Review of Sociology* 9:333–354.

Secretary of Health and Human Services. 1990. *Alcohol and Health: Seventh Special Report to the U.S. Congress*. Rockville, MD: National Institute of Alcohol Abuse and Alcoholism, Government Printing Office.

Shibutani, Tamotsu. 1986. *Social Processes.* Berkeley: University of California Press.

Skipper, James K., Jr., William L. McWhorter, Charles H. McCaghy, and Mark Lefton, eds. 1981. *Deviance: Voices From the Margin.* Belmont, CA: Wadsworth.

Stephens, Richard C. 1987. *Mind-Altering Drugs: Use, Abuse, and Treatment.* Beverly Hills, CA: Sage Publications.

Sutherland, Edwin H. 1937. *The Professional Thief.* Chicago: University of Chicago Press.

———. 1950. "The Sexual Psychopath Laws." *Journal of Criminal Law, Criminology, and Police Science,* 40: 540–549

Sutherland, Edwin H. and Donald R. Cressey. 1978. *Criminology,* 10th ed. Philadelphia: Lippincott.

Szasz, Thomas. 1987. *Insanity: The Idea and Its Consequences.* New York: Wiley.

Troiden, Richard R. 1989. "The Formation of Homosexual Identities." *Journal of Homosexuality,* 17: 43–73.

Turner, Ralph H. 1972. "Deviance Avowal as Neutralization of Commitment," *Social Problems* 19:308–322.

Wilson, James Q. and Richard J. Herrnstein. 1985. *Crime and Human Nature.* New York: Simon and Schuster.

Yamaguchi, Kazuo and Denise B. Kandel. 1985. "On the Resolution of Role Incompatibility: A Life Event History Analysis of Family Roles and Marijuana Use." *Amerian Journal of Sociology,* 90: 1284–1325.

Zinberg, Norman. 1984. *Drug, Set, and Setting: The Basis for Controlled Intoxicant Use.* New Haven, CT: Yale University Press.

CHAPTER 3—URBANIZATION, URBANISM, AND DEVIANT ATTITUDES

Adler, Patricia A. 1985. *Wheeling and Dealing: An Ethnography of an Upper-Level Drug Dealing and Smuggling Community.* New York: Columbia University Press.

Akers, Ronald L. 1985. *Deviant Behavior: A Social Learning Approach,* 3rd ed. Belmont, CA: Wadsworth.

Attorney General's Commission on Pornography. 1986. Final Report. Washington, DC: Department of Justice, Government Printing Office.

Baker, Robert and Sandra Ball, eds. 1969. *Violence and the Media.* Washington, DC: Government Printing Office.

Bandura, Albert. 1973. *Aggression: A Social Learning Approach.* Englewood Cliffs, NJ: Prentice-Hall.

Baron, Larry and Murray A. Straus. 1989. *Four Theories of Rape in American Society: A State-Level Analysis.* New Haven, CT: Yale University Press.

Becker, Howard S. 1973. *Outsiders: Studies in the Sociology of Deviance,* enlarged ed. New York: Free Press.

Benda, Brent B. 1987. "Crime, Drug Abuse, Mental Illness, and Homelessness." *Deviant Behavior,* 8: 361–375.

Brantingham, Paul and Patricia. 1984. *Patterns in Crime.* New York: Macmillan.

Buendia, Hernando Gomez. 1990. *Urban Crime: Global Trends and Policies.* New York: United Nations Publications.

Choldin, Harvey M. 1978. "Urban Density and Pathology," *Annual Review of Sociology*, 4: 91–113.

Citizen's Crime Commission. 1975. Tokyo: *One City Where Crime Doesn't Pay! A Study of the reasons for Tokyo's low urban crime rate and what can be learned to help America's crime crisis*. Philadelphia: The Citizen's Crime Commission.

Clinard, Marshall B. 1976. "The Problem of Crime and Its Control in Developing Countries." Pp. 47–71 in *Crime in Papua New Guinea*, edited by David Biles. Canberra: Australian Institute of Criminology.

———. 1978. *Cities with Little Crime: The Case of Switzerland*. Cambridge: Cambridge University Press.

———. 1990. *Corporate Corruption*. New York: Greenwood/Praeger.

Cogan, John F. 1982. "The Decline in Black Teenage Employment: 1950–70." *American Economic Review*, 72: 621–638.

Comstock, George. 1980. *Television in America*. Beverly Hills, CA: Sage.

DeFleur, Melvin L. 1983. *Social Problems*. Boston: Houghton Mifflin.

DeFleur, Melvin L. and Sandra Ball-Rokeach. 1982. *Theories of Mass Communication*, 4th ed. New York: Longman.

Dornbusch, Sanford M. 1989. "The Sociology of Adolescence." *Annual Review of Sociology*, 15: 233–259.

Elliott, Delbert S. and David Huizinga. 1983. "Social Class and Delinquent Behavior in a National Youth Pancl," *Criminology*, 21: 149–177.

Feagin, Joe R. 1985. "The Global Context of Metropolitan Growth: Houston and the Oil Industry," *American Journal of Sociology*, 90: 1204–1230.

Feldman, M. Philip. 1977. *Criminal Behavior: A Psychological Analysis*. New York: Wiley.

Feshbach, Seymour and Robert D. Singer. 1971. *Television and Aggression*. San Francisco: Jossey Bass.

Fischer, Claude S. 1975. "The Effect of Urban Life on Traditional Values," *Social Forces*, 53: 420–432.

———. 1984. *The Urban Experience,* 2nd ed. New York: Harcourt Brace Jovanovich.

Friday, Paul C. and Jerald Hage. 1976. "Youth Crime in Post-Industrial Societies: An Integrated Perspective," *Criminology*, 14: 347–367.

Gibbs, Jack P. 1982. "Law as a Means of Social Control." Pp. 83–113 in *Social Control: Views From the Social Sciences*. Beverly Hills, CA: Sage.

Glassner, Barry and Julia Loughlin. 1987. *Drugs in Adolescent Worlds: Burnouts to Straights*. New York: St. Martin's Press.

Grinnell, Jr., Richard M. and Cheryl A. Chambers. 1979. "Broken Homes and Middle-Class Delinquency: A Comparison," *Criminology* 17: 395–400.

Harris, Louis. 1987. *Inside America*. New York: Vintage.

Herman, Nancy J. 1987. " 'Mixed Nutters' and 'Looney Tuners': The Emergence, Development, Nature, and Functions of Two Informal, Deviant Subcultures of Chronic, Ex-Psychiatric Patients." *Deviant Behavior*, 8: 235–258.

Hirschi, Travis and Michael Gottfredson. 1980. "Introduction: The Sutherland Tradition in Criminology." Pp. 7–19 in *Understanding Crime: Current Theory and Research*, edited by Travis Hirschi and Michael Gottfredson. Beverly Hills, CA: Sage.

Hodges, William F., Helen K. Buchsbaum, and Carol W. Tierney. 1983. "Parent-Child Relationships and Adjustment in Preschool Children in Divorced and Intact Families." *Journal of Divorce*, 7: 43–58.

Hogan, Dennis P. and Evelyn M. Kitagawa. 1985. "The Impact of Social Status, Family Structure, and Neighborhood on the Fertility of Black Adolescents," *American Journal of Sociology*, 90: 825–855.

Hunter, Albert. 1974. *Symbolic Communities: The Persistence and Change of Chicago's Local Communities.* Chicago: University of Chicago Press.

Johnson, Elmer H., ed. 1983. *International Handbook of Contemporary Developments in Criminology*, Volumes I and II. Westport, CT: Greenwood Press.

Kaplan, Robert M. and Robert D. Singer. 1976. "Television Violence and Viewer Aggression: A Reexamination of the Evidence," *Journal of Social Issues*, 32: 35–70.

Kobrin, Solomon. 1951. "The Conflict of Values in Delinquency Areas," *American Sociological Review*, 16: 653–661.

Liebert, Robert M., John M. Neale, and Emily S. Davidson. 1973. *The Early Window: Effects of Television on Children and Youth.* New York: Pergamon.

Loeber, Rolf and Madga Stouthamer-Loeber. 1986. "Family Factors as Correlates and Predictors of Juvenile Conduct Problems and Delinquency." Pp. 29–149 in *Crime and Justice: An Annual Review of Research*, Vol. 7, edited by Michael Tonry and Norval Morris. Chicago: University of Chicago Press.

Lowery, Shearon A. and Melvin L. DeFleur. 1988. *Milestones in Mass Communication Research: Media Effects*, 2nd ed. New York: Longman.

Lowry, Carol R. and Shirley A. Settle. 1985. "Effects of Divorce on Children: Differential Impact of Custody and Visitation Patterns." *Family Relations*, 34: 455–463.

Mayhew, Bruce H. and Roger L. Levinger. 1976. "Size and Density of Interaction in Human Aggregates," *American Journal of Sociology*, 82: 86–110.

Meier, Robert F. 1982. "Perspectives on the Concept of Social Control." *Annual Review of Sociology*, 8:35–55

———, ed. 1984. *Major Forms of Crime.* Beverly Hills, CA: Sage.

Miller, Eleanor M. 1986. *Street Woman.* Philadelphia: Temple University Press.

Moore, Joan. 1987. "L.A. Gangs: Throat of a Nightmare." *Los Angeles Times*, December 9: Part II, p. 7.

Murray, Charles. 1984. *Losing Ground: American Social Policy, 1950–1980.* New York: Basic Books.

Peek, Charles W. and George D. Lowe. 1977. "Wirth, Whiskey, and WASPs: Some Consequences of Community Size for Alcohol Use," *The Sociological Quarterly*, 18: 209–222.

Pins, Kenneth. 1990. "Census Report Shows Breakup of Black Family," *Des Moines Register*, July 12: p. 1A

Population Reference Bureau. 1987. "Where is the Metropolitan U.S.?" *Interchange*, no volume: 1–4.

President's Commission on Law Enforcement and Administration of Justice. 1967. *The Challenge of Crime in a Free Society.* Washington, DC: Government Printing Office.

Rankin, Joseph H. 1983. "The Family Context of Delinquency," *Social Problems*, 30: 466–479.

Reiss, Albert J., Jr. 1986. "Why Are Communities Important in Understanding Crime?" Pp. 1–33 in *Communities and Crime.* Edited by Albert J. Reiss, Jr. and Michael Tonry. Chicago: University of Chicago Press.

Reiss, Albert J., Jr. and Michael Tonry, eds. 1986. *Communities and Crime.* Chicago: University of Chicago Press.

Reiss, Ira L. and Gary R. Lee. 1988. *Family Systems in America*, 4th ed. New York: Holt, Rinehart and Winston.

Rosecrance, John. 1988. *Gambling Without Guilt: The Legitimation of an American Pastime*. Pacific Grove, CA: Brooks/Cole.

Sampson, Robert J. 1986. "Crime in Cities: The Effects of Formal and Informal Social Control." Pp. 271–311 in *Communities and Crime*. Edited by Albert J. Reiss, Jr. and Michael Tonry. Chicago: University of Chicago Press.

Schur, Edwin M. 1973. *Radical Non-Intervention*. Englewood Cliffs, NJ: Prentice-Hall.

Sennett, Richard. 1977. *The Fall of Public Man: On the Social Psychology of Capitalism*. New York: Knopf.

Shelley, Louise I. 1981. *Crime and Modernization: The Impact of Industrialization and Urbanization on Crime*. Carbondale, IL: University of Southern Illinois Press.

Short, James F., Jr., 1990. *Delinquency and Society*. Englewood Cliffs, NJ: Prentice Hall.

Spates, James L. and John J. Maciones. 1987. *The Sociology of Cities*, 2nd ed. Belmont, CA: Wadsworth.

Stack, Steven. 1982. "Suicide: A Decade Review of the Sociological Literature." *Deviant Behavior*, 4: 41–66.

Stephens, Richard S. 1987. *Mind-Altering Drugs: Use, Abuse, and Treatment*. Newbury Park, CA: Sage.

Tittle, Charles R. 1989. "Urbanness and Unconventional Behavior: A Partial Test of Claude Fischer's Subcultural Theory." *Criminology*, 27: 273–306.

Tittle, Charles R., Wayne J. Villemez, and Douglas Smith. 1978. "The Myth of Social Class and Criminality: An Empirical Assessment," *American Sociological Review*, 43: 643–656.

—— and Raymond Paternoster. 1988. "Geographic Mobility and Criminal Behavior." *Criminology*, 25: 301–343.

—— and Robert F. Meier. 1990. "Specifying the SES/Delinquency Relationship." *Criminology*, 28: 271–299.

Tuch, Steven A. 1987. "Urbanism, Region, and Tolerance Revisited: The Case of Racial Prejudice," *American Sociological Review*, 52: 504–510.

Vaillant, George E., Jane R. Bright, and Charles MacArthur. 1970. "Physicians' Use of Mood-Altering Drugs." *New England Journal of Medicine*, 282: 365–370.

Wilson, Thomas C. 1985. "Urbanism and Tolerance: A Test of Some Hypotheses Drawn from Wirth and Stouffer," *American Sociological Review*, 50: 117–123.

Wilson, William Julius. 1987. *The Truly Disadvantaged: The Inner City, The Underclass and Public Policy*. Chicago: University of Chicago Press.

Winick, Charles. 1961. "Physician Narcotic Addicts," *Social Problems*, 9: 174–186.

Wirth, Louis. 1938. "Urbanism as a Way of Life." *American Journal of Sociology*, 44: 1–24.

CHAPTER 4—GENERAL THEORIES OF DEVIANCE

Aday, David P., Jr. 1990. *Social Control at the Margins: Toward a General Understanding of Deviance*. Belmont, CA: Wadsworth.

Agnew, Robert. 1985. "A Revised Strain Theory of Delinquency." *Social Forces*, 64: 151–167.

Akers, Ronald L. 1985. *Deviant Behavior: A Social Learning Perpsective*, 3rd ed. Belmont, CA: Wadsworth.

Akers, Ronald L., Marvin D. Krohn, Lonn Lonza-Kaduce, and Marcia Radosevich. 1979. "Social Learning and Deviant Behavior: A Specific Test of a General Theory." *American Sociological Review*, 44: 635–655.

Andrews, D. A. 1980. "Some Experimental Investigations of the Principles of Differential Association Through Deliberate Manipulations of the Structure of Service Systems." *American Sociological Review*, 45: 448–462.

Becker, Howard S. 1973. *Outsiders: Studies in the Sociology of Deviance*, enlarged ed. New York: Free Press.

Brantingham, Paul and Patricia. 1984. *Patterns of Crime*. New York: Macmillan.

Brownfield, David H. 1987. "A Reassessment of Cultural Deviance Theory: The Use of Underclass Measures." *Deviant Behavior*, 8: 343–359.

Bulmer, Martin. 1984. *The Chicago School of Sociology: Institutionalization, Diversity, and the Rise of Sociological Research*. Chicago: University of Chicago Press.

Cooley, Charles H. 1918. *Social Process*. New York: Scribner.

Davis, Nanette J. 1980. *Sociological Constructions of Deviance: Perspectives and Issues,* 2nd ed. Dubuque, IA: William C. Brown.

Downs, William R. 1987. "A Panel Study of Normative Structure, Adolescent Alcohol Use, and Peer Alcohol Use." *Journal of Studies on Alcohol*, 48: 167–175.

Federal Bureau of Investigation. 1987. *Crime in the United States, 1986*. Washington, DC: Department of Justice, Government Printing Office.

Gibbons, Don C. and Joseph W. Jones. 1975. *The Study of Deviance: Perspectives and Problems*. Englewood Cliffs, NJ: Prentice-Hall.

Higgins, Paul C. and Richard R. Butler. 1982. *Understanding Deviance*. New York: McGraw-Hill.

Kornhauser, Ruth Rosner. 1978. *Social Sources of Delinquency: An Appraisal of Analytic Models*. Chicago: University of Chicago Press.

Lemert, Edwin M. 1951. *Social Pathology*. New York: McGraw-Hill.

Liska, Allen E. 1987. *Perspectives on Deviance*, 2nd ed. Englewood Cliffs, NJ: Prentice-Hall.

Matsueda, Ross L. 1988. "The Current State of Differential Association Theory." *Crime and Delinquency*, 34: 277–306.

Matsueda, Ross L. and Karen Heimer. 1987. "Race, Family Structure, and Delinquency: A Test of Differential Association and Social Control Theories." *American Sociological Review*, 52: 826–840.

Matza, David. 1969. *Becoming Deviant*. Englewood Cliffs, NJ: Prentice-Hall.

McCaghy, Charles H. 1985. *Deviant Behavior: Crime, Conflict, and Interest Groups*, 2nd ed. New York: Macmillan.

Meier, Robert F. 1976. "The New Criminology: Continuity in Criminological Theory." *Journal of Criminal Law and Criminology*, 67: 461–467.

———. 1982. "Perspectives on the Concept of Social Control." *Annual Review of Sociology*, 8:35–55

Messner, Steven F. 1988. "Merton's 'Social Structure and Anomie': The Road Not Taken." *Deviant Behavior*, 9: 33–53.

Mills, C. Wright. 1943. "The Professional Ideology of Social Pathologists." *American Journal of Sociology*, 49: 165–180.

Randall, Susan and Vicki McNickle Rose. 1984. "Forcible Rape." Pp. 47–72 in *Major Forms of Crime*. Edited by Robert F. Meier. Beverly Hills, CA: Sage.

Rubington, Earl and Martin S. Weinberg. 1981. *The Study of Social Problems: Five Perspectives*, 3rd ed. New York: Oxford University Press.

Schwendinger, Herman and Julia Schwendinger. 1974. *Sociologists of the Chair*. New York: Basic Books.

Scull, Andrew. 1988. "Deviance and Social Control." Pp. 667–693 in Neil J. Smelser, ed., *Handbook of Sociology*. Beverly Hills, CA: Sage.

Simon, Rita J. 1975. *Women and Crime*. Lexington, MA: Lexington Books.

Smith, Samuel. 1911. *Social Pathology*. New York: Macmillan.

Sutherland, Edwin H. 1947. *Criminology*, 4th ed. Philadelphia: Lippincott.

Sutherland, Edwin H. and Donald R. Cressey. 1978. *Criminology*, 10th ed. Philadelphia: Lippincott.

Thomas, W.I. and Florian Znaniecki. 1918. *The Polish Peasant in Europe and America*. New York: Alfred A. Knopf.

Traub, Stuart H. and Craig B. Little, eds. 1985. *Theories of Deviance*, 3rd ed. Itasca, IL: F. E. Peacock.

Whyte, William F. 1943. *Street Corner Society*. Chicago: University of Chicago Press.

Wilson, James Q. and Richard J. Herrnstein. 1985. *Crime and Human Nature*. New York: Simon and Schuster.

Wrong, Dennis. 1961. "The Oversocialized Conception of Man in Modern Sociology." *American Sociological Review*, 26: 183–193.

CHAPTER 5—LABELING, CONTROL, AND CONFLICT THEORIES OF DEVIANCE

Akers, Ronald L. 1968. "Problems in the Sociology of Deviance: Social Definitions and Behavior." *Social Forces*, 46: 455–465.

Arnold, William R. and Terrance M. Brungardt. 1983. *Juvenile Misconduct and Delinquency*. Boston: Houghton Mifflin.

Attorney General's Commission on Pornography. 1986. Final Report. Washington, DC: Government Printing Office.

Becker, Howard S. 1973. *Outsiders: Studies in the Sociology of Deviance*, enlarged ed. New York: Free Press.

Beirne, Piers and Richard Quinney, eds. 1982. *Marxism and Law*. New York: Wiley.

Biernacki, Patricia. 1986. *Pathways from Heroin Addiction: Recovery Without Treatment*. Philadelphia: Temple University Press.

Block, Alan A. and William J. Chambliss. 1981. *Organizing Crime*. New York: Elsevier.

Blumer, Herbert. 1969. *Symbolic Interactionism*. Englewood Cliffs, NJ: Prentice-Hall.

Box, Steven. 1981. *Deviance, Reality and Society*, 2nd ed. New York: Holt, Rinehart and Winston.

Cain, Maureen and Alan Hunt, eds. 1979. *Marx and Engels on Law*. New York: Academic Press.

Chambliss, William J. 1976. "Functional and Conflict Theories of Crime: The Heritage of Emile Durkheim and Karl Marx." Pp. 1–28 in *Whose Law? Whose*

Order? A Conflict Approach to Criminology. Edited by William J. Chambliss and Milton Mankoff. New York: Wiley.

Coleman, James W. 1989. *The Criminal Elite: The Sociology of White Collar Crime,* 2nd ed. New York: St. Martin's Press.

Colvin, Mark and John Pauly. 1983. "A Critique of Criminology: Toward an Integrated Structural-Marxist Theory of Delinquency Production." *American Journal of Sociology,* 89: 513–551.

Cressey, Donald R. 1971. *Other People's Money.* Belmont, CA: Wadsworth, originally published 1953.

Davies, Christie. 1982. "Sexual Taboos and Social Boundaries." *American Journal of Sociology,* 87: 1032–1063.

Dotter, Daniel L. and Julian B. Roebuck. 1988. "The Labeling Approach Re-examined: Interactionism and the Components of Deviance." *Deviant Behavior,* 9: 19–32.

Durkheim, Emile. 1933. *The Division of Labor in Society.* New York: Macmillan; originally published 1893.

Erikson, Kai T. 1962. "Notes on the Sociology of Deviance." *Social Problems,* 9: 307–314.

Fenton, Steve. 1984. *Durkheim and Modern Sociology.* Cambridge: Cambridge University Press.

Friday, Paul C. 1977. "Changing Theory and Research in Criminology." *International Journal of Criminology and Penology,* 5: 159–170.

Garfinkel, Harold. 1967. *Studies in Ethnomethodology.* Englewood Cliffs, NJ: Prentice-Hall.

Gibbs, Jack P. 1981. "The Sociology of Deviance and Social Control." Pp. 483–522 in *Social Psychology: Sociological Perspectives.* Edited by Morris Rosenberg and Ralph H. Turner. New York: Basic Books.

Goffman, Erving. 1963. *Asylums.* New York: Doubleday Anchor.

Gouldner, Alvin W. 1980. *The Two Marxisms: Contradictions and Anomalies in the Development of Theory.* New York: Seabury Press.

Gove, Walter R. 1970. "Societal Reaction as an Explanation of Mental Illness: An Investigation." *American Sociological Review,* 35: 873–884.

———. 1982. "Labeling Theory's Explanation of Mental Illness: An Update of Recent Evidence." *Deviant Behavior,* 3: 307–327.

Greenberg, David F. 1988. *The Construction of Homosexuality.* Chicago: University of Chicago Press.

Hamilton, V. Lee and Steve Rytina. 1980. "Social Consensus on Norms of Justice: Should the Punishment Fit the Crime?" *American Journal of Sociology,* 85: 1117-1144.

Hirschi, Travis. 1969. *Causes of Delinquency.* Berkeley: University of California Press.

———. 1984. "A Brief Commentary on Akers' 'Delinquent Behavior, Drugs, and Alcohol: What is the Relationship?'" *Today's Delinquent,* 3: 49–52.

Horton, John. 1981. "The Rise of the Right: A Global View." *Social Justice,* 15: 7–17.

Inverarity, James M., Pat Lauderdale, and Barry C. Feld 1983. *Law and Society: Sociological Perspectives on Criminal Law.* Boston: Little, Brown.

Langevin, Ron, ed. 1985. *Erotic Preference, Gender Identity, and Aggression in Men: New Research Studies.* Hillsdale, NJ: Lawrence Erlbaum Associates.

Lemert, Edwin M. 1951. *Social Pathology*. New York: McGraw-Hill.

———. 1972. *Human Deviance, Social Problems and Social Control*, 2nd ed. Englewood Cliffs, NJ: Prentice-Hall.

Liazos, Alexander. 1972. "The Poverty of the Sociology of Deviance: Nuts, Sluts, and Preverts." *Social Problems*, 20: 103–120.

Kornhauser, Ruth Rosner. 1978. *Social Sources of Delinquency: An Appraisal of Analytic Models*. Chicago: University of Chicago Press.

Krisberg, Barry. 1975. *Crime and Privilege*. Englewood Cliffs, NJ: Prentice-Hall.

Mankoff, Milton. 1971. "Societal Reaction and Career Deviance: A Critical Analysis." *The Sociological Quarterly*, 12: 204–218.

Matsueda, Ross L. 1982. "Testing Control Theory and Differential Association: A Causal Modeling Approach." *American Sociological Review*, 47: 489–504.

Matsueda, Ross L. and Karen Heimer. 1987. "Race, Family Structure, and Delinquency: A Test of Differential Association and Social Control Theories." *American Sociological Review*, 52: 826–840.

Meier, Robert F. 1983. "Shoplifting: Behavioral Aspects." Pp. 1497–1500 in *Encyclopedia of Crime and Justice*. Edited by Sanford H. Kadish. New York: Free Press.

Nye, F. Ivan. 1958. *Family Relationships and Delinquent Behavior*. New York: Wiley.

O'Malley, Pat. 1987. "Marxist Theory and Marxist Criminology." *Crime and Social Justice*, 29: 70–87.

Osgood, D. Wayne, Lloyd D. Johnston, Patrick M. O'Malley, and Jerald G. Bachman. 1988. "The Generality of Deviance in Late Adolescence and Early Adulthood." *American Sociological Review*, 53: 81–93.

Platt, Tony. 1974. "Prospects for a Radical Criminology in the United States." *Crime and Social Justice*, 1: 2–10.

Plummer, Ken. 1979. "Misunderstanding Labelling Perspectives." Pp. 85–121 in *Deviant Interpretations*. Edited by David Downes and Paul Rock. Oxford: Martin Robertson.

Quinney, Richard. 1979. *Criminology: Analysis and Critique of Crime in America*, 2nd ed. Boston: Little, Brown.

———. 1980. *Class, State and Crime*, 2nd ed. New York: Longman.

Reckless, Walter. 1973. *The Crime Problem*, 4th ed. New York: Appleton.

Reiman, Jeffery H. 1984. *The Rich Get Richer and the Poor Get Prison: Ideology, Class, and Criminal Justice*, 2nd ed. New York: Wiley.

Rubington, Earl and Martin S. Weinberg. 1987. *Deviance: The Interactionist Perspective*, 5th ed. New York: Macmillan.

Scheff, Thomas J. 1984. *Being Mentally Ill*, rev. ed. New York: Aldine.

Schur, Edwin M. 1971. *Labeling Deviant Behavior*. New York: Harper and Row.

———. 1979. *Interpreting Deviance*. New York: Harper and Row.

———. 1980. *The Politics of Deviance: Stigma Contests and the Uses of Power*. Englewood Cliffs, NJ: Prentice-Hall.

Schutz, Alfred. 1967. *The Phenomenology of the Social World*. Evanston, IL: Northwestern University Press.

Schwendinger, Herman and Julia. 1974. *Sociologists of the Chair*. New York: Basic Books.

Seeman, Melvin and Carolyn S. Anderson. 1983. "Alienation and Alcohol: The Role of Work Mastery and Community in Drinking Behavior." *American Sociological Review*, 48: 60–77.

Shover, Neal. 1985. *Aging Criminals.* Beverly Hills, CA: Sage.

Simon, David and D. Stanley Eitzen. 1987. *Elite Deviance,* 2nd ed. Boston: Allyn and Bacon.

Spitzer, Steven. 1975. "Toward a Marxian Theory of Deviance." *Social Problems,* 22: 638–651.

Stafford, Mark C. and Richard R. Scott. 1986. "Stigma, Deviance, and Social Control: Some Conceptual Issues." Pp. 77–91 in *The Dilemma of Difference: A Multidisciplinary View of Stigma.* Edited by Stephen C. Ainlay, Gaylene Becker, and Lerita M. Coleman. New York: Plenum.

Sykes, Gresham M. and David Matza. 1957. "Techniques of Neutralization: A Theory of Delinquency." *American Sociological Review,* 22: 664–670.

Takagi, Paul. 1974. "A Garrison State in a 'Democratic' Society." *Crime and Social Justice,* 1: 27–33.

Taylor, Ian, Paul Walton, and Jock Young. 1973. *The New Criminology: For A Social Theory of Deviance.* London: Routledge and Kegan Paul.

Thio, Alex. 1973. "Class Bias in the Sociology of Deviance." *The American Sociologist,* 8: 1–12.

Toby, Jackson. 1957. "Social Disorganization and Stake in Conformity: Complementary Factors in the Predatory Behavior of Hoodlums." *Journal of Criminal Law, Criminology and Police Science,* 48: 12–17.

Traub, Stuart H. and Craig B. Little, eds. 1985. *Theories of Deviance,* 3rd ed. Itasca, IL: F. E. Peacock.

Turk, Austin. 1969. *Criminality and Legal Order.* Chicago: Rand McNally.

———. 1984. "Political Crime." Pp. 119–135 in *Major Forms of Crime.* Edited by Robert F. Meier. Beverly Hills, CA: Sage.

Vaillant, George E., Jane R. Bright, and Charles MacArthur. 1970. "Physicians' Use of Mood-Altering Drugs." *New England Journal of Medicine,* 282: 365–370.

Vold, George B. 1958. *Theoretical Criminology.* New York: Oxford University Press.

Wiatrowski, Michael D., David B. Griswold, and Mary K. Roberts. 1981. "Social Control Theory and Deliquency." *American Sociological Review,* 46: 525–541.

Wilsnack, Richard W., Sharon C. Wilsnack, and Albert D. Klassen. 1987. "Antecedents and Consequences of Drinking and Drinking Problems in Women: Patterns from a U.S. National Survey." Pp. 85–158 in *Alcohol and Addictive Behavior: Nebraska Symposium on Motivation,* 1986. Edited by P. Clayton Rivers. Lincoln: University of Nebraska Press.

CHAPTER 6—CRIMES OF INTERPERSONAL VIOLENCE

Abadinsky, Howard. 1988. *Law and Justice.* Chicago: Nelson-Hall.

Amir, Menachim. 1971. *Patterns of Forcible Rape.* Chicago: University of Chicago Press.

Archer, Dane and Rosemary Gartner. 1984. *Violence and Crime in Cross-National Perspective.* New Haven, CT: Yale University Press.

Athens, Lonnie H. 1989. *The Creation of Dangerous Violent Criminals.* London: Routledge.

Ball-Rokeach, Sandra J. 1973. "Values and Violence: A Test of the Subculture of Violence Thesis." *American Sociological Review,* 38: 736–749.

Baron, Larry and Murray A. Straus. 1989. *Four Theories of Rape in American Society: A State-Level Analysis.* New Haven, CT: Yale University Press.

Bensing, Robert C. and Oliver Schroeder, Jr. 1960. *Homicide in an Urban Community.* Springfield, IL: Charles C Thomas.

Berk, Richard, Harold Brackman, and Selma Lesser. 1977. *A Measure of Justice: An Empirical Study of Changes in the California Penal Code, 1955–1971.* New York: Academic Press.

Bierne, P. and Richard Quinney, eds. 1982. *Marxism and Law.* New York: Wiley.

Blau, Judith R. and Peter M. Blau. 1982. "The Cost of Inequality: Metropolitan Structure and Violent Crime." *American Sociological Review*, 47: 114–129.

Blumstein, Alfred, Jacqueline Cohen, Jeffrey A. Roth, and Christy Visher, eds. 1986. *Criminal Careers and "Career Criminals,"* Volume 1. Washington, DC: National Academy Press.

Braithwaite, John. 1985. "White Collar Crime." *Annual Review of Sociology*, 11: 1–25.

Brantingham, Paul and Patricia Brantingham. 1984. *Patterns in Crime.* New York: Macmillan.

Brownmiller, Susan. 1975. *Against Our Will: Men, Women, and Rape.* New York: Simon and Schuster.

Bullock, Henry Allen. 1955. "Urban Homicide in Theory and Fact." *Journal of Criminal Law, Criminology, and Police Science*, 45: 565–575.

Bureau of Justice Statistics. 1987. *Criminal Victimization in the United States, 1985.* Washington, DC: Department of Justice, Government Printing Office.

Chambliss, William J. 1964. "A Sociological Analysis of the Law of Vagrancy." *Social Problems*, 11: 67–77.

——— and Robert Seidman. 1982. *Law, Order, and Power*, 2nd ed. Reading, MA: Addison-Wesley.

Clinard, Marshall B. and Daniel J. Abbott. 1973. *Crime in Developing Countries: A Comparative Perspective.* New York: Wiley.

——— and Richard Quinney. 1973. *Criminal Behavior Systems: A Typology*, 2nd ed. New York: Holt, Rinehart and Winston.

Cohen, Murray L., Ralph Garofalo, Richard Boucher, and Theoharis Seghorn. 1975. "The Psychology of Rapists." Pp. 113–140 in *Violence and Victims.* Edited by Stefan A. Pasternack. New York: Spectrum Publications.

Curtis, Lynn A. 1974. "Victim Precipitation and Violent Crime." *Social Problems*, 21: 594–605.

Daly, Martin and Margo Wilson. 1988. *Homicide.* Hawthorne, NY: Aldine de Gruyter.

Deming, Mary Beard and Ali Eppy. 1981. "The Sociology of Rape." *Sociology and Social Research*, 65: 357–380.

Dixon, Jo and Alan J. Lizotte. 1987. "Gun Ownership and the 'Southern Subculture of Violence'." *American Journal of Sociology*, 93: 383–405.

Dunford, Franklyn W., David Huizinga, and Delbert S. Elliott. 1990. "The Role of Arrest in Domestic Assault: The Omaha Police Experiment." *Criminology*, 28: 183–206.

Erlanger, Howard S. 1974. "The Empirical Status of the Subculture of Violence Thesis." *Social Problems*, 22: 280–292.

———. 1979. "Estrangement, Machismo, and Gang Violence." *Social Science Quarterly*, 60: 235–248.

Federal Bureau of Investigation. 1990. *Crime in the United States, 1989.* Washington, DC: Department of Justice, Government Printing Office.

Ferracuti, Franco, Renato Lazzari, and Marvin E. Wolfgang, eds., *Violence in Sardinia* (Rome: Mario Bulzoni 1970).

—— and Graeme Newman. 1974. "Assaultive Offenses," in Daniel Glaser, ed., *Handbook of Criminology* (Chicago: Rand McNally, 1974), pp. 194–195.

Finkelor, David. 1982. "Sexual Abuse: A Sociological Perspective." *Child Abuse and Neglect*, 6: 95–102.

——. 1984. *Child Sexual Abuse: New Theory and Research.* New York: Free Press.

Garbarino, James. 1989. "The Incidence and Prevalence of Child Maltreatment." Pp. 219–261 in Lloyd E. Ohlin and Michael Tonry, eds., *Family Violence.* Chicago: University of Chicago Press.

Gelles, Richard J. 1985. "Family Violence." *Annual Review of Sociology*, 11: 347–367.

Gelles, Richard J. and Murray Straus. 1979. "Violence in the American Family." *Journal of Social Issues*, 35: 15–39.

——. 1979. *Family Violence.* Beverly Hills, CA: Sage.

Gibson, Lorne, Rick Linden, and Stuart Johnson. 1980. "A Situational Theory of Rape." *Canadian Journal of Criminology*, 25: 51–65.

Green, Edward and Russell P. Wakefield. 1979. "Patterns of Middle and Upper Class Homicide." *Journal of Criminal Law and Criminology*, 70: 172–181.

Groth, A. Nicholas. 1979. *Men Who Rape: The Psychology of the Offender.* New York: Plenum Press.

Hagan, John. 1980. "The Legislation of Crime and Delinquency: A Review of Theory, Method, and Research." *Law and Society Review*, 14: 603–628.

Hall, Jerome. 1952. *Theft, Law and Society*, 2nd ed. Indianapolis: Bobbs Merrill.

Hartnagel, Timothy F. 1980. "Subculture of Violence: Further Evidence." *Pacific Sociological Review*, 23: 217–242.

Helfer, R. and H. Kempe. 1974. *The Battered Child.* Chicago: University of Chicago Press.

Hepburn, John R. and Harwin L. Voss. 1970. "Patterns of Criminal Homicide in Chicago and Philadelphia." *Criminology*, 8: 21–45.

——. 1973. "Violent Behavior in Interpersonal Relationships." *The Sociological Quarterly*, 14: 419–429.

Hills, Stuart L. 1980. *Demystifying Social Deviance.* New York: McGraw-Hill.

Holmstrom, Lynda Lytle and Ann Wolbert Burgess. 1990. "Rapists' Talk: Linguisitic Strategies to Control the Victim." Pp. 556–576 in Clifton D. Bryant, ed., *Deviant Behavior: Readings in the Sociology of Deviant Behavior.* New York: Hemisphere.

Jacobs, Joanne. 1990. "Some Are Trying to Broaden Definition of Rape." *Des Moines Register*, December 18, p. 9A.

Johnson, Elmer H., ed. 1983. *International Handbook of Contemporary Developments in Criminology*, 2 Volumes. Westport, CT: Greenwood Press.

Luckenbill, David F. 1984. "Murder and Assault." Pp. 19–45 in *Major Forms of Crime.* Edited by Robert F. Meier. Beverly Hills, CA: Sage.

Lundsgaarde, Henry P. 1977. *Murder in Space City: A Cultural Analysis of Houston Homicide Patterns.* New York: Oxford University Press.

Meier, Robert F. 1989. *Crime and Society.* Boston: Allyn and Bacon.

Messner, Steven F. 1989. "Economic Discrimination and Societal Homicide Rates: Further Evidence on the Cost of Inequality." *American Sociological Review*, 54: 597–611.

McClintock, F. H. 1963. *Crimes of Violence*. London: Macmillan.

Newman, Graeme. 1979. *Understanding Violence*. New York: Harper and Row.

Pagelow, Mildred Daley. 1989. "The Incidence and Prevalence of Criminal Abuse of Other Family Members." Pp. 263–313 in Lloyd E. Ohlin and Michael Tonry, eds., *Family Violence*. Chicago: University of Chicago Press.

Phillips, David P. 1983. "The Impact of Mass Media Violence on U.S. Homicides." *American Sociological Review*, 48: 560–568.

Pillemer, Karl A. and Rosalie S. Wolf, eds. 1986. *Elder Abuse: Conflict in the Family*. Dover, MA: Auburn House.

Pittman, David J. and William Handy. 1964. "Patterns in Criminal Aggravated Assault." *Journal of Criminal Law, Criminology, and Police Science*, 55: 462–470.

Pokorny, Alex D. 1965. "A Comparison of Homicides in Two Cities." *Journal of Criminal Law, Criminology, and Police Science*, 56: 479–487.

Randall, Susan and Vicki McNickle Rose. 1984. "Forcible Rape." Pp. 47–72 in *Major Forms of Crime*. Edited by Robert F. Meier. Beverly Hills, CA: Sage.

Renvoize, Jean. 1982. *Incest: A Family Pattern*. London: Routledge and Kegan Paul.

Sarafino, Edward P. 1979. "An Estimate of Nationwide Incidence of Sexual Offenses Against Children." *Child Welfare*, 58: 127–134.

Secretary of Health and Human Services. 1990. *Alcohol and Health: Seventh Special Report to the U.S. Congress*. Rockville, MD: National Institute of Alcohol Abuse and Alcoholism, Government Printing Office.

Schwartz, Martin. D. 1987. "Gender and Injury in Spousal Assault." *Social Forces*, 20: 61–75.

Schwendinger, Julia R. and Herman Schwendinger. 1981. "Rape, the Law, and Private Property." *Crime and Delinquency*, 28: 271–291.

Sherman, Lawrence W. and Richard A. Berk. 1984. "Deterrent Effects of Arrest for Domestic Violence." 49: 261–272.

Simon, Rita J. and Sandra Baxter. 1989. "Gender and Violent Crime." Pp. 171–197 in Neil Alan Weiner and Marvin E. Wolfgang, eds., *Violent Crime, Violent Criminals*. Newbury Park, CA: Sage.

Straus, Murray, Richard J. Gelles, and Suzanne Steinmetz. 1980. *Behind Closed Doors: Violence in the American Family*. Garden City, NY: Doubleday/Anchor.

Thomas, Charles W. and Donna M. Bishop. 1987. *Criminal Law: Understanding Basic Principles*. Newbury Park, CA: Sage.

Voss, Harwin L. and John R. Hepburn. 1968. "Patterns of Criminal Homicide in Chicago." *Journal of Criminal Law, Criminology and Police Science*, 59: 499–508.

Warr, Mark. 1985. "Fear of Rape among Urban Women." *Social Problems*, 32: 238–250.

Weinberg, S. Kiron. 1955. *Incest Behavior*. New York: Citadel Press.

Weiner, Neil Alan. 1989. "Violent Criminal Careers and 'Violent Criminal Careers': An Overview of the Research Literature." Pp. 35–138 in Neil Alan Weiner and Marvin E. Wolfgang, eds., *Violent Crime, Violent Criminals*. Newbury Park, CA: Sage.

Widom, Cathy Spatz. 1989. "The Intergenerational Transmission of Violence." Pp. 137–201 in Neil Alan Weiner and Marvin E. Wolfgang, eds., *Violent Crime, Violent Criminals*. Newbury Park, CA: Sage.

Wilbanks, William. 1983. "The Female Homicide Offender in Dade County, Florida." *Criminal Justice Review*, 8: 9–14.

———. 1984. *Murder in Miami*. Washington, DC: American University Press.

Wilson, James Q. and Richard J. Hernnstein. 1985. *Crime and Human Nature*. New York: Simon and Schuster.

Wolfgang, Marvin E. 1958. *Patterns of Criminal Homicide*. Philadelphia: University of Pennsylvania Press.

——— and Franco Ferracuti. 1982. *The Subculture of Violence*. Beverly Hills, CA: Sage; originally published 1967.

——— and Margaret A. Zahn. 1983. "Homicide: Behavioral Aspects." Pp. 849–855 in *Encyclopedia of Crime and Justice*, Vol 2. Edited by Sanford H. Kadish. New York: Free Press.

Zahn, Margaret A. 1975. "The Female Homicide Victim." *Criminology*, 13: 400–415.

CHAPTER 7—ECONOMIC AND POLITICAL CRIMINALITY

Abadinsky, Howard. 1981. *Organized Crime*. Boston: Allyn and Bacon.

Adams, Stuart. 1989. "Ripping Off Books." pp. 32–33 in *Degrees of Deviance: Student Accounts of Their Deviant Behavior*, edited by Stuart Henry. Brookfield, VT: Avebury.

Amaya, Atilio Ramirez, Miguel Angel Amaya, Carlos Alberto Avilez, Josefina Ramirez, and Miguel Angel Reyes. 1987. "Justice and the Penal System in El Salvador." *Crime and Social Justice*, 30: 1–27.

Anderson, Annelise Graebner. *The Business of Organized Crime: A Cosa Nostra Family*. Stanford, CA: Hoover Institution Press.

Anderson, Robert T. 1965. "From Mafia to Cosa Nostra." *American Journal of Sociology*, 81: 302–310.

Attorney General's Commission on Pornography. 1986. *Final Report*. Washington, DC: Government Printing Office.

Baumer, Terry L. and Dennis P. Rosenbaum. 1984. *Combatting Retail Theft: Programs and Strategies*. Toronto: Butterworths.

Bayh, Birch. 1975. *Our Nation's Schools: A Report Card—'A' in School Violence and Vandalism*. Washington, DC: Government Printing Office.

Bequai, August. 1978. *Computer Crime*. Lexington, MA: Lexington Books.

Blankenburg, Erhard. 1976. "The Selectivity of Legal Sanctions: An Empirical Investigation of Shoplifting." *Law and Society Review*, 11: 1110–1130.

Block, Alan A. and William J. Chambliss. 1981. *Organizing Crime*. New York: Elsevier.

Braithwaite, John and Gilbert Geis. 1982. "On Theory and Action for Corporate Crime Control." *Crime and Delinquency*, 28: 292–314.

Braithwaite, John. 1984. *Corporate Crime in the Pharmaceutical Industry*. London: Routledge and Kegan Paul.

Broedeur, Paul. 1974. *Expendable Americans*. New York: Viking.

Bureau of Justice Statistics. 1990. *Crime and the Nation's Households, 1989.* Washington, DC: Department of Justice, Government Printing Office.

Cameron, Mary Owen. 1964. *The Booster and the Snitch: Department Store Shoplifting.* New York: Free Press.

Carlin, Jerome E. 1966. *Lawyer's Ethics.* New York: Russell Sage.

Chambliss, Bill. 1972. *Box Man: A Professional Thief's Journey.* (The story of Harry King, as told to and edited by Bill Chambliss). New York: Harper and Row.

Chambliss, William J. 1978. *On the Take: From Petty Crooks to Presidents.* Bloomington: Indiana University Press.

Clark, John P. and Richard C. Hollinger. 1983. *Theft by Employees in Work Organizations.* Lexington, MA: Lexington Books.

Clinard, Marshall B. and Peter C. Yeager. 1980. *Corporate Crime.* New York: Free Press.

Clinard, Marshall B. 1990. *Corporate Corruption: The Abuse of Power.* New York: Praeger.

Cohen, Albert K. 1977. "The Concept of Criminal Organization." *British Journal of Criminology,* 17: 97–111

Coleman, James W. 1989. *The Criminal Elite: The Sociology of White Collar Crime,* 2nd ed. New York: St. Martin's Press.

Cressey, Donald R. 1969. *Theft of a Nation.* New York: Harper and Row.

———. 1971. *Other People's Money.* Belmont, CA: Wadsworth, originally published 1953.

———. 1972. *Criminal Organization.* New York: Harper and Row.

Cullen, Francis T., William J. Maakestad, and Gray Cavender. 1984. "The Ford Pinto Case and Beyond." Pp. 107–130 in *Corporations as Criminals.* Edited by Ellen Hochstedler. Beverly Hills, CA: Sage.

Douglas, Jack D. 1974. "Watergate: Harbinger or the American Prince." *Theory and Society,* 1: 89–97.

Eisenstadter, Werner J. 1969. "The Social Organization of Armed Robbery." *Social Problems,* 17: 67–68.

Ermann, M. David, Mary B. Williams, and Claudio Guiterrez, eds., 1990. *Computers, Ethics, and Society.* New York: Oxford University Press.

Fisse, Brent and John Braithwaite. 1983. *The Impact of Publicity on Corporate Offenders.* Albany: State University of New York Press.

Gandossey, Robert P. 1985. *Bad Business: The OPM Scandal and the Seduction of the Establishment.* New York: Basic Books.

Gasser, Robert Louis. 1963. "The Confidence Man." *Federal Probation,* 27: 47–54.

Geis, Gilbert. 1977. "The Heavy Electrical Equipment Antitrust Cases of 1961." Pp. 117–132 in *White-Collar Crime: Offenses in Business, Politics and the Professions.* Edited by Gilbert Geis and Robert F. Meier. New York: Free Press.

———. 1984. "White-Collar and Corporate Crime." Pp. 137–166 in *Major Forms of Crime.* Edited by Robert F. Meier. Beverly Hills, CA: Sage.

Gibbons, Don C. 1965. *Changing the Lawbreaker.* Englewood Cliffs, NJ: Prentice-Hall.

Greenberg, Peter S. 1982. "Fun and Games with Credit Cards." Pp. 377–379 in *Contemporary Criminology.* Edited by Leonard D. Savitz and Norman Johnston. New York: Wiley.

Haller, Mark H. 1990. "Illegal Enterprise: A Theoretical and Historical Interpretation." *Criminology,* 28: 207–235.

Harris, Louis. 1987. *Inside America.* New York: Vintage.

Hayno, David M. 1977. "The Professional Poker Player: Career Identification and the Problem of Respectability." *Social Problems*, 24: 556–565.

Hepburn, John. 1984. "Occasional Property Crime." Pp. 73–94 in *Major Forms of Crime.* Edited by Robert F. Meier. Beverly Hills, CA: Sage.

Hollinger, Richard C. and John P. Clark. 1982. "Formal and Informal Social Controls over Employee Deviance." *The Sociological Quarterly*, 23: 333–343.

Holzman, Harold R. 1982. "The Serious Habitual Property Offender as 'Moonlighter': An Empirical Study of Labor Force Participation Among Robbers and Burglars." *Journal of Criminal Law and Criminology*, 73: 1774–1792.

Homer, Frederic D. 1974. *Guns and Garlic: Myths and Realities of Organized Crime.* West Lafayette, IN: Purdue University Press.

Ianni, Francis A. J. 1972. *A Family Business: Kinship and Social Control in Organized Crime.* New York: Russell Sage.

———. 1975. *Black Mafia: Ethnic Succession in Organized Crime.* New York: Simon and Schuster.

——— and Elizabeth Reuss-Ianni. 1983. "Organized Crime: Overview," Pp. 1094–1106 in *Encyclopedia of Crime and Justice*, Vol. 3. Edited by Sanford Kadish. New York: Free Press.

Inciardi, James A. 1975. *Careers in Crime.* Chicago: Rand McNally.

———. 1984. "Professional Theft." Pp. 221–243 in *Major Forms of Crime.* Edited by Robert F. Meier. Beverly Hills, CA: Sage.

Jaworski, Leon. 1977. *The Right and the Power: The Prosecution of Watergate.* New York: Pocket Books.

Kempf, Kimberly. 1987. "Specialization and the Criminal Career." *Criminology*, 25: 399–420.

Klemke, Lloyd W. 1982. "Exploring Juvenile Shoplifting." *Sociology and Social Research*, 67: 59–75.

Klockars, Carl B. 1974. *The Professional Fence.* New York: Free Press.

Knapp Commission. 1977. "Official Corruption and the Construction Industry." Pp. 225–232 in *Official Deviance: Readings in Malfeasance, Misfeasance, and Other Forms of Corruption.* Edited by Jack D. Douglas and John M. Johnson. Philadelphia: Lippincott.

Kruissink, M. 1990. *The Halt Program: Diversion of Juvenile Vandals.* The Hague: Netherlands. Research and Documentation Center, Dutch Ministry of Justice.

Lanza-Kaduce, Lonn 1980. "Deviance among Professionals: The Case of Unnecessary Surgery." *Deviant Behavior*, 1: 333–359.

Lemert, Edwin M. 1958. "The Behavior of the Systematic Check Forger." *Social Problems*, 6: 141–149.

———. 1972. "An Isolation Closure Theory of Naive Check Forgery," in Edwin M. Lemert, *Human Deviance, Social Problems and Social Control*, 2nd ed. (Englewood Cliffs, NJ: Prentice-Hall, 1972), p. 139.

Light, Ivan. 1977. "Numbers Gambling among Blacks: A Financial Institution." *American Sociological Review*, 42: 892–904.

Maurer, David W. 1949. *The Big Con.* New York: Pocket Books, 1949).

———. 1964. *Whiz Mob.* New Haven, CT: College and University Press.

Meier, Robert F. and James F. Short, Jr. 1982. "The Consequences of White-Collar Crime." Pp. 23–49 in *White-Collar Crime: An Agenda for Research.* Edited by Herbert Edelhertz and Thomas A. Overcast. Lexington, MA: Lexington Books.

Meier, Robert F. 1983. "Shoplifting: Behavioral Aspects." Pp. 1497–1500 in *Encyclopedia of Crime and Justice*. Edited by Sanford H. Kadish. New York: Free Press.

McCaghy, Charles H. Peggy C. Giodano, and Trudy Knicely Hensen. 1977. "Auto Theft: Offender and Offense Characteristics." *Criminology*, 15: 367–385.

Minor, William J. 1975. "Political Crime, Political Justice and Political Prisoners." *Criminology*, 12: 385–398.

Mintz, Morton. 1985. *At Any Cost: Corporate Greed, Women, and the Dalkon Shield.* New York: Pantheon.

Mizurchi, Mark S. 1987. "Why Do Corporations Stick Together? An Interorganizational Theory of Class Cohesion." Pp. 205–218 in *Power Elites and Organizations*. Edited by G. William Domhoff and Thomas R. Dyc. Newbury Park, CA: Sage.

Morash, Merry. 1984. "Organized Crime." Pp. 191–220 in *Major Forms of Crime*. Edited by Robert F. Meier. Beverly Hills, CA: Sage.

Morris, Norval and Gordon Hawkins. 1971. *The Honest Politician's Guide to Crime Control*. Chicago: University of Chicago Press.

Nagel, Ilene H. and John L. Hagan. 1982. "The Sentencing of White-Collar Criminals in Federal Courts: A Socio-Legal Exploration of Disparity." *Michigan Law Review*, 80: 1427–1465.

Plate, Thomas. 1975. *Crime Pays!* New York: Ballantine.

Polsky, Ned. 1964. "The Hustlers." *Social Problems*, 12: 9–17.

Pontell, Henry N., Paul D. Jesilow, and Gilbert Geis. 1982. "Policing Physicians: Practitioner Fraud and Abuse in a Government Medical Program." *Social Problems*, 30: 117–125.

Prus, R. and C. R. D. Sharper. 1979. *The Road Hustler: The Career Contingencies of Professional Card and Dice Hustlers*. Toronto: Gage.

Reuter, Peter. 1983. *Disorganized Crime: Illegal Markets and the Mafia*. Cambridge, MA: MIT Press.

Rhodes, Robert P. 1984. *Organized Crime: Crime Control vs. Civil Liberties*. New York: Random House.

Richards, Pamela. 1979. "Middle-Class Vandalism and Age-Status Conflicts." *Social Problems*, 26: 482–497.

Robin, Gerald. 1974. "White-Collar Crime and Employee Theft." *Crime and Delinquency*, 20: 251–262.

Roebuck, Julian B. and Mervyn L. Cadwallader. 1961. "The Negro Armed Robber as a Criminal Type: The Construction and Application of a Typology." *Pacific Sociological Review*, 4: 21–26.

Roebuck, Julian B. and Ronald C. Johnson. 1963. "The 'Short Con' Man," *Crime and Delinquency*, 10: 235–248.

Roebuck, Julian B. 1983. "Professional Criminal: Professional Thief." Pp. 1260–1263 in *Encyclopedia of Crime and Justice*, Vol. 3. Edited by Sanford H. Kadish. New York: Free Press.

Rowan, Roy. 1986. "The Biggest Mafia Bosses," *Fortune*, November 10: 24–38.

Schafer, Stephen. 1974. *The Political Criminal: The Problem of Morality and Crime*. New York: Free Press.

Schwartz, Richard D. and Jerome H. Skolnick. 1964. "Two Studies of Legal Stigma." in *The Other Side: Perspectives on Deviance*. Edited by Howard S. Becker. New York: Free Press.

Shapiro, Susan P. 1984. *Wayward Capitalists: Target of the Securities and Exchange Commission*. New Haven, CT: Yale University Press.

———. 1990. "Collaring the Crime, Not the Criminal: Reconsidering the Concept of White-Collar Crime." *American Sociological Review*, 55: 346–365.

Simon, Carl P. and Ann D. Witte. 1982. *Beating the System: The Underground Economy*. Boston: Auburn House.

Simon, David R. and D. Stanley Eitzen. 1987. *Elite Deviance*, 2nd ed. Boston: Allyn and Bacon.

Simpson, Sally S. 1987. "Cycles of Illegality: Antitrust Violations in Corporate America." *Social Forces*, 65: 943–963.

Smith, Dwight C. 1975. *The Mafia Mystique*. New York: Basic Books.

Steffensmeier, Darrell J. 1986. *The Fence: In the Shadow of Two Worlds*. Totowa, NJ: Rowan and Littlefield.

Sutherland, Edwin H. 1937. *The Professional Thief*. Chicago: University of Chicago Press.

———. 1949. *White-Collar Crime*. New York: Dryden Press.

———. 1983. *White-Collar Crime: The Uncut Version.*, Introduction by Gilbert Geis and Colin Goff. New Haven, CT: Yale University Press.

Tien, James M. Thomas F. Rich, and Michael F. Cahn. 1986. Electronic Fund Transfer Systems Fraud: *Computer Crime. Public Systems Evaluation, Inc., 1985*. Washington, DC: Department of Justice, Government Printing Office.

Turk, Austin T. 1982. *Political Criminality: The Defiance and Defense of Authority*. Beverly Hills, CA: Sage Publications.

Tyler, Gus. 1981. "The Crime Corporation." Pp. 273–290 in *Current Perspectives in Criminal Behavior*, 2nd ed. Edited by Abraham S. Blumberg. New York: Knopf.

Vaughan, Diane and Giovanna Carlo. 1975. "The Appliance Repairman: A Study of Victim-Responsiveness." *Journal of Research in Crime and Delinquency*, 12: 153–161.

Vaughan, Diane. 1982. "Transaction Systems and Unlawful Organizational Behavior." *Social Problems*, 29: 373–379.

Vera Institute of Justice. 1981. *Felony Arrests*, rev. ed. New York: Longman.

Wade, Andrew L. 1967. "Social Processes in the Act of Juvenile Vandalism." Pp. 94–109 in *Criminal Behavior Systems: A Typology*. Edited by Marshall B. Clinard and Richard Quinney. New York: Holt, Rinehart and Winston.

Walker, Andrew. 1981. "Sociology and Professional Crime." Pp. 153–178 in *Current Perspectives in Criminal Behavior*, 2nd ed. Edited by Abraham S. Blumberg. New York: Knopf.

Wattenberg, William W. and James Balistieri. 1952. "Automobile Theft: A 'Favored-Group' Delinquency." *American Journal of Sociology*, 57: 575–579.

Wertheimer, Albert I. and Henri R. Manasse, Jr. 1976. "Pharmacist Practice Deviance." *Social Science and Medicine*, 10: 232.

Wheeler, Stanton, David Weisburd, and Nancy Bode. 1982. "Sentencing the White-Collar Offender: Rhetoric and Reality." *American Sociological Review*, 47: 641–659.

Williams, Kristen M. and Judith Lucianovic. 1979. *Robbery and Burglary: A Study of The Characteristics of the Persons Arrested and the Handling of Their Cases in Court*. Washington, DC: Institute for Law and Social Research.

CHAPTER 8—DRUG USE AND ADDICTION

Adler, Patricia A. 1985. *Wheeling and Dealing: An Ethnography of an Upper-Level Drug Dealing and Smuggling Community*. New York: Columbia University Press.

Advisory Council on the Misuse of Drugs. 1982. *Treatment and Rehabilitation: Report of the Advisory Council on the Misuse of Drugs*. London: Her Majesty's Stationery Office.

Agar, Michael. 1973. *Ripping and Running: A Formal Ethnography of Urban Heroin Addicts*. New York: Seminar Press.

―――― and Richard C. Stephens. 1975. "The Methadone Street Scene: The Addict's View." *Psychiatry*, 38: 381–387.

Anglin, M. Douglas, Yih-Ing Hser, and William McGlothlin. 1987. "Sex Differences in Addict Careers. 2. Becoming Addicted," *American Journal of Drug and Alcohol Abuse*, 13: 59–71.

Ball, John C. and M. P. Lau. 1966. "The Chinese Narcotic Addict in the United States," *Social Forces*, 45: 68–72.

―――― and Carl D. Chambers, eds. 1970. *The Epidemiology of Opiate Addiction in the United States*. Springfield, IL: Charles C Thomas.

――――, Lawrence Rosen, John A. Flueck, and David N. Nurco. 1982. "Lifetime Criminality of Heroin Addicts in the United States." *Journal of Drug Issues*, 12: 225–239.

Becker, Howard S. 1973. *Outsiders: Studies in the Sociology of Deviance*, enlarged ed. New York: Free Press.

Bennett, Trevor. 1986. "A Decision-Making Approach to Opioid Addiction." Pp. 83–102 in *The Reasoning Criminal: Rational Choice Perspectives on Offending*. Edited by Derek B. Cornish and Ronald V. Clarke. New York: Springer-Verlag.

Biernacki, Patricia. 1986. *Pathways from Heroin Addiction: Recovery Without Treatment*. Philadelphia: Temple University Press.

Blackwell, Judith Stephenson. 1983. "Drifting, Controlling, and Overcoming: Opiate Users Who Avoid Becoming Chronically Dependent." *Journal of Drug Issues*, 13: 219–235.

Blum, Richard H., with Eva Blum and Emily Garfield. 1976. *Drug Education: Results and Recommendations*. Lexington, MA: Lexington Books.

Brown, George F., Jr. and Lester P. Silverman. 1980. "The Retail Price of Heroin: Estimation and Applications." Pp. 25–33 in *Quantitative Explorations in Drug Abuse Policy*. Edited by Irving Leveson. New York: Spectrum.

Chambers, Carl D. and Michael T. Harter. 1987. "The Epidemiology of Narcotic Abuse among Blacks in the United States, 1935–1980." Pp. 191–223 in *Chemical Dependencies: Patterns, Costs, and Consequences*. Edited by Carl D. Chambers, James A. Inciardi, David M. Peterson, Harvey A. Siegal, and O.Z. White. Athens: Ohio University Press.

Chein, Isador, Donald L. Gerard, Robert S. Lee, and Eva Rosenfeld. 1964. *The Road to H*. New York: Basic Books.

Cohen, Sidney. 1984. "Recent Developments in the Use of Cocaine." *Bulletin on Narcotics*, (April - June, 1984): 9.

――――. 1987. "Causes of the Cocaine Outbreak." Pp. 3–9 in *Cocaine: A Clinician's Handbook*. Edited by Arnold M. Washton and Mark S. Gold. New York: Guilford Press.

Coombs, Robert H. 1981. "Drug Abuse as a Career." *Journal of Drug Issues*, 4: 369–387.

Courtwright, David T. 1982. *Dark Paradise*. Cambridge, MA: Harvard University Press.

Cox, Terrance, Michael R. Jacobs, A. Eugene Leblanc, and Joan A. Marshman. 1983. *Drugs and Drug Abuse: A Reference Text*. Toronto: Addiction Research Foundation.

Cuskey, Walter R. and Richard B. Wathey. 1982. *Female Addiction: A Longitudinal Study*. Lexington, MA: Lexington Books.

Dembo, Richard and Michael Miran. 1976. "Evaluation of Drug Prevention Programs by Youths in a Middle-Class Community." *The International Journal of Addictions*, 11: 881–903.

Dupont, Robert L. and M. H. Greene. 1973. "The Dynamics of a Heroin Addiction Epidemic." *Science*, 181: 716–722.

Estroff, Todd Wilk. 1987. "Medical and Biological Consequences of Cocaine Abuse." Pp. 23–32 in *Cocaine: A Clinician's Handbook*. Edited by Arnold M. Washton and Mark S. Gold. New York: Guilford Press.

Faupel, Charles E. and Carl B. Klockars. 1987. "Drugs-Crime Connections: Elaborations from the Life Histories of Hard-Core Heroin Addicts." *Social Problems*, 34: 54–68.

Finestone, Harold. 1957. "Cats, Kicks, and Color," *Social Problems*, 5: 3–13.

Galliher, John F. and Allyn Walker. 1977. "The Puzzle of the Social Origins of the Marihuana Tax Act of 1937," *Social Problems*, 24: 366–376.

Gonzales, Laurence. 1987. "Addiction and Rehabilitation." *Playboy*, 34: 149–152, 182–198.

Goode, Erich. 1970. *The Marijuana Smokers*. New York: Basic Books.

———. 1984. *Drugs in American Society*. 2nd ed. New York: Knopf.

Goodstadt, Michael S., Margaret A. Sheppard, and Godwin C. Chan. 1982. "Relationships Between Drug Education and Drug Use: Carts and Horses." *Journal of Drug Issues*, 12: 431–441.

Grabowski, John, Maxine L. Stitzer, and Jack E. Henningfield. 1984. *Behavioral Intervention Techniques in Drug Abuse Treatment*. Washington, DC: National Institute on Drug Abuse.

Graevan, David B. and Kathleen A. Graenan 1983. "Treated and Untreated Addicts: Factors Associated with Participation in Treatment and Cessation of Heroin Use." *Journal of Drug Issues*, 13: 207–218.

Griffin, M. L., R. D. Weiss, and S. M. Mirin. 1989. "A Comparison of Male and Female Cocaine Abusers," *Archives of General Psychiatry*, 46: 122–126.

Grinspoon, Lester and James B. Bakalar. 1979. "Cocaine." Pp. 241–247 in *Handbook on Drug Abuse*. Edited by Robert I. Dupont, Avram Goldstein, and John O'Donnell. Washington, DC: National Institute of Drug Abuse.

Hanson, Bill, George Beschner, James M. Walters, and Elliott Bovelle. 1985. *Life With Heroin*. Lexington, MA: Lexington Books.

Haas, Ann Pollinger and Herbert Hendin. 1987. "The Meaning of Chronic Marijuana Use among Adults: A Psychosocial Perspective." *Journal of Drug Issues*, 17: 333–348.

Hser, Yih-Ing, M. Douglas Anglin, and William McGlothlin. 1987. "Addict Careers. 1. Initiation of Use," *American Journal of Drug and Alcohol Abuse*, 13: 33–57.

Hunt, Leon Gibson and Carl D. Chambers. 1976. *The Heroin Epidemics: A Study of Heroin Use in the United States, 1965–1975*. New York: Spectrum Publications.

Inciardi, James A. 1977. *Methadone Diversion: Experience and Issues*. Washington, DC: National Institute on Drug Abuse.

———. 1986. *The War on Drugs*. Palo Alto, CA: Mayfield.

——— and Anne E. Pottieger. 1986. "Drug Use and Crime among Two Cohorts of Women Narcotics Users: An Empirical Assessment." *Journal of Drug Issues*, 16: 91–106.

——— and Duane C. McBride. 1990. "Legalizing Drugs: A Gormless, Naive Idea." The Criminologist, 15 (September–October): 1,3–4.

Johnson, Bruce D., Paul J. Goldstein, Edward Preble, James Schmidler, Douglas S. Lipton, Barry Spunt, and Thomas Miller. 1985. *Taking Care of Business: The Economics of Crime by Heroin Abusers*. Lexington, MA: Lexington Books.

Johnston, Lloyd D., Patrick M. O'Malley, and Jerald G. Bachman. 1987. *National Trends in Drug Use and Related Factors among American High School Seniors and Young Adults, 1975–1986*. Rockville, MD: U.S. Department of Health and Human Services, National Institute of Drug Abuse.

Johnston, Lloyd D. 1988. "Decline in Cocaine Use among High School Seniors." Press release, January 13, 1988. Ann Arbor, MI: University of Michigan News and Information Services.

Joint Committee on Narcotic Drugs. 1960. *Drug Addiction: Crime or Disease?*. Interim and Final Reports of the Joint Committee of the American Bar Association and American Medical Association on Narcotic Drugs. Bloomington, IN: Indiana University Press.

Jones, Paul. 1989. "Marijuana Smoking" pp. 59–61 in *Degrees of Deviance: Student Accounts of Their Deviant Behavior*, edited by Stuart Henry. Brookfield, VT: Avebury.

Jorquez, James S. 1983. "The Retirement Phase of Heroin Using Careers." *Journal of Drug Issues*, 13: 343–365.

Journal. 1983. Symposium on the British Drug System. *British Journal of Addictions*, Vol. 78: entire issue.

Kaplan, John. 1983. *The Hardest Drugs: Heroin and Public Policy*. Chicago: University of Chicago Press.

Kaufman, Joel Sidney Wolfe, and the Public Citizen Health Research Group. 1983. *Over the Counter Pills that Don't Work*. New York: Pantheon.

Kramer, John C., Vitezslav S. Fishman, and Don C. Littlefield. 1967. "Amphetamine Abuse Patterns and Effects of High Doses Taken Intravenously," *Journal of the American Medical Association*, 201: 305–309.

Kreek, Mary Jeanne. 1979. "Methadone in Treatment: Physiological and Pharmacological Issues." Pp. 57–86 in *Handbook on Drug Abuse*. Edited by Robert I. Dupont, Avram Goldstein, and John O'Donnell. Washington, DC: National Institute of Drug Abuse.

Levison, Peter K. Dean R. Gerstein, and Deborah R. Maloff. 1983. *Commonalities in Substance Abuse and Habitual Behavior*. Lexington, MA: Lexington Books.

Lewis, Virginia and Daniel Glaser. 1974. "Lifestyles Among Heroin Users," *Federal Probation*, 38: 21–28.

Lindesmith, Alfred R. and John H. Gagnon. 1964. "Anomie and Drug Addiction." Pp. 162–178 in *Anomie and Deviant Behavior: A Discussion and Critique*. Edited by Marshall B. Clinard. New York: Free Press.

Lindesmith, Alfred R. 1968. Addiction and Opiates. Chicago: Aldine.

———. 1975. "A Reply to McAuliffe and Gordon's 'A Test of Lindesmith's Theory of Addiction'." *American Journal of Sociology*, 81: 149–150.

McAuliffe, William E. and Robert A. Gordon. 1974. "A Test of Lindesmith's Theory of Addiction: Frequency of Euphoria Among Long-Term Addicts." *American Journal of Sociology*, 79: 795–840.

McBride, Robert B. 1983. "Business as Usual: Heroin Distribution in the United States." *Journal of Drug Issues*, 13: 147–166.

Merton, Robert K. 1968. *Social Theory and Social Structure*, enlarged ed. New York: Free Press.

Milby, Jesse B. 1981. *Addictive Behavior and Its Treatment*. New York: Springer.

Morgan, H. Wayne. 1981. *Drugs in America: A Social History, 1800–1980*. Syracuse, NY: Syracuse University Press.

Musto, David F. 1973. *The American Disease: The Origins of Narcotics Control*. New Haven, CT: Yale University Press.

National Commission on Marijuana and Drug Abuse. 1972. *Marihuana: A Signal of Misunderstanding*. Washington, DC: Government Printing Office.

National Institute of Drug Abuse. 1987. "WHO Reports Growing AIDS Epidemic in Europe." *NIDA Notes*, 2 (June, 1987): 2–3.

Newcomb, Michael D. and Peter M. Bentler. 1990. "Antecedents and Consequences of Cocaine Use: An Eight-Year Study From Early Adolescence to Young Adulthood." Pp. 158–181 in Lee N. Robins and Michael Rutter, eds., *Straight and Devious Pathways From Childhood to Adulthood*. Cambridge: Cambridge University Press.

Newman, Robert G., Sylvia Bashkow, and Margot Cates. 1973. "Arrest Histories Before and After Admission to a Methadone Maintenance Program." *Contemporary Drug Problems*, 2: 417–430.

Olin, William. 1980. *Escape From Utopia: My Ten Years in Synanon*. Santa Cruz, CA: Unity Press.

Peele, Stanton. 1990. *The Diseasing of America: Addiction Treatment Out of Control*. Lexington, MA: Lexington Books.

Petersen, Robert C. 1984. "Marijuana Overview." Pp. 1–17 in *Correlates and Consequences of Marijuana Use*. Edited by Meyer D. Glantz. Rockville, MD: National Institute of Drug Abuse, U.S. Department of Health and Human Services.

Platt, Jerome J. 1986. *Heroin Addiction*. Melbourne, FL: Krieger.

Poundstone, William. 1983. *Big Secrets: The Uncensored Truth About All Sorts of Stuff You are Never Supposed to Know*. New York: Quill.

Ray, Oakley. 1983. *Drugs, Society, and Human Nature*. 3rd ed. St. Louis: C.V. Mosby.

Redlinger, John and Jerry B. Michael. 1970. "Ecological Variations in Heroin Abuse," *The Sociological Quarterly*, 11: 219–229.

Reznikov, Diane. 1987. "States Spend $1.3 Billion on Substance Abuse Services, Treat 305,000 Drug Abuses in FY 1985." *NIDA Notes*, 2: (March): 11–12.

Rosecran, Jeffrey S. and Henry I. Spitz. 1987. "Cocaine Reconceptualized: Historical Overview." Pp. 5–16 in *Cocaine Abuse: New Directions in Treatment and Research*. Edited by Henry I. Spitz and Jeffrey S. Rosecran. New York Brunner/ Mazel.

Rosenbaum, Marsha. 1981. *Women on Heroin*. New Brunswick, NJ: Rutgers University Press.

Rowe, Dennis. 1987. "UN Drug Meet Draws 138 Nations." *C. J. International*, 3: 9.

Schaef, Ann Wilson. 1987. *When Society Becomes an Addict*. New York: Harper and Row.

Schlaadt, Richard G. and Peter T. Shannon. 1990. *Drugs*, 3rd ed. Englewood Cliffs, NJ: Prentice-Hall.

Schur, Edwin M. 1983. *Labeling Women Deviant: Gender, Stigma, and Social Control*. New York: Random House.

Seidenberg, Robert. 1976. "Advertising and Drug Acculturation." Pp. 19–25 in *Socialization in Drug Use*. Edited by Robert H. Coombs, Lincoln J. Fry, and Patricia G. Lewis. Cambridge, MA: Schenkman.

Siegal, Harvey Alan. 1987. "Current Patterns of Psychoactive Drug Use: Epidemiological Observations." Pp. 45–113 in *Chemical Dependencies: Patterns, Costs, and Consequences*. Edited by Carl D. Chambers, James A. Inciardi, David M. Peterson, Harvey A. Siegal, and O.Z. White. Athens: Ohio University Press.

Smart, Lisa. 1989. "College Kids as Coke Characters." Pp. 69–72 in *Degrees of Deviance: Student Accounts of Their Deviant Behavior*, edited by Stuart Henry. Brookfield, VT: Avebury.

Smith, David E., Donald R. Wesson, and Richard B. Seymour. "The Abuse of Barbiturates and Other Sedative-Hypnotics." Pp. 233–240 in *Handbook on Drug Abuse*. Edited by Robert I. Dupont, Avram Goldstein, and John O'Donnell. Washington, DC: National Institute of Drug Abuse.

Stephens, Richard C. and Emily Cottrell. 1972. "A Follow-Up Study of 200 Narcotic Addicts Committed for Treatment under the Narcotic Addict Rehabilitation Act (NARA)." *British Journal of Addiction*, 67: 45–53.

Stephens, Richard C. 1987. *Mind-Altering Drugs: Use, Abuse, and Treatment*. Newbury Park, CA: Sage.

———. 1991. *The Street Addict Role: A Theory of Heroin Addiction*. Albany: State University of New York Press.

Sullivan, Barbara. 1990. "Society's Hang Ups Wear Other Labels Now." *Des Moines Register*, October 16: 1T.

Tennant, Forest. 1990. "Outcomes of Cocaine-Dependence Treatment." Pp. 314–415 in *Problems of Drug Dependence, 1989*, edited by Louis S. Harris. Rockville, MD: National Institute on Drug Abuse.

Trebach, Arnold S. 1982. *The Heroin Solution*. New Haven, CT: Yale University Press.

———. 1987. *The Great Drug War*. New York: Macmillan.

Troyer, Ronald J. and Gerald E. Markle. 1983. *Cigarettes: The Battle Over Smoking*. New Brunswick, NJ: Rutgers University Press.

Vaillant, George E., Jane R. Bright, and Charles MacArthur. 1970. "Physicians' Use of Mood-Altering Drugs." *New England Journal of Medicine*, 282: 365–370.

Volkman, Rita and Donald R. Cressey. 1963. "Differential Association and the Rehabilitation of Drug Addicts." *American Journal of Sociology*, 69: 129–142.

Waldorf, Dan. 1973. *Careers in Dope*. Englewood Cliffs, NJ: Prentice-Hall.

———. 1983. "Natural Recovery from Opiate Addiction: Some Social-Psychological Processes of Untreated Recovery," *Journal of Drug Issues*, 13: 237–280.

Welti, Charles V. 1987. "Fatal Reactions to Cocaine." Pp. 33–54 in *Cocaine: A Clinician's Handbook*. Edited by Arnold M. Washton and Mark S. Gold. New York: Guilford Press.

Willie, Rolf. 1983. "Processes of Recovery from Heroin Dependence: Relationship to Treatment, Social Changes, and Drug Use." *Journal of Drug Issues*, 13: 333–342.

Wilson, Lana. 1989. "The Bong and the Weed." Pp. 57–59 in *Degrees of Deviance: Student Accounts of Their Deviant Behavior*, edited by Stuart Henry. Brookfield, VT: Avebury.

Winick, Charles. 1959–60. "The Use of Drugs by Jazz Musicians." *Social Problems*, 7: 240–254.

———. 1961. "Physician Narcotic Addicts." *Social Problems*, 9: 174–186.

———. 1965. "Epidemiology of Narcotics Use." Pp. 3–18 in *Narcotics*. Edited by David M. Wilner and Gene G. Kassebaum. New York: McGraw-Hill.

Yablonsky, Lewis. 1965. *The Tunnel Back: Synanon*. New York: Macmillan.

CHAPTER 9—DRUNKENNESS AND ALCOHOLISM

Bahr, Howard M. 1973. *Skid Row: An Introduction to Dissaffiliation*. New York: Oxford University Press.

Beckman, Linda S. 1975. "Women Alcoholics: A Review of Social and Psychological Studies." *Journal of Studies on Alcohol*, 36: 797–824.

Berreman, Gerald D. 1956. "Drinking Patterns of the Aleuts." *Quarterly Journal of Studies on Alcohol*, 17: 503–514.

Biegel, Allan and Stuart Chertner. 1977. "Toward a Social Model: An Assessment of Social Factors Which Influence Problem Drinking and Its Treatment." Pp. 197–233 in *Treatment and Rehabilitation of the Chronic Alcoholic*. Edited by Benjamin Kassin and Genri Geglieter. New York: Plenum Press.

Biernacki, Patricia. 1986. *Pathways from Heroin Addiction: Recovery Without Treatment*. Philadelphia: Temple University Press.

Bradstock, M. Kirsten, James S. Marks, Michelle R. Forman, Eileen M. Gentry, Gary C. Hogelin, Nancy J. Binkin, and Frederick L. Trowbridge. 1987. "Drinking-Driving and Health Lifestyle in the United States: Behavioral Risk Factor Surveys." *Journal of Studies on Alcohol*, 48: 147–152.

Bromet, Evelyn and Rudolf Moos. 1976. "Sex and Marital Status in Relation to the Characteristics of Alcoholics." *Journal of Studies on Alcohol*, 37: 1302–1312.

Bureau of Justice Statistics. 1985. *Jail Inmates in 1985*. Washington, DC: Department of Justice, Government Printing Office.

Burkett, Steven R. 1980. "Religiosity, Beliefs, Normative Standards and Adolescent Drinking." *Journal of Studies on Alcohol*, 41: 662–671.

Burkett, Steven R. and William T. Carrithers. 1980. "Adolescents' Drinking and Perceptions of Legal and Informal Sanctions." *Journal of Studies on Alcohol*, 41: 839–853.

Caetano, Raul. 1984. "Ethnicity and Drinking in Northern California: A Comparison among Whites, Blacks, and Hispanics." *Alcohol and Alcoholism*, 19: 31–44.

———. 1987. "Public Opinion about Alcoholism and Its Treatment." *Journal of Studies on Alcohol*, 48: 153–160.

———. 1989. "Drinking Patterns and Alcohol Problems in a National Sample of U.S. Hispanics." Pp. 147–162 in *The Epidemiology of Alcohol Use and Abuse among U.S. Minorities*. National Institute of Alcohol Abuse and Alcoholism. Washington, DC: Government Printing Office.

Cahalan, Don. 1982. "Epidemiology: Alcohol Use in American Society." Pp. 96–118 in Gomberg, White and Carpenter, *Alcohol, Science and Society Revisited*. Edited by Edith Lisansky Gomberg, Helene Raskin White, and John A. Carpenter. Ann Arbor: University of Michigan Press.

Chafetz, Morris E. 1983. "Is Compulsory Treatment of the Alcoholic Effective?" Pp. 294–302 in *Alcoholism: Introduction to Theory and Treatment*, 2nd ed. Edited by David A. Ward. Dubuque, IA: Kendall-Hunt.

Cherpitel, Cheryl J. Stephens and Haydee Rosovsky. 1990. "Alcohol Consumption and Casualties: A Comparison of Emergency Room Populations in the United States and Mexico." *Journal of Studies on Alcohol*, 51: 319–326.

Clark, Walter B. 1981. "The Contemporary Tavern." Pp. 425–470 in *Research Advances in Alcohol and Drug Problems*, Vol. 6. Edited by Y. Israel, Г. B. Glaser, H. Kalant, R. Popham, W. Schmidt, and R. Smart. New York: Plenum Press.

Conrad, Peter and Joseph W. Schnieder. 1980. *Deviance and Medicalization: From Badness to Sickness*. St. Louis: C. V. Mosby.

Cospers, Ronald and Florence Hughes. 1983. "So-Called Heavy Drinking Occupations: Two Empirical Tests." *Journal of Studies on Alcohol*, 43: 110–118.

Cressey, Donald R. 1955. "Changing Criminals: The Application of the Theory of Differential Association." *American Journal of Sociology*, 61: 116–120.

Donavan, John E., Richard Jessor, and Lee Jessor. 1983. "Problem Drinking in Adolescence and Young Adulthood: A Follow-Up Study." *Journal of Studies on Alcohol*, 44: 109–115.

Downs, William R. 1987. "A Panel Study of Normative Structure, Adolescent Alcohol Use, and Peer Alcohol Use." *Journal of Studies on Alcohol*, 48: 167–175.

Eckhardt, Michael J., Thomas C. Harford, Charles T. Kaelber, Elizabeth S. Parker, Laura S. Rosenthal, Ralph S. Ryback, Gian C. Salmoiraghi, Ernestine Vanderveen, and Kenneth R. Warren. 1981. "Health Hazards Associated with Alcohol Consumption." *Journal of the American Medical Association*, 246: 648–666.

Fagin, Ronald W., Jr. and Armand L. Mauss. 1978. "Padding the Revolving Door: An Initial Assessment of the Uniform Alcoholism and Intoxication Treatment Act in Practice." *Social Problems*, 26: 232–246.

Federal Bureau of Investigation. 1987. *Crime in the United States, 1986*. Washington, DC: Department of Justice, Government Printing Office.

Field, Eugene. 1897. *The Colonial Tavern*. Providence, RI: Preston and Rounds.

Fingarette, Herbert. 1988. *Heavy Drinking: The Myth of Alcoholism as a Disease*. Berkeley: University of California Press.

Firebaugh, W. C. 1928. *The Inns of Greece and Rome*. Chicago: F. M. Morris.

Glascow, Russell E., Robert C. Klesges, and Michael W. Vasey. 1983. "Controlled Smoking for Chronic Smokers: An Extension and Replication." *Addictive Behaviors*, 8: 143–150.

Gusfield, Joseph R. 1963. *Symbolic Crusade: Status Politics and the American Temperance Movement*. Urbana: University of Illinois Press.

Hasin, Deborah S., Bridget Grant, and Thomas C. Harford. 1990. "Male and Female Differences in Liver Cirrhosis Mortality in the United States, 1961–1985" *Journal of Studies on Alcohol*, 51: 123–129.

Herd, Denise. 1985. "Migration, Cultural Transformation, and the Rise of Black Liver Cirrhosis Mortality." *British Journal of Addiction*, 80: 397–410.

————. 1989. "The Epidemiology of Drinking Patterns and Alcohol-Related Problems among U.S. Blacks." In *The Epidemiology of Alcohol Use and Abuse among U.S. Minorities*. National Institute of Alcohol Abuse and Alcoholism. Washington, DC: Government Printing Office.

————. 1990. "Subgroup Differences in Drinking Patterns among Black and White Men: Results from a National Survey." *Journal of Studies on Alcohol*, 51: 221–232.

Hill, Shirley Y., Stuart R. Steinhauer, and Joseph Zubin. 1987. "Biological Markers for Alcoholism: A Vulnerability Model Conceptualization." Pp. 207–256 in *Alcohol and Addictive Behavior: Nebraska Symposium on Motivation, 1986*. Edited by P. Clayton Rivers. Lincoln: University of Nebraska Press.

Hilton, Michael E. 1988. "Trends in U.S. Drinking Patterns: Further Evidence From the Past 20 Years." *British Journal of Addictions*, 83: 269–278.

Hilton, Michael E. and Walter B. Clark. 1987. "Changes in American Drinking Patterns and Problems, 1967–1984." *Journal of Studies on Alcohol*, 48: 515–522.

Hilton, Michael E. and B. M. Johnstone. 1988. "Symposium on International Trends in Alcohol Consumption." *Contemporary Drug Problems*.

Hingson, Ralph, Tom Mangione, Allen Meyers, and Norman Scotch. 1982. "Seeking Help for Drinking Problems: A Study in the Boston Metropolitan Area." *Journal of Studies on Alcohol*, 43: 273–288.

Hingson, Ralph and Jonathan Howland. 1987. "Alcohol as a Factor for Injury or Death Resulting from Accidental Falls: A Review of the Literature." *Journal of Studies on Alcohol*, 48: 212–219.

Horgan, M.M., M. D. Sparrow, and R. Brazeau. 1986. *Alcoholic Beverage Taxation and Control Policies*, 6th ed. Ottawa: Brewer's Association of Canada.

Jacobs, James B. 1989. *Drunk Driving: An American Dilemma*. Chicago: University of Chicago Press.

Jacob, Theordore. 1987. "Alcoholism: A Family Interaction Perspective." Pp. 159–206 in *Alcohol and Addictive Behavior: Nebraska Symposium on Motivation, 1986*. Edited by P. Clayton Rivers. Lincoln: University of Nebraska Press.

Jellinek, E. M. 1960. *The Disease Concept of Alcoholism*. Highland Park, NJ: Hillhouse Press.

Jessor, Richard and Shirley L. Jessor. 1978. *Problem Behavior and Psychosocial Development: A Longitudinal Study of Youth*. New York: Academic Press.

Johnston, Lloyd D., Patrick M. O'Malley, and Jerald G. Bachman. 1986. *Drug Use among American High School Students, College Students, and Other Young Adults: National Trends Through 1985*. Rockville, MD: National Institute on Drug Abuse.

————. 1989. *Drug Use, Drinking, and Smoking: National Survey Results From High School, College and Young Adult Populations, 1975–1988*. Rockville, MD: National Institute on Drug Abuse.

Kandel, Denise B. 1980. "Drug and Drinking Behavior Among Youth." *Annual Review of Sociology*, 6: 235–285.

Keller, Mark and Vera Efron. 1955. "The Prevalence of Alcoholism." *Quarterly Journal of Studies on Alcohol*, 16: 628–637.

————. 1961. *Selected Statistical Tables on Alcoholic Beverages, 1950–1960, and on Alcoholism, 1930–1960*. New Brunswick, NJ: Quarterly Journal of Studies on Alcohol.

Keller, Mark. 1982. "On Defining Alcoholism: With Comment on Some Other Relevant Words." Pp. 119–133 in *Alcohol, Science and Society Revisited*. Edited by Edith Lisansky Gomberg, Helene Raskin White, and John A. Carpenter. Ann Arbor: University of Michigan Press.

Kitano, H. H. L., H. Hatanaka, W. T. Yeung, and S. Sue. 1985. "Japanese-American Drinking Patterns." Pp. 335–357 in *The American Experience with Alcohol: Contrasting Cultural Perspectives*. Edited by L. A. Bennett and G. M. Ames. New York: Plenum.

Klatsky, A. L., A. B. Siegelaub, C. Landy, and G. D. Friedman. 1985. "Racial Patterns of Alcoholic Beverage Use." *Alcoholism: Clinical and Experimental Research*, 7: 372–377.

Kurtz, Ernest. 1982. "Why A.A. Works: The Intellectual Significance of Alcoholics Anonymous." *Journal of Studies on Alcohol*, 43: 38–80.

Larkin, John R. 1965. *Alcohol and the Negro: Explosive Issues*. Zebulon, NC: Record Publishing Company.

LeMasters, Ersel E. 1975. *Blue-Collar Aristocrats: Life-Styles at a Working Class Tavern*. Madison: University of Wisconsin Press.

Lemert, Edwin M. 1954. *Alcohol and the Northwest Coast Indians. University of California Publications on Culture and Society*, Vol. 2, No. 6. Berkeley: University of California Press.

———. 1972. "Alcohol, Values, and Social Control." Pp. 112–122 in *Human Deviance, Social Problems and Social Control*, 2nd ed. Edited by Edwin M. Lemert. Englewood Cliffs, NJ: Prentice-Hall.

———. 1982. "Drinking Among American Indians." Pp. 80–95 in *Alcohol, Science and Society Revisited*. Edited by Edith Lisansky Gomberg, Helene Raskin White, and John A. Carpenter. Ann Arbor: University of Michigan Press.

Lender, Mark Edward and James Kirby Martin. 1982. *Drinking in America: A History*. New York: Free Press.

Lex, B. W. 1985. "Alcohol Problems in Special Populations." Pp. 89–187 in *The Diagnosis and Treatment of Alcoholism*, 2nd ed. Edited by J. H. Mendelson and N. K. Mello. New York: McGraw-Hill.

Linsky, Arnold S., John P. Colby, Jr., and Murray A. Straus. 1986. "Drinking Norms and Alcohol-Related Problems in the United States." *Journal of Studies on Alcohol*, 47: 384–393.

Linsky, Arnold S., John P. Colby, Jr., and Murray A. Straus. 1987. "Social Stress, Normative Constraints, and Alcohol Problems in the United States." *Social Science and Medicine*, 24: 875–883.

Lolli, Giorgio, Emilio Serianni, Grace M. Golder, and Peirpaolo Luzzatto-Fegis. 1958. *Alcohol in Italian Culture*. New York: Free Press.

Longclaws, Lyle, Gordon E. Barnes, Linda Grieve and Ron Dumoff. 1980. "Alcohol and Drug Use Among Broken Head Ojibwa." *Journal of Studies on Alcohol*, 41: 21–36.

Lowman, C., T. C. Hardford, and C. T. Kaelber. 1983. "Alcohol Use among Black Senior High School Students." *Alcohol Health and Research World*, 7: 37–46.

MacAndrew, Craig and Robert B. Edgerton. 1969. *Drunken Comportment: A Social Explanation*. Chicago: Aldine.

Mangin, William. 1957. "Drinking Among Andean Indians." *Quarterly Journal of Studies on Alcohol*, 18: 55–66.

Marchant, Ward. 1990. "Secular Approach to Alcoholism: Groups Offer Alternatives to AA." *Santa Barbara News Press*, October 20: 6.

Martin, Casey and Sally Casswell. 1987. "Types of Male Drinkers: A Multivariate Study." *Journal of Studies on Alcohol*, 48: 109–118.

McCarthy, Dennis, Sherry Morrison, and Kenneth C. Mills. 1983. "Attitudes, Beliefs, and Alcohol Use." *Journal of Studies on Alcohol*, 44: 328–341.

McCord, William and Joan McCord. 1961. *Origins of Alcoholism*. Stanford, CA: Stanford University Press.

Meilman, Philip W., Janet E. Stone, Michael S. Gaylor, and John H. Turco. 1990. "Alcohol Consumption by College Undergraduates: Current Use and 10-year Trends." *Journal of Studies on Alcohol*, 51: 38–395.

Miller, W. R. and R. K. Hester. 1980. "Treating the Problem Drinker: Modern Approaches." Pp. 11–141 in *The Addictive Behaviors: Treatment of Alcoholism, Drug Abuse, Smoking, and Obesity*. Oxford: Pergamon Press.

Mulford, Harold A. 1964. "Drinking and Deviant Behavior, USA." *Quarterly Journal of Studies on Alcohol*, 25: 634–650.

Mulkern, V. and Spence. R. 1984. *Alcohol Abuse/Alcoholism among Homeless Persons: A Review of the Literature. Final Report*. Washington, DC: Government Printing Office.

National Highway Traffic Safety Administration. 1988. *Drunk Driving Facts*. Washington, DC: National Center for Statistics and Analysis, Government Printing Office.

National Institute of Drug Abuse. 1988. National Household Survey on Drug Abuse: Main Findings, 1985. DHHS Pub. No. (ADM)88–1586. Rockville, MD: National Instititute of Drug Aubse.

Nordstrom, Goran and Mats Berglund. 1987. "A Prospective Study of Successful Long-Term Adjustment in Alcohol Dependence: Social Drinking Versus Abstinence." *Journal of Studies on Alcohol,* 48: 95–103.

Nusbaumer, Michael R., Armand L. Mauss, and David C. Pearson. 1982. "Draughts and Drunks: The Contributions of Taverns and Bars to Excessive Drinking in America." *Deviant Behavior*, 3: 329–358.

Orcutt, James D. 1976. "Ideological Variations in the Structure of Deviant Types: A Multivariate Comparison of Alcoholism and Heroin Addiction." *Social Forces*, 55: 419–437.

Pandina, Robert J. 1982. "Effects of Alcohol on Psychological Processes." Pp. 38–62 in *Alcohol, Science and Society Revisited*. Edited by Edith Lisansky Gomberg, Helene Raskin White, and John A. Carpenter. Ann Arbor: University of Michigan Press.

Patrick, Charles H. 1952. *Alcohol, Culture and Society*. Durham, NC: Duke University Press.

Peele, Stanton. 1990. *The Diseasing of America: Addiction Treatment Out of Control*. Lexington, MA: Lexington Books.

Pendery, Mary L., Irving M. Maltzman, and L. Jolyon West. 1982. "Controlled Drinking by Alcoholics." *Science*, 217: July, 169–175.

Phillips, David P. 1979. "Suicide, Motor Vehicle Fatalities, and Mass Media: Evidence Toward a Theory of Suggestion." *American Journal of Sociology*, 84: 1150–1173.

Pittman, David J. and C. Wayne Gordon. 1958. *The Revolving Door*. New York: Free Press.

Plaut, Thomas F. A. 1967. *Alcohol Problems: A Report to the Nation by the Cooperative on the Study of Alcoholism*. New York: Oxford University Press.

Popham, Robert E. 1962. "The Urban Tavern: Some Preliminary Remarks." *Addictions,* 9: 17–26.

Robinson, David. 1972. "The Alcohologist's Addition: Some Implications of Having Lost Control Over the Disease Concept of Alcoholism." *Quarterly Journal of Studies on Alcohol,* 33: 1028–1042.

Room, Robin. 1976. "Ambivalence as a Sociological Explanation: The Case of Cultural Explanations of Alcohol Problems." *American Sociological Review,* 41: 1047–1065.

———. 1983. "Alcohol and Crime: Behavioral Aspects." Pp. 34–44 in *Encyclopedia of Crime and Justice,* Vol. 1. Edited by Sanford H. Kadish. New York: Free Press.

Roman, Paul M. 1981. "From Employee Alcoholics to Employee Assistance: Deemphasis on Prevention and Alcohol Problems in Work-Based Programs." *Journal of Studies on Alcohol,* 42: 244–272.

Ross, H. Laurence. 1982. *Deterring the Drunk Driver: Legal Policy and Social Control.* Lexington, MA: Lexington Books.

———. 1985. "Britain's Christmas Crusade against Drinking and Driving." *Journal of Studies on Alcohol,* 48: 476–482.

Sadoun, Roland, Giorgio Lolli, and Milton Silverman. 1965. *Drinking in French Culture.* New Brunswick, NJ: Rutgers Center of Alcohol Studies.

Secretary of Health and Human Services. 1981. *Alcohol and Health: Fourth Special Report to the U.S. Congress.* Rockville, MD: National Institute of Alcohol Abuse and Alcoholism, Government Printing Office.

———. 1987. *Alcohol and Health: Sixth Special Report to the U.S. Congress.* Rockville, MD: National Institute of Alcohol Abuse and Alcoholism, Government Printing Office.

———. 1990. *Alcohol and Health: Seventh Special Report to the U.S. Congress.* Rockville, MD: National Institute of Alcohol Abuse and Alcoholism, Government Printing Office.

Snyder, Charles R. 1978. *Alcohol and the Jews.* Carbondale, IL: Southern Illinois University Press; originally published 1958.

Sobell, Mark B. and Linda C. Sobell. 1973. "Alcoholics Treated by Individualized Behavior Therapy." *Behavior Research and Therapy,* 11: 599–618.

Spradley, James A. 1970. *You Owe Yourself a Drunk: An Ethnography of Urban Nomads.* Boston: Little, Brown. p. 117

Sterne, Muriel W. 1967. "Drinking Patterns and Alcoholism Among American Negroes." Pp. 71–74 in *Alcoholism.* Edited by David J. Pittman. New York: Harper and Row.

Stivers, Richard. 1976. *A Hair of the Dog: Irish Drinking and American Stereotype.* University Park, PA: Pennsylvania State University Press.

Straus, Robert. 1982. "The Social Costs of Alcohol." Pp. 137–147 in *Alcohol, Science and Society Revisited.* Edited by Edith Lisansky Gomberg, Helene Raskin White, and John A. Carpenter. Ann Arbor: University of Michigan Press.

Sue, S., H. H. L. Kitano, H. Hatanka, and W. T. Yeung. 1985. "Alcohol Consumption among Chinese in the United States." Pp. 359–371 in *The American Experience with Alcohol: Contrasting Cultural Perspectives.* Edited by L. A. Bennett and G. M. Ames. New York: Plenum.

Sugerman, A. Arthur. 1982. "Alcoholism: An Overview of Treatment Models and Methods." Pp. 262–278 in Gomberg, White and Carpenter.

Taylor, John R., Terri Combs-Orme, and David A. Taylor. 1983. "Alcohol and Mortality: Diagnostic Considerations." *Journal of Studies on Alcohol*, 44: 17–25.

Traml, Vladimir G. 1975. "Production and Consumption of Alcoholic Beverages in the USSR." *Journal of Studies on Alcohol*, 36: 285–320.

Trebach, Arnold. 1987. *The Great Drug War*. New York: Macmillan.

Trice, Harrison M. and J. Richard Wahl. 1958. "A Rank Order Analysis of the Symptoms of Alcoholism." *Quarterly Journal of Studies on Alcohol*, 19: 636–648.

Trice, Harrison M. 1959. "The Affiliation Motive and Readiness to Join Alcoholics Anonymous." *Quarterly Journal of Studies on Alcohol*, 20: 313–321.

———. 1966. *Alcoholism in America*. New York: McGraw-Hill.

Trice, Harrison M. and Paul M. Roman. 1972. *Spirits and Demons at Work: Alcohol and Other Drugs on the Job*. Ithaca, NY: Industrial and Labor Relations Paperback.

Umanna, Ifekandu. 1967. "The Drinking Culture of a Nigerian Community: Onitsha." *Quarterly Journal of Studies on Alcohol*, 28: 529–537.

U.S. General Accounting Office. 1985. *Homelessness: A Complex Problem and the Federal Response*. GAO/HRD-85-40. Gaithersburg, MD: Government Accounting Office, Government Printing Office.

Waldorf, Dan. "Natural Recovery From Opiate Addiction: Some Social-Psychological Processes of Untreated Recovery." *Journal of Drug Issues*, 13: 237–280.

Wallack, Lawrence, Warren Breed, and John Cruz. 1987. "Alcohol on Prime-Time Television." *Journal of Studies on Alcohol*, 48: 33–38.

Wanberg, Kenneth W. and John L. Horn. 1970. "Alcoholism Patterns in Men and Women." *Quarterly Journal of Studies on Alcoholism*, 31: 40–61.

Ward, David A. 1983. "Conceptions of Alcoholism." Pp. 4–13 in *Alcoholism: Introduction to Theory and Treatment*, 2nd ed. Edited by David A. Ward. Dubuque, IA: Kendall-Hunt.

Washburne, Chandler. 1961. *Primitive Drinking: A Study of the Uses and Functions of Alcohol in Preliterate Societies*. New Haven, CT: College and University Press.

Weed, Frank J. 1990. "The Victim-Activist Role in the Anti-Drunk Driving Movement." *The Sociological Quarterly*, 31: 459–473.

Welte, John W. and Grace M. Barnes. 1987. "Alcohol Use among Adolescent Minority Groups." *Journal of Studies on Alcohol*, 48: 329–336.

Williams, G. D., D. Doernberg, F. Stinson, and J. Noble. 1986. "National, State, and Regional Trends in Apparent Per Capita Consumption of Alcohol." *Alcohol Health and Research World*, 10: 60–63.

Wilsnack, Richard W., Sharon C. Wilsnack, and Albert D. Klassen. 1984. "Women's Drinking and Drinking Problems: Patterns from a 1981 Survey." *American Journal of Public Health*, 74: 1231–1238.

———. 1987. "Antecedents and Consequences of Drinking and Drinking Problems in Women: Patterns from a U.S. National Survey." Pp. 85–158 in *Alcohol and Addictive Behavior: Nebraska Symposium on Motivation, 1986*. Edited by P. Clayton Rivers. Lincoln: University of Nebraska Press.

Wiseman, Jacqueline P. 1981. "Sober Comportment: Patterns and Perspectives on Alcohol Addition." *Journal of Studies on Alcohol*, 42: 106–126.

Wolfgang, Marvin E. 1958. *Patterns of Criminal Homicide*. Philadelphia: University of Pennsylvania Press.

CHAPTER 10—HETEROSEXUAL DEVIANCE

Abramson, Paul R. and Haruo Hayashi. 1984. "Pornography in Japan: Cross-Cultural and Theoretical Considerations." Pp. 173–183 in *Pornography and Sexual Aggression*. Edited by Neil M. Malamuth and Edward Donnerstein. New York: Academic Press.

Aday, Jr., David P. 1990. *Social Control at the Margins: Toward a General Understanding of Deviance*. Belmont, CA: Wadsworth.

Angel, Ronald and Marta Tienda. 1982. "Determinants of Extended Household Structure: Cultural Pattern or Economic Necessity?" *American Journal of Sociology*, 87: 1360–1383.

Attorney General's Commission on Pornography. 1986. Final Report. Washington, DC: Government Printing Office.

Baron, Larry and Murray A. Straus. 1984. "Sexual Stratification, Pornography, and Rape in the United States." Pp. 186–209 in *Pornography and Sexual Aggression*. Edited by Neil M. Malamuth and Edward Donnerstein. New York: Academic Press.

Barry, Kathleen. 1984. *Female Sexual Slavery*. New York: New York University Press.

Bartell, Gilbert D. 1971. *Group Sex*. New York: Peter H. Wyden.

Bell, Robert R. and Dorthyann Peltz. 1976. "Extramarital Sex among Women." Paper cited in *Social Deviance: a Substantive Analysis*, rev. ed., p. 74. By Robert R. Bell. Homewood, IL: Dorsey.

Bellis, David J. 1990. "Fear of AIDS and Risk Reduction among Heroin-Addicted Female Street Prostitutes: Personal Interviews with 72 Southern California Subjects." *Journal of Alcohol and Drug Education*, 35:26–37.

Bryan, James H. 1965. "Apprenticeships in Prostitution." *Social Problems*, 12: 278–297.

———. 1966. "Occupational Ideologies and Individual Attitudes of Call Girls." *Social Problems*, 13: 437–447.

Buunk, Bram P. and Barry van Driel. 1989. *Variant Lifestyles and Relationships*. Newbury Park, CA: Sage.

Cohen, Bernard. 1980. *Deviant Street Networks: Prostitution in New York City*. Lexington, MA: Lexington Books.

Commission on Obscenity and Pornography. 1970. Final Report. New York: Bantam Books (originally published by Government Printing Office, 1970).

Crockenberg, Susan B. and Barbara A. Soby. 1989. "Self-Esteem and Teenage Pregnancy." pp. 125–164 in *The Social Importance of Self-Esteem*. Edited by Andrew M. Mecca, Neil J. Smelser, and John Vasconcellos. Berkeley: University of California Press.

Davis, Nanette J. 1981. "Prostitutes," Pp. 305–313 in *Deviance: The Interactionist Perspective*, 4th ed. Edited by Earl Rubington and Martin S. Weinberg. New York: Macmillan.

Davis, Kingsley. 1937. "The Sociology of Prostitution." *American Sociological Review*, 2: 744–755.

D'Emilio, John and Estelle B. Freedman. 1988. *Intimate Matters: A History of Sexuality in America*. New York: Harper and Row.

DeLamater, John. 1981. "The Social Control of Sexuality." *Annual Review of Sociology*, 7: 263–290.

Diana, Lewis. 1985. *The Prostitute and Her Clients*. Springfield, IL: Charles C Thomas.

Donnerstein, Edward, Daniel Linz, and Steven Penrod. 1987. *The Question of Pornography: Research Findings and Policy Implications*. New York: Free Press.

Downs, Donald Alexander. 1989. *The New Politics of Pornography*. Chicago: University of Chicago Press.

Eysenck, Hans J. 1972. "Obscenity—Officially Speaking." *Penthouse*, 3 (11): 95–102.

Fang, B. 1976. "Swinging: In Retrospect." *Journal of Sex Research*, 12: 220–237.

Federal Bureau of Investigation. 1987. *Crime in the United States*, 1986. Washington, DC: Department of Justice, Government Printing Office.

Forsyth, Craig J. and Lee Fournet. 1987. "A Typology of Office Harlots: Mistresses, Party Girls, and Career Climbers." *Deviant Behavior*, 8: 319–328.

Frost, Janet. 1989. "Affairs." Pp. 25–27 in *Degrees of Deviance: Student Accounts of Their Deviant Behavior*, edited by Stuart Henry. Brookfield, VT: Avebury.

Gagnon, John H. and William Simon. 1970a. *Sexual Conduct: The Social Sources of Human Sexuality*. Chicago: Aldine.

———. 1970b. "Perspectives on the Sexual Scene." Pp. 1–12 in *The Sexual Scene*. Edited by John H. Gagnon and William Simon. Chicago: Aldine.

Gebhard, Paul H., John H. Gagnon, Wardell B. Pomeroy, and Cornelia V. Christenson. 1965. *Sex Offenders*. New York: Harper and Row.

Gilmartin, Brian D. 1978. *The Gilmartin Report*. Secaucus, NJ: Citadel Press.

Goldstein, M. J., H. S. Kant, and J. J. Hartman. 1974. *Pornography and Sexual Deviance*. Berkeley: University of California Press.

Gorchos, Harvey L., Jean S. Gorchos, and Joel Fischer, eds. 1986. *Helping the Sexually Oppressed*. Englewood Cliffs, NJ: Prentice-Hall.

Gray, Diana. 1973. "Turning Out: A Study of Teenage Prostitution." *Urban Life and Culture*, 1: 401–425.

Harris, Louis. 1987. *Inside America*. New York: Vintage.

Heyl, Barbara Sherman. 1979. *The Madam as Entrepreneur: Career Management in House Prostitution*. New Brunswick, NJ: Transaction Books.

Hobson, Barbara Meil. 1987. *Uneasy Virtue: The Politics of Prostitution and the American Reform Tradition*. New York: Basic Books

Holzman, Harold R. and Sharon Pines. 1982. "Buying Sex: The Phenomenology of Being a John." *Deviant Behavior*, 4:

Hunt, Morton. 1974. *Sexual Behavior in the 1970s*. Chicago: Playboy Press.

Jackman, Norman R., Richard O'Toole, and Gilbert Geis. 1963. "The Self-Image of the Prostitute." *The Sociological Quarterly*, 4: 150–161.

James, Jennifer. 1977. "Prostitutes and Prostitution." Pp. 368–428 in *Deviants: Voluntary Actors in a Hostile World*. Edited by Edward Sagarin and Fred Montanino. Morristown, NJ: General Learning Press.

Jencks, Richard J. 1985. "Swinging: A Test of Two Theories and a Proposed New Model." *Journal of Sex Research*, 14: 517–527.

Jenness, Valerie. 1990. "From Sex as Sin to Sex as Work: COYOTE and the Reorganization of Prostitution as a Social Problem." *Social Problems*, 37: 403–420.

Kantner, J. F. and M. Zelnick. 1972. "Sexual Experience of Young Unmarried Women in the United States." *Family Planning Perspectives*, 4: 9–18.

Karlen, Arno. 1988. *Threesomes: Studies in Sex, Power, and Intimacy*. New York: William Morrow.

Kinsey, Alfred C. Ward B. Pomeroy, and Charles E. Martin. 1948. *Sexual Behavior in the Human Male*. Philadelphia: Saunders.

————, Ward B. Pomeroy, Charles C. Martin, and Paul H. Gebhard. 1953. *Sexual Behavior in the Human Female*. Philadelphia: Saunders.

Klassen, Albert D., Colin J. Williams, and Eugene E. Levitt. 1989. *Sex and Morality in the U.S.* Middletown, CT: Wesleyan University Press.

Kunstel, Marcia and Joseph Albright. 1987. "Prostitution Thrives on Young Girls." *C.J. International*, 3: 9–11.

Little, Craig B. 1983. *Understanding Deviance and Control: Theory, Research, and Control*. Itasca, IL: F. E. Peacock.

Malamuth, Neil M. and Edward Donnerstein, eds., 1984. *Pornography and Sexual Aggression*. New York: Academic Press.

Miller, Eleanor M. 1986. *Street Woman*. Philadelphia: Temple University Press.

Murphey, Michael. 1987. "She Sells Sex on the Dark Side of the Street." *Spokane Spokesman Review*, November 11: A7.

Muedeking, George D. 1977. "Pornography and Society." Pp. 463–502 in *Deviants: Voluntary Actors in a Hostile World*. Edited by Edward Sagarin and Fred Montanino. Morristown, NJ: General Learning Press.

Palson, Charles and Rebecca Palson. 1972. "Swinging in Wedlock." *Society*, 9: 28–37.

Plummer, Ken. 1982. "Symbolic Interactionism and Sexual Conduct: An Emergent Perspective." Pp. 223–241 in *Human Sexual Relations: Towards A Redefinition of Sexual Politics*. Edited by Mike Brake. New York: Pantheon.

Polsky, Ned. 1967. *Hustlers, Beats, and Others*. Chicago: Aldine.

Potter, Gary W. 1989. "The Retail Pornography Industry and the Organization of Vice." *Deviant Behavior*, 10: 233–251.

Reiss, Ira L. 1970. "Premarital Sex as Deviant Behavior: An Application of Current Approaches to Deviance." *American Sociological Review*, 35: 78–88.

————. 1967. *The Social Context of Premarital Sexual Permissiveness*. New York: Holt, Rinehart and Winston.

Reynolds, Helen. 1986. *The Economics of Prostitution*. Springfield, IL: Charles C Thomas.

Sagarin, Edward. 1977. "Sex Deviance: A View From Middle America." Pp. 429–462 in *Deviants: Voluntary Actors in a Hostile World*. Edited by Edward Sagarin and Fred Montanino. Morristown, NJ: General Learning Press.

Skolnick, Jerome H. 1975. *Justice Without Trial: Law Enforcement in Democratic Society*, 2nd ed. New York: Wiley.

Smith, M. Dwayne and Carl Hand. 1987. "The Pornography/Aggression Linkage: Results From a Field Study," *Deviant Behavior*, 8: 389–399.

Smith, Thomas W. 1990. "The Sexual Revolution?" *Public Opinion Quarterly*, 54:

Spanier, Graham P. 1983. "Married and Unmarried Cohabitation in the United States, 1980." *Journal of Marriage and the Family*, 45: 277–288.

Stack, Carol B. 1974. *All Our Kin: Strategies for Survival in a Black Community*. New York: Harper Colophon.

Valentine, Bettylou. 1978. *Hustling and Other Hard Work*. New York: Free Press.

Weisberg, D. Kelly. 1985. *Children of the Night: A Study of Adolescent Prostitution*. Lexington: Lexington Books.

Winick, Charles and Paul M. Kinsie. 1971. *The Lively Commerce: Prostitution in the United States*. New York: Quadrangle.

Zelnick, M. and J.F. Kantner. 1980. "Sexual Activity, Contraceptive Use and Pregnancy among Metropolitan-Area Teenagers, 1971–1979." *Family Planning Perspectives*, 12: 230–237.

CHAPTER 11—HOMOSEXUALITY

Adam, Barry D. 1987. *The Rise of a Gay and Lesbian Movement.* Boston: Twayne.

Akers, Ronald L. 1985. *Deviant Behavior: A Social Learning Approach*, 3rd ed. Belmont, CA: Wadsworth.

American College Health Association. 1987. "AIDS: What Everyone Should Know." Rockville, MD: American College Health Association.

Baum, Andrew and Lydia Temoshok. 1990. "Psychosocial Aspects of Acquired Immunodeficiency Syndrome." Pp. 1–16 in Lydia Temoshok and Andrew Baum, eds., *Psychosocial Perspectives on AIDS: Etiology, Prevention, and Treatment.* Hillsdale, NJ: Laurence Erlbaum Associates.

Bauman, Laurie J. and Karolyn Siegel. 1990. "Misperception among Gay Men of the Risk of AIDS Associated with Their Sexual Behavior." Pp. 81–101 in Lydia Temoshok and Andrew Baum, eds., *Psychosocial Perspectives on AIDS: Etiology, Prevention, and Treatment.* Hillsdale, NJ: Laurence Erlbaum Associates.

Bell, Alan P. and Martin S. Weinberg. 1978. *Homosexualities: A Study of Diversity among Men and Women.* New York: Simon and Schuster.

———, Martin S. Weinberg, and Sue Kiefer Hammersmith. 1981. *Sexual Preference: Its Development in Men and Women.* Bloomington: Indiana University Press.

Berger, Gregory, Lori Hank, Tom Ravzi, and Lawrence Simkins. 1987. "Detection of Sexual Orientation by Heterosexuals and Homosexuals." *Journal of Homosexuality*, 13: 83–100.

Berger, Raymond M. 1986. "Gay Men." Pp. 162–180 in *Helping the Sexually Oppressed.* Edited by Harvey L. Gochros, Jean S. Gochros, and Joel Fischer. Englewood Cliffs, NJ: Prentice-Hall.

———. 1990. "Men Together: Understanding the Gay Couple," *Journal of Homosexuality,* 19: 31–49.

Brannock, JoAnn C. and Beata E. Chapman. 1990. "Negative Sexual Experiences with Men among Heterosexual Women and Lesbians," *Journal of Homosexuality,* 19: 105–110.

Blumstein, Philip W. and Pepper Schwartz. 1974. "Lesbianism and Bisexuality." Pp. 278–295 in *Sexual Deviance and Sexual Deviants.* Edited by Erich Goode and Richard T. Troiden. New York: Morrow.

———. 1983. *American Couples.* New York: Morrow.

Boston Lesbian Psychologies Collective, eds. 1987. *Lesbian Psychologies: Explorations and Challenges.* Urbana: University of Illinois Press.

Britton, Dana M. 1990. "Homophobia and Homosociality: An Analysis of Boundary Maintenance." *The Sociological Quarterly*, 31: 423–439.

Browning, Christine. 1987. "Therapeutic Issues and Intervention Strategies with Young Adult Lesbian Clients: A Developmental Approach." *Journal of Homosexuality*, 14: 45–52.

Coates, Thomas J., Ron D. Stall, and Colleen C. Hoff. 1990. "Changes in Sexual Behavior among Gay and Bisexual Men Since the Beginning of the AIDS Epidemic." Pp. 103–137 in Lydia Temoshok and Andrew Baum, eds., *Psychosocial Perspectives on AIDS: Etiology, Prevention, and Treatment.* Hillsdale, NJ: Laurence Erlbaum Associates.

Coleman, Eli. 1981–2. "Developmental Stages in the Coming Out Process." *Journal of Homosexuality*, 7: 31–43.

Cronin, Denise M. 1974. "Coming Out among Lesbians." Pp. 265–277 in *Sexual Deviance and Sexual Deviants*. Edited by Erich Goode and Richard T. Troiden. New York: Morrow.

Dank, Barry. 1971. "Coming Out in the Gay World." *Psychiatry*, 34: 192–198.

———. 1972. "Why Homosexuals Marry Women." *Medical Aspects of Human Sexuality*, 6: 14–23.

Davies, Christie. 1982. "Sexual Taboos and Social Boundaries." *American Journal of Sociology*, 87: 1032–1063.

Dooley, Janne. 1986. "Lesbians." Pp. 181–190 in *Helping the Sexually Oppressed*. Edited by Harvey L. Gochros, Jean S. Gochros, and Joel Fischer. Englewood Cliffs, NJ: Prentice-Hall.

Dover, Kenneth J. 1978. *Greek Homosexuality*. Cambridge, MA: Harvard University Press.

Eckholm, Erik and John Tierney. 1990. "AIDS in Africa: A Killer Rages On." *New York Times*, September 16, p. 1.

Ficarrotto, Thomas J. 1990. "Racism, Sexism, and Erotophobia: Attitudes of Heterosexuals Toward Homosexuals." *Journal of Homosexuality,* 19: 111–116.

Ford, Clellan S. and Frank A. Beach. 1951. *Patterns of Sexual Behavior*. New York: Harper and Row.

Fumento, Michael. 1990. T*he Myth of Heterosexual AIDS*. New York: Basic Books.

Gallup Poll. 1986. "Sharp Decline Found in Support for Legalizing Gay Relations." *The Gallup Report*, 254.

Geis, Gilbert. 1972. *Not the Law's Business?* Rockville, MD: National Institute of Mental Health.

Gilligan, Carol. 1982. *In a Different Voice: Psychological Theory and Women's Development*. Cambridge, MA: Harvard University Press.

Glassner, Barry and Carol Owen. 1976 "Variations in Attitudes Toward Homosexuality." *Cornell Journal of Social Relations*, 11: 161–176.

Goode, Erich and Richard T. Troiden, eds. 1974. *Sexual Deviance and Sexual Deviants*. New York: Morrow.

Goffman, Erving. 1963. *Stigma: Notes on the Management of Spoiled Identity*. Englewood Cliffs, NJ: Prentice-Hall.

Green, Richard. 1987. *The "Sissy Boy Syndrome" and the Development of Homosexuality*. New Haven, CT: Yale University Press.

Greenberg, David F. 1988. *The Construction of Homosexuality*. Chicago: University of Chicago Press.

Hammersmith, Sue Kiefer. 1987. "A Sociological Approach to Counseling Homosexual Clients and Their Families." *Journal of Homosexuality*, 14: 173–190.

Harry, Joseph. 1982. *Gay Children Grown Up: Gender Culture, and Gender Deviance*. New York: Praeger.

———. 1984. "Sexual Orientation as Destiny." *Journal of Homosexuality*, 10: 111–124.

———. 1990. "A Probability Sample of Gay Men," *Journal of Homosexuality,* 19: 89–104.

Hedblom, Jack H. 1972. "The Female Homosexual: Social and Attitudinal Dimensions." Pp. 50–65 in *The Homosexual Dialectic*. Edited by James A. McCaffrey. Englewood Cliffs, NJ: Prentice-Hall.

Herdt, Gilbert. 1981. *The Guardians of the Flute*. New York: McGraw-Hill.

Herek, G. M. 1984. "Beyond Homophobia: A Social Psychological Perspective on Attitudes Toward Lesbians and Gay Men." *Journal of Homosexuality*, 10: 1–21.

Hetrick, Emery S. and A. Damien Martin. 1987. "Developmental Issues and Their Resolution for Gay and Lesbian Adolescents." *Journal of Homosexuality*, 14: 25–43.

Humphreys, Laud. 1972. *Out of the Closets: The Sociology of Homosexual Liberation.* Englewood Cliffs, NJ: Prentice-Hall.

———. 1975. *Tearoom Trade: Impersonal Sex in Public Places*, enlarged ed. Chicago: Aldine.

——— and Brian Miller. 1980. "Identities in the Emerging Gay Culture." Pp. 142–156 in *Homosexual Behavior: A Modern Reappraisal.* Edited by Judd Marmor. New York: Basic Books.

Jay, Karla and Allen Young. 1979. *The Gay Report: Lesbians and Gay Men Speak Out About Sexual Experiences and Lifestyles.* New York: Summit Books.

Johnston, Jill. 1973. *Lesbian Nation: The Feminist Solution.* New York: Simon and Schuster.

Katz, Jonathan, ed. 1976. *Gay American History: Lesbians and Gay Men in the U.S.A.* New York: Cromwell.

———. 1983. *Gay/Lesbian Almanac: A New Documentary.* New York: Harper and Row.

Kelly, Robert J. 1990. "AIDS and the Societal Reaction." Pp. 47–61 in *Perspectives on Deviance: Dominance, Degradation, and Denigration.* Edited by Robert J. Kelly and Donal E. J. MacNamara. Cincinnati: Anderson.

Kinsey, Alfred C., Ward B. Pomeroy, and Charles E. Martin. 1948. *Sexual Behavior in the Human Male.* Philadelphia: Saunders.

———, Ward B. Pomeroy, Charles E. Martin and Paul H. Gebhard. 1953. *Sexual Behavior in the Human Female.* Philadelphia: Saunders.

Klassen, Albert D., Colin J. Williams, and Eugene E. Levitt. 1989. *Sex and Morality in the U.S.* Middletown, CT: Wesleyan University Press.

Langevin, Ron. 1985. "Introduction." Pp. 1–13 in *Erotic Preference, Gender Identity, and Aggression in Men: New Research Studies.* Edited by Ron Langevin. Hillsdale, NJ: Lawrence Erlbaum Associates.

Lewin, Ellen and Terrie A. Lyons. 1982. "Everything In Its Place: The Coexistence of Lesbianism and Motherhood." Pp. 249–273 in *Homosexuality: Social, Psychological, and Biological Issues.* Edited by William Paul, James D. Weinrich, John C. Gonsiorek, and Mary E. Hotvedt. Beverly Hills, CA: Sage.

Lindner, Robert. 1963. "Homosexuality and the Contemporary Scene." Pp. 60–82 in *The Problem of Homosexuality in Modern Society.* Edited by Hendrix Ruitenbeek. New York: Dutton.

Luckenbill, David F. 1986. "Deviant Career Mobility: The Case of Male Prostitutes." *Social Problems*, 33: 283–293.

Lynch, Frederick R. 1987. "Non-Ghetto Gays: A Sociological Study of Suburban Homosexuals." *Journal of Homosexuality*, 13: 13–42.

Magee, Brian. 1966. *One in Twenty: A Study of Homosexuality in Men and Women.* New York: Stein and Day.

McCaghy, Charles H. and James K. Skipper, Jr. 1969. "Lesbian Behavior as an Adaptation to the Occupation of Stripping." *Social Problems*, 17: 262–270.

McDonald, A. P. 1976. "Homophobia: Its Roots and Meanings." *Homosexual Counseling Journal*, 3: 23–33.

McDonald, Gary J. 1982. "Individual Differences in the Coming Out Process for Gay Men: Implications for Theoretical Models." *Journal of Homosexuality*, 8: 47–60.

McWhirter, David P. and Andrew M. Mattison. 1984. *The Male Couple: How Relationships Develop*. Englewood Cliffs, NJ: Prentice-Hall.

Miller, Neil. 1989. *In Search of Gay America*. New York: Atlantic Monthly Press.

Money, John. 1988. *Gay, Straight, and In-Between: The Sociology of Eriotic Orientation*. New York: Oxford.

Muchmore, Wes and William Hanson. 1989. "A Gay Man's Guide to Coming Out." Pp. 72–74 in *The Alyson Almanac*. Boston: Alyson Publications.

Paul, Jay P. 1990. Review of Richard Green, "The 'Sissy Boy Syndrome' and the Development of Homosexuality," *Journal of Homosexuality*, 19: 14–147.

Paul, William and James D. Weinrich. 1982. "Whom and What We Study: Definition and Scope of Sexual Orientation." Pp. 23–28 in *Homosexuality: Social, Psychological, and Biological Issues*. Edited by William Paul, James D. Weinrich, John C. Gonsiorek, and Mary E. Hotvedt. Beverly Hills, CA: Sage.

Peplau, Letitia Anne, Christine Padesky, and Mykol Hamilton. 1982. "Satisfaction in Lesbian Relationships." *Journal of Homosexuality*, 8: 23–35.

Plummer, Kenneth. 1975. *Sexual Stigma: An Interactionist Account*. London: Routledge and Kegan Paul.

———, ed. 1981. *The Making of the Modern Homosexual*. London: Hutchinson.

———. 1989. "Lesbian and Gay Youth in England," *Journal of Homosexuality*, 17: 195–223.

Price, Monroe. 1989. *Shattered Mirrors: Our Search for Identity and Community in the AIDS Era*. Cambridge, MA: Harvard University Press.

Quadland, Michael C and William D. Shattis. 1987. "AIDS, Sexuality, and Sexual Control." *Journal of Homosexuality*, 13: 13–42.

Rechy, John. 1963. *City of the Night*. New York: Grove Press.

Reiss,, Ira L. 1986. *Journey into Sexuality: An Exploratory Voyage*. Englewood Cliffs, NJ: Prentice-Hall.

Reiss, Jr., Albert J. 1961. "The Social Integration of Queers and Peers." *Social Problems*, 9: 102–120.

Reitzes, Donald C. and Juliette K. Diver. 1982. "Gay Bars as Deviant Community Organizations: The Management of Interactions with Outsiders." *Deviant Behavior*, 4: 1–18.

Risman, Barbara and Pepper Schwartz. 1988. Sociological Research on Male and Female Homosexuality." *Annual Review of Sociology*, 14: 125–147.

Ross, H. Laurence. 1971. "Modes of Adjustment of Married Homosexuals." *Social Problems*, 18: 385–393.

Saghir, Marcel T. and Eli Robins. 1973. *Male and Female Homosexuality: A Comparative Investigation*. Baltimore: Williams and Wilkins.

———. 1980. "Clinical Aspects of Female Homosexuality." Pp. 286–315 in *Homosexual Behavior: A Modern Reappraisal*. Edited by Judd Marmor. New York: Basic Books.

Schofield, Michael. 1965. *Sociological Aspects of Homosexuality: A Comparative Study of Three Types of Homosexuals*. Boston: Little, Brown.

Schur, Edwin M. 1984. *Labeling Women Deviant: Gender, Stigma, and Social Control*. New York: Random House.

Schwanberg, Sandra L. 1990. "Attitudes Toward Homosexuality in American Health Care Literature, 1983–1987," *Journal of Homosexuality*, 19: 117–136.

Shilts, Randy. 1987. *And the Band Played On: Politics, People and the AIDS Epidemic.* New York: St. Martin's Press.

Silverstein, Charles. 1981. *Man To Man: Gay Couples in America.* New York: William Morrow.

Simon, William and John H. Gagnon. 1967. "The Lesbians: A Preliminary Overview." Pp. 247–282 in *Sexual Deviance.* Edited by William Simon and John H. Gagnon. New York: Harper and Row.

Simpson, Ruth. 1976. *From the Closet to the Courts: The Lesbian Tradition.* Baltimore: Penguin Books.

Spector, Malcolm. 1977. "Legitimizing Homosexuality." *Society,* 14: 52–56.

Stephan, G. Edward and Douglas R. McMullin. 1982. "Tolerance of Sexual Nonconformity: City Size as a Situational and Early Learning Determinant." *American Sociological Review,* 47: 411–415.

St. Lawrence, Janet S., Brenda A. Husfeldt, Jeffrey A. Kelly, Harold V. Hood, and Steve Smith, Jr. 1990. "The Stigma of AIDS: Fear of Disease and Prejudice Toward Gay Men," *Journal of Homosexuality,* 19: 85–101.

Surgeon General. 1986. *Surgeon General's Report on Acquired Immune Deficiency Syndrome.* Washington, DC: U.S. Public Health Service, Government Printing Office.

Troiden, Richard R. 1979. "Becoming Homosexual: A Model of Gay Identity Acquisition." *Psychiatry,* 42: 362–373.

———. 1988. *Gay and Lesbian Identity: A Sociological Analysis.* New York: General Hall.

———. 1989. "The Formation of Homosexual Identities." *Journal of Homosexuality,* 17: 43–73.

Warren, Carol A. B. 1972. *Identity and Community in the Gay World.* New York: Wiley.

Weinberg, Martin S. and Colin J. Williams. 1974. Male *Homosexuals: Their Problems and Adaptations.* New York: Oxford University Press.

——— and Colin J. Williams. 1975. "Gay Baths and the Social Organization of Impersonal Sex." *Social Problems,* 23: 124–136.

Westwood, Gordon. 1960. *A Minority: A Report on the Life of the Male Homosexual in Great Britain.* London: Longmans Green.

Whitam, Frederick L. and Robin M. Mathy. 1985. *Male Homosexuality in Four Societies: Brazil, Guatemala, the Philippines, and the United States.* New York: Praeger.

CHAPTER 12—MENTAL DISORDERS

Abrams, Richard and Michael Alan Taylor. 1983. "The Genetics of Schizophrenia: A Reassessment Using Modern Criteria." *American Journal of Psychiatry,* 140: 171–175.

Aday, Jr., David P. 1990. *Social Control at the Margins: Toward a General Understanding of Deviance.* Belmont, CA: Wadsworth.

American Psychiatric Association. 1980. *Diagnostic and Statistical Manual,* 3rd ed. Washington, DC: American Psychiatric Association.

Bastide, Roger. 1972. *The Sociology of Mental Disorder,* Jean McNeil, trans. New York: David McKay.

Bittner, Egon. 1967. "Police Discretion in Apprehending the Mentally IL" *Social Problems*, 14: 282–285.

Blazer, Dan, Dana Hughes, and Linda K. George. 1987. "Stressful Life Events and the Onset of a Generalized Anxiety Syndrome." *American Journal of Psychiatry*, 114: 1178–1183.

Brenner, Harvey. 1973. *Mental Illness and the Economy*. Cambridge, MA: Harvard University Press.

Brown, George W. and Tirril Harris. 1978. *Social Origins of Depression: A Study of Psychiatric Disorder in Women*. New York: Free Press.

Cameron, Norman. 1947. *The Psychology of Behavior Disorders*. Boston: Houghton Mifflin.

Carstairs, G. M. 1959. "The Social Limits of Eccentricity: An English Study." in Opler, Culture and Mental Health.

Catalano, Ralph and David Dooley. 1977. "Economic Predictors of Depressed Mood and Stressful Life Events in a Metropolitan Community." *Journal of Health and Social Behavior*, 18: 292–307.

Catalano, Ralph, David Dooley, and Robert Jackson. 1981. "Economic Predictors of Admissions to Mental Health Facilities in a Nonmetropolitan Area." *Journal of Health and Social Behavior*, 22: 284–297.

Conrad, Peter and Joseph W. Schneider. 1980. *Deviance and Medicalization: From Badness to Sickness*. St. Louis: Mosby.

Dohrenwend, Bruce P. and Barbara Snell Dohrenwend. 1969. *Social Status and Psychological Disorder*. New York: Wiley-Interscience.

Dohrenwend, Bruce P. and Barbara Snell Dohrenwend. 1980. "Psychiatric Disorders and Susceptibility to Stress." Pp. 183–197 in *The Social Consequences of Psychiatric Illness*. Edited by Lee N. Robins, Paula J. Clayton, and John K. Wing. New York: Brunner/Mazel.

Dohrenwend, Bruce P., Barbara Snell Dohrenwend, M. S. Gould, B. Link, R. Neugebauer, and R. Wunsch-Hitzig. 1980. *Mental Illness in the United States: Epidemiological Estimates*. New York: Praeger.

Dunham, H. Warren. 1965. *Community and Schizophrenia: An Epidemiological Analysis*. Detroit: Wayne State University Press.

Eaton, Joseph W. and Robert J. Weil. 1955. *Culture and Mental Disorders*. New York: Free Press.

Eaton, William W. 1974. "Residence, Social Class, and Schizophrenia." *Journal of Health and Social Behavior*, 15: 289–299.

Fein, Sara and Kent S. Miller. 1972. "Legal Processes and Adjudication in Mental Incompetency Proceedings." *Social Problems*, 20: 57–64.

Goffman, Erving. 1961. *Asylums*. Garden City, New York: Anchor Doubleday.

Golann, Stuart and William H. Fremouw, eds. 1976. *The Right to Treatment for Mental Patients*. New York: Irvington Publishers.

Goleman, D. 1985. "New Psychiatric Syndromes Spur Protest." *New York Times*, November 19: C-1, 16.

Gottesman, I.I. and James Shields. 1972. *Schizophrenia and Genetics: A Twin Study Vantage Point*. New York: Academic Press.

Gove, Walter R. 1970. "Societal Reaction as an Explanation of Mental Illness: An Evaluation." *American Sociological Review*, 35: 873–884.

———. 1972. "The Relationship Between Sex Roles, Marital Status, and Mental Illness." *Social Forces*, 51: 34–44.

————. 1982. "The Current Status of the Labelling Theory of Mental Illness." Pp. 285–297 in *Deviance and Mental Illness*. Edited by Walter R. Gove. Beverly Hills, CA: Sage.

————. 1987. "Sociobiology Misses the Mark: An Essay on Why Biology But Not Sociobiology is Very Relevant to Sociology." *The American Sociologist*, 18: 258–277.

———— and Michael Hughes. 1989. "A Theory of Mental Illness: An Attempted Integration of Biological, Psychological, and Sociological Variables." Pp. 61–76 in Steven F. Messner, Marvin D. Krohn, and Allen E. Liska, eds., *Theoretical Integration in the Study of Deviance and Crime: Problems and Prospects*. Albany: State University of New York Press.

Greenley, James. 1972. "The Psychiatric Patient's Family and Length of Hospitalization." *Social Problems*, 13: 25–37.

Grusky, Oscar and Melvin Pollner, eds. 1981. *The Sociology of Mental Illness: Basic Studies*. New York: Holt, Rinehart and Winston.

Herman, Nancy J. 1987. " 'Mixed Nutters' and 'Looney Tuners': The Emergence, Development, Nature, and Functions of Two Informal, Deviant Subcultures of Chronic, Ex-Psychiatric Patients." *Deviant Behavior*, 8: 235–258.

Heston, Lenord. 1966. "Psychiatric Disorders in Foster Home Reared Children of Schizophrenic Mothers." *British Journal of Psychiatry*, 112: 819–825.

————. 1970. "The Genetics of Schizophrenia and Schizoid Disease." *Science*, 167: 249–256.

Hollingshead, August B. and Frederick Redlich. 1958. *Social Class and Mental Illness*. New York: Wiley.

Horney, Karen. 1937. *The Neurotic Personality in Our Time*. New York: Norton.

Isaac, Rael Jean and Virginia C. Armat. 1990. *Madness in the Streets: How Psychiatry and the Law Abandoned the Mentally Ill*. New York: Free Press.

Jaco, E. Gartly. 1959. "Mental Health of the Spanish-American in Texas." Pp. 467–489 in *Culture and Mental Health*. Edited by Marvin K. Opler. New York: Macmillan.

Johnson, Ann Braden. 1990. *Out of Bedlam: The Truth About Deinstitutionalization*. New York: Basic Books.

Joint Commission on Mental Illness and Health. 1961. *Action for Mental Health, Final Report*. New York: Basic Books.

Kallman, Franz. 1938. *The Genetics of Schizophrenia*. Locust Valley, NY: J.J. Augustine.

————. 1946. "The Genetic Theory of Schizophrenia." *American Journal of Psychiatry*, 103: 309–322.

Kennedy, John F. 1967. "The Role of the Federal Government in the Prevention and Treatment of Mental Disorders." Pp. 297–300 in *The Sociology of Mental Disorders: Analyses and Readings in Psychiatric Sociology*. Edited by S. Kirson Weinberg. Chicago: Aldine.

Kessler, Ronald C., Roger L. Brown and Clifford L. Broman. 1981. "Sex Differences in Psychiatric Help-Seeking: Evidence from Four Large-Scale Surveys." *Journal of Health and Social Behavior*, 22: 49–64.

Kessler, Ronald C. 1982. "A Disaggregation of the Relationship Between Socioeconomic Status and Psychological Distress." *American Sociological Review*, 47: 752–764.

Kiesler, Charles A. and Amy E. Sibulkin. 1987. *Mental Hospitalization: Myths and Facts about a National Crisis*. Newbury Park, CA: Sage.

Kirkpatrick, Brian and Robert W. Buchanan. 1990. "The Neural Basis of the Deficit Syndrome of Schizophrenia." *Journal of Nervous and Mental Disease*, 178: 545–555.

Leaf, Philip J. and Martha Livingston Bruce. 1987. "Gender Differences in the Use of Mental Health-Related Services: A Re-examination." *Journal of Health and Social Behavior*, 28: 171–183.

Lemert, Edwin M. 1972. *Human Deviance, Social Problems and Social Control*, 2nd ed. Englewood Cliffs, NJ: Prentice-Hall.

Lipkowitz, Marvin H. and Sudharam Idupuganti. 1983. "Diagnosing Schizophrenia in 1980: A Survey of U.S. Psychiatrists." *American Journal of Psychiatry*, 140: 52–55.

Little, Craig B. 1983. *Understanding Deviance and Control: Theory, Research, and Social Policy*. Itasca, IL: F. E. Peacock.

Luborsky, Lester, Paul Crits-Christoph, A. Thomas McLellan, George Woody, William Piper, Bernard Liberman, Stanley Imber, and Paul Pilkonis. 1986. "Do Therapists Vary Much in Their Success: Findings From Four Outcome Studies." *American Journal of Orthopsychiatry*, 56: 501–512.

Mechanic, David. 1972. "Social Class and Schizophrenia: Some Requirements for a Plausible Theory of Social Influence." *Social Forces*, 50: 305–309.

———. 1989. *Mental Health and Social Policy*, 3rd ed. Englewood Cliffs, NJ: Prentice-Hall.

Menninger, Karl. 1946. *The Human Mind*. New York: Knopf.

Monahan, John. 1981. *The Prediction of Dangerousness*. Beverly Hills, CA: Sage.

Murphey, Jane M. 1982. "Cultural Shaping and Mental Disorders." in Gove, 1982.

Murphey, H. B. M. 1959. "Culture and Mental Disorder in Singapore." Pp. 291–316 in *Culture and Mental Health*. Edited by Marvin K. Opler. New York: Macmillan.

Myers, Jerome K., Jacob J. Lindenthal, and Max P. Pepper. 1974. "Social Class, Life Events, and Psychiatric Symptoms: A Longitudinal Study." Pp. 191–205 in *Stressful Life Events: Their Nature and Effects*. Edited by Barbara Snell Dohrenwend and Bruce P. Dohrenwend. New York: Wiley.

Opler, Marvin K. 1959. "Cultural Differences in Mental Disorders: An Italian and Irish Contrast in the Schizophrenics—USA." Pp. 425–442 in *Culture and Mental Health*. Edited by Marvin K. Opler. New York: Macmillan.

Paykel, E. S. 1974. "Life Stress and Psychiatric Disorder: Applications of the Clinical Approach." Pp. 135–149 in *Stressful Life Events: Their Nature and Effects*. Edited by Barbara Snell Dohrenwend and Bruce P. Dohrenwend. New York: Wiley.

Perucci, Robert. 1974. *Circle of Madness: On Being Insane and Institutionalized in America*. Englewood Cliffs, NJ: Prentice-Hall.

Phillips, Derek L. 1968. "Rejection: A Possible Consequence of Seeking Help for Mental Disorders." Pp. 213–226 in *The Mental Patient: Studies in the Sociology of Deviance*. Edited by Stephen P. Spitzer and Norman K. Denzin. New York: McGraw-Hill.

Redlich, Frederick C. 1957. "The Concept of Health in Psychiatry." Pp. 145–146 in *Explorations in Social Psychiatry*. Edited by Alexander H. Leighton, John A. Clausen, and Robert N. Wilson. New York: Basic Books.

Rogler, Lloyd R. and August B. Hollingshead. 1965. *Trapped*. New York: Wiley.

Romney, David M. 1990. "Thought Disorder in the Relatives of Schizophrenics: A Meta-Analytic Review of Selected Published Studies." *Journal of Nervous and Mental Disease*, 178: 481–486.

Rosenhan, David L. 1973. "On Being Sane In Insane Places." *Science*, 179: 250–258.

Rotenberg, Mordechai. 1975. "Self-Labeling Theory: Preliminary Findings among Mental Patients." *British Journal of Criminology*, 15: 360–375.

Roth, Martin and Jerome Kroll. 1986. *The Reality of Mental Illness*. Cambridge, England: Cambridge University Press.

Rothman, David. 1971. *The Discovery of the Asylum: Social Order and Disorder in the New Republic*. Boston: Little, Brown.

Sarbin, Theodore R. and James C. Mancuso. 1980. *Schizophrenia: Medical Diagnosis or Moral Verdict?* New York: Pergamon Press.

Scheff, Thomas J. 1974. "The Labeling Theory of Mental Illness." *American Sociological Review*, 39: 444–452.

———. 1975. *Labeling Madness*. Englewood Cliffs, NJ: Prentice-Hall.

———. 1984. *Being Mentally Ill: A Sociological Theory*, 2nd ed. New York: Aldine.

Scull, Andrew. 1977. *Decarceration: Community Treatment and the Deviant—A Radical View*. Englewood Cliffs, NJ: Prentice-Hall.

Shibutani, Tamotsu. 1986. *Social Processes*. Berkeley: University of California Press.

Srole, Leo, Thomas S. Langner, Stanley T. Michael, Marvin K. Opler, and Thomas A. C. Rennie. 1962. *Mental Health in the Metropolis: The Midtown Manhattan Study*. New York: McGraw-Hill.

Srole, Leo, Thomas S. Langner, Stanley T. Michael, Price Kirkpatrick, Marvin K. Opler, and Thomas A. C. Rennie. 1977. *Mental Health in the Metropolis*, Book Two, revised ed. Edited by Leo Srole and Anita K. Fisher. New York: Harper and Row.

Stanton, Alfred H. and Morris S. Schwartz. 1954. *The Mental Hospital*. New York: Basic Books.

Stone, Alan A. 1975. *Mental Health and the Law: A System in Transition*. Rockville, MD: National Institute of Mental Health, Government Printing Office.

———. 1982. "Psychiatric Abuse and Legal Reform: Two Ways to Make A Bad Situation Worse." *International Journal of Law and Psychiatry*, 5: 9–28.

Szasz, Thomas S. 1974. *The Myth of Mental Illness*, rev. ed. New York: Harper and Row.

———. 1976. *Schizophrenia: The Sacred Symbol of Psychiatry*. New York: Basic Books.

———. 1981. "Crime, Punishment and Psychiatry." Pp. 342–363 in *Current Perspectives on Criminal Behavior*, 2nd ed. Edited by Abraham S. Blumberg. New York: Knopf.

———. 1987. *Insanity: The Idea and Its Consequences*. New York: Wiley.

Thoits, Peggy A. 1987. "Gender and Marital Status Differences in Control and Distress: Common Stress Versus Unique Stress Explanations." *Journal of Health and Social Behavior*, 28: 7–22.

Torrey, E. Fuller. 1974. *The Death of Psychiatry*. Radnor, PA: Chilton Books.

Townsend, J. Marshall. 1975. "Cultural Conceptions, Mental Disorders, and Social Roles: A Comparison of Germany and America." *American Sociological Review*, 40: 739–752.

Van Praag, Herman M. 1990. "The DSM-IV (Depression) Classification: To Be or Not To Be?" *Journal of Nervous and Mental Disease*, 178: 147–149.

Warheit, George J., Charles E. Holzer, III, Robert A. Bell and Sandra A. Arey. 1976. "Sex, Marital Status and Mental Health: A Reappraisal." *Social Forces*, 55: 459–470.

Waxler, Nancy E. 1974. "Culture and Mental Illness: A Social Labeling Perspective." *Journal of Nervous and Mental Disease*, 159: 379–395.

Weinberg, S. Kirson and H. Warren Dunham. 1960. *The Culture of the State Mental Hospital*. Detroit: Wayne State University Press.

Williams, Ann W., John E. Ware, Jr., and Cathy A. Donald. 1981. "A Model of Mental Health, Life Events and Social Supports Applicable to General Populations." *Journal of Health and Social Behavior*, 22: 324–336.

Wheaton, Blair. 1983. "Stress, Personal Coping Resources, and Psychiatric Symptoms." *Journal of Health and Social Behavior*, 24: 208–229.

Zigler, Edward and Marion Glick. 1986. *A Development Approach to Adult Psychopathology*. New York: Wiley.

Zimmerman, M. 1988. "Why are we rushing to publishing DSM-IV?" *Archives of General Psychiatry*, 45: 1135–1138.

CHAPTER 13—SUICIDE

Akers, Ronald L. 1985. *Deviant Behavior: A Social Learning Approach*, 3rd ed. Belmont, CA: Wadsworth.

Ansel, Edward L. and Richard McGee. 1971. "Attitudes Toward Suicide Attempters." *Bulletin of Suicidology*, 8: 22–29.

Blackstone, William. 1765–1769. *Commentaries on the Laws of England*, Vol. IV.

Bould, Sally, Beverly Sanborn, and Laura Reif. 1989. *Eighty-Five Plus: The Oldest Old*. Belmont, CA: Wadsworth.

Breed, Warren. 1963. "Occupational Mobility and Suicide among White Males." *American Sociological Review*, 28: 179–188.

Brealt, Kevin D. 1986. "Suicide in America: A Test of Durkheim's Theory of Religious and Family Integration, 1933–1980." *American Journal of Sociology*, 92: 628–656.

———. 1988. "Beyond the Quick and Dirty: Problems Associated with Analyses Based on Small Samples or Large Ecological Aggregates: Reply to Girard." *American Journal of Sociology*, 93: 1479–1486.

Bridge, T. Peter, Steven G. Potkin, William W. K. Zung, and Beth J. Soldo. 1977. "Suicide Prevention Centers: Ecological Study of Effectiveness." *The Journal of Nervous and Mental Disease*, 164: 18–24.

Bush, James A. 1978. "Similarities and Differences in Precipitating Events Between Black and Anglo Suicide Attempts." *Suicide and Life-Threatening Behavior*, 8: 243–249.

Daly, Martin and Margo Wilson. 1988. *Homicide*. Hawthorne, NY: Aldine de Gruyter.

Davidson, L. and W. Rokay. 1986. "Characteristics of Suicide Attempts During a Series of Suicides." Paper presented at the Epidemic Intelligence Service conference, Centers for Disease Control, Atlanta, GA. Cited in McGinnis, 1987.

Davis, Robert and James F. Short, Jr. 1978. "Dimensions of Black Suicide: A Theoretical Model." *Suicide and Life-Threatening Behavior*, 8: 161–173.

Davis, Robert A. 1979. "Black Suicide in the Seventies: Current Trends." *Suicide and Life-Threatening Behavior*, 9: 131–140.

———. 1981. "Female Labor Force Participation, Status Integration, and Suicide, 1950–1969." *Suicide and Life-Threatening Behavior*, 11: 111–123.

Douglas, Jack D. 1967. *The Social Meanings of Suicide*. Princeton, NJ: Princeton University Press.

Dublin, Louis I. and Bessie Bunzel. 1933. *To Be or Not To Be*. New York: Harrison Smith and Robert Haas.

Dublin, Louis I. 1963. *Suicide: A Sociological and Statistical Study*. New York: Ronald Press.

Durkheim, Emile. 1951. *Suicide*. Translated by John A. Spaulding and George Simpson. New York: Free Press, 1951; originally published 1895.

Farber, Maurice L. 1968. *Theory of Suicide*. New York: Funk and Wagnalls.

———. 1979. "Suicide in France: Some Hypotheses." S*uicide and Life-Threatening Behavior*, 9: 154–162.

Farberow, Norman L. 1977. "Suicide," in Edward Sagarin and Fred Montanino, eds., *Deviants: Voluntary Actors in an Involuntary World*, (Morristown, NJ: General Learning Press, 1977), pp. 503–505.

———. ed. 1980. *The Many Faces of Suicide: Indirect Self-Destructive Behavior*. New York: McGraw-Hill.

Faris, Robert E. L. 1948. *Social Disorganization*. New York: Ronald Press.

Ferracuti, Franco. 1957. "Suicide in a Catholic Country." Pp. 57–74 in *Clues to Suicide*. Edited by Edwin S. Schneidman and Norman L. Farberow. New York: McGraw-Hill.

Gibbs, Jack P. 1971. "Suicide." Pp. 271–312 in *Contemporary Social Problems*, 3rd ed. Edited by Robert K. Merton and Robert Nisbet. New York: Harcourt Brace Jovanovich.

———. 1982. "Testing the Theory of Status Integration and Suicide Rates." *American Sociological Review*, 47: 227–237.

Gibbs, Jack P. and Walter T. Martin. 1958. "Theory of Status Integration and Its Relationship to Suicide." *American Sociological Review*, 23: 140–147.

———. 1964. *Status Integration and Suicide: A Sociological Study*. Eugene: University of Oregon Books.

Gould, Madelyn S, David Shaffer, and Mark Davies. 1990. "Truncated Pathways from Childhood to Adulthood: Attrition in Follow-Up Studies Due to Death." Pp. 3–9 in *Straight and Devious Pathways from Childhood to Adulthood*. Edited by Lee N. Robins and Michael Rutter. Cambridge: Cambridge University Press.

Hacker, Andrew. 1983. *U/S: A Statistical Portrait of the American People*. Baltimore: Penguin Books.

Hankoff, L. D. 1979. "Judaic Origins of the Suicide Prohibition." Pp. 3–20 in *Suicide: Theory and Research*. Edited by L. D. Hankoff and Bernice Einsidler. Littleton, MA: PSG Publishing.

Hawton, Keith and Jose Catalan. 1982. *Attempted Suicide: A Practical Guide to its Nature and Management*. Oxford, England: Oxford University Press.

Hawton, Keith. 1986. *Suicide and Attempted Suicide among Children and Adolescents*. Beverly Hills, CA: Sage.

Headley, Lee A., ed. 1983. *Suicide in Asia and the Near East*. Berkeley: University of California Press.

Hendin, Herbert. 1964. *Suicide in Scandanavia.* New York: Grune and Stratton.

——. 1969. *Black Suicide.* New York: Basic Books.

Henry, Andrew F. and James F. Short, Jr. 1954. *Suicide and Homicide.* New York: Free Press.

Hoelter, Jon W. 1979. "Religiosity, Fear of Death, and Suicide Acceptability." *Suicide and Life-Threatening Behavior,* 9: 163–172.

Humphrey, Derek. 1987. "Letter to the Editor: The Case for Rational Suicide," *Suicide and Life-Threatening Behavior,* 17: 335–338.

Iga, Mamora. 1981. "Suicide of Japancse Youth." *Suicide and Life-Threatening Behavior,* 11: 17–30.

Jacobs, Jerry. 1967. "A Phenomenological Study of Suicide Notes." *Social Problems,* 15: 60–73.

——. 1970. "The Use of Religion in Constructing the Moral Justification of Suicide." Pp. 229–252 in *Deviance and Respectability: The Social Construction of Moral Meanings.* Edited by Jack D. Douglas. New York: Basic Books.

Jacobson, Gerald F. and Stephen H. Portuges. 1978. "Relation of Marital Separation and Divorce to Suicide: A Report." *Suicide and Life-Threatening Behavior,* 8: 217–225.

Jobes, David A., Alan L. Berman, and Arnold R. Josselsen. 1986. "The Impact of Psychological Autopsies on Medical Examiners' Determination of Manner of Death." *Journal of Forensic Science,* 31: 177–189.

——. 1987. "Improving the Validity and Reliability of Medical-Legal Certifications of Suicide." *Journal of Suicide and Life-Threatening Behavior,* 17: 310–325.

Johnson, David, Starla D. Fitch, Jon P. Alston, and William Alex McIntosh. 1980. "Acceptance of Conditional Suicide and Euthanasia among Adult Americans." *Suicide and Life-Threatening Behavior,* 10: 157–166.

Kirk, Alton R. and Robert A. Zucker. 1979. "Some Sociopsychological Factors in Attempted Suicide among Urban Black Males." *Suicide and Life-Threatening Behavior,* 9: 76–86.

Kitagawa, Evelyn M. and Philip M. Hauser. 1973. *Differential Mortality in the United States: A Study in Socioeconomic Epidemiology.* Cambridge, MA: Harvard University Press.

Kobler, Arthur L. and Ezra Stotland. 1964. *The End of Hope: A Socio-Clinical Study of Suicide.* New York: Free Press.

Kosky, Robert, Sven Silburn, and Stephen R. Zubrick. 1990. "Are Children and Adolescents Who Have Suicidal Thoughts Different From Those Who Attempt Suicide?" *Journal of Nervous and Mental Disease,* 178: 38–43.

Leenaars, Antoon A. 1988. *Suicide Notes: Predictive Clues and Patterns.* New York: Human Sciences Press.

Labovitz, Sanford. 1968. "Variations in Suicide Rates." Pp. 57–73 in *Suicide.* Edited by Jack P. Gibbs. New York: Harper and Row.

Levi, Ken. 1982. "Homicide and Suicide: Structure and Process." *Deviant Behavior,* 3: 91–115.

Li, Wen L. 1972. "Suicide and Educational Attainment in a Transitional Society." *The Sociological Quarterly,* 13: 253–258.

Litman, Robert E. 1972. "Experiences in a Suicide Prevention Center." Pp. 217–230 in *Suicide and Attempted Suicide.* Edited by Jan Waldenstrom, Trage Larsson, and Nils Ljungstedt. Stockholm: Nordiska Bokhandelns Forlag.

——. 1987. "Mental Disorders and Suicidal Intention." *Suicide and Life-Threatening Behavior,* 17: 85–92.

Malinowski, Bronislaw. 1926. *Crime and Custom in Savage Society*. London: Routledge and Kegan Paul.

Maris, Ronald W. 1969. *Social Forces in Urban Suicide*. Homewood, IL: Dorsey Press.

———. 1981. *Pathways to Suicide: A Survey of Self-Destructive Behaviors*. Baltimore: Johns Hopkins University Press.

Markides, Kyriakos S. 1981. "Death-Related Attitudes and Behavior among Mexican Americans: A Review." *Suicide and Life-Threatening Behavior*, 11: 75–85.

Marshall, James. 1981. "Political Integration and the Effect of War on Suicide." *Social Forces*, 59: 771–785.

Meneese, William B. and Barbara A. Yutrzenka. 1990. "Correlates of Suicidal Ideation among Rural Adolescents." *Suicide and Life-Threatening Behavior*, 20: 206–212.

McGinnis, J. Michael. 1987. "Suicide in America—Moving Up the Public Health Agenda." *Suicide and Life-Threatening Behavior*, 17: 18–32.

Miller, H. L., D. W. Coombs, J. D. Leeper, and S. N. Barton. 1984. "An Analysis of the Effects of Suicide Prevention Facilities on Suicide Rates in the United States." *American Journal of Public Health*, 74: 340–343.

Nelson, Franklyn L. Norman L. Farberow, and Douglas R. MacKinnon. 1978. "The Certification of Suicide in Eleven Western States: An Inquiry into the Validity of Reported Suicide Rates." *Suicide and Life-Threatening Behavior*, 8: 75–88.

Osgood, Nancy J. and Barbara A Brant. 1990. "Suicide Behavior in Long-Term Care Facilities." *Suicide and Life-Threatening Behavior*, 20: 113–122.

Paerregaard, Grethe. 1980. "Suicide in Denmark: A Statistical Review for the Past 150 Years." *Suicide and Life-Threatening Behavior*, 10: 150–156.

Pescosolido, Bernice A. 1990. "The Social Context of Religious Integration and Suicide: Pursuing the Network Explanation." *The Sociological Quarterly*, 31: 337–357.

Phillips, David P. 1974. "The Influence of Suggestion on Suicide: Substantive and Theoretical Implications of The Werther Effect." *American Sociological Review*, 39: 340–354.

———. 1979. "Suicide, Motor Vehicle Fatalities, and the Mass Media: Evidence toward a Theory of Suggestion." *American Journal of Sociology*, 84: 1150–1174.

———. 1980. "Airplane Accidents, Murder, and the Mass Media: Toward a Theory of Imitation and Suggestion." *Social Forces*, 58: 1001–1024.

Phillips, David P. and John S. Wills. 1987. "A Drop in Suicides Around Major National Holidays," *Suicide and Life-Threatening Behavior*, 17: 1–12.

Pope, Whitney. 1976. *Durkheim's "Suicide": A Classic Analyzed*. Chicago: University of Chicago Press.

Retterstol, Nils. 1975. "Suicide in Norway." Pp. 77–94 in *Suicide in Different Cultures*. Edited by Norman L. Farberow. Baltimore: University Park Press.

Ringel, Erwin. 1977. "The Presuicidal Syndrome." *Suicide and Life-Threatening Behavior*, 6: 131–149.

Roa, A. Venkoba. 1983. "India." Pp. 210–237 in *Suicide in Asia and the Near East*. Edited by Lee A. Headley. Berkeley: University of California Press.

Robins, Lee N., Patricia A. West, and George E. Murphey. 1977. "The High Rate of Suicide in Older White Men: A Study of Testing Ten Hypotheses." *Social Psychiatry*, 12: 1–20.

Rudd, M. David. 1990. "An Integrative Model of Suicide Ideation." *Suicide and Life-Threatening Behavior*, 20: 16–30.

Sainsbury, Peter. 1963. "Social and Epidemiological Aspects of Suicide with Special Reference to the Aged." Pp. 155–178 in *Processes of Aging*, Vol II. Edited by Richard H. Williams, Clark Tibbitts, and Wilma Donahue. New York: Atherton Press.

Sanborn, Charlotte J. 1990. "Gender Socialization and Suicide: American Association of Suicidology Presidential Address, 1989." *Suicide and Life-Threatening Behavior*, 20: 148–155.

Sawyer, John B. and Elizabeth M. Jameton. 1979. "Chronic Callers to a Suicide Prevention Center." *Suicide and Life-Threatening Behavior*, 9: 97–104.

Schneideman, Edwin S. ed. 1976. *Suicidology: Contemporary Developments*. New York: Grune and Stratton.

Schneidman, Edwin S. and Norman L. Farberow, eds. 1957. *Clues to Suicide*. New York: McGraw-Hill.

———. 1961. "Statistical Comparisons Between Attempted and Committed Suicides." Pp. 19–47 in *The Cry for Help*. Edited by Norman L. Farberow and Edwin S. Schneidman. New York: McGraw-Hill.

Secretary of Health and Human Services. 1990. *Alcohol and Health: Seventh Special Report to the U.S. Congress*. Rockville, MD: National Institute of Alcohol Abuse and Alcoholism, Government Printing Office.

Seiden, Richard H. and Raymond P. Freitas. 1980. "Shifting Patterns of Deadly Violence." *Suicide and Life-Threatening Behavior*, 10: 195–209.

Stack, Steven. 1980. "Interstate Migration and the Rate of Suicide." *International Journal of Social Psychiatry*, 26: 17–26.

———. 1982. "Suicide: A Decade Review of the Sociological Literature." *Deviant Behavior*, 4: 41–66.

Stafford, Mark C. and Jack P. Gibbs. 1985. "A Major Problem with the Theory of Status Integration and Suicide." *Social Forces*, 63: 643–660.

———. 1988. "Change in the Relation Between Marital Integration and Suicide Rates." *Social Forces*.

Stephans, B. Joyce. 1987. "Cheap Thrills and Humble Pie: The Adolescence of Female Suicide Attempters." *Suicide and Life-Threatening Behavior*, 17: 107–118.

Stengel, Erwin. 1964. *Suicide and Attempted Suicide*. Baltimore: Penguin.

Tatai, Kechinosuke. 1983. "Japan." Pp. 12–58 in *Suicide in Asia and the Near East*. Edited by Lee A. Headley. Berkeley: University of California Press.

Trout, Deborah. 1980. "The Role of Social Isolation in Suicide." *Suicide and Life-Threatening Behavior*, 10: 10–23.

U.S. Department of Health and Human Services. 1985. *Suicide Surveillance Summary: 1970–1980*. Atlanta: Centers for Disease Control.

West, Donald J. 1965. *Murder Followed by Suicide*. London: William Heinemann.

Westermark, Edward A. 1908. *Origin and Development of Moral Ideas*, Vol. II. London: Macmillan.

Wilson, Michele. 1981. "Suicide Behavior: Toward an Explanation of Differences in Female and Male Rates." *Suicide and Life-Threatening Behavior*, 11: 131–140.

Wold, Carl I. 1970. "Characteristics of 26,000 Suicide Prevention Center Patients." *Bulletin of Suicidology*, 7: 24–28.

Wolfgang, Marvin E. 1958. *Patterns of Criminal Homicide*. Philadelphia: University of Pennsylvania Press.

CHAPTER 14—PHYSICAL DISABILITIES

Albrecht, Gary L. 1976. "Socialization and the Disability Process." Pp. 3–38 in *The Sociology of Physical Disability and Rehabilitation.* Edited by Gary L. Albrecht. Pittsburg: University of Pittsburg Press.

Becker, Howard S. 1973. *Outsiders: Studies in the Sociology of Deviance,* enlarged ed. New York: Free Press.

Becker, Gaylene. 1980. *Growing Old in Silence.* Berkeley: University of California Press.

Cahnman, Werner J. 1968. "The Stigma of Obesity." *The Sociological Quarterly,* 9: 283–299.

Colvin, Robert H. and Susan B. Olson. 1983. "A Descriptive Analysis of Men and Women Who Have Lost Significant Weight and Are Highly Successful at Maintaining the Loss." *Addictive Behaviors,* 8: 287–295.

Connors, Mary E. and Craig L. Johnson. 1987. "Epidemiology of Bulimia and Bulimic Behaviors." *Addictive Behaviors,* 12: 165–179.

Davis, Fred. 1961. "Deviance Disavowal: The Management of Strained Interaction by the Visibly Handicapped." *Social Problems,* 9: 120–132.

———. 1972. *Illness, Interaction and the Self.* Belmont, CA: Wadsworth.

Deegan, Mary Jo. 1985. "Multiple Minority Groups: A Case Study of Physically Disabled Women." Pp. 37–55 in *Women and Disability: The Double Handicap.* New Brunswick, NJ: Transaction Books.

DeJong, William. 1980. "The Stigma of Obesity: The Consequences of Naive Assumptions Concerning the Causes of Physical Deviance." *Journal of Health and Social Behavior,* 21: 75–87.

Edgerton, Robert B. 1967. *The Cloak of Competence: Stigma in the Lives of the Mentally Retarded.* Berkeley: University of California Press.

———. 1979. *Mental Retardation.* Cambridge, MA: Harvard University Press.

Eisenberg, M. G., C. Griggins, and R. J. Duval. eds. 1982. *Disabled People as Second-Class Citizens.* New York: Springer.

Elliott, Gregory C., Herbert L. Ziegler, Barbara M. Altman, and Deborah R. Scott. 1982. "Understanding Stigma: Dimensions of Deviance and Coping." *Deviant Behavior,* 3: 275–300.

Evans, Daryl Paul. 1983. *The Lives of Mentally Retarded People.* Boulder, CO: Westview Press.

Fox, Renee. 1977. "The Medicalization and Demedicalization of American Society." *Daedalus,* 106: 12–19.

Friedson, Eliot. 1965. "Disability as Social Deviance." Pp. 70–82 in *Sociology and Rehabilitation.* Edited by Marvin B. Sussman. Washington, DC: American Sociological Association.

Garrity, Thomas F. 1973. "Vocational Adjustment after First Myocardial Infarction: Comparative Assessment of Several Variables Suggested in the Literature." *Social Science and Medicine,* 7: 705–717.

Goffman, Erving. 1963. *Stigma: Notes on the Management of Spoiled Identity.* Englewood Cliffs, NJ: Prentice-Hall.

Haber, Lawrence D. and Richard T. Smith. 1971. "Disability and Deviance: Normative Adaptations of Role Behavior." *American Sociological Review,* 36: 87–97.

Hamli, Katherine A., James R. Falk, and Estelle Schwartz. 1981. "Binge-Eating and Vomiting: A Survey of a College Population." *Psychological Medicine*, 11: 697–706.

Hawkins, R. C. and P.F. Clement. 1980. "Development and Construct Validation of a Self Report Measure of Binge Eating Tendencies." *Addictive Behaviors*, 5: 219–226.

Hays, Diane and Catherine E. Ross. 1987. "Concern with Appearance, Health Beliefs, and Eating Habits." *Journal of Health and Social Behavior*, 28: 120–130.

Higgins, Paul C. 1980. *Outsiders in a Hearing World*. Beverly Hills, CA: Sage Publications.

Howards, Irving, Henry P. Brehm, and Saad Z. Nagi. 1980. *Disability: From Social Problem to Federal Program*. New York: Praeger.

Humphries, Laurie L., Sylvia Wrobel, and H. Thomas Wiegert. 1982. "Anorexia Nervosa." *American Family Physician*, 26: 199–204.

Hyman, Marvin. 1971. "Disability and Patients' Perceptions of Preferential Treatment." *Journal of Chronic Diseases*, 24: 329–342

Kaplan, H. I. and H. S. Kaplan. 1957. "The Psychosomatic Concept of Obesity." *Journal of Nervous and Mental Disease*, 125: 181–201.

Koestler, Frances A. 1976. *The Unseen Minority: A Social History of Blindness in the United States*. New York: David McKay.

Krause, Elliott A. 1976. "The Political Sociology of Rehabilitation." Pp. 201–221 in *The Sociology of Physical Disability and Rehabilitation*. Edited by Gary L. Albrecht. Pittsburg: University of Pittsburg Press.

Kubler-Ross, Elizabeth. 1969. *On Death and Dying*. New York: Macmillan.

Laslet, Barbara and Carol A. B. Warren. 1975. "Losing Weight: The Organizational Promotion of Behavior Change." *Social Problems*, 23: 79.

Lemert, Edwin M. 1951. *Social Pathology*. New York: McGraw-Hill.

Levitin, Teresa E. 1975. "Deviants as Active Participants in the Labeling Process: The Visibly Handicapped." *Social Problems*, 22: 548–557.

Maddox, George L., Kurt W. Back, and Veronica R. Liederman. 1968. "Overweight as Social Deviance and Social Disability." *Journal of Health and Social Behavior*, 9: 287–298.

Mayer, Jean. 1968. *Overweight: Causes, Cost, and Control*. Englewood Cliffs, NJ: Prentice-Hall.

McLorg, Penelope A. and Diane E. Taub. 1987. "Anorexia Nervosa and Bulimia: The Development of Deviant Identities." *Deviant Behavior*, 8: 177–189.

Mercer, Jane R. 1973. *Labeling the Mentally Retarded*. Berkeley: University of California Press.

Millman, M. 1980. *Such a Pretty Face: Being Fat in America*. New York: W.W. Norton.

Mitchell, James E., Richard L. Pyle, Elke D. Eckert, Dorothy Hatsukami, and Elizabeth Soll. 1990. "Bulimia Nervosa in Overweight Individuals." *Journal of Nervous and Mental Disease*, 178: 324–327.

Morgan, Carolyn Stout, Marilyn Affleck, and Orin Solloway. 1990. "Gender Role Attitudes, Religiosity, and Food Behavior: Dieting and Bulimia in College Women." *Social Science Quarterly*, 71: 142–151.

Myerson, Lee. 1971. "Physical Disability as a Social Psychological Problem." Pp. 205–210 in *The Other Minorities*. Edited by Edward Sagarin. Boston: Ginn.

Nagi, Saad Z. 1969. *Disability and Rehabilitation*. Columbus, OH: Ohio State University Press.

Nonbeck, Michael E. 1973. *The Meaning of Blindness: Attitudes Toward Blindness and Blind People.* Bloomington: Indiana University Press.

Parsons, Talcott. 1951. *The Social System.* New York: Free Press.

Powers, Pauline S., Richard G. Schulman, Alice A. Gleghorn, and Mark E. Prange. 1987. "Perceptual and Cognitive Abnormalities in Bulimia." *American Journal of Psychiatry*, 144: 1456–1460.

Richardson, Stephen A., Norman Goodman, Albert H. Hastorf, and Sanford M. Dornbush. 1961. "Cultural Uniformity in Reaction to Physical Disabilities." *American Sociological Review*, 26: 241–247.

Roth, William. 1981. *The Handicapped Speak.* Jefferson, NC: McFarland.

Ruderman, Audrey J. 1983. "Obesity, Anxiety, and Food Consumption." *Addictive Behaviors*, 8: 235–242.

Sagarin, Edward. 1975. *Deviants and Deviance.* New York: Praeger.

Safilios-Rothschild, Constantina. 1970. *The Sociology and Social Psychology of Disability and Rehabilitation.* New York: Random House.

———. 1976. "Disabled Persons' Self-Definitions and Their Implications for Rehabilitation." Pp. 39–56 in *The Sociology of Physical Disability and Rehabilitation.* Edited by Gary L. Albrecht. Pittsburg: University of Pittsburg Press.

Schur, Edwin M. 1984. *Labeling Women Deviant: Gender, Stigma, and Social Control.* New York: Random House.

Scott, Robert A. 1969. *The Making of Blind Men: A Study of Adult Socialization.* New York: Russell Sage.

———. 1980. "Introduction." Pp. 2–10 in *Outsiders in a Hearing World.* By Paul C. Higgins. Beverly Hills, CA: Sage Publications.

Stafford, Mark C. and Richard R. Scott. 1986. "Stigma, Deviance, and Social Control: Some Conceptual Issues." Pp. 77–91 in *The Dilemma of Difference: A Multidisciplinary View of Stigma.* Edited by Stephen A. Ainlay, Gaylene Becker, and Lerita M. Coleman. New York: Plenum Press.

Taub, Diane E. and Penelope A. McLorg. 1990. "The Sociocultural Context of Anorexia Nervosa and Bulimia Nervosa: A Review and Discussion." Unpublished paper, Department of Sociology, Southern Illinois University.

Thomas, David. 1982. *The Experience of Handicap.* London: Methuen.

Topliss, Eda. 1982. *Social Responses to Handicap.* New York: Longman.

Turner, R. Jay and Morton Beiser. 1990. "Major Depression and Depressive Symptomatology among the Physcially Disabled." *Journal of Nervous and Mental Disease*, 178: 343–350.

Vandereycken, Walter and Eugene L. Lowenkopf. 1990. "Anorexia Nervosa in 19th Century America." *Journal of Nervous and Mental Disease*, 178: 531–535.

Warren, Carol A. B. 1974. "The Use of Stigmatized Labels in Conventionalizing Deviant Behavior." *Sociology and Social Research*, 58: 303–311.

Wolinsky, Fredric D. and Sally R. Wolinsky. 1981. "Expecting Sick-Role Legitimation and Getting It." *Journal of Health and Social Behavior*, 22: 229–242.

Wright, Beatrice A. 1960. *Physical Disability—A Psychological View.* New York: Harper and Row.

Zincand, H., R. J. Cadoret, and R. B. Widman. 1984. "Incidence and Detection of Bulimia in a Family Practice Population." *Journal of Family Practice*, 18: 555–560.

NAME INDEX

SUBJECT INDEX

AA. *See* Alcoholics Anonymous (AA)
Absolutist definition of deviance, 7, 9
Abstinence syndrome, 192
Abuse. *See* Child abuse; Spouse abuse
Actus reus, 128–129
Adaptation. *See* Social adaptation
Adaptations to strain, 93–94
Addict crime, 217–218
"Addict" personality, 55–56
Addiction. *See also* Heroin
    becoming an opiate user, 205–210
    cost of, 217
    criminal law and, 218–220
    criminal sanctions against, 217–220
    glossary of terms used by addicts, 201
    meaning of, 199–200
    process of, 192, 206–208
    self-help programs, 225–227
    social control of, 217–229
    sociological theory of, 210–211
    subculture of, 208–210
    treatment for, 220–225
Adolescents
    pregnancy of, 64–65, 78–79
    as prostitutes, 296–297
    suicide of, 405–407
    urbanization and, 64–65
African Americans. *See* Blacks
Age
    mental disorders and, 368
    murder and, 134
    suicide and, 399–401
AIDS
    disease and its transmission, 345–347
    effects of, 348–350
    homosexuality and, 320–322, 345–350
    impact of, 347–350
    incidence of, 346, 347
    political impact of, 349–350
    prostitution and, 301
    sexual behavior, impact on, 349
AIDS-Related Complex (ARC), 345–346
Alcohol use
    alcohol substitution, 260
    alcoholism and problem drinking, 237, 248–253
    companions and excessive drinking, 256
    controversy on whether alcoholics can safely return
        to drinking, 269–270
    crime related to, 250–251
    deviant nature of, 232–233
    drunk driving, 251–253
    education about consequences of, 260
    ethnic differences in, 241, 243–248
    group and subcultural factors in excessive drinking,
        253–259
    health-related consequences of, 234–235
    heavy drinker, 237–238
    homeless and skid row drinking, 256–257
    legal regulation of, 260
    models of alcoholism, 261–265
    occupation and excessive drinking, 257–258
    physiological and behavioral aspects of, 233–235
    prevalence of, 235–237
    prevention of alcohol abuse, 260–261

    problem drinker, 238–239
    Prohibition, 259–260
    psychological effects of, 235
    public drinking houses, 241–243
    public policy on, 265–269
    religious differences in excessive drinking, 241,
        258–259
    sex differences in, 241, 254–255
    as social and group activity, 239–248
    social background and, 241
    social control of, 259–265
    social or controlled drinker, 237–238
    types of drinkers, 237–239
    urbanization and, 67
    variations in, 241
Alcohol-related crime, 250–251
Alcoholics Anonymous (AA), 225, 266–269
Alcoholism
    Alcoholics Anonymous, 266–269
    behavioral model of, 262
    biological model of, 262–263
    chronic alcoholics, 237, 239
    combined perspectives on, 264–265
    community-based treatment programs, 265–266
    companions and excessive drinking, 256
    controversy on whether alcoholics can safely return
        to drinking, 269–270
    costs of, 249–250
    crime related to, 250–251
    definition of alcoholics, 237, 239
    drunk driving, 251–253
    extent of, 248–249
    family interaction model of, 262
    group and subcultural factors in excessive drinking,
        253–259
    homeless and skid row drinking, 256–257
    male-limited susceptibility to, 263
    medical model of, 263–264
    milieu-limited susceptibility to, 263
    models of, 261–265
    occupation and excessive drinking, 257–258
    psychoanalytic model of, 261
    psychoses resulting from, 357–358
    public policy on, 265–269
    religious differences in excessive drinking, 241,
        258–259
    sex differences in, 254–255
    social control of, 259–265
Altruistic suicide, 403–404
American Indians, and alcohol use, 241, 245–246
Amphetamines, 190–191
Anomic suicide, 405
Anomie
    adaptation to strain, 93–94
    evaluation of theory of, 95–97
    reformulations of, 94–95
    theory of, 92–97
Anorexia, 437, 448
Anxiety, 369–370
ARC. *See* AIDS-Related Complex (ARC)
Asian Americans, and alcohol use, 241, 244–245
Assault
    age differences in, 134
    child abuse, 139–142